Biomechanical Basis of Human Movement

Biomechanical Basis of Human Movement

Joseph Hamill, Ph. D.
University of Massachusetts at Amherst
and Kathleen M. Knutzen, Ph. D.
Western Washington University

Williams & Wilkins

BALTIMORE • PHILADELPHIA • HONG KONG
LONDON • MUNICH • SYDNEY • TOKYO
A WAVERLY COMPANY
1995

Executive Editor: Donna Balado
Developmental Editor: Lisa Stead
Production Coordinator: Peter J. Carley
Project Editor: Robert D. Magee

Copyright © 1995
Williams & Wilkins
Rose Tree Corporate Center-Building II
Suite 5025, 1400 North Providence Road
Media, PA 19063-2043 USA

Accurate indications, adverse reactions, and dosage schedules for drugs are provided in this book, but it is possible they may change. The reader is urged to review the package information data of the manufacturers of the medications mentioned.

Printed in the United States of America

Library of Congress Cataloging in Publication Data

Hamill, Joe
    Biomechanical basis of human movement / Joe Hamill and Kathleen
Knutzen
      p. cm.
    Includes index.
    ISBN 0-683-03863-X
    1. Human mechanics.   I. Knutzen, Kathleen.   II. Title.
QP303.H354   1995   94-22325
612.7'6—dc20

94 95 96 97 98
1 2 3 4 5 6 7 8 9 10

Preface

Biomechanics is a quantitative field of study within the discipline of Exercise Science. This book is intended as an introductory textbook but will stress this quantitative rather than qualitative nature of biomechanics. It is hoped that while stressing the quantification of human movement the book will also acknowledge those with a limited background in mathematics. The quantitative examples are presented in a detailed, logical manner that highlight topics of interest. The goal of this book, therefore, is to provide an introductory text in biomechanics that integrates basic anatomy, physics, calculus, and physiology for the study of human movement. We decided to use this approach because numerical examples are meaningful and easily clear up misconceptions concerning the mechanics of human movement.

Several features of this text serve as effective means for students to comprehend biomechanics as a field of study. This book is organized into three major sections: Part I—Foundations of Human Movement; Part II—Functional Anatomy; and Part III—Mechanical Analysis of Human Motion. The chapters are ordered to provide a logical progression of material essential toward the understanding of biomechanics and the study of human movement.

Part I, Foundations of Human Movement, includes Chapters 1 through 4. Chapter 1 "Basic Movement Terminology," presents the terminology and nomenclature generally used in biomechanics. Chapter 2, "Skeletal Considerations for Movement," covers the skeletal system with particular emphasis on joint articulation. Chapter 3, "Muscular Considerations for

Movement," discusses the organization of the muscular system. Finally, in Chapter 4, "Neurological Considerations For Movement," the control and activation systems for human movement are presented.

Part II, Functional Anatomy, includes Chapters 5 through 7, deals with specific regions of the body: the upper extremity, lower extremity, and trunk, respectively. Each chapter integrates the general information presented in Part I relative to each region.

Part III, Mechanical Analysis of Human Motion includes Chapters 8 through 12, in which quantitative mechanical techniques for the analyses of human movement are presented. Chapter 8 and 9 present the concepts of linear and angular kinematics. Included in these two chapters are the equations used to describe the motion of projectiles. Conventions for the study of angular motion in the analysis of human movement are also detailed. A portion of each chapter is devoted to a review of the research literature on human locomotion, which is used throughout the section to illustrate the quantitative techniques presented.

Chapters 10 and 11 present the concepts of linear and angular kinetics, including discussions on the forces and torques that act on the human body during daily activities. Included here is a discussion of the inertial characteristics of the segments of the body.

Chapter 12, "Types of Mechanical Analysis," is the culminating chapter in the mechanics section, in which three methods of analyzing human movement are presented with computations and examples of each method.

Although the book follows a progressive order, the major sections are generally self-contained. Therefore, instructors may delete or de-emphasize certain sections. Parts I and II, Foundations of Human Movement and Functional Anatomy, for example, could be used in a traditional kinesiology course, while Part III—Mechanical Analysis of Human Motion, could be used for a biomechanics course.

A second feature of the book is the means used to reinforce the principles presented. Each chapter contains a list of chapter objectives to enable the student to focus on key points in the material. Each concept is accompanied by problems, which are solved to highlight the concept. A chapter summary then outlines the major concepts presented. A glossary is presented at each chapter's end defining terms found in each chapter, as a source of re-enforcement and reference. Both an up-to-date ref-erence list of the research literature cited in the chapter and a list of additional readings for students who wish to study the material in more depth are included.

Each chapter contains a series of problems to challenge students and to help them digest and inte-grate the material presented. Finally, four appen-dices present information on bony landmarks, liga-ments, muscles, units of measurement, and a review of trigonometric functions.

A third feature of this book is the examples of human movement presented. While illustrations of the principles of human movement are easily seen in most sports examples, applications from ergonomics, orthopedics, and exercise are presented as well, with references from the current biome-chanics literature. In doing so, the full continuum of human movement potential is considered in examples.

Acknowledgments

To those who reviewed the drafts of this book and who made a substantial contribution to its development, the authors wish to express our sincere appreciation. These individuals are: Nelson Ng, Ed. D.; John Sigg, Ph. D.; Janet Dufek, Ph. D.; Dave Barlow, Ph. D.; Steve McCaw; and Richard Hinrichs, Ph. D.

Finally, we wish to thank Lisa Stead, Development Editor for Williams & Wilkins, and Tim Hengst, Medical Illustrator, for their substantial contribution to this book. A special thanks also to Andrew Stephens for the photography used throughout.

Joseph Hamill
Kathleen M. Knutzen

Contents

PART 1

Foundations of Human Movement

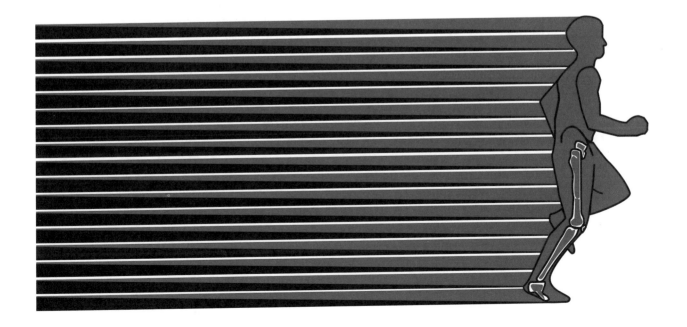

CHAPTER 1

Basic Movement Terminology

After reading this chapter, the student will be able to:

1. Define mechanics, biomechanics, and kinesiology and differentiate between their uses in the analysis of human movement.
2. Define and provide examples of linear and angular motion.
3. Define kinematics and kinetics.
4. Explain the difference between a relative and absolute reference system.
5. Define sagittal, frontal, and transverse planes along with corresponding frontal, sagittal, and longitudinal axes. Provide examples of human movements that occur in each plane.
6. Explain "degree of freedom" and provide examples of degrees of freedom associated with numerous joints in the body.
7. Describe the location of segments or landmarks using correct anatomical terms such as medial, lateral, proximal, and distal.
8. Identify segments by their correct name, define all of segmental movement descriptors, and provide specific examples in the body.
9. Describe the segmental movements occurring in a multijoint activity or sport skill.

Introduction

To begin or renew the study of kinesiology and biomechanics using this textbook requires a fresh mind. Remember that human movement is the theme and the focus of study in both biomechanics and kinesiology. A thorough understanding of various aspects of human movement may facilitate better teaching, successful coaching, more observant therapy, knowledgeable exercise prescription, or new research ideas. Movement is used to interact with the environment, whether it be simply to take a walk in a park, to strengthen muscles in a bench press, to compete in the high jump at a collegiate track meet, or to stretch or rehabilitate a joint that has been injured. Movement, or motion, involves a change in place, position, or posture relative to some point in the environment.

This textbook focuses on developing knowledge in the area of human movement in order to feel comfortable observing human movement and solving movement problems. There are many different approaches to the study of movement, such as observing movement using only the human eye, or collecting data on movement parameters using laboratory equipment. Different observers of activities also have different concerns: a coach may be interested in the final outcome of a tennis serve, while a therapist may be interested in identifying where in the serve the athlete is placing the stress on the medial elbow that has a tendinitis condition. Some applications of biomechanics and kinesiology will only require a cursory view of a movement, such as visual inspection of the forearm position in the jump shot. Other applications, such as evaluating the forces applied by the hand on the basketball during the shot will require some advanced knowledge and the use of sophisticated equipment and techniques.

Elaborate equipment is not needed to apply the material in this text, but it will be necessary to understand and interpret numerical examples collected using such intricate instrumentation. There will be qualitative examples in this text where movement characteristics will be described. A qualitative analysis is a non-numeric evaluation of the motion based on direct observation. These examples can be applied directly to a particular movement situation using visual observation or video.

There will also be quantitative information presented in this text. A quantitative analysis is a numerical evaluation of the motion based on data collected during the performance. For example, movement characteristics can be presented to describe the forces, or the temporal and spatial components of the activity. The application of this material to a practical setting such as teaching a sport skill is more difficult, since it is more abstract and often cannot be visually observed. However, the quantitative information is very important since it substantiates what is seen visually in a qualitative analysis. It also directs the instructional technique because a quantitative analysis will identify the source of a movement. For example, a front handspring can be qualitatively evaluated through visual observation by focusing on such things as whether the legs are together and straight, the back arched, the landing solid, and whether it was too fast or slow. But it is through the quantitative analysis that the source of the movement, the magnitude of the torque generated about the ground and the center of gravity, can be identified. Torque cannot be observed qualitatively, but knowing it is the source of the movement will help to qualitatively assess the effects of the torque, the success of the handspring.

This chapter will introduce terminology that will be used numerous times throughout the text. The chapter will begin by defining and introducing the various areas of study for movement analysis. This will be the first exposure to the areas that will be presented in much greater depth later in the text. Next, this chapter will establish a working vocabulary for movement description by presenting reference systems, anatomical descriptors, segment names, and names for all of the major movements of the body. By the end of this chapter, you should be able to describe a movement or skill using correct anatomical terminology and references. Numerous examples are provided to assist you with this.

For example, a common locomotor activity such as walking can be studied using different approaches. In exercise prescription, it may be important to know what muscles are used in walking and when in the walking cycle the muscles are used. It may also be important to understand the changes in muscular usage occurring when an individual walks up or down a hill or walks with weights on the ankles. To teach physical education, it may be important to understand the sequence of joint motions that comprise the walking pattern so that the pattern can be emphasized for students with developmental impairments. Also, a race-walking unit may be included in the curriculum, in which case a thorough understanding of the joint sequences would be helpful. In addition, understanding basic force concepts as they relate to walking substantiates the benefit of using walking, rather than running, as a fitness activity, since walking reduces by half the forces that may cause an injury. A completely different aspect of walking, concentrating on one injured joint, may be the focus of a physical therapist. For example, the hip joint may be creating an abnormal walking pattern such as a limp. Knowing this, the focus may be shifted to developing strength and flexibility surrounding the hip joint so that the individual can resume a normal walking pattern. Lastly, a researcher may be interested in measuring walking to evaluate forces at the feet, forces in the joints, muscle forces involved in the movement patterns of walking, and measuring the speeds of the body and the limbs. These research measurements may serve purposes such as designing a new walking shoe, evaluating an artificial limb, or describing the efficiency of various walking patterns. This textbook will explore the tools that can be used to conduct and understand any of these movement evaluations.

Core Areas of Study

Biomechanics vs Kinesiology

Among people who study human movement, there is often disagreement over the use of the terms "kinesiology" and "biomechanics". The word "kinesiology" can be used in two ways. First, kinesiology is the scientific study of human movement, and can be an umbrella term used to describe any form of anatomical, physiological, psychological, or mechanical human movement evaluation. A second use of the term kinesiology is to describe the content of a class where human movement is evaluated by examining its source and characteristics. Consequently, kinesiology has been used by several disciplines to describe many different content areas. Some departments of physical education have gone so far as to adopt kinesiology as the department name. A class in kinesiology may consist primarily of functional anatomy at one university and strictly biomechanics at another.

Historically, a kinesiology course has been part of college curricula as long as there have been physical education programs. The course originally focused on the musculoskeletal system, movement efficiency from the anatomical standpoint, and joint and muscular actions during simple and complex movements. A typical student activity in the kinesiology course was to identify discrete phases in an activity, describe the segmental movements occurring in each phase, and then identify the major muscular contributors to each joint movement. Thus, if one were completing a kinesiologic analysis of the movement of rising from a chair, the movements would be hip extension, knee extension, and plantar flexion via the hamstrings, quadriceps femoris, and triceps surae muscle groups, respectively. Most kinesiological analyses are considered qualitative because they involve observing a movement and providing a breakdown of the skills and identification of the muscular contributions to the movement.

The content from the study of kinesiology is currently incorporated into many biomechanics courses, and is used as a precursor to the introduction of the more quantitative biomechanical content. In this text, biomechanics will be used as an umbrella term to describe that content previously covered in courses in kinesiology, as well as content developed as a result of growth of the area of biomechanics (FIGURE 1–1).

In the 1960s and 1970s, biomechanics was being developed as an area of study within the undergraduate and graduate curricula across North America. The content of biomechanics was extracted from an area of physics, mechanics, the study of motion and the effect of forces on an object. Mechanics is used by engineers to design and build structures such as bridges, or machines such as airplanes, since it provides the tools for analyzing the strength of structures and ways of predicting and measuring the movement of a machine. It was a natural transition to take the tools of mechanics and apply them to living organisms. Thus, biomechanics, the study of the application of mechanics to biological systems, evolved.

Biomechanics evaluates the motion of a living organism and the effect of force–either a push or pull–on a living organism. The biomechanical approach to movement analysis can be qualitative, with movement observed and described, or quantitative, meaning that some measurement of the movement will be performed. The use of the term "biomechanics" in this text will incorporate qualitative components as well as a more specific, quantitative approach in which the motion characteristics of a human or an object will be described using such parameters as speed and direction, how the motion is created through application of forces both inside and outside the body, and the optimal body positions and actions for efficient, effective human or object motion. To evaluate biomechanically the motion of rising from a chair, for example, one would attempt to measure and identify joint forces acting at the hip, knee, and ankle, as well as the force between the foot and the floor, all of which act together to produce the movement up out of the chair. The components of a biomechanical and kinesiologic movement analysis are presented in FIGURE 1–1. Let's examine some of these components individually. A glossary of movement analysis terms is provided at the end of this chapter.

Anatomy vs Functional Anatomy

Anatomy, the science of the structure of the body, is the base of the pyramid from which expertise about human movement will be developed. It is very helpful to develop a strong understanding of regional gross anatomy so that for a specific region such as the shoulder, the bones, location of muscles, nerve innervation of those muscles, blood supply to those muscles, and other significant structures such as ligaments can be identified. A knowledge of anatomy can be put to good use if, for

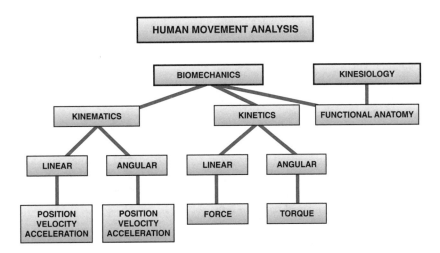

FIGURE 1–1. Types of movement analysis. Movement can be analyzed by assessing the anatomical contributions to the movement (functional anatomy), describing the motion characteristics (kinematics), or determining the cause of the motion (kinetics).

example, one is trying to assess an injury. Let's say that a patient has a pain on the inside of her elbow. Using a knowledge of anatomy, the medial epicondyle of the humerus would be recognized as the prominent bony structure of the medial elbow. It would also be known that the muscles that pull the hand and fingers toward the forearm into flexion attach on this epicondyle. Thus, familiarity with anatomy might lead to a diagnosis of medial epicondylitis, possibly caused by overuse of the hand flexor muscles.

Functional anatomy is the study of the body components needed to achieve or perform a human movement or function. Using functional anatomy to analyze a dumbbell lateral raise of the arm, the deltoid, trapezius, levator scapulae, rhomboid, and supraspinatus muscles would be identified as contributors to upward rotation and elevation of the shoulder girdle as well as the abduction of the arm. Knowledge of functional anatomy will be useful in a variety of situations to set up an exercise or weight-training program, assess the injury potential in a movement or sport activity, or when training techniques and drills for athletes are established. The prime consideration of a functional anatomy perspective is not the muscle's location, but rather the movement produced by the muscle or muscle group.

Linear vs Angular Motion

When looking at a human movement or an object being propelled by a human, two different types of motion are present. First is linear motion, often termed translation or translational motion, which is movement along a straight or curved pathway. Examples focusing only on the linear motion in the activity are the examination of the speed of a sprinter, the path of a baseball, the bar movement in a bench press, or the movement of the foot during a football punt. The focus in these activities is on the direction, path, and speed of the movement of the body or object. FIGURE 1–2 illustrates two different focal points for a linear movement analysis.

The center of mass of the body, segment, or object is usually the point monitored in a linear analysis (see FIGURE 1–2). The center of mass is the point about which the mass of the object is balanced, and it represents the point where the total effect of gravity acts on the object. However, any point can be selected and evaluated for linear motion characteristics. In skill analysis, for example, it is often very helpful to monitor the motion of the top of the head to gain an indication of certain trunk motions. An examination of the head in running is a prime example. Does the head move up and down? Side to side? If so, it is an indication that the central mass of the body is

FIGURE 1–2. Linear motion examples. Examining the motion of the center of gravity or the path of a projected object are examples of how linear motion analyses are applied.

moving in those directions as well. The path of the hand or racket is very important in throwing or racket sports, so visually monitoring the linear movement of the hand or racket throughout the execution of the skill is beneficial. In an activity such as sprinting, the linear movement of the whole body is the most important component to analyze since the object of the sprint is to move the body quickly from one point to another.

The second type of motion is angular motion, which is motion around some point where different regions of the same body segment or object do not move through the same distance. As illustrated in FIGURE 1–3, swinging around a high bar is an angular motion because the whole body rotates around the contact point with the bar. To make one full revolution around the bar, the feet travel through a much greater distance than the arms because they are farther away from the point of turning. It is typical in biomechanics to examine the linear motion characteristics of an activity and then follow up with a closer look at the angular motions that create and contribute to the linear motion.

All linear movements of the human, or objects propelled by humans, occur as a consequence of angular contributions. The only exceptions to this rule are movements such as skydiving or free falling where the body is held in a position to let gravity create the movement lin-

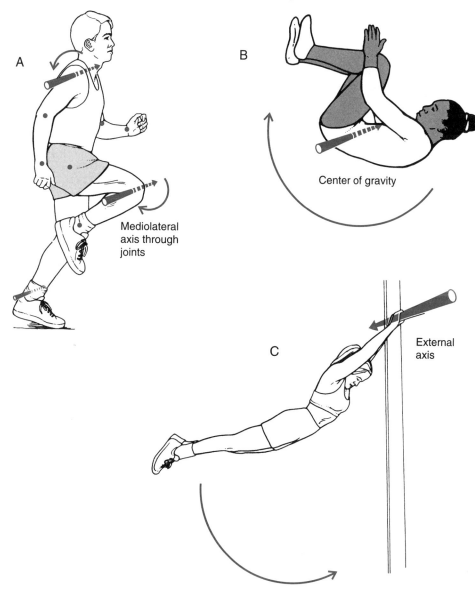

FIGURE 1–3. Angular motion examples. Angular motion of the body, an object, or segment can take place around an axis running through a joint (A), through the center of gravity (B), or about an external axis (C).

early downward, or cases where an external pull or push moves the body or an object. It is important to identify the angular motions and the sequence of angular motions that make up a skill or human movement, because the angular motions will determine the success or failure of the linear movement.

Angular motions occur around an imaginary line called the axis of rotation. Angular motion of a segment, such as the arm, occurs about an axis running through the joint. Thus, lowering the body into a deep squat entails angular motion of the thigh around the hip joint, angular motion of the leg around the knee joint, and angular motion of the foot around the ankle joint. Angular motion can also occur around an axis through the center of mass. Examples of this type of angular motion are performing a somersault in the air, and spinning vertically when figure skating. Finally, angular motion can occur around a fixed external axis, such as swinging around a high bar, rotating over the foot in a run or walk, or swinging on the end of the pole in the pole vault.

To be proficient in human movement analysis, the identification of angular motion contributions to the linear motion of the body or an object is important. This is apparent in a simple activity such as kicking a ball for maximum distance. The intent of the kick is to make solid contact between a foot travelling at a high linear speed and moving in the proper direction and a ball to send the ball in the desired linear direction. The linear motion of interest is the actual path and movement of the ball after it leaves the foot. To create the high speeds and the right path, the angular motions in the kicking leg are sequential and draw speed from each other so that the velocity of the foot is determined by the summation of the individual velocities of the connecting segments. The kicking leg moves into a preparatory phase and draws back through angular motions of the thigh, leg, and foot. The leg whips back underneath the thigh very quickly as the thigh starts to move forward to initiate the kick. In the power phase of the kick, the thigh moves vigorously forward, and rapidly extends the leg and foot forward at very fast angular speeds. As contact is made with the ball, the foot is moving very fast because the velocities of the thigh and leg have been transferred to the foot. Through a skilled observation of human movement, the relationship between angular and linear motion, shown in this kicking example, will serve as a foundation for techniques used to correct or facilitate a movement pattern or skill.

Kinematics vs Kinetics

Biomechanical analysis can be conducted from one of two perspectives. The first, kinematics, is concerned with motion characteristics, and examines motion from a spatial and temporal perspective without reference to the forces causing motion. A kinematic analysis involves the description of movement to determine how fast an object is moving, how high it goes, or how far it travels. Thus, position, velocity, and acceleration are the components of interest in a kinematic analysis. Examples of linear kinematic analysis are the examination of the projectile characteristics of a high jumper and a study of the performance of elite swimmers. Examples of angular kinematic analysis are an observation of the joint movement sequence for a tennis serve, and an examination of the segmental velocities and accelerations in a vertical jump. FIGURE 1–4 presents both an angular and linear example of the kinematics of the golf swing. By examining an angular or linear movement kinematically, we can identify segments of a movement needing improvement, obtain ideas and technique enhancements from elite performers, break a skill down into identifiable parts, and further our understanding of human movement.

Pushing on a table may or may not move the table, depending upon the direction and strength of the push. A push or pull between two objects that may or may not result in motion is termed a force. Kinetics is the area of study that examines the forces acting on a system such as the human body or any object. A kinetic movement analysis area attempts to define the forces causing a movement. A kinetic movement analysis is more difficult than a kinematic analysis to both comprehend and evaluate, since forces cannot be seen (FIGURE 1–5). Only the effects of forces can be observed!

Watch someone lift a 200 pound barbell in a squat. How much force has been applied? Since the force cannot be seen, there is no way of accurately evaluating the force unless it can be measured using recording instruments. A likely estimate of the force would be at least 200 pounds since that is the weight of the bar. The estimate may be off by a significant amount if the weight of the body lifted and the speed of the bar is not considered. The forces produced are very important since they are responsible for creating all of our movements and for maintaining positions or postures having no movement. The assessment of these forces represents the greatest technical challenge in this field since it requires equipment and expertise. Thus, for the novice movement analyst, concepts relating to maximizing or minimizing force production in the body will be more important than evaluating the actual forces themselves.

A kinetic analysis can provide the teacher, therapist, coach, or researcher with valuable information about how the movement is produced or how a position is

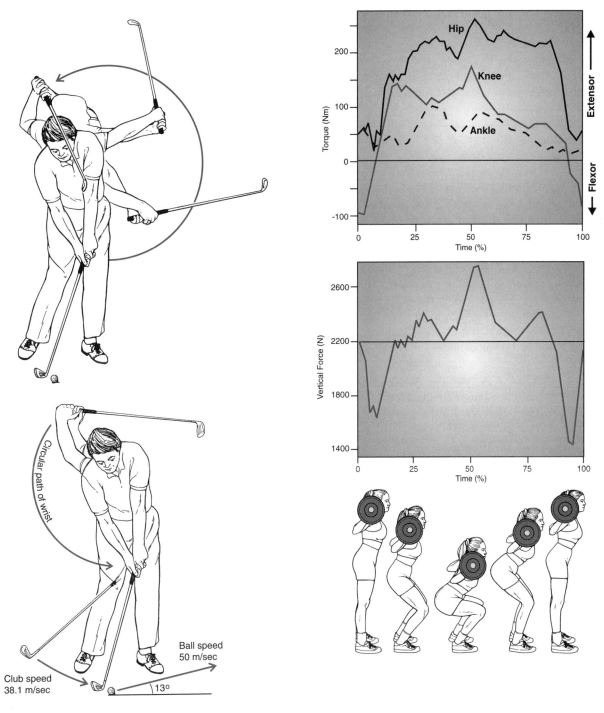

FIGURE 1–4. Examples of kinematic movement analysis. A kinematic analysis is one which focuses on the amount and type of movement, the direction of the movement, and the speed or change in speed of the body or an object. The golf example above is presented from two of these perspectives: the angular components of the golf swing (top) and the linear direction and speed of the club and ball (bottom).

FIGURE 1–5. Examples of kinetic movement analysis. A kinetic analysis is one which focuses on the cause of movement. The lifting example above demonstrates how lifting can be analyzed by looking at the vertical forces on the ground that produce the lift (linear) and the torques produced at the three lower extremity joints which generate the muscular force required for the lift (Redrawn from Lander, 1984).

maintained. This information can direct conditioning and training for a sport or movement. For example, kinetic analyses performed by researchers have identified weak and strong positions in various joint movements. Thus, we know that starting an arm curl exercise with the weights hanging down and the forearm straight is the weakest position. If the same exercise was started with the elbow slightly bent, more weight could be lifted. The area of kinetics also identifies the important parts of a skill in terms of movement production. For example, what is the best technique for maximizing a vertical jump? By measuring the forces produced against the ground that are used to propel upward, researchers have agreed that the vertical jump incorporating a very quick drop, stop, and pop action upward produces more effective forces at the ground than the slow, deep, gather jump.

Lastly, the area of kinetics has played a crucial role in identifying aspects of a skill or movement that make the performer prone to injury. Why do 43% of participants and 76% of instructors of high-impact aerobics incur an injury (1)? The answer was clearly identified through a kinetic analysis that found forces in typical high impact aerobic exercises to be in the magnitude of 4–5 times body weight (2). For an individual weighing 667.5 newtons (150 lbs), repeated exposure to forces in the range of 2670–3337.5 newtons (600–750 lbs) partially contributes to injury of the musculoskeletal system.

To fully understand all aspects of a movement, examination of both the kinematic and kinetic components must be performed. It is also important to study the kinematic and kinetic relationships since any acceleration of a limb, object, or human body is a result of a force applied at some point, at a particular time, of a given magnitude, and for a particular duration. While it is of some use to merely describe the motion characteristics kinematically, one must also explore the kinetic sources before a thorough comprehension of a movement or skill is possible.

Statics vs Dynamics

Examine the posture used to sit at a desk and work at a computer. Are there forces present? Yes, even though there is no movement, there are forces present between the back and the chair, the foot and the ground, as well as muscular forces acting in the neck to counteract gravity in order to keep the head up and looking at the screen. Forces are present without motion and are being produced continuously to maintain positions and postures that do not involve movement. Principles from the area of *statics* are used to evaluate the sitting posture. Statics is the examination of systems that are not moving, or are

moving at a constant speed such that they are considered to be in equilibrium. Equilibrium is a balanced state in which there is no acceleration because the forces causing a person, or object, to begin moving, speed up, or slow down are neutralized by opposite forces that cancel them out.

Statics is also very useful for determining stresses on anatomical structures in the body, identifying the magnitude of muscular forces, and identifying the magnitude of force that would result in the loss of equilibrium and create movement in the system. How much force from the deltoid would be required to hold the arm out to the side? Why is it easier to hold an arm at the side if you lower or raise the arm so that it is no longer perpendicular to the body? What is the effect of an increased curvature, or swayback, on forces coming through the lumbar vertebrae? These are the types of questions one might answer using a static analysis. Since the static case involves no change in the kinematics of the system, a static analysis is usually performed using kinetic techniques to identify the forces and the site of the force applications responsible for maintaining a posture, position, or constant speed in an object. However, kinematic analyses can be applied in statics to substantiate whether there is equilibrium through the absence of acceleration.

To leave the computer work station and get up out of the chair, forces must be produced in the lower extremity and on the ground to produce the rising motion. *Dynamics* is the area of mechanics used to evaluate this movement since it examines systems in accelerated motion using both the kinematic and kinetic approach to movement analysis. A dynamic analysis of an activity such as running would incorporate a kinematic analysis in which the linear motion of the total body and the angular motion of the segments would be described and related to the kinetic analysis, which would then describe forces applied to the ground and across the joints in order to produce the running actions. Since this textbook will deal with numerous examples involving motion of the human or a human-propelled object, the area of dynamics will be addressed in detail later in specific chapters on linear and angular kinematics and kinetics.

Anatomical Movement Descriptors
Segment Names

To flex the arm, would a lift be performed at the elbow with weights in the hand or would the whole arm be raised in front of you? Whatever interpretation is placed on the segment name "arm" will determine the type of movement performed. It is important to correctly identify segment names and use them consistently when analyzing movement. The correct interpretation of flexing the arm would be to raise the whole arm, since the arm refers to the humerus, not the radius and the ulna. A review of segment names is worthwhile in preparation for more extensive use of them in the study of biomechanics.

The head, neck, and trunk are segments composing the main part of the body and the axial portion of the skeleton. This portion of the body is large, accounting for more than 50% of a person's weight, and it usually moves much more slowly than the other parts of the body. Because of its large size and slow speed, the trunk is a good segment to visually observe when one is learning to analyze movement or if one wants to follow the total body activity.

The upper and lower extremities are termed the appendicular portion of the skeleton. Generally speaking, as one moves away from the trunk, the segments become smaller, move faster, and are more difficult to observe due to their size and speed. In the upper extremity, the humerus is termed the arm, the radius and the ulna constitute the forearm, and the carpals, metacarpals, and phalanges are termed the hand. Thus, in the example above, the movement of arm flexion would be a movement of raising the upper extremity in front, while forearm flexion would describe a movement at the elbow joint. The movements of the arm will typically be described as they occur in the shoulder joint, the forearm movements will be described in relation to either elbow or radioulnar joint activity, and the hand movements will be described relative to the wrist joint activity. See FIGURE 1–6 for a graphical representation of the segments.

In the lower extremity, the segment name thigh describes the femur, leg describes the tibia and fibula segment, and foot describes the tarsals, metatarsals, and phalanges. Additionally, the movement of the thigh will typically be described as it occurs at the hip joint, the leg movement as described by actions at the knee joint, and the foot movements as determined by ankle joint activity.

Anatomical Terms

The description of a segmental position or joint movement is typically expressed relative to a designated starting position. The anatomical starting position has been a standardized reference used for many years by anatomists, biomechanists, and the medical profession. In this position, the body is in an erect stance with the head facing forward, arms at the side of the trunk with palms facing forward, and the legs together with the feet pointing forward. Some biomechanists prefer to use the fundamental starting position that is similar to the

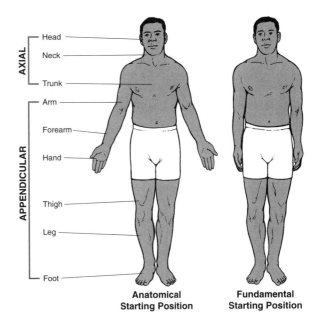

FIGURE 1–6. Anatomical vs fundamental starting position. The anatomical and fundamental starting positions serve as a reference point for the description of joint movements.

anatomical position except that the arms are in a more relaxed posture at the sides with the palms facing in toward the trunk. Whatever starting position used, all segmental movement descriptions are made relative to this starting position. Both of these starting positions are illustrated in FIGURE 1–6.

The starting position is also called the zero position, or origin, for description of most joint movements. For example, when standing, there is zero movement at the hip joint. If the thigh is lifted or rotated in or out, the amount of movement is described relative to the fundamental or anatomical starting position. Most zero positions appear to be quite obvious since there is usually a straight line between the two segments so that no relative angle is formed between them. A relative angle is the included angle between the two segments. Zero position in the trunk occurs when the trunk is vertical and lined up with the lower extremity. The zero position at the knee is found in the standing posture when there is no angle between the thigh and the leg. One not so obvious zero position is at the ankle joint, and is the position assumed in stance with the foot at a right angle to the leg.

Movement description or anatomical location can best be presented using terminology universally accepted and understood. Movement terms should become a part of a working vocabulary, regardless of the level of application of kinesiology required. The development of a solid knowledge of the movement characteristics of the various phases of a human movement or a sport skill can improve the effectiveness of teaching a skill, assist in correcting flaws in a performance, identify important movements and segments for emphasis in conditioning, and identify aspects of the skill that may be associated with injury. The experienced coach or teacher can determine the most relevant movements in a skill, and will use a specific vocabulary of terms to instruct students or athletes. A standardized set of terms is most helpful in this situation.

The term medial refers to a position relatively closer to the midline of the body or object, or a movement that moves toward the midline. In the anatomical position, the little finger and the big toe would be considered to be on the medial side of the extremity, since they are on the side of the limb closest to the midline of the body. Also, pointing the toes in is considered a medial movement because it moves toward the midline. The term lateral describes a position relatively farther away from the midline or a movement away from the midline. The thumb and the little toe are on the lateral side of the hand and foot, respectively, since they are farther away from midline. Likewise, if toes are pointed out, it is considered a lateral movement, since the movement is away from the midline. Landmarks are also commonly designated as medial or lateral based on their relative position to the midline, such as medial and lateral condyles, epicondyles, and malleoli.

Proximal and distal positions are used to describe the relative position with respect to a designated reference point, with proximal representing a position closer or nearest to the reference point, and distal being a point further away from the reference. The elbow joint is proximal and the wrist joint is distal in relation to the shoulder joint. The ankle joint is proximal and the knee joint is distal in relationship to the point of heel contact with the ground. Both proximal and distal must be expressed in relation to some reference point.

A segment or anatomical landmark could be located on the superior aspect of the body, placing it above a particular reference point or closer to the top of the head, or it could be located inferiorly, where it would be lower than a reference segment or landmark. Consequently, the head is positioned superior to the trunk, the trunk is superior to the thigh, and so on. The greater trochanter is located on the superior aspect of the femur, while the medial epicondyle of the humerus is located on the inferior end of the humerus.

The location of an object or a movement relative to the front or back is termed anterior and posterior, respec-

tively. Thus, the quadriceps are located on the anterior aspect of the thigh, while the hamstrings are located on the posterior aspect of the thigh. Anterior is also synonymous with the term ventral when describing a location on the human body, while posterior refers to the dorsal surface or position on the human.

To describe activity or the location of a segment or landmark positioned on the same side as a particular reference point, one would use the term ipsilateral. Actions, positions, or landmark locations on the opposite side can be designated using the term contralateral. Thus, when a person lifts his right leg in a forward direction, there is extensive muscular activity in the iliopsoas muscle of that same leg, the ipsilateral leg, and there is also extensive activity in the gluteus medius of the contralateral leg in order to maintain balance and support. In walking, as the ipsilateral lower limb is swinging forward, the other limb, the contralateral limb, is pushing on the ground to propel the walker forward.

Movement Description

Basic Movements There are six basic movements occurring in varying combinations in the joints of the body. The first two movements, flexion and extension, are movements found in almost all of the synovial, or freely moveable, joints in the body, including the toes, ankle, knee, hip, trunk, shoulder, elbow, wrist, and fingers. Flexion is a bending movement where the relative angle between two adjacent segments decreases. Extension is a straightening movement where the relative angle between two adjacent segments increases as the joint returns back to the zero, anatomical position. Numerous examples of both flexion and extension are provided in FIGURE 1–7. A person can also perform hyperflexion if the flexion movement goes beyond 180 degrees of flexion or more than half of a circle. This can only happen when the arm moves forward and up in flexion through 180 degrees until it is at the side of the head, and then hyperflexes as it continues to move past the head toward the back. Hyperextension can occur in many different joints as the extension movement continues past the original zero position. It is common to see hyperextension movements in the trunk, arm, thigh, and hand.

In a toe-touch movement, there is flexion at the vertebral, shoulder, and hip joints. The return to the standing position would involve the opposite movements of vertebral extension, hip extension, and shoulder extension. The power phase of the jump shot is produced via smooth timing of lower extremity movements of hip extension, knee extension, and ankle extension coordinated with the movements of shoulder flexion, elbow extension, and

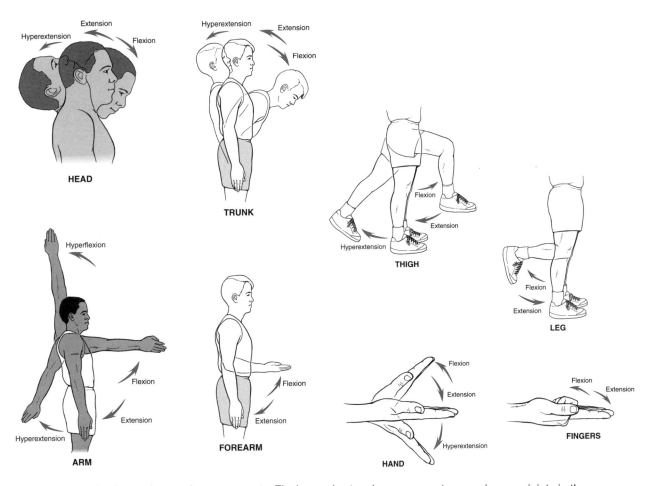

FIGURE 1–7. Flexion and extension movements. Flexion and extension movements occur in many joints in the body including: vertebral, shoulder, elbow, wrist, metacarpophalangeal, interphalangeal, hip, knee, and metatarsophalangeal.

wrist flexion in the shooting limb. This example illustrates the importance of the lower extremity extension movements to the production of power. Lower extremity extensions often serve to produce upward propulsion, working against the downward pull of gravity. It is opposite in the shoulder joint where flexion movements are primarily used to develop propulsion upward against gravity in order to raise the limb.

The second pair of movements, abduction and adduction, are not as common as flexion and extension, and only occur in the metatarsophalangeal, hip, shoulder, wrist, and metacarpophalangeal joints. Many of these movements are presented in FIGURE 1–8. Abduction is a movement away from the midline of the body or the segment. Raising an arm or leg out to the side or the spreading of the fingers or toes is an example of abduction. Hyperabduction can occur in the shoulder joint as the arm moves more than 180 degrees from the side all the way up past the head. Adduction is the return movement of the segment back toward the midline of the body or segment. Consequently, bringing the arms back to the trunk, bringing the legs together, or closing the toes or fingers are examples of adduction. Hyperadduction occurs frequently in the arm and thigh as the adduction continues past the zero position so that the limb crosses the body. These side-to-side movements are commonly used to maintain balance and stability during the performance of both upper- and lower-extremity sport skills. Controlling or preventing abduction and adduction movements of the thigh are especially crucial to the maintenance of pelvic and limb stability during walking and running.

The last two basic movements are segment rotations, illustrated in FIGURE 1–9. Rotations can be either medial, also known as internal, or lateral, also known as external. Rotations are designated as right and left for the

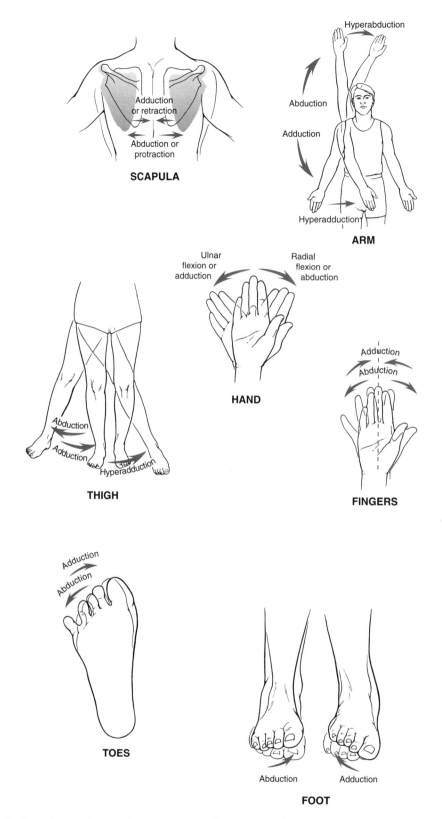

FIGURE 1–8. Abduction and adduction movements. Abduction and adduction movements occur in the sternoclavicular, shoulder, wrist, metacarpophalangeal, hip, intertarsal, and metatarsophalangeal joints.

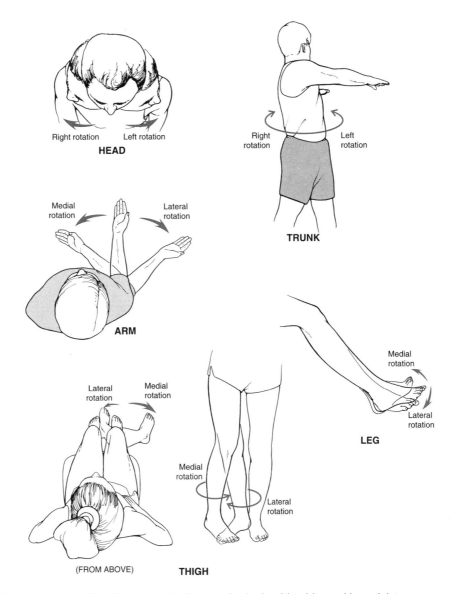

FIGURE 1–9. Rotation movements. Rotation occurs in the vertebral, shoulder, hip, and knee joints.

head and trunk only. When in the fundamental starting position, medial or internal rotation refers to the movement of a segment around a vertical axis running through the segment, so that the anterior surface of the segment moves toward the midline of the body while the posterior surface moves away from the midline. Lateral or external rotation is the opposite movement in which the anterior surface moves away from the midline, and the posterior surface of the segment moves toward the midline. Since the midline runs through the trunk and head segments, the rotations in these segments are described as left or right, from the perspective of the performer. Right rota-

tion is the movement of the anterior surface of the trunk so that it faces right while the posterior surface faces left, and left rotation is the opposite movement so that the anterior trunk faces left and the posterior trunk faces right. Rotations occur in the vertebrae, shoulder, hip, and knee joints. The rotation movements are very important in the power phase of sport skills involving the trunk, arm, or thigh. For the skill of throwing, the throwing arm laterally rotates in the preparatory phase and medially rotates in the power and follow-through phases. The trunk complements the arm action with right rotation in the preparatory phase (right-handed thrower) and left rotation in the power and follow-through phase. Likewise, the right thigh laterally rotates in the preparatory phase and medially rotates until it comes off the ground in the power phase.

Specialized Movement Descriptors   There are specialized movement names assigned to a variety of segmental movements (FIGURE 1–10). While most of these segmental movements are technically among the six basic movements described above, the specialized movement name is the terminology commonly used by movement professionals. Right and left lateral flexion is a specialized movement name that applies only to the movement of the head or trunk. When the trunk or head tilts sideways, the movement is termed lateral flexion. If the right side of the trunk

or head moves so that it faces down, the movement is termed right lateral flexion, and vice versa.

The shoulder girdle has specialized movement names that can best be described by observing the movements of the scapula. The raising of the scapula, as in a shoulder shrug, is termed elevation, while the opposite lowering movement is depression. If the two scapulae move away from each other, as in rounding the shoulders, the movement is termed protraction or abduction. The return movement where the scapulae move towards each other with the shoulders back, is called retraction or adduction. Finally, the scapulae can swing out so that the bottom of the scapula moves away from the trunk and the top of the scapula moves toward the trunk. This movement is termed upward rotation, and the return movement, when

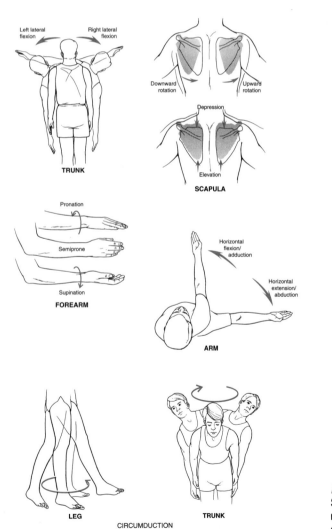

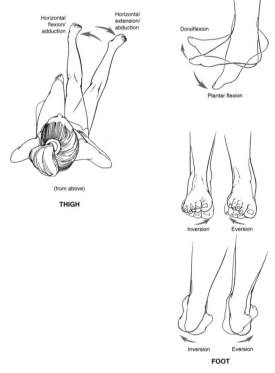

FIGURE 1–10. Examples of specialized movements. Some joint movements are designated using specialized movement names, even though they may, technically, be one of the six basic movements.

the scapula swings back down into the resting position, is called downward rotation.

In the arm and the thigh, a combination of flexion and adduction is termed horizontal adduction, and a combination of extension and abduction is called horizontal abduction. Horizontal adduction, sometimes called horizontal flexion, is the movement of the arm or thigh across the body, toward the midline, using a movement horizontal to the ground. Horizontal abduction, or horizontal extension, is a movement of the arm or thigh away from the midline of the body in the horizontal direction. These movements are present in a wide variety of sport skills. The arm action of the discus throw is a good example of the use of horizontal abduction in the preparatory phase and horizontal adduction of the arm in the power and follow-through phase. Many soccer skills utilize horizontal adduction of the thigh to bring the leg up and across the body for a shot or pass.

In the forearm, the movements of pronation and supination occur as the distal end of the radius rotates over and back on the ulna at the radioulnar joints. Supination is the movement of the forearm in which the palm rotates to face forward from the fundamental starting position. Pronation is the movement in which the palms face backward. Supination and pronation joint movements have also been referred to as external and internal rotation, respectively. As the forearm moves from a supinated position to a pronated position, the forearm passes through the semiprone position where the palms face the midline of the body with the thumbs forward. The actions of forearm pronation and supination are used with arm rotation movements to increase the range of motion, add spin, enhance power, and change direction during the force application phases in racket sports, volleyball, and throwing.

At the wrist joint, the movement of the hand toward the thumb is called radial flexion while the opposite movement of the hand toward the little finger is called ulnar flexion. These specialized movement names are easier to remember because they do not depend on forearm or arm position as do the interpretation of abduction and adduction, and they can easily be interpreted if the location of the radius (thumb side) and the ulna (little finger side) is known. Ulnar and radial flexion are important movements in racket sports for control and stabilization of the racket. Also, in volleyball, the movement of ulnar flexion is a valuable component of the forearm pass as it helps to maintain the extended arm position and increases the contact area of the forearms.

In the foot, the movements of plantar flexion and dorsiflexion are specialized names for foot extension and flexion, respectively. Plantar flexion is the movement in which the bottom of the foot moves down and the angle formed between the foot and the leg increases. This movement can be created by raising the heel so the weight is shifted up on the toes, or by placing the foot flat on the ground in front and moving the leg backward so that the body weight is behind the foot. Dorsiflexion is the movement of the foot up toward the leg that decreases the relative angle between the leg and the foot. This movement may be created by putting weight on the heels and raising the toes, or by keeping the feet flat on the floor and lowering with weight centered over the foot. Any foot-leg angle greater than 90 degrees is termed a plantar flexed position while any foot-leg angle less than 90 degrees is termed dorsiflexion.

The foot has another set of specialized movements called inversion and eversion that occur in the intertarsal and metatarsal articulations. Inversion of the foot takes place when the medial border of the foot lifts so that the sole of the foot faces inward toward the other foot. Eversion is the opposite movement of the foot where the lateral aspect of the foot lifts so that the sole of the foot faces outward away from the other foot.

Often there is confusion over the use of the terms inversion and eversion and the popularized use of pronation and supination as descriptors of foot motion. Inversion and eversion movements are not the same as pronation and supination; in fact, they are only a part of the pronation and supination movements. Pronation of the foot is actually a combined set of movements consisting of dorsiflexion at the ankle joint, eversion in the tarsals, and abduction of the forefoot. Supination is created through ankle plantar flexion, tarsal inversion, and forefoot adduction. It is important to note that pronation and supination are dynamic movements of the foot and ankle occurring when the foot is on the ground during a run or walk. These two movements are determined by structure and laxity of the foot, body weight, playing surfaces, and shoes.

The final specialized movement, circumduction, is a movement that can be created in any joint or segment that has the potential to move in two directions, so that the segment can be moved in a conic fashion as the end of the segment moves in a circular path. An example of circumduction would be placing the arm out in front and drawing an imaginary "O" in the air. Circumduction is not a simple rotation, but a combination of four movements created in sequential combination. The movement of the arm in the creation of the imaginary "O" is actually a combination arm action of flexion, adduction, extension, and abduction. Circumduction movements are also possible in the foot, thigh, trunk, head, and hand. The movements of all of the major segments are reviewed in TABLE 1–1.

Reference Systems

Relative vs Absolute

To observe and describe any type of motion accurately, a reference system must first be established. The use of joint movements relative to the fundamental or anatomical starting position is an example of a simple reference system, which was previously used in this chapter to describe movement of the segments. To improve upon the precision of a movement analysis, a movement can be evaluated with respect to a different starting point or position.

A reference system is necessary in order to specify position of the body, a segment, or an object to describe motion or identify whether any motion has occurred. The reference frame or system is arbitrarily established, and may be placed in or out of the body. The reference frame is placed at a designated spot and usually consists of 2 to 3 imaginary lines called axes that intersect at right angles at a common point termed the origin. Any position can be described by identifying the distance the object is from each of the three axes. It is very important to identify the frame of reference used in the description of motion.

An example of a reference system placed outside the body is the starting line in a 100 m race. The center of an anatomical joint such as the shoulder can be used as a reference system within the body. The arm could be described as moving through a 90 degree angle if abducted until it was at right angles to the trunk. However, if the ground is used as a frame of reference, the same arm abduction movement could be described with respect to the ground, such as movement to a height of 1.6 m from the ground.

When angular motion is described, the joint positions, velocities, and accelerations can be described using either an absolute or a relative frame of reference. An absolute reference frame is one in which the three axes intersect in the center of the joint and movement of the

Table 1–1 Movement Review

Segment	Joint	Df	Movements
Head	Intervertebral	3	Flexion, extension, hyperextension, R/L lateral flexion, R/L rotation, circumduction
	Atlantoaxial (3 jts)	1 each	R/L rotation
Trunk	Intervertebral	3	Flexion, extension, hyperextension, R/L rotation R/L lateral flexion, circumduction
Arm	Shoulder	3	Flexion, extension, hyperextension, abduction, adduction, hyperabduction, hyperadduction, horizontal abduction, horizontal adduction, med/lat rotation, circumduction
Arm/Shoulder	Sternoclavicular	3	Elevation, depression,
Girdle	Acromioclavicular	3	abduction, adduction (protraction, retraction), upward/downward rotation
Forearm	Elbow	1	Flexion, extension, hyperextension
	Radioulnar	1	Pronation, supination
Hand	Wrist	2	Flexion, extension, hyperextension, radial flexion, ulnar flexion, circumduction
Fingers	Metacarpophalangeal	2	Flexion, extension, hyperextension, abduction, adduction, circumduction
	Interphalangeal	1	Flexion, extension, hyperextension
Thumb	Carpometacarpal	2	Flexion, extension, abduction, adduction, opposition, circumduction
	Metacarpophalangeal	1	Flexion, extension
	Interphalangeal	1	
Thigh	Hip	3	Flexion, extension, hyperextension, abduction, adduction, hyperadduction, horizontal adduction, horizontal abduction, med/lat rotation, circumduction
Leg	Knee	2	Flexion, extension, hyperextension, med/lat rotation
Foot	Ankle	1	Plantar flexion, dorsiflexion
	Intertarsal	3	Inversion, eversion
Toes	Metatarsophalangeal	2	Flexion, extension, abduction, adduction, circumduction
	Interphalangeal	1	Flexion, extension

segment is described with respect to that joint. The absolute positioning of an abducted arm perpendicular to the trunk is 0 or 360 degrees when described relative to the axes running through the shoulder joint. A relative reference frame is one in which the movement of a segment is described relative to the adjacent segment. See FIGURE 1–11 for an example of an absolute and relative reference system applied to the lower extremity. In the same arm example with abduction perpendicular to the trunk, the relative positioning of the arm with respect to the trunk is 90 degrees. The reference frame should be clearly identified so that the results can be interpreted accordingly, and since different reference systems are used by different researchers, the reference system and reference point need to be identified before comparing and contrasting results between studies. For example, some researchers label a fully extended forearm as a 180 degree position and others label the position 0 degrees. After 30 degrees of flexion at the elbow joint, the final position would be 150 degrees or 30 degrees, respectively, for the two systems described above. There can be considerable confusion when trying to interpret an article using a different reference system than the authors.

Planes/Axes

The universally used method of describing human movements in three dimensions is based on a system of planes and axes. Three imaginary planes are positioned through the body at right angles so they intersect at the center of mass of the body. Movement is said to occur in a specific plane if it is actually along the plane or parallel to it. Movement in a plane always occurs about an axis running perpendicular to the plane. Refer to the illustra-

tion presented in FIGURE 1–12. Stick a pin piece of cardboard and spin the paper around the pin. movement of the cardboard takes place in the plane, and the pin represents the axis of rotation. The cardboard can spin around the pin while holding the pin horizontally, vertically, or sideways, representing movement of the cardboard in all three of the planes. This example can be applied to describe imaginary lines running through the hip joint in the same three pin directions, representing thigh flexion about an imaginary line running side to side, abduction about an imaginary line running front to back, rotation about an imaginary line running up and down. These planes allow one to describe fully a motion and to contrast an arm movement straight out in front of the body with one straight out to the side of the body. The planes and axes placed on the human body for motion description are presented in FIGURE 1–13.

The sagittal plane bisects the body into right and left halves. Movements occurring in the sagittal plane occur about a frontal or "z" axis running side to side. These movements can occur about an axis running through a joint, through the center of the body at the center of mass, or through an external point of contact such as a high bar. FIGURE 1–14 provides examples of these movements. Most two-dimensional analyses in biomechanics are concerned with motion in the sagittal plane. Examples of sagittal plane movements around a joint can be demonstrated by performing flexion and extension movements,

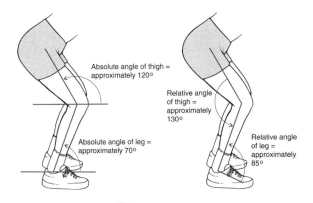

FIGURE 1–11. Absolute vs relative reference frame. It is important to designate the reference frame you are using in movement description. An absolute reference frame measures the segment angle with respect to the distal joint. A relative reference frame measures the angle formed by the two segments.

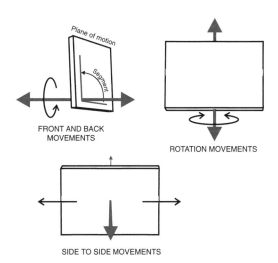

FIGURE 1–12. The plane and axis. Movement takes place in a plane about an axis running perpendicular to the plane.

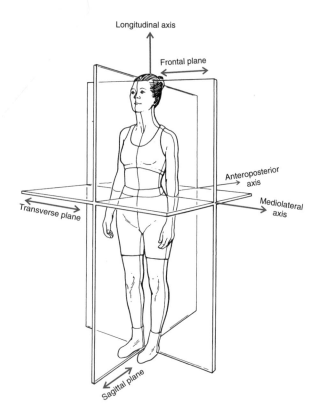

Longitudinal axis

Frontal plane

Anteroposterior
axis

Mediolateral
axis

Transverse plane

Sagittal plane

FIGURE 1–13. Planes and axes on the human body. The three cardinal planes that originate at the center of gravity are the sagittal plane, which divides the body into right and left halves, the frontal plane, dividing the body into front and back halves, and the transverse plane, dividing the body into top and bottom halves. Movement takes place in or parallel to the planes about a mediolateral axis (sagittal plane), an anteroposterior axis (frontal plane), or a longitudinal axis (transverse plane).

rior and posterior from the plane. Like the sagittal plane movements, frontal plane movements can occur about a joint, about the center of mass of the body, or about an external point of contact (FIGURE 1–15). Characteristic joint movements in the frontal plane include thigh abduction/adduction, finger and hand abduction/adduction, lateral flexion of the head and trunk, and inversion and eversion of the foot. Frontal plane motions of the whole body about the center of mass are not as common as movements in the other planes, but, aerials would be an example. Frontal plane motion about an external point of contact can be seen often in dance and ballet especially, as the dancers move laterally from a pivot point, or in gymnastics where the body will rotate sideways over the hand, such as in the cartwheel. To view frontal plane movements, the best position is in the front or back of the body to focus on the joint or the point about which the whole body is going to rotate.

The transverse or horizontal plane bisects the body to create upper and lower halves. Movements occurring in this plane are primarily rotations occurring about a longitudinal, vertical, or "y" axis. Examples of movements occurring in the transverse plane about vertical joint axes are rotations at the vertebral, shoulder, and hip joints. Pronation and supination of the forearm at the radioulnar joints is also a transverse plane movement. The axis for all of these movements is an imaginary line running vertically down through the vertebrae, the shoulder, the radioulnar, or the hip joints. An example of transverse plane movement about the body's center of mass would be spinning vertically around the body while in the air (FIGURE 1–16). This is a very common movement in gymnastics, dance, and ice skating. There are also numerous examples from dance, skating, or gymnastics in which the athlete performs transverse plane movements about an external axis running through a pivot point between the foot and the ground. All spinning movements that have the whole body turning about the ground or the ice are examples. While transverse plane motions are very vital aspects of most successful sport skills, these movements are difficult to follow visually since the best viewing position is either above or below the movement to be perpendicular to the plane of motion. Consequently, rotation motions are evaluated by following the linear movement of some point on the body if vertical positioning cannot be achieved.

Most human movements take place in two or more planes at the various joints. In running, for example, the lower extremity appears to move predominantly in the sagittal plane as the lower limbs swing forward and back through the cycle. Upon closer examination of the limbs

such as raising the arm in front, bending the trunk forward and back, lifting and lowering the leg in front, and raising up on the toes. Sagittal plane movements involving the whole body rotating around the center of mass include somersaults, backward and forward handsprings, and flexing to a pike position in a dive. Sagittal plane movements of the body about an external support include planting the foot and rotating the body over the foot, and rotating over the hands in a vault. To obtain the most accurate view of any motion in a plane, a position perpendicular to the plane of movement to allow viewing along the axis of rotation is preferred. Therefore, all of the activities offered as examples of sagittal plane movements are best viewed from the side of the body to allow focus on a frontal axis of rotation.

The frontal or coronal plane bisects the body to create front and back halves. The axis about which frontal plane movements occur is the sagittal or "x" axis running ante-

SAGITTAL PLANE MOVEMENTS ABOUT AXIS JOINTS

SAGITTAL PLANE MOVEMENTS ABOUT THE CENTER OF GRAVITY

SAGITTAL PLANE MOVEMENTS ABOUT AN EXTERNAL AXIS

FIGURE 1–14. Movements in the sagittal plane. Sagittal plane movements are typically flexions and extensions or some forward or backward turning exercise. The movements can take place about a joint axis, the center of gravity, or an external axis.

and joints, one will find movements in all of the planes. At the hip joint, for example, the thigh will perform flexion and extension in the sagittal plane, it abducts and adducts in the frontal plane, and rotates internally and externally in the transverse plane. If the human movements were more confined to single plane motion, we would look like robots as we performed our skills or joint motions. Examine the three-dimensional motion for an overhand throw presented in FIGURE 1–17. Note the positioning for viewing motion in each of the three planes.

The movement in a plane can also be described as a single degree of freedom (df). This terminology is commonly used to describe the type and amount of motion structurally allowed by the anatomical joints. One df for a joint indicates that the joint allows the segment to move through one plane of motion. A joint with 1 df is also termed uniaxial, since there is one axis, perpendicular to the plane of motion, about which movement occurs. A 1 df joint, the elbow, only allows flexion and extension in the sagittal plane.

Conventionally, most joints are considered to have 1, 2, or 3 degrees of freedom offering movement potential that is uniaxial, biaxial, or triaxial, respectively. The shoulder is an example of a 3 df triaxial joint because it allows the arm to move in the frontal plane via abduction and adduction, in the sagittal plane via flexion and extension, and in the transverse plane via rotation.

Three-degrees-of-freedom joints in the body include the vertebrae, shoulder, and hip joint; 2 df joints include the knee, metacarpophalangeal, wrist, and the thumb carpometacarpal joints; 1 df joints include the atlantoaxial, interphalangeal, elbow, radioulnar, and ankle joints. Three degrees of freedom does not always imply great mobility, but it does indicate that the joint allows movement in all three planes of motion. The shoulder is much more mobile than the hip, even though they both are triaxial joints and are capable of performing the same movements. The trunk movements, although classified as having 3 df, are quite restricted if one evaluates movement at a single vertebral level. For example, the lumbar and cervical areas of the vertebrae allow the trunk to flex

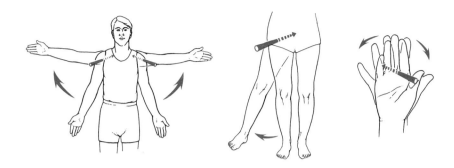

FRONTAL PLANE MOVEMENTS ABOUT JOINT AXES

FRONTAL PLANE MOVEMENT ABOUT THE CENTER OF GRAVITY

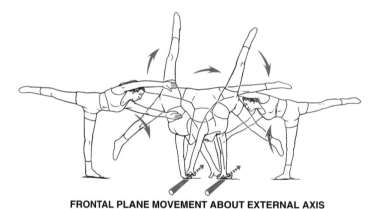

FRONTAL PLANE MOVEMENT ABOUT EXTERNAL AXIS

FIGURE 1–15. Movements in the frontal plane. Segmental movements in the frontal plane about anteroposterior joint axes are abduction and adduction, or some specialized movement in the side-to-side direction. Frontal plane movements about the center of gravity or an external point involve sideways movement of the body, which is more difficult than movement to the front or back.

and extend, but this plane of movement is limited in the middle thoracic portion of the vertebrae. Likewise, the rotation actions of the trunk occur primarily in the thoracic and cervical regions since the lumbar region has limited movement potential in the horizontal plane. It is only the combination of all of the vertebral segments that allows the 3 df motion produced by the spine.

There are also additional gliding movements occurring across the joint surfaces that may be interpreted as

adding more degrees of freedom to those defined in the literature. For example, the knee joint is considered to have 2 df for the flexion/extension movements in the sagittal plane and the rotations in the transverse plane. However, the knee joint also demonstrates linear translation and it is well known that there is movement in the joint in the frontal plane as the joint surfaces glide over one another to create side-to-side translation movements. While these movements have been measured and are

TRANSVERSE PLANE MOVEMENTS ABOUT JOINT AXES

TRANSVERSE PLANE MOVEMENTS ABOUT AN EXTERNAL AXIS

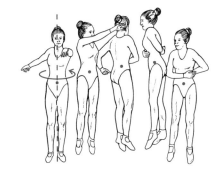

TRANSVERSE PLANE MOVEMENTS ABOUT THE CENTER OF GRAVITY

FIGURE 1–16. Movements in the transverse plane. Transverse plane movements are usually rotations occurring about a longitudinal axis running through a joint, the center of gravity, or an external contact point.

fairly significant, they have not been established as an additional degree of freedom for the joint. The degrees of freedom for the majority of the joints in the body are included in the movement review chart presented in TABLE 1–1.

A kinematic chain is derived from combining degrees of freedoms at various joints to produce a skill or movement. The chain is the summation of the degrees of freedom in adjacent joints that identifies the total degrees of freedom available or necessary for the performance of a movement. For example, kicking a ball might involve an 11 df link system relative to the trunk, as 3 df may be used at the hip, 2 df at the knee, 1 df at the ankle, 3 df in the tarsals, and 2 df in the toes.

Examples of Joint Movement Characteristics

Single Joint Movement Examples

Knowledge of the movement characteristics of a specific joint is valuable for the implementation of a stretching, strengthening, or rehabilitation program. The safest method for stretching or strengthening a segment's musculature is to move the limb through the range of motion of the segment in one plane of motion using only 1 df in the joint. While simple planar movements do not transfer directly to the complicated patterns in a sport skill that

will typically use diagonal patterns through two to three planes, the uniaxial movement is the commonly used approach for conditioning and rehabilitation. This is because movement through one plane allows better isolation of muscle groups that are arranged in functional compartments on the front, back, and sides of the joints. It is also safer, since movement through more than one plane can create stress in the joint and surrounding musculature.

Two segments and their movement characteristics will be presented here. More detailed presentation is available for all segments in the chapters focusing on the musculoskeletal components of the body segments and joints. The active mobility characteristics of each joint will be examined. The range of motion produced actively indicates the joint mobility associated with a voluntary muscle contraction. Another common technique, passive range of motion, is measured by moving the limb through a range of motion using external force provided by a therapist rather than muscular effort from the participant. This technique assesses the movement potential in the joint. Another important distinction is the difference between segmental position and movement. For example, to bend forward at the waist and hold the position, both the movement to that position and the final position would be termed trunk flexion. Returning to the standing

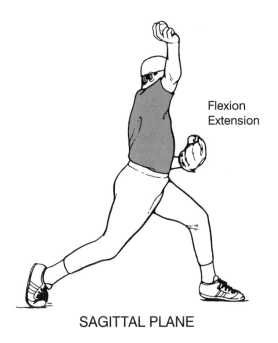

Flexion
Extension

SAGITTAL PLANE

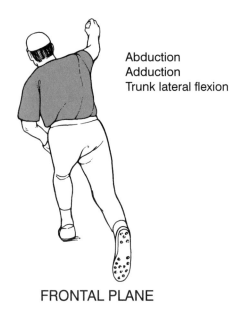

Abduction
Adduction
Trunk lateral flexion

FRONTAL PLANE

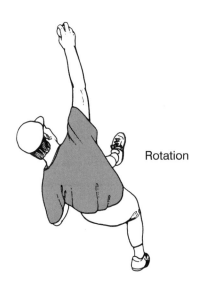

Rotation

TRANSVERSE PLANE

FIGURE 1–17. Movements in all three planes. Most human movements will employ movement in all three planes. The release phase of the overhand throw shown above, illustrates movements which occur in all three planes. Notice that the sagittal plane movements are viewed from the side, the frontal plane movements from the rear, and the transverse plane movements from above.

position but stopping before reaching an erect posture would still be a trunk flexion position, but the movement to that position is trunk extension. Positions are used in static analyses to determine such things as forces on the joints, while movements are used in dynamic analyses to interpret kinetic and kinematic components of motion.

The first segment example presented is the forearm, a segment that is confined to movement in two planes with respect to the anatomical position. A kinematic description of the forearm movements using the information on the anatomical position frame of reference, angular motion relative to this position, joint movement terminology, and the three planes of motion is provided. The sagittal plane movements of flexion and extension occur at the elbow joint that has 1 df (FIGURE 1–18). The forearm can be flexed through approximately 140 to 160 degrees before it is slowed by considerable soft tissue resistance. The segment extends back to the anatomical position and, for some individuals, it can go beyond the anatomical position into hyperextension (5 to 10 degrees) if the structural characteristics of the posterior elbow joint allow. The forearm can also move in the transverse plane through the movements of pronation and supination that are produced at the superior and inferior radioulnar joints (1 df). The forearm can be pronated or supinated so that the hand is rotated a full 180 degrees from the anatomical position. The forearm movements are important components to many sport skills involving the upper extremity. In the typical overhand pattern, the forearm will flex and pronate in the preparatory phase and extend with sometimes more pronation in both the power and follow-through phases. The movement of forearm extension is a significant component to the force application phase of many sport activities. The movement of forearm pronation is important in both force application and in the achievement of correct direction or placement of the hand or racket.

While the forearm movements are contained within two primary planes, the sagittal and the transverse, the forearm can be oriented three-dimensionally in space by the arm so that the segment itself can move through any of the three planes or diagonally across two planes with reference to the total body. This is a good example of the kinematic chain, illustrating the contribution of the three-dimensional shoulder joint to forearm movements. It is best to first identify planes and degrees of freedom of motion of the forearm with respect to the anatomical position before attempting to describe the characteristics of the forearm when it is positioned in space by the arm.

Movement characteristics of the most mobile segment in the body, the arm, are complex. The arm moves at the shoulder joint in three different planes of action (3 df). From the fundamental starting position, the arm elevates forward through flexion in the sagittal plane up to a full 180 degrees where it is beside the head. Actually, the segment can only move through approximately 150 to 160 degrees of motion before assisting movements such as trunk extension or lateral bending are necessary to achieve the last few degrees (FIGURE 1–19). Any hyperflexion visually apparent as the arm goes beyond 180 degrees of movement is not a result of pure arm movement but rather created by adjacent joint movements. The arm extends back down to the starting position and can continue on to approximately 50 to 60 degrees of hyperextension.

In the frontal plane, the arm can abduct 160 to 170 degrees out to the side before trunk movements assist to bring the arm up to 180 degrees or beyond into hyperabduction. On the return movement, the arm can be

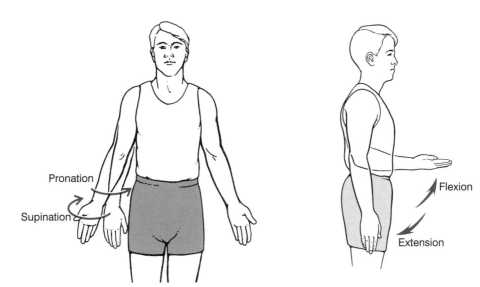

FIGURE 1–18. Single joint movement: the forearm. Flexion and extension of the forearm occur in the sagittal plane about a mediolateral axis running through the elbow joint. Pronation and supination of the forearm take place in the transverse plane about a longitudinal axis running through the radioulnar joints.

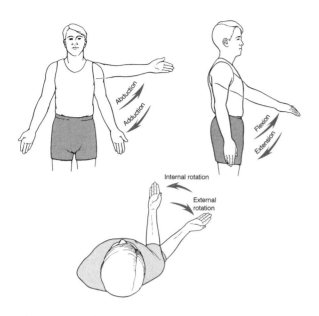

FIGURE 1–19. Single joint movement: the arm. Flexion and extension of the arm take place in the sagittal plane about a mediolateral axis running through the joint, and are best viewed from the side. Abduction and adduction occur in the frontal plane about an anteroposterior axis, making the front or back viewing position the best. Rotations occur in the transverse plane about the longitudinal axis and are best viewed from above.

adducted into 50 to 75 degrees of hyperadduction. In the transverse plane, the arm can rotate externally approximately 90 degrees and internally approximately 90 degrees.

In many sport activities, the arm will move through two to three planes of motion during skill performance. For example, as a right-handed batter swings, the left arm will horizontally abduct and possibly rotate externally away from the body as the right arm horizontally adducts and possibly internally rotates toward the body. For the overhand pattern seen in throwing and many racket sports, the throwing arm will move through multiple planes of motion, exhibiting abduction, horizontal abduction, and external rotation in the backswing; extension or adduction, horizontal adduction, and internal rotation in the force application phase; and extension or adduction and internal rotation in the follow through phase. Since the arm is capable of 3 df, it is a complex joint in terms of movement potential and characteristics. Adding to the complexity is the fact that all arm movements are assisted by movements in the shoulder girdle; consequently, arm movements cannot be interpreted in isolation. The relationship between arm and shoulder-girdle movements will be clarified in a later chapter.

Multiple Joint Movement Examples

Most human movements or sport skills are complicated to observe or analyze because they encompass multiple segments moving individually using 1 to 2 df. The contribution and involvement of adjacent segments is very important; for example, a soccer kick cannot be analyzed by just monitoring the foot movement. The leg, thigh, trunk, and even the arms also need to be analyzed in terms of their influence on the foot motion and position. The following sections will present total body movement characteristics for a variety of sport skills ranging from simple activities involving a limited number of segments and movement through predominately one plane to complex activities involving multiple segments and multiple planes of motion. All of the examples will be analyzed with respect to the anatomical starting position.

The squat is a simple activity involving positioning in the upper extremity, slight movement of the trunk, and significant movement of three lower extremity segments: the thigh, leg, and foot. The different phases of the squat exercise are presented in FIGURE 1–20. Refer to this figure when following the movement description provided below. With the person standing erect and a bar across the shoulders, the trunk is maintained in the neutral position or moved into a slightly flexed position, the head is hyperextended or maintained in the neutral position, the arms are abducted and externally rotated, the shoulder girdles are elevated and upwardly rotated, the forearms flexed and pronated, the hand is maintained in the neutral position and the fingers are flexed around the bar. Before beginning descent, the lower extremity segments are all in the neutral position except for the thigh, which is slightly abducted to bring the feet to a shoulder-width stance for stability. In descent, the upper extremity and the trunk are maintained in the starting position as much as possible. The trunk will have the tendency to flex, and this movement should be controlled so that stress on the lumbar spine is kept to a minimum. The lower extremity progresses slowly into increasing amounts of thigh flexion, leg flexion, and dorsiflexion until the bottom position is achieved. The return movement is the reverse movements of thigh extension, leg extension, and plantar flexion until the neutral starting or standing position is achieved (3,4).

Wheel chair propulsion is an activity just the opposite of the squat, using the upper extremity, specifically the shoulder girdle, arm, forearm, and hand, to generate the movements that propel the wheelchair. Propelling a regular wheelchair is different from a racing or sports chair, due to the differences in design. The regular wheelchair

1   2   3   4   5

START   DESCENT   ASCENT

FIGURE 1–20. Multiple joint movement: the squat. The performance of the squat exercise uses multiple joints in the body. Although the lower extremity joints are primarily responsible for the movement, the upper extremity and trunk are involved in stabilization. The lift begins with the bar resting on the upper back (#1 above), is slowly lowered into flexion (descent—#1–3), and then extended back to the starting position (ascent #3–5).

has a narrow wheelbase and the wheel is aligned vertically. The force applied to the rim is vertical. In a track or racing chair, the wheelbase is much wider and the wheel slants medially, making the width at the top much less than the bottom. By bringing the top closer to the upper body, pushing is diagonal and much more efficient. The seat in the chair of the racing style is usually higher in the front, bringing the knees up. This facilitates trunk flexion, makes for a more aerodynamic position, and places the arms in a more efficient position for pushing. To take a closer look at the movements that generate wheelchair propulsion in a track or racing chair, refer to FIGURE 1–21.

The propulsion of the chair comes from a push generated by the upper extremity. The push begins with the upper extremity positioned behind the body via shoulder-girdle adduction and elevation, arm hyperextension, forearm flexion, and hand radial flexion. The trunk is also flexed, allowing the arms to hyperextend even further. The head is hyperextended to look in the direction of travel. The push is initiated from this position through the movements of shoulder-girdle abduction, arm flexion, forearm extension, and hand ulnar flexion. Force on the hand rim is applied through approximately 200 degrees

for racing. (Pushing for short propulsive thrusts or in a regular wheelchair will be for a smaller distance ranging from 90 to 110 degrees.) When the push is completed, the upper extremity is close to the anatomical position, with only slight hyperextension of the arm, slight flexion of the forearm, and ulnar flexion of the hand.

The next phase is the recovery phase, where the arm is brought back to the start of the push phase by sliding the hand along the hand rim. The movements here are shoulder-girdle adduction and elevation, arm hyperextension, forearm flexion, and hand radial flexion. The trunk may also assist in this movement by slightly extending (5).

An activity involving the total body is the forward pike dive off the springboard, illustrated in FIGURE 1–22. This skill is primarily confined to movement in one plane, but involves more phases and segments in the activity. Beginning with the foot contact phase before takeoff, the springboard is deflected downward while the trunk flexes slightly with the head in the neutral position, the arms hyperextend, the shoulder girdles adduct, the forearm is maintained in the semiprone position, and the hands remain in the neutral position. The lower extremity moves into hip flexion, knee flexion, and dorsiflexion.

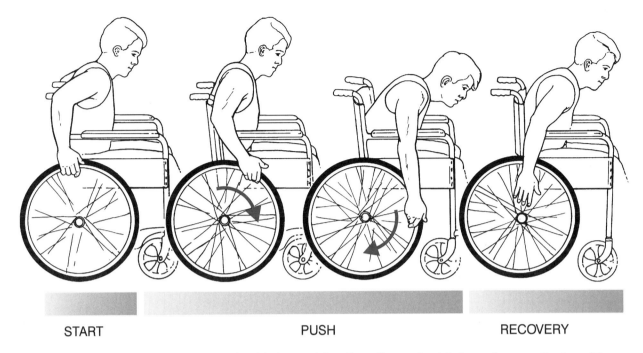

START   PUSH   RECOVERY

FIGURE 1–21. Multiple joint movement: wheelchair propulsion. Propelling a wheelchair requires coordination of the movements in the upper extremity. The hands rest on the rim of the wheel and push forward through 120–220 degrees in the power phase. There is a quick recovery as the hand is brought back up to begin the push again.

Before the board reaches maximum deflection, the upward movement of the body starts as the trunk extends, the arms begin to flex forward, the shoulder girdles follow with upward rotation, elevation, and abduction, the forearms flex, and the hands are maintained in the same position. The arms continue to flex, they internally rotate, the forearms extend and begin to pronate until the arms are above the head and the diver leaves the board. The lower body pushes on the board with hip extension, knee extension, and plantar flexion that are initiated before maximum deflection and continue until the diver leaves the board.

Once in the air, the diver maintains the arm in the previously described takeoff position at the side of the head, while the trunk flexes and the thigh flexes in order to assume the pike position. In the pike position, the knees are maintained in the neutral position and the ankles are plantar flexed. To prepare for entry into the water, all joint positions are maintained except for the extension of the thigh and the trunk back into the neutral position. Even though the pike dive involves many body segments, it is still simple to analyze because the movement occurs primarily in one plane, and both upper and lower limbs perform the same movements at the same time in the event (6,7).

The final movement example, baseball pitching, is a complex activity for analysis because it involves numerous segments, multiple planes of activities, multiple phases, and the right and left extremities do not have the same motion characteristics. Again, there are many individual variations for the skill of baseball pitching. A common variation will be presented in this example of a right-handed pitcher (FIGURE 1–23).

In the preparatory phase when the pitcher performs the windup and cocking of the arm, the motion begins as the front or striding leg pushes on the ground using the movements of thigh extension, leg extension, and plantar flexion. The same leg is then quickly raised and moved toward the body via thigh flexion, horizontal adduction of the thigh, and internal rotation of the thigh. The other limb segments are moved to a position of leg flexion and a neutral foot position. As the weight shifts to the back or driving limb, the thigh rotates externally and slightly flexes, the leg flexes slightly, and the foot dorsiflexes. The trunk rotates approximately 90 degrees to the right as the head maintains a position looking toward home plate through rotation to the left. Both arms flex upward, the shoulder girdles upwardly rotate, elevate, and abduct, the elbows are slightly flexed and there may be slight wrist flexion if the ball and mitt are brought above the head.

PREPARATORY   TAKE-OFF   PIKE   ENTRY

FIGURE 1–22. Multiple joint movement: pike dive. Diving involves the majority of joints in the body, both upper and lower extremity and the trunk. The phases of the pike dive, shown above, begin with the preparatory phase, the takeoff, the pike, and the entry. The approach movements to the final plant have been excluded in the present analysis.

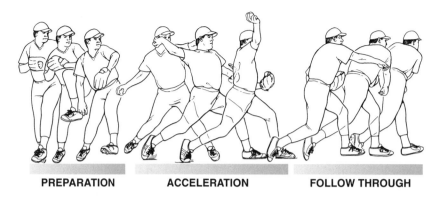

PREPARATION   ACCELERATION   FOLLOW THROUGH

FIGURE 1–23. Multiple joint movement: pitch. The overhand pitch is a very complex activity because numerous joints are involved, and the right and left sides of the body are performing different movements. The pitch is characterized by a preparatory phase which brings the arm behind the body, a force application, or acceleration phase where the ball is brought rapidly forward and then released, and the follow-through phase where the segment motions are slowed down.

The movement toward the plate in the acceleration phase is again initiated with the front or striding limb as the thigh horizontally abducts and externally rotates, the knee extends, and the foot is slightly dorsiflexed. Upon contact with the ground, the striding limb flexes at the hip and knee joints as weight is shifted over onto it. The foot remains slightly dorsiflexed after contact. Meanwhile, the contralateral or driving limb extends and medially rotates at the hip joint, while the leg extends and the foot plantar flexes. Shortly after both thighs begin to rotate toward the plate, the trunk also begins a rotation to the left accompanied by left lateral flexion that continues until the trunk faces home plate.

Also accompanying the striding and rotation action of the lower limbs is a horizontal abduction and external rotation action in both arms. These movements bring the front, glove hand to a position that is pointing at home plate and the contralateral ball hand to a position behind the back. In both limbs, the forearms are usually pronated and extended, and the wrists are in a neutral position with fingers flexed around both the glove or the ball, depending upon the hand. Once the striding leg is planted and the driving leg and the trunk have nearly completed their movements, the final acceleration of the ball arm is completed as it moves through a rapid horizontal adduction and internal rotation. This movement forces the shoulder girdle into a rapid abduction with upward rotation and elevation. At the same time the forearm is rapidly extending and pronating, and the hand is either maintained in the neutral position or moved into slight ulnar flexion or flexion. It is important to note that all high velocity

throwing motions involve forearm pronation regardless of the type of pitch thrown. The contralateral glove arm usually adducts down to the side with the forearm flexed and the hand in a neutral position.

The final follow-through phase of pitching has the purpose of slowing down (negatively accelerating) the joint movements while also placing the pitcher in some form of defensive position. The front lower limb flexes at the thigh and leg, and dorsiflexes at the foot to absorb the weight and forward momentum of the body. The contralateral driving limb swings through with thigh and leg flexion and the foot plantar flexed until it makes contact with the ground. The trunk flexes and continues on with some rotation to the left until motion is stopped. The throwing arm continues across the body through arm horizontal adduction, extension, and internal rotation, and accompanying shoulder girdle abduction, depression, and downward rotation until it also is stopped. The forearm of the throwing arm extends and pronates while the hand will also flex until motion stops. The glove arm is maintained in a position similar to that described at the end of the acceleration phase (8,9,10).

Ideally by studying some examples of the movement characteristics of simple segments, simple sport skills, and a complex skill such as a baseball pitch, expertise in movement identification and kinematic analysis can be developed. The examples presented in this section may vary from techniques seen elsewhere in the literature. Also, the examples may not include every joint movement present in the skill, but they do offer a starting point for the development of movement analysis skills. Finally, it was not the intent of this chapter to identify the source, or the kinetics, of the joint movements. Examination of the source of the movement will be presented later in a kinetics section after kinematic components have been explored.

Chapter Summary

Biomechanics, the study of motion and the effect of forces on an object, is a useful tool for studying human motion. Human motion can be quantitatively assessed biomechanically, using kinematic or kinetic applications that analyze a skill or movement by identifying its components or by assessing the forces creating the motion, respectively. To provide a specific description of a movement, it is helpful to define movements with respect to a starting point or to one of the three planes of motion: sagittal, frontal, or transverse.

Anatomical movement descriptors should be used to describe the segmental movements. This requires acknowledgement of the starting position (fundamental or anatomical), a standardized use of segment names (arm, forearm, hand, thigh, leg, foot), and the correct use of movement descriptors (flexion, extension, abduction, adduction, rotation).

The study and analysis of movement parameters of both single joint and multiple joints provides useful information for understanding a basic movement or a sport skill. By identifying the sequence of simple joint movements combining to produce a sport skill or complex movement, a clearer understanding of conditioning requirements and technique emphases can be obtained.

Review Questions

1. Define biomechanics. What would be the focus of a biomechanical analysis of running?
2. Define kinesiology. What would be the focus of a kinesiologic analysis of long jumping?
3. Provide one example of a sagittal, frontal, and transverse plane movement about a joint, about the center of mass, and about an external axis.
4. List the degrees of freedom and the name of the movements for each of the following joints:
   a. shoulder
   b. elbow
   c. wrist
   d. hip
   e. knee
   f. ankle
5. How do the joint movements differ between the following movements?
   a. curl up vs straight-leg raise
   b. raise up on toes vs heel lower
   c. going up stairs vs coming down stairs
   d. push up vs pull up
6. Compare the movements of the hip and shoulder in terms of differences and similarities.
7. Describe an activity that uses trunk rotation. Where is the trunk rotation used?
8. Describe an activity using pronation and supination of the forearm. Where is the pronation/supination used?

Additional Questions:
1. Describe the upper-extremity movements that occur to produce the following skills/actions:
   a. freestyle swimming stroke
   b. combing hair
   c. rowing
   d. bench press
   e. volleyball spike
2. Describe the lower-extremity movements contributing to and producing the following human movements:
   a. kicking a ball
   b. walking
   c. deadlift
   d. throwing
   e. rising from a chair

Additional Reading

Grieve, D.W.: Dynamic characteristics of man during crouch and stoop lifting. *In* Biomechanics IV. Edited by R.C. Nelson and C.A. Morehouse. Baltimore. University Park Press, 1974.

O'Shea, P.: The parallel squat. National Strength and Conditioning Association Journal. *7:*1, 1985.

Roozbazar, A.: Biomechanics of lifting. *In* Biomechanics IV. Edited by R.C. Nelson and C.A. Morehouse. Baltimore, University Park Press, 1974.

Vetter, W.L., Helfet, D.L., Spear, K., and Matthews, L.S.: Aerobic dance injuries. The Physician and Sports Medicine. *13:*2, 1985.

References

1. Richie, D.H., Kelso, S.F., and Belluci, P.A.: Aerobic dance injuries: A retrospective study of instructors and participants. The Physician and Sports Medicine. *13:*2, 1985.
2. Ulibarri, V.D., Fredericksen, R., and Soutas-Little, R.W.: Ground reaction forces in selected aerobics movements. Biomechanics in Sport. New York, Bioengineering Division of the American Society of Mechanical Engineering, 1987 p. 19–21.
3. Lander, J., Bates, B.T., and Devita, P.: Biomechanics of the squat exercise using a modified center of mass bar. Medicine and Science in Sports and Exercise. *18:*4, 1986.
4. McLaughlin, T.M., Lardner, T., and Dillman, C.J.: Kinetics of the parallel squat. The Research Quarterly. *49:*2, 1978.
5. Lamontagne, M.: Biomechanical study of wheelchair propulsion. *In* Biomechanics X1-A: International Series on Biomechanics, *7-A.* Edited by G. deGroot, A.H. Hollander, P.A. Huijing, and G.J. van Ingen Schenau. Amsterdam, Free University Press, 1987.
6. Miller, D.I., and Munro, C.F.: Body segment contributions to height achieved during flight of a springboard dive. Medicine and Science in Sports and Exercise. *16:*3, 1984.
7. Miller, D.I., and Munro, C.F.: Greg Louganis' springboard takeoff: I: Temporal and joint position analysis. International Journal of Sport Biomechanics. 1, 1985.
8. Jacobs, P.: The overhand baseball pitch: A kinesiological analysis and related strength-conditioning programming. National Strength and Conditioning Association Journal: *9:*1, 1987.
9. Jobe, F.W., Radovich Moynes, D., Tibone, J.E., and Perry, J.: An EMG analysis of the shoulder in pitching. The American Journal of Sports Medicine. *12:*3, 1984.
10. Pappas, A.M., Zawacki, R.M., and Sullivan, T.J.: Biomechanics of baseball. The American Journal of Sports Medicine, *13:*4, 1985.

Glossary

Acceleration:	The time rate of change in velocity. A term used to describe whether an object is speeding up, slowing down, or maintaining the same speed.
Angular Motion:	Motion around an axis of rotation where different regions of the same object do not move through the same distance.
Axis of Rotation:	The imaginary line about which an object rotates.
Biomechanics:	The study of motion and the effect of forces on biological systems.
Center of Mass:	The point at which all of the body's mass is concentrated; the balance point of the body.
Dynamics:	The branch of mechanics in which the system being studied undergoes an acceleration.
Equilibrium:	A balanced state in which there is no movement or no change in the movement, because the sum of all of the forces and torques acting on the object is zero.
Follow-Through Phase:	The sequence of joint movements used to decelerate the segments following the force application phase.
Force:	An interaction between two objects in the form of a push or pull that may or may not produce motion.
Force Application Phase:	Also referred to as the acceleration phase, it is the sequence of joint movements used to accelerate an object, segment, or the whole body, while at the same time maintaining accuracy in the directional component of the movements.
Functional Anatomy:	The study of the body components needed to achieve a human movement or function.
Kinematics:	Area of study that examines the spatial and temporal components of motion (position, velocity, acceleration).
Kinesiology:	The scientific study of human movement.
Kinetics:	Area of study that examines the forces that act on a system.
Linear Motion:	Motion in a straight or curved line in which different regions of the same object move the same distance.
Mass:	The amount of matter of which an object is composed.
Movement:	A change in place, position, or posture, occurring over time and relative to some point in the environment.
Plane of Motion:	A two-dimensional flat surface running through an object. Motion occurs in the plane or parallel to the plane.
Position:	The location of an object or a point on the object with respect to a designated reference point in the environment.

Preparatory Phase:	Joint movements that precede a purposeful action such as throwing, kicking, or jumping. The purpose of the preparatory phase is to place the musculoskeletal system in an advantageous position for optimal performance in the succeeding force application phase.
Qualitative Analysis:	A non-numeric description or evaluation of movement based on direct observation.
Quantitative Analysis:	A numeric description or evaluation of movement based on data collected during the performance of the movement.
Statics:	A branch of mechanics in which the system being studied undergoes no acceleration.
Torque:	A rotation resulting from a force that is not applied through the center of mass.
Velocity:	The time-rate change in position. A term used to describe how fast an object is moving.

CHAPTER 2

Skeletal Considerations for Movement

I. *Introduction*
- A. Functions of the Skeletal System
  1. levers
  2. support
  3. other functions
- B. Types of Bones
  1. long bones
  2. short bones
  3. flat bones
  4. irregular bones
  5. sesamoid bones

II. *Biomechanical Characteristics of Bone*
- A. Bone Tissue
  1. constituency
  2. resorption and deposit of bone
  3. physical activity vs bone remodeling
  4. lack of activity vs bone remodeling
  5. bone deposit in soft tissue
  6. osteoporosis
- B. Architecture of Bone
  1. compact bone
  2. spongy bone
- C. Strength and Stiffness of Bone
  1. anisotropic characteristics
  2. viscoelastic characteristics
  3. elastic response
  4. plastic response
  5. strength
  6. stiffness
  7. stress and strain
  8. normal and shear stress or strain

- D. Types of Load
  1. compression forces
  2. tension forces
  3. shear forces
  4. bending forces
  5. torsional forces
  6. injury vs loading
  7. muscular activity vs loading
  8. stress fractures

III. *Bony Articulations*
- A. The Diarthrodial or Synovial Joint
  1. Characteristics of the Diarthrodial Joint
     a) articular endplate b) articular cartilage
     c) fibrocartilage d) joint capsule
     e) synovial membrane and fluid f) ligaments
  2. stability of the diarthrodial joint
  3. degeneration of the diarthrodial joint
- B. Types of Diarthrodial Joints
  1. simple, compound, and complex joints
  2. close-packed vs loose-packed positions
  3. plane or gliding joint
  4. hinge joint
  5. pivot joint
  6. condylar joint
  7. ellipsoid joint
  8. ball-and-socket joint
- C. Other Types of Joints
  1. synarthrodial or fibrous joints
  2. amphiarthrodial or cartilaginous joints

IV. *Summary*

Student Objectives

After reading this chapter, the student will be able to:

1. List the functions of the skeletal system, the types of bones found in the skeletal system, and describe the role each type of bone plays in human movement or support.
2. Describe the process of bone resorption and deposit.
3. Describe the characteristics of compact and spongy bone, and the mechanical properties of bone under load.
4. Describe the characteristics of the load-deformation curve and the stress-strain curve, and identify how one would determine strength or stiffness of a material from these curves.
5. Identify the elastic region, yield point, plastic region, and failure point on a stress-strain curve.
6. Define the following types of loads that bone must absorb and provide an example to illustrate each load on the skeletal system:
   a) compression; b) tension; c) shear
   d) bending; e) torsion
7. Describe some common injuries to the skeletal system and explain the load causing the injury.
8. Describe all of the components of the diarthrodial joint, factors that contribute to joint stability, and examples of injury to the diarthrodial joint.
9. List the seven different types of diarthrodial joints, providing examples for each one.
10. Describe the characteristics of the synarthrodial and amphiarthrodial joint, and provide an example of each.

Introduction

The skeletal framework determines our shape and body size, and while one can roughly predict adult height by doubling the height of a 2 year old, the skeletal system and frame can also be greatly influenced by nutrition, activity level, and postural habits. Although the general size and shape of the bones are inherited, structural adaptations in shape, size, and landmarks can be induced by weight bearing and forces exerted by tendons, ligaments, and muscles (1).

In the developing or immature skeleton, the influence of weight bearing and muscular forces will have a more substantial effect on the formation of the size and shape of the bones than the same forces will have on a mature skeleton. For this reason, it is important to pay careful attention to the types of activities and habitual postures in which a pre-adolescent child is engaging.

A common example of skeletal alteration in the immature skeleton is the presence of idiopathic scoliosis, a lateral curvature in the spine, present in approximately 15 to 20% of females, aged 10 to 12. It is termed idiopathic because the forces causing the lateral curve have not been identified. Is it because young females spend too much time weight bearing on one limb? Is it related to a posture that includes hyperextended knees and swayback? It is easy to speculate on possible causes but the scientific substantiation of this disorder still evades us. We do know that the skeletal system is a malleable system that can be shaped and formed through activity. It is important to understand how the skeletal system responds in order to institute programs that will promote skeletal health in which individuals are free from skeletal injury.

Functions of the Skeletal System

The skeletal system performs many functions: leverage, support, protection, storage, and blood cell formation. Two of these functions, leverage and support, are important for human movement.

Levers The skeletal system provides the levers and the axes of rotation about which the muscular system generates the movements. A lever is a simple machine that magnifies force or speed of movement. The levers are primarily the long bones of the body, and the axes are the joints where the bones meet.

The manner in which portions of the skeleton contribute to movement is determined by the shape of the bones, the structural arrangement of the bones, and the characteristics of the articulations connecting the bones. For example, if the forearm is moved into a position of maximum flexion so that the forearm is up against the arm and then extended until it stops, hyperextension is limited and the forearm stops because of the shape of the bones forming the elbow joint. Conversely, if the arm is moved into flexion so that it is above the head, followed by extension back down, the arm will continue into a hyperextended position because the structure of the shoulder joint will allow the movement.

The skeleton also restricts motion in other areas of the body, as shown in the two examples provided in FIGURE 2–1. A movement up on the toes will wedge the talus up against the calcaneus and restrict the amount of plantar

FIGURE 2–1. Skeletal obstructions are responsible for limiting joint movement in many different areas of the body. (A) In the elbow joint, the olecranon process and the olecranon fossa meet to restrict the amount of forearm hyperextension. (B) In the foot, the talus and calcaneus make contact to restrict the amount of plantar flexion that can take place.

flexion. Dancers consider this a serious limitation, since they desire maximum plantar flexion for many dance and ballet maneuvers. Some dancers will even undergo corrective surgery that involves shaving the back side of the talus in order to allow for more freedom of movement in plantar flexion. Dancers have been known to have this operation 2 or 3 times during their careers. Understanding skeletal structure and shape provides information on the movement potential in each joint that can be used to work within the limits of the system.

Support   A second important function of the skeletal system is the provision of a support structure, used to maintain upright posture. The skeleton can maintain a posture, while at the same time being capable of accommodating large external forces such as those involved in jumping. The bones constituting the skeletal system increase in size from top to bottom as more of the body weight is assumed by the skeletal structure; thus the bones of the lower extremity, and the lower vertebrae and pelvic bones are larger than their upper extremity or upper torso counterparts. A visual comparison of the humerus and femur, or the cervical vertebrae and lumbar vertebrae, will provide good examples of size increases moving down in the body.

Other Functions   There are three additional bone functions not specifically related to human movement: protection, storage, and blood cell formation. The bones protect the brain and internal organs. Bone also stores fat and minerals. Finally, blood cell formation, called hematopoiesis, takes place within the cavities of bone.

Types of Bones

The skeletal system consists of different types of bones categorized according to shape, function, and the proportion of spongy and compact bone tissue. Spongy bone is a high porosity bone that is a high energy absorber and compact bone is dense and offers strength and stiffness to the skeleton. Five different types of bones are illustrated in FIGURE 2–2 and include the long, short, flat, irregular, and sesamoid bone.

Long Bones   These bones are longer than they are wide. The long bones in the body are the clavicle, humerus, radius, ulna, femur, tibia, fibula, metatarsals, metacarpals, and the phalanges.

The long bone has a shaft, the diaphysis, which is a thick layer of compact bone surrounding the bone marrow cavity (FIGURE 2–3). The shaft widens toward the end into the section called the metaphysis. The end of the

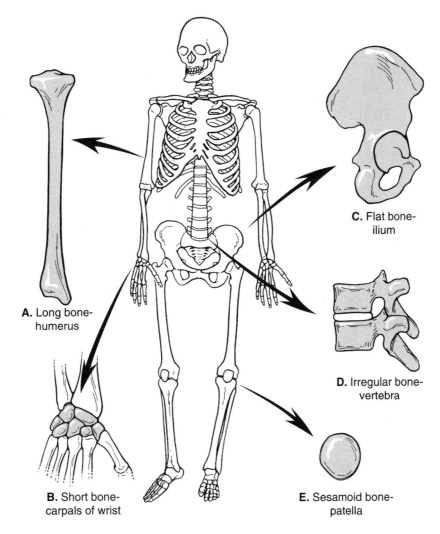

A. Long bone-
humerus

B. Short bone-
carpals of wrist

C. Flat bone-
ilium

D. Irregular bone-
vertebra

E. Sesamoid bone-
patella

FIGURE 2–2. There are different types of bones in the skeleton that serve specific functions. (A) Long bones serve as levers; (B) short bones offer support and shock absorption; (C) flat bones protect and offer large muscular attachment sites; (D) irregular bones have specialized functions; and (E) sesamoid bones alter the angle of muscular insertion.

long bone is called the epiphysis, separated from the shaft by a cartilaginous disc in the immature skeleton. The ends consist of a thin outer layer of compact bone covering spongy inner bone. A thin white membrane, the periosteum, covers the outside of the bone.

Long bones offer the body support and also provide the interconnected set of levers and linkages that allow us to create movement. The length of the long bone is generally formed by compressive forces, while muscle attachment sites and protuberances are formed by tensile forces. Most long bones are beam-shaped in order to handle and minimize the bending loads imposed upon them.

Short Bones The short bones, such as the carpals and tarsals, consist primarily of spongy bone covered with a thin layer of compact bone. These bones play an important role in shock absorption and the transmission of forces.

Flat Bones A third type of bone, the flat bone, is represented by the ribs, ilium, sternum, and scapula. These bones consist of two layers of compact bone, with spongy bone and marrow in between. Flat bones protect internal structures and offer broad surfaces for muscular attachment.

Irregular Bones Irregular bones, such as those found in the skull, pelvis, and vertebrae, consist of spongy bone with a thin compact bone exterior. These bones are

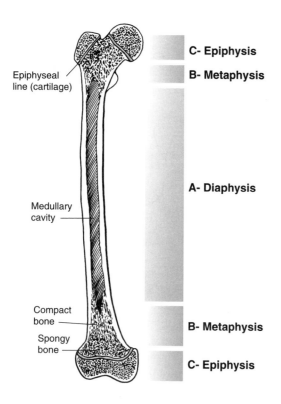

Epiphyseal
line (cartilage)

Medullary
cavity

Compact
bone

Spongy
bone

C- Epiphysis

B- Metaphysis

A- Diaphysis

B- Metaphysis

C- Epiphysis

FIGURE 2–3. The long bone has a shaft, or (A) diaphysis, which broadens out into the (B) metaphysis and the (C) epiphysis. Layers of compact bone make up the diaphysis. The metaphysis and epiphysis are made up of spongy bone with a thin layer of compact bone on the outside.

termed "irregular" because of the specialized shape and function for each bone. The irregular bones perform a variety of different functions including: supporting weight, dissipating loads, protecting the spinal cord, contributing to movement, and providing sites for muscular attachment.

Sesamoid Bones   The last type of bone, the sesamoid bone, is a short bone embedded within a tendon or joint capsule. The patella is a sesamoid bone at the knee joint because it is embedded in the tendon of the quadriceps. Other sesamoid bones can be found at the base of the first metatarsal in the foot where the bones are embedded in the distal tendon of the flexor hallucis brevis muscle, and at the thumb where the bones are embedded in the tendon of the flexor pollicis brevis muscle. The role of the sesamoid bone is to alter the angle of insertion of the muscle.

Biomechanical Characteristics of Bone

Bone Tissue

Constituency   Osseous tissue is strong and one of the body's hardest structures because of its combination of inorganic and organic elements. The minerals, calcium and phosphate, along with collagen, constitute the organic element in bone and make up approximately 60 to 70% of bone tissue. Water constitutes approximately 25 to 30% of the weight of bone tissue (2).

Bone tissue is a viscoelastic material whose mechanical properties are affected by its deformation rate. The ductile properties of bone are provided by the collagenous material in bone. The collagen content gives bone the ability to withstand tensile loads. Bone is also a brittle material and its strength depends on the loading mechanism. The brittleness of bone is provided by the mineral constituents that provide bone with the ability to withstand compressive loads. These properties will be discussed later in the chapter.

Resorption and Deposit of Bone   Bone is a highly adaptive material that is very sensitive to disuse, immobilization, or vigorous activity and high levels of loading. Bone tissue is self-repairing and can alter its properties and configuration in response to mechanical demand. This was first determined by the German anatomist, Julius Wolff, who provided us with the theory of bone development, termed "Wolff's Law," which states: "Every change in the form and function of a bone or of their function alone is followed by certain definitive changes in their internal architecture, and equally definite secondary alteration in their external conformation, in accordance with mathematical laws" (3).

During growth and in life, the bones are subjected to externally applied loads and muscular forces, to which the bone responds. Bone is remodeled and repaired, and is a dynamic and active tissue in which large volumes of bone are removed through bone resorption and replaced through bone deposit. This process is not the same in all bones or even in a single bone. For example, the bone in the distal part of the femur is replaced every 5 to 6 months, whereas the bone in the shaft is replaced much more slowly. In young adults, the bone deposits equal the bone resorption, and the total bone mass is fairly constant. However, through exercise, the bone mass can be increased even up through young adults. This is one of the major benefits of physical activity.

Bone deposits will exceed bone resorption when there is an injury or when greater strength is required. Thus, weight lifters will develop thickenings at the insertion of very active muscles, bones will be more dense in sites where the stresses are greatest, and there will be a change in the shape of bone during fracture healing. The tennis arm of professional tennis players have cortical thicknesses that are 35% greater than the contralateral arm (3).

Physical Activity vs Bone Remodeling   Bones require mechanical stress in order to grow and strengthen; thus physical activity is an important component in the development and maintenance of skeletal integrity and strength. Bone tissue must experience daily stimulus to maintain health in the bone. The daily applied loading history, comprising the number of loading cycles and the stress magnitude, influences the density of the bone. Intermittent loading for 100 cycles a day has been shown to produce a significant increase in the bone cross section (4). Body weight and activity level are examples of factors that regulate bone density in the weight-bearing bones. If the activity level is increased, there will be a moderate increase in bone mass (4).

Lack of Activity vs Bone Remodeling   Bone loss following a decrease in the activity level may be significant (4). Astronauts, subjected to reduced activity and the loss of body weight influences, experience significant bone loss in relatively short periods of time. Some of the changes occurring to bone after outer space travel include: less rigidity, more bending displacement, a decrease in bone length and cortical cross section, and a slowing in bone formation (5).

Bone Deposit in Soft Tissue   There is a condition known as myositis ossificans in which bone deposits are laid down in soft tissue and near bone as a safety response to repeated trauma or a hematoma (bruise) to an area. The body first responds to the repeatedly bruised area by developing fibrous tissue that eventually develops into cartilage, and then into bone. The anterior portion of the thigh and the hip joint are regions of the body where this condition will commonly develop. Football players who are repeatedly hit in the thigh are susceptible to this type of injury. It has also been seen in the hip region of soccer players who repeatedly fall on the hip (6). Recognizing that the body will develop a fibrous, cartilaginous, and osseus proliferation at the specific site of repeated trauma, some performers in the martial arts will deliberately attempt to obtain this response in the hands and feet.

Osteoporosis   Bone resorption will exceed bone deposits in a condition known as osteoporosis. The symptoms of osteoporosis often begin to appear in the elderly, especially postmenopausal women, but actually, the condition begins to develop in early years when bone mineral density decreases. When bone deposits cannot keep up with bone resorption, there is a decrease in the bone mineral mass, resulting in reduced bone density. There is also an accompanying loss of trabecular integrity. The loss of bone mineral density means loss of the stiffness in bone, and the loss of trabecular integrity creates a weaker structure. Both of these losses create the potential for a much higher incidence of fracture (7), ranging from 2 to 3.7% in non-osteoporotic individuals, and increasing almost two times to 5 to 7% in osteoporotic individuals (8).

The exact causes of osteoporosis are not fully understood, but the condition has been shown to be related to hormonal factors, nutritional imbalances, and the lack of exercise. Normal bone volume is 1.5 to 2.0 liters, and the cortical diameter of bone is at its maximum between the ages of 30 and 40 years of age for both men and women (9,10). After age 30, there is a 0.2 to 0.5% yearly loss in the mineral weight of bone (4), accelerating after menopause in females to bone loss that is 50% more than in males (10). It is speculated that a substantial proportion of this bone loss may be related to the accompanying reduction in activity level (4).

Through mild or moderate exercise, the bone mineral content in the elderly can be increased (11). In one study, bone mineral content in runners, aged 50 to 72, was shown to be greater than in control subjects. There was also a decreased rate of bone loss with age: 4% over 2 years for the runners and 6 to 7% over 2 years for the control group (12). However, when the runners quit running or moved to walking as an alternative exercise, the bone resorption and loss increased substantially to 10 to 13% (12). Therefore, it is suggested that a substitute activity for running is one which provides high intensity loads and low repetitions, such as can be found in weight lifting.

Lifestyle and activity habits seem to play an important role in the maintenance of bone health. In one study, the incidence of osteoporosis was 47% in a sedentary population, compared to only 23% in hard physical labor occupations (11). It is clear that the elderly may benefit from some form of weight-bearing exercise, even though the exercise intensity and durations have not been determined.

Estrogen levels in anorexic women and amenorrheic female athletes have also been related to the presence of osteoporosis in this population. There is speculation that the occurrence of stress fractures in the femoral neck of female runners may be related to a noted loss of bone mineral density due to osteoporosis (7). Elite female athletes in a variety of sports have experienced a loss of bone, usually associated with bouts of heavy training and associated menstrual irregularity. Some of these female athletes have lost so much bone mass, their skeletal characteristics resemble those of elderly women.

Architecture of Bone

Compact Bone   The architectural arrangement of bony tissue is remarkably well suited for the mechanical demands imposed upon the skeletal system during physical activity. A midsection of the femoral head is presented in FIGURE 2–4, and illustrates the internal architecture of the long bone. The osseus tissue on the exterior of a bone is made up of compact or cortical bone that is very dense and has a porosity less than 15% (13).

The compact bone consists of a system of hollow tubes, lamellae, that are placed inside one another to form the haversian system of canals. The arrangement of these weight-bearing pillars and the density of the compact bone provide strength and stiffness to our skeletal system. Compact bone can withstand high levels of weight bearing or muscle tension in the longitudinal direction before it will fail and fracture (1).

Compact bone is especially capable of absorbing tensile loads if the collagen fibers are parallel to the load. Typically, the collagen is arranged in layers going in different directions in longitudinal, circumferential, and oblique configurations. This offers resistance to tensile forces in different directions, because the more layers there are, the more strength and stiffness the bone will have. Also, where muscles, ligaments, and tendons attach to the skeleton, the collagen fibers are arranged parallel to the insertion of the soft tissue, thereby offering greater tensile strength for these attachments.

A thick layer of compact bone is found in the shafts of long bones, the diaphyses, where strength is necessary to respond to the high loads imposed down the length of the bone during weight bearing or in response to muscular tension. Thin layers of compact bone are found on the ends of the long bones, the epiphyses, and also covering the short or irregular bones.

Spongy Bone   The bone tissue in the interior of bone is spongy, or cancellous bone, except for the shaft of the long bones. This bone is spongy and lattice-like with a porosity greater than 70% (13). The spongy bone structure, although quite rigid, is weaker and less stiff than the compact bone. The small flat pieces of bone making up the spongy bone are called trabeculae (FIGURE 2–4).

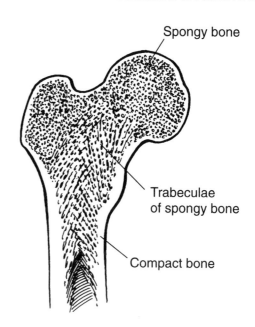

FIGURE 2–4. This midsection of the proximal end of the femur shows both compact and spongy bone. The dense compact bone lines the outside of the bone, continuing down to form the shaft of the bone. Spongy bone is found in the ends, and is distinguishable by its lattice-like appearance. Note the curvature in the trabeculae, which forms to withstand the stresses.

The trabeculae adapt to the direction of the imposed stress on the bone, providing strength while maintaining low weight in the structure (14). Collagen runs along the axis of the trabecular bone, providing spongy bone with both tensile and compressive resistance.

The high porosity gives spongy bone high energy storage capacity so that it becomes a crucial element in the energy absorption and stress distribution when loads are applied to the skeletal structure (2). Spongy bone is not as strong as compact bone, and there is a high incidence of fracture in the spongy bone of elderly, believed to be caused by a loss of compressive strength due to mineral loss (osteoporosis).

Strength and Stiffness of Bone

The behavior of any material under different loading conditions is determined by its strength and stiffness. When an external force is applied to a bone or any other material there is an internal reaction. The strength can be evaluated by examining the relationship between the load imposed (external force) and the amount of deformation (internal reaction) occurring in the material, known as the load-deformation curve.

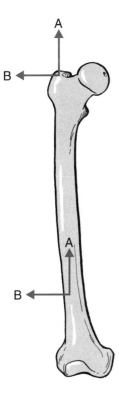

FIGURE 2–5. Bone is considered anisotrophic because it will respond differently if forces are applied in different directions. (A) Bone can handle large forces applied in the longitudinal direction. (B) Bone is not as strong in handling forces applied transversely across its surface.

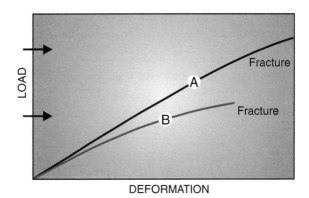

FIGURE 2–6. Bone is considered viscoelastic because it will respond differently when loaded at different rates. (A) When loaded quickly, bone responds with more stiffness, and can handle a greater load before fracturing. (B) When loaded slowly, bone is not as stiff or strong, fracturing under lower loads.

Anisotropic Characteristics Bone tissue is an anisotropic material, indicating that the behavior of bone will vary depending on the direction of the load application. This is illustrated in FIGURE 2–5. In general, bone tissue can handle the greatest loads in the longitudinal direction and the least amount of load when applied across the surface of the bone (1). Bone is stronger withstanding loads in the longitudinal direction because it has been habitually loaded in that direction.

Viscoelastic Characteristics Bone is also viscoelastic, meaning it will respond differently depending upon the rate at which the load is applied and the duration of the load. At a higher speed of loading, bone can handle greater loads before it fails, or fractures. As shown in FIGURE 2–6, the bone loaded slowly fractures at a load that is approximately half of the load handled by bone at a fast rate of loading.

Elastic Response When a load is first applied, a bone will deform through a change in length or angular shape.

Bone deforms no more than approximately 3% (15). This is considered the elastic range of the load-deformation curve because, when the load is removed, the bone will recover and return to its original shape or length.

A stress-strain or load-deformation curve is presented in FIGURE 2–7. An examination of this curve can be

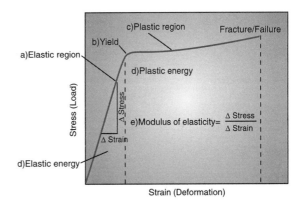

FIGURE 2–7. The stress/strain and the load/deformation curve illustrate the strength and performance characteristics of a material when subjected to a load. When a load is applied, there is an initial (A) elastic response which eventually reaches a (B) yield point, entering into the (C) plastic response when the material permanently deforms or fractures. The strength of a material is determined by the (D) energy or area under the curve. The stiffness of a material, called modulus of elasticity, is determined by the (E) slope of the curve during the elastic response phase.

used to determine whether a material is stiff, ductile, brittle, strong, or weak. The curve shown would be representative of a material that is strong and ductile.

Plastic Response   With continued loading the bone tissue reaches its yield point, after which the outer fibers of the bone tissue will begin to yield, experiencing microtears and debonding of the material in the bone. This is termed the plastic or nonelastic phase in the load-deformation curve. The bone tissue begins to permanently deform and eventually fracture if loading continues in the nonelastic phase. Thus, when the load is removed, the bone tissue does not return to its original length and stays permanently elongated.

Strength   The strength of bone or any other material is defined by the failure point or the load sustained before failure. Strength can also be assessed in terms of energy storage, the area under the load-deformation, or stress-strain curve.

Stiffness   Stiffness, or modulus of elasticity of a material, is determined by the slope of the load-deformation curve (FIGURE 2–7) during the elastic response range and is representative of the material's resistance to load as the structure deforms. This is a response in most materials, including bone, tendons, and ligaments. The stress-strain curve for ductile, brittle, and bone material is shown in FIGURE 2–8. A stiff material will respond with

minimal deformation to the increased load. If the material fails at the end of the elastic phase, it is termed a brittle material. Glass is an example of a brittle material. Bone is not as stiff as glass or metal, and unlike these does not respond in a linear fashion because it yields and deforms non-uniformly during the loading phase (2).

The greater the load imposed upon the bone, the greater the deformation. Furthermore, if the load exceeds the material's elastic limits, there will be permanent deformation and failure of the material. If a material continues to elongate and deform a great deal in the plastic phase, it is termed a ductile material. Skin is an example of material that will deform a considerable amount before failure. Bone is a material having properties that respond in both a brittle and ductile manner.

A variety of materials have been plotted according to strength and stiffness in FIGURE 2–9. Examples of material considered stiff and weak are glass and copper, material stiff and strong are steel and iron, material flexible and strong are fiberglass and silk, and material flexible and weak are oak, lead, and a spider web. Bone is considered to be a flexible and weak material (16).

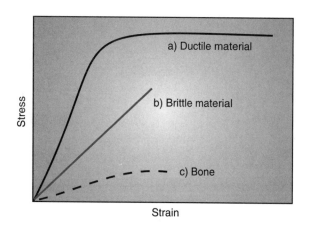

FIGURE 2–8. These stress/strain curves illustrate the differences in the behavior between (A) ductile material, (B) brittle material, and (C) bone, which has both brittle and ductile properties. When a load is applied, a brittle material will respond linearly and fail or fracture before undergoing any permanent deformation. The ductile material will enter the plastic region and deform considerably before failure or fracture. Bone will deform slightly before failure.

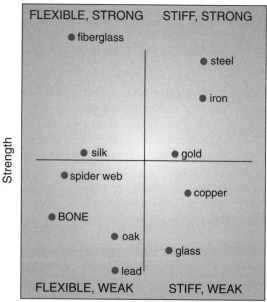

FIGURE 2–9. The strength and stiffness of a variety of different materials are plotted in four quadrants representing material which is (A) flexible and weak; (B) stiff and weak; (C) stiff and strong; and (D) flexible and strong. Note that bone is categorized as being flexible and weak along with other materials such as a spider web and oak wood. (Adapted from P. Shipman et al., 1985)

Stress and Strain   Another way of evaluating the behavior of bone or any other material when subjected to loading is to measure the stress, or load per cross-sectional area, and the strain, or deformation with respect to the original length of the material. A stress-strain curve can be produced which, like the load-deformation curve, illustrates the mechanical behavior of the material and can be used to examine strength and stiffness of the material (FIGURE 2.7).

The load-deformation curve of a particular material looks exactly like the stress-strain curve for the same material and is interpreted in the same manner described previously using the load-deformation curve. The only difference in the curves are the units used to plot each curve. The load-deformation curve is plotted with absolute load and deformation values, while the stress-strain curve is plotted with relative values computed with respect to material length and cross section. The benefit of producing the stress-strain curve is that the standardization with respect to unit area and length allows one to compare different materials.

Normal and Shear Stress or Strain   Stress and strain can occur perpendicular to the plane of a cross section of the loaded object, termed normal stress and strain, or parallel to the plane of the cross section, termed shear stress and strain. For example, normal strain involves a change in the length of an object, whereas shear strain is characterized by a change in the original angle of the object. An example of both normal strain and shear strain is the response of the femur to weight bearing that shortens in length due to normal strain and bends anteriorly due to shear strain imposed by the body weight (1). Normal stress and shear stress, developed in response to tension applied to the tibia, are presented in FIGURE 2–10. Normal and shear strain, developed in response to compression of the femur, are also illustrated.

Types of Loads

The skeletal system is subjected to a variety of different types of forces so that bone is loaded in various directions. There are loads produced by weight bearing, by gravity, by muscular forces, and by external forces. The loads are applied in different directions producing forces that are one of five different types: compression, tension, shear, bending, or torsion.

Injury to the skeletal system can be produced by a single high-magnitude force application of one of these types of load, or by repeated application of a low-magnitude load over an extended period of time. The latter type of injury to the bone is termed a stress fracture, fatigue

fracture, or bone strain. An x-ray photo of a stress fracture to the metatarsal is shown in FIGURE 2–11. These fractures occur as a consequence of cumulative microtraumas imposed upon the skeletal system, when loading of the system is so frequent that the bone repair process cannot keep up with the breakdown of bone tissue. The development of a stress fracture will be discussed in greater detail at the end of this section.

Compression Forces   A compression force presses the ends of the bones together, and is produced by muscles, weight bearing, gravity, or some external loading down the length of the bone. The compressive stress and strain inside the bone causes shortening and widening, and the bone absorbs maximal stress on a plane perpendicular to the compressive load. (Refer to FIGURE 2–12 for a visual illustration of the effects of a compressive force.) Compressive forces are very necessary for the development and growth in the bone. The stress and strain produced by compressive or other types of forces are responsible for facilitating the deposit of osseus material.

If a large compressive force is applied, and if the loads surpass the stress limits of the structure, a fracture will occur. There are numerous sites in the body susceptible to compressive fractures or injury. A compressive force is responsible for patellar pain and the softening and destruction of the cartilage underneath the patella, known as chondromalacia patella. As the knee joint moves through a range of motion, the patella moves up and down in its groove. The load between the patella and the femur increases and decreases to a point at which the compressive patellofemoral force is greatest at approximately 50 degrees of flexion and lowest at full extension or hyperextension of the knee joint. The high-compressive force in flexion, primarily on the lateral patellofemoral surface, is the source of the destructive process that breaks down the cartilage and underlying surface of the patella (17).

Compression is also the source of fractures to the vertebrae (18). Fractures to the cervical area have been reported in activities such as water sports, gymnastics, wrestling, rugby, ice hockey, and football. Normally the cervical spine is slightly extended with a curve anteriorly convex. If the head is lowered, the cervical spine will flatten out to approximately 30 degrees of flexion. If force is applied against the top of the head when it is in this position, the cervical vertebrae are loaded down their length with a compressive force, creating a dislocation or fracture-dislocation of the facets of the vertebrae. When spearing or butting with the head in flexion was outlawed

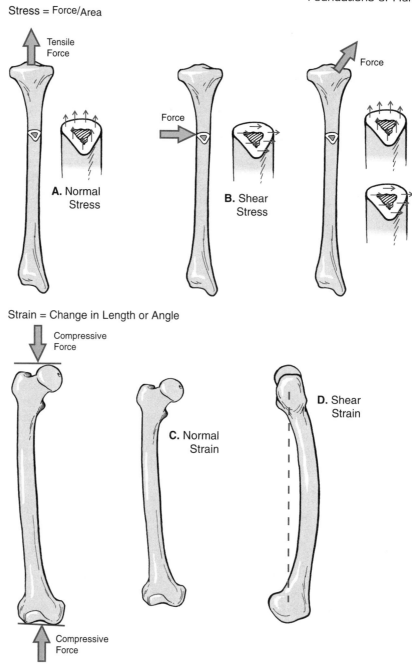

FIGURE 2–10. Stress, force per unit area, can occur perpendicular to the plane (normal stress) as shown in (A), or parallel to the plane (shear stress) as shown in (B). Strain, deformation of the material, is labeled (C) normal strain, in which the length varies, and (D) shear strain, in which the angle changes.

in football, the number of cervical spine injuries was dramatically reduced (18).

Compression fractures have also been reported in the lumbar vertebrae for weight lifters, football lineman, or gymnasts who load the vertebrae while the spine is held in hyperlordotic or swayback position (19). An x-ray photo

of a fracture to the lumbar vertebrae is presented in FIGURE 2–13, demonstrating the shortening and widening effect of the compressive force. Finally, compression fractures are also common in individuals with osteoporosis.

Spondylolysis, a stress fracture of the pars interarticularis section of the vertebrae may result. Specific lifts in

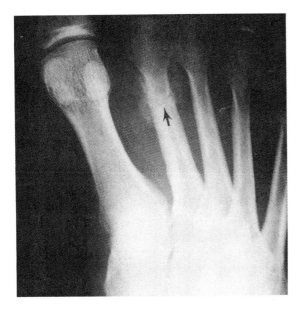

FIGURE 2–11. Stress fractures occur in response to overloading of the skeletal system so that cumulative microtraumas occur in the bone. A stress fracture to the second metatarsal, as shown in the x-ray photo above, is caused by running on hard surfaces or in stiff shoes. It is also associated with persons with high arches and can be created by fatigue of the surrounding muscles. (From Fu, H.F. and Stone, D.A.: Sports Injuries: Baltimore, Williams and Wilkins, 1994)

weight training that have a higher incidence of this fracture are the clean and jerk and the snatch from the Olympic lifts, and the squat and deadlift from powerlifting (19,20). It is also seen in gymnasts and is associated with extreme extension positions in the lumbar region of the vertebrae. This injury will be discussed in greater detail in Chapter 7, when the trunk is reviewed.

A compressive force at the hip joint can increase or decrease the injury potential of the femoral neck. The hip joint must absorb compressive forces of approximately 3 to 7 times body weight during walking (1,2). Compressive forces are up to 15 to 20 times body weight in jumping (1). In a normal standing posture, the hip joint assumes approximately one third of the body weight if both limbs are on the ground (2). This creates large compressive forces on the inferior portion of the femoral neck and a large pulling, or tensile, force on the superior portion of the neck. FIGURE 2–14 shows how this happens as the body pushes down on the femoral head, pushing the bottom of the femoral neck together and pulling the top of the femoral neck apart as it creates bending.

The hip abductors, specifically the gluteus medius, contract to counteract the body weight during stance. They also produce a compressive load on the superior aspect of the femoral neck that reduces the tensile forces

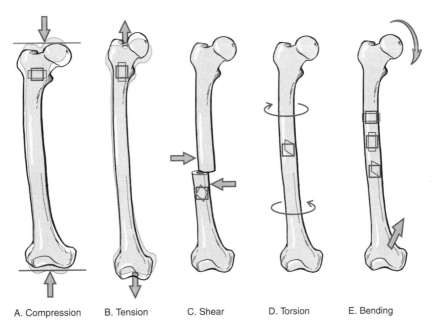

A. Compression   B. Tension   C. Shear   D. Torsion   E. Bending

FIGURE 2–12. The skeletal system is subjected to a variety of different loads which alter the stresses within the bone. The solid square drawn in the femur above indicates the original state of the bone tissue. The colored area illustrates the effect of the force applied to the bone. (A) A compressive force causes shortening and widening; (B) a tensile force causes narrowing and lengthening; (C) a shear force and (D) a torsional force create angular distortion; and (E) the bending force includes all of those changes seen in compression, tension, and shear.

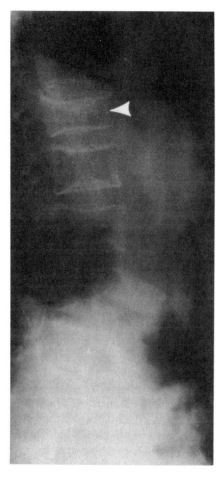

FIGURE 2–13. The lumbar vertebrae can incur a compressive fracture, where the body of the vertebrae is shortened and widened. This type of fracture has been associated with loading of the vertebrae while maintaining a hyperlordotic position. (From Nordin, M. and Frankel, V.H.: Basic Biomechanics of the Musculoskeletal System, 2nd Ed. Philadelphia, Lea and Febiger, 1989.)

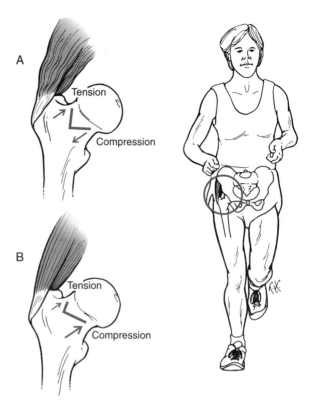

FIGURE 2–14. During standing, or in the stance phase of walking and running, there is a bending force applied to the femoral neck. This force creates a large compressive force on the inferior neck and a tensile force on the superior neck (see A above). If the gluteus medius contracts, the compressive force is increased and the tensile force is decreased (B above). This reduces the injury potential, since injury is more likely to occur in tension.

and injury potential in the femoral neck, since bone will usually fracture sooner with a tensile force (2) (FIGURE 2–14). It is proposed that runners develop femoral neck fractures because the gluteus medius fatigues and cannot maintain its reduction of the high tensile force producing the fracture (1,21). A femoral neck fracture can also be produced by a strong co-contraction of the hip muscles, specifically the abductors and adductors, creating excessive compressive forces on the superior neck.

Tension Forces   A tension force is usually applied to the bony surface and pulls or stretches the bone apart so that it tends to lengthen and narrow (FIGURE 2–12). The maximum stress, like compression, is perpendicular to

the plane of the applied load. The source of the tensile force is usually muscle. When muscle applies a tensile force to the system through the tendon, the collagen in the bone tissue will arrange itself in line with the tensile force of the tendon. (Refer to FIGURE 2–15 for an example of the collagen alignment at the tibial tuberosity.) This figure also illustrates the influence of the tensile forces on the development of apophyses, showing how the tibial tuberosity is formed by tensile forces.

Failure of the bone usually occurs at the site of the muscle insertion. Tensile forces can also create ligament avulsions, which occur more frequently in children. Additionally, ligament avulsions are common on the lateral ankle as a result of ankle sprain.

Avulsion fractures occur when the tensile strength of the bone is not sufficient to prevent the fracture. This is typical of some of the injuries occurring in the high-velocity throwing motion of a Little Leaguer's pitching

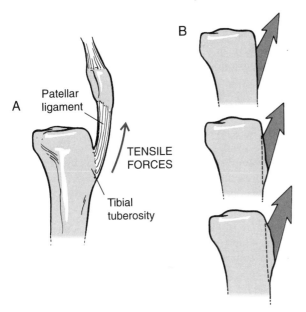

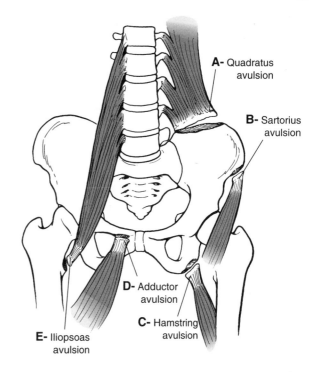

FIGURE 2–15. (A) When tensile forces are applied to the skeletal system, the bone is strengthened in the direction of the pull as collagen fibers align with the pull of the tendon or ligament. (B) Tensile forces are also responsible for the development of apophyses, which are bony outgrowths such as processes, tubercles, or

FIGURE 2–16. Avulsion fractures can occur as a result of tension applied by a tendon or a ligament. Sites of avulsion fracture injuries in the pelvic region are presented above and include the following sites: (A) anterior superior spine; (B) anterior inferior spine; (C) ischial tuberosity; (D) pubic bone; and (E) lesser trochanter.

arm. The avulsion fracture in this case is commonly on the medial epicondyle as a result of tension generated in the wrist flexors.

Two other common tension-produced fractures are at the fifth metatarsal due to the tensile forces generated by the peroneal muscle group, and at the calcaneus where the forces are generated by the triceps surae muscle group. The tensile force on the calcaneus can also be produced in the stance phase of gait as the arch is depressed and the plantar fascia covering the plantar surface of the foot is tightened, exerting a tensile force on the calcaneus. Some sites of avulsion fractures for the pelvic region, presented in FIGURE 2–16, include: the anterior superior and inferior spines, the lesser trochanter, the ischial tuberosity, and the pubic bone.

Tension forces are generally responsible for sprains and strains. For example, the typical ankle inversion sprain occurs when the foot is rolled over to the outside, stretching the ligaments. Tensile forces are also identified with shin splints as the tibialis anterior pulls on its attachment site and on the interosseus membrane.

Another site in the body exposed to high-tensile forces is the tibial tuberosity that transmits very high tensile forces when the quadriceps femoris muscle group is active. This tensile force, if of sufficient magnitude and

duration, may create a tendinitis condition in the older participant. In the younger participant, however, the damage usually occurs at the site of tendon-bone attachment and can result in inflammation, bony deposits, or an avulsion fracture of the tibial tuberosity. Osgood Schlatter Disease is the name of the condition characterized by inflammation and formation of bony deposits at the tendon-bone junction.

Bone responds to the demands placed upon it as described by the previously mentioned Wolff's Law (3). Therefore, different bones and different sections in a bone will respond to tension and compressive forces differently. For example, the tibia and femur participate in weight bearing in the lower extremity and are strongest when loaded with a compressive force. The fibula, which does not participate significantly in weight bearing, but is a site for muscle attachment, is strongest when tensile forces are applied (1).

An evaluation done of the differences that can be found in the femur uncovered greater tensile strength capabilities in the middle third of the shaft that is loaded through a bending force in weight bearing. In the femoral neck, the bone can withstand large compressive forces, and at the attachment sites of the muscles, there is great tensile strength (1).

Shear Forces   A shear force is one applied parallel to the surface within an object, creating deformation internally in an angular direction (FIGURE 2–12). Maximum shear stress acts on the surface parallel to the plane of the applied force. Shear stresses are created when a bone is subjected to compressive forces, tension forces, or both. FIGURE 2–17 shows how a shear stress is developed with the application of a compressive or tensile force. Note the change in shape of the diamond. As the diamond undergoes distortion through compression or tension, there is a shear force applied across the surface.

Bone fails more rapidly when exposed to a shear force than a tensile or compressive force. This is because bone is anisotropic and responds differently when loaded from different directions.

Shear forces are responsible for disc problems in the vertebrae. A shear force can produce spondylolisthesis, in which the vertebrae slip anteriorly over one another. In the lumbar vertebrae, the shear force across the vertebrae increases with increased swayback or hyperlordosis (19). The pull of the psoas muscle on the lumbar vertebrae also creates an increased shear force on the vertebrae. This injury will be discussed in greater detail in a later chapter on the trunk.

Examples of fractures due to shear forces are commonly found in the femoral condyles or the tibial plateau. The mechanism of injury for both of these is usually a hyperextension in the knee through some fixation of the foot and a valgus or medial force to the thigh or shank. In the adult, this shear force can create a fracture as well as injury to the collateral or cruciate ligaments (22). In the developing child, this shear force can create epiphyseal fractures, such as in the distal femoral epiphysis. The mechanism of injury and the resulting epiphyseal damage are presented in FIGURE 2–18. The effects of such a fracture can be very significant since this epiphysis is the fastest growing in the body and accounts for approximately 37% of the bone growth in length (23).

It is common for bone to be loaded with different types of forces at one time. FIGURES 2–19 and 2–20 contain an examination of the multiple loads absorbed by the tibia in walking and running, respectively (2). In walking, there is a compressive stress at heel strike, created by body weight, contact with the ground, and muscular contraction. A tensile stress dominates in midsupport as a result of muscular contraction. Compressive stress develops in preparation for propulsion, as the force on the ground and the muscular contractions increase. A shear force is also present in the propulsive phase of sup-

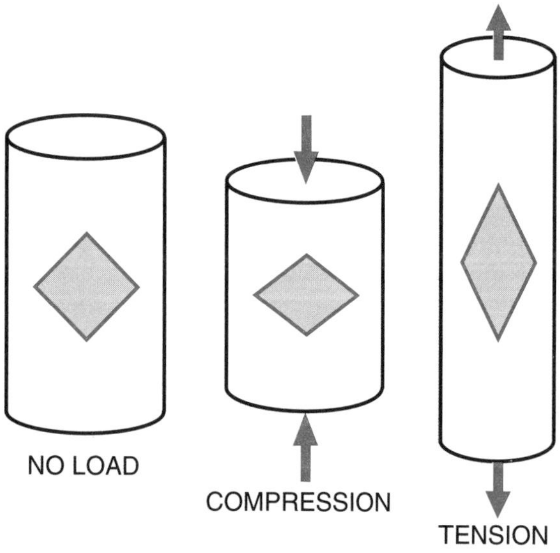

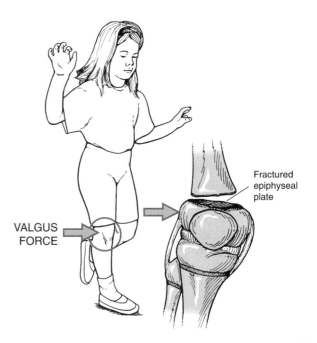

NO LOAD

COMPRESSION

TENSION

FIGURE 2–17. Shear stress and strain accompanies both compression and tension loads.

VALGUS FORCE

Fractured epiphyseal plate

FIGURE 2–18. An epiphyseal fracture of the distal femoral epiphysis is usually created by a shear force. This is commonly produced by a valgus force applied to the thigh or shank, with the foot fixed and the knee hyperextended.

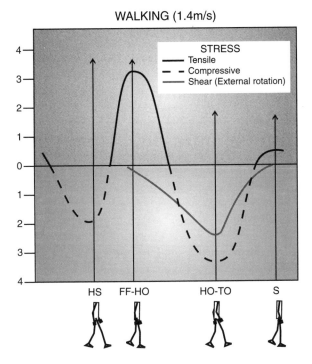

FIGURE 2–19. Tensile, compressive, and shear stresses on the adult tibia during walking. HS = heel strike; FF = foot flat; HO = heel off; TO = toe off; S = swing (Nordin and Frankel, 1989).

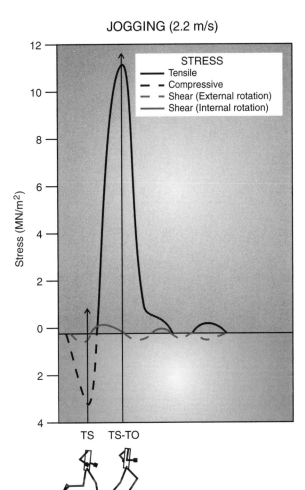

FIGURE 2–20. Tensile, compressive, and shear stresses on the adult tibia during running. TS = toe strike; TO = toe off. (From Slaby, F. and Jacobs, E.R.: Radiographic Anatomy. Philadelphia, Harwal Publishing, 1990).

port, and is believed to be related to torsion created through external rotation of the tibia (2).

In running, the stress increases substantially, and the patterns of stress are different than those seen in walking. There are similarities in the foot-strike phase as a compressive stress is created due to contact with the ground, body weight, and muscular contraction. This is followed by a large tensile stress continuing on through the toe-off phase and into the swing phase. The pattern of shear stress is also different, and is representative of torsion created in response to internal and external rotation of the tibia (2).

Compressive, tensile, and shear forces applied simultaneously to the bone are important in the development of the strength of the bones. FIGURE 2–21 illustrates both the compressive and tensile stress lines in the tibia and femur during running. Bone strength develops along these stress lines.

Bending Forces   A bending force is one applied to an area having no direct support offered by the structure. When a bone is subjected to a bending force and deformation occurs, one side of the bone will form a convex-

ity where tensile forces are present, and the other side of the bone will form a concavity where compressive forces are present (FIGURE 2–12). Typically, the bone will fail and fracture on the convex side in response to the high tensile forces since bone can withstand greater compressive forces than tensile (2). The magnitude of the compressive and tensile forces produced by bending becomes greater farther away from the axis of the bone; thus they are greater on the outer portions of the bone.

During normal stance, there is bending produced in both the femur and the tibia. The femur bends both anteriorly and laterally due to its shape and the manner of the force transmission due to weight bearing. Weight bearing

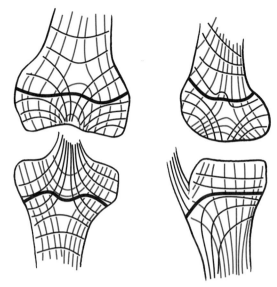

FIGURE 2–21. The lines of compressive stress (solid line) and tension stress (dotted line) are shown for both the distal femur and proximal tibia during the stance phase of running.

Injury-producing bending loads are produced by three- or four-point force applications. The three-point force application usually involves force applied perpendicular to the bone at the ends of the bone, with a force applied in the opposite direction in the middle. The bone will break in the middle as is the case of a ski boot fracture shown in FIGURE 2–22. This fracture is produced as the skier falls over the top of the boot with the ski and boot pushing in the other direction. The bone will usually fracture on the posterior side since that is where the convexity and the tensile forces are applied (2).

Ski boot fractures have been significantly reduced because of improvements in bindings, skis that turn easier, well-groomed slopes, and the change in skiing technique that puts the weight forward. The reduction of tibial fractures through the improvement of equipment and technique has led to an increase in the number of knee injuries for the same reasons (24).

The three-point bending force is also responsible for injuries to the finger that is jammed and forced into hyperextension (1), and to the knee or lower extremity when the foot is fixed in the ground and the lower body bends. Just by eliminating the long cleats in the shoes of football players, and playing on good, resurfaced fields, this type of injury can be reduced by half (20).

Three-point bending force applications are also used in bracing. FIGURE 2–23 presents two brace applications using the three-point force application to correct a postural deviation or stabilize a region.

produces an anterior bend in the tibia. Although these bending forces are not injury producing, if one examines the strength of the tibia and femur, the bone will be stronger in those regions where the bending force is greatest (1).

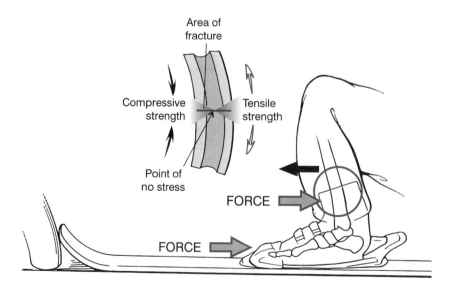

FIGURE 2–22. The ski boot fracture is created by a three-point bending load, and occurs when the ski is abruptly stopped. A compressive force is created on the anterior tibia and a tensile force on the posterior tibia. The tibia usually fractures on the posterior side.

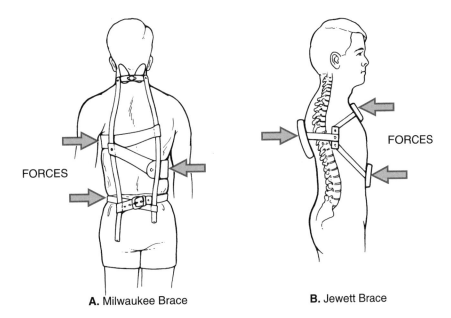

A. Milwaukee Brace   **B.** Jewett Brace

FIGURE 2–23. Three-point bending loads are used in many braces. (A) The Milwaukee brace was used for the correction of lateral curvature of the spine and applied a three-point bending force to the spine. (B) The Jewett brace applies a three-point bending force to the thoracic spine to create spinal extension in that region.

A four-point bending load is applied through the application of two equal and opposite pairs of forces at each end of the bone. In the case of four-point bending, the bone will break at its weakest point (2). This is illustrated in FIGURE 2–24 with the application of a four-point bending force to the femur. The femur fractures in a site that was weak.

Torsional Forces   A torsion force applied to a bone is a twisting force, creating a shear stress over the entire material (FIGURE 2–12). The magnitude of the stress increases with distance from the axis of rotation, and maximum shear stress acts both perpendicular and parallel to the axis of the bone. A torsional loading also produces both tensile and compressive forces at an angle across the structure.

Fractures resulting from torsional force will occur in the humerus when poor throwing techniques create a twist on the arm (1), and in the lower extremity when the foot is planted and the body is changing direction. A spiral fracture is produced as a result of the application of a torsional force. An example of the mechanism of a spiral fracture to the humerus in a pitcher is shown in FIGURE 2–25. The fracture usually begins on the outside of the bone and parallel to the middle of the bone. Torsional loading of the lower extremity is also responsible for cartilage and ligament injuries at the knee joint (20).

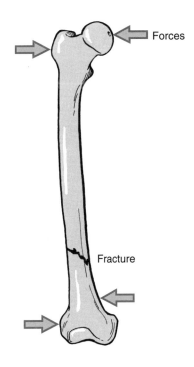

FIGURE 2–24. A four-point bending load applied to a structure will create a fracture or failure at the weakest point. A hypothetical example using the femur is shown above.

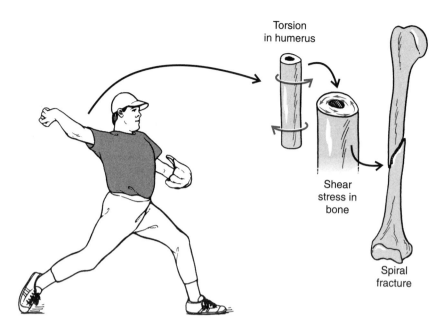

FIGURE 2–25. A torsional force applied to bone creates a shear stress across the surface. An example of torsion applied to the humerus is shown above.

Injury vs Loading   Whether or not a bone incurs an injury as a result of an applied force is determined by the critical strength limits of the material and the loading history of the bone. These limits are influenced primarily by the loading of bone which can be increased or decreased by physical activity and conditioning, immobilization, and skeletal maturity of the individual. The rate of loading is also important, for response and tolerance of bone is rate sensitive. At high rates of loading, when bone tissue cannot deform fast enough, an injury can occur.

Muscular Activity vs Loading   Muscular activity can also influence the loads that can be managed by the bones. The muscles alter the forces applied to the bone by creating compressive and tensile forces. These muscular forces may reduce tensile forces or redistribute the forces on the bone. Since most bones can handle larger amounts of compressive forces, the total amount of load can increase due to the muscular contribution. However, if the muscles fatigue during an exercise bout, their ability to alleviate the load on the bone diminishes. The altered stress distribution or increase in tensile forces leaves the athlete, or performer, susceptible to injury.

Stress Fractures   The typical stress fracture injury will occur during a load application that produces a shear or tensile strain, and results in lacerations, fractures, ruptures, or avulsions. Bone tissue can also develop a stress fracture in response to compressive or tensile loading that overloads the system, either through an excessive magnitude of force applied one or a few times, or through too great of frequency of application of a low or moderate level force (21,25,26). The relationship between the magnitude and the frequency of applications of load on bone is presented in FIGURE 2–26. The tolerance of bone to injury is a function of the load and the cycles of loading.

A stress fracture occurs when bone resorption weakens the bone too much and the bone deposit does not occur rapidly enough to strengthen the area. The cause of stress fractures in the lower extremity can be attributed to muscle fatigue, which reduces shock absorption and allows redistribution of forces to specific focal points in the bone. In the upper extremity, stress fractures are created by repetitive muscular forces pulling on the bone. Stress fractures account for 10% of all injuries to athletes (26).

Examples of injuries to the skeletal system are presented in TABLE 2–1. The activity associated with the injury, the type of load causing the injury, and the mechanism of injury are summarized.

Bony Articulations
The Diarthrodial or Synovial Joint

Movement potential of a segment is determined by the structure and function of the diarthrodial or synovial joint.

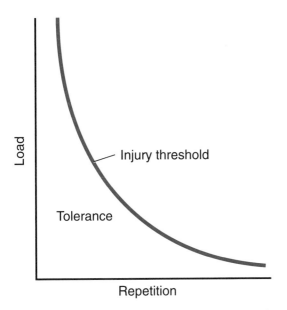

FIGURE 2–26. Injury can occur when a high load is applied a small number of times or when low loads are applied numerous times. It is important to remain within the injury tolerance area.

The diarthrodial joint provides us with a low friction articulation capable of withstanding significant wear and tear. The characteristics of all diarthrodial joints are similar, such that the knee joint has similar structures to the finger joints, and so forth. Because of this similarity, it is worthwhile to look at the various components of the diarthrodial joint to gain general knowledge about joint function, support, and nourishment. Refer to FIGURE 2–27 as we examine the characteristics of the diarthrodial joint.

Characteristics of the Diarthrodial Joint   Lining the ends of the bones is the articular endplate, a thin layer of compact bone over spongy bone. Over the top of the endplate is the articular cartilage, a tough connective tissue lining the ends of the bones. The articular cartilage is hyaline cartilage, an avascular substance consisting of 60 to 80% water and a solid matrix composed of collagen and proteoglycan gel. Cartilage has no blood supply and is nourished by the fluid within the joint (27).

The properties of cartilage make it well suited to resisting shear forces because it responds to load in a viscoelastic manner. It deforms instantaneously to a low or moderate load and, if rapidly loaded, it will respond with stiffness and deform over a longer period of time. The force distribution across the area in the joint determines the stress in the cartilage, and the distribution of the force depends on the cartilage thickness.

Cartilage is 1 to 7 mm in thickness, depending on the stress and the incongruity of the joint surfaces (28). For example, in the ankle and the elbow joints, the cartilage is very thin, while at the hip and knee joints, it is thick. The cartilage is thin in the ankle because of the ankle architecture. There is a substantial area of force distribution that imposes less stress on the cartilage. Conversely, the knee joint is exposed to lower forces, but the area of force distribution is smaller, imposing more stress on the cartilage. Some of the thickest cartilage in the body is found on the underside of the patella, and is approximately 5 mm in thickness (29).

Cartilage is very important to the stability and function of the joint since it distributes loads over the surface and reduces the contact stresses by half (15). It allows a movement to occur between two bones with minimal friction and wear.

In the tibiofemoral, acromioclavicular, temporomandibular, sternoclavicular, and distal radioulnar joints there is an additional cartilage, the fibrocartilage or meniscus, a fibrous tissue connected to the capsule. This cartilage in the joint offers additional load transmission, stability, improved fit of the surfaces, protection of the joint edges, and lubrication of the joint.

Collagen fibers in the meniscus are arranged to withstand load bearing. For example, in the knee, the medial meniscus transmits 50% of the compression load. Removal of just a small part of the meniscus has been shown to increase the contact stress by as much as 350% (30). Ten years ago, a meniscus tear would have meant removal of the whole meniscus, but today orthopedists trim up the meniscus and remove only minimal amounts in order to maintain as much shock absorption and stability in the joint as possible.

Meniscus tears usually occur during a change of direction movement with the weight all on one limb. This creates a compression and tension on the meniscus that tears the fibrocartilage. There is no pain associated with the actual tear but the peripheral attachment sites are the site of the irritation and resulting sensitivity.

Another important characteristic of the diarthrodial joint is the capsule, a white connective fibrous tissue made primarily of collagen. It protects the joint. Thickenings in the capsule, known as ligaments, are common where additional support is needed. The capsule basically defines the joint, creating the interarticular portion of the joint, or inside of the joint, that has a joint cavity and a reduced atmospheric pressure (15). Although soft tissue loads are difficult to compute, the capsule sustains some of the load imposed upon the joint (31).

Table 2–1 Injuries to the Skeletal System

Injury to Bone Type of Injury	Activity Examples	Load Causing the Injury	Mechanism of Injury
Tibial stress fracture	Dancing Running Basketball	Compression	Poor conditioning, stiff footwear, non-yielding surfaces, hypermobile foot (overpronation).
Medial epicondyle fracture	Gymnastics	Tension Compression	Too much work on floor exercise and tumbling.
Stress fracture of big toe	Sprinting Fencing Rugby	Tension	Toe extensors create bowstring effect on big toe when up on toes; primarily in individuals with hallux valgus.
Stress fracture of femoral neck	Running Gymnasts	Compression	Muscle fatigue, high-arched foot.
Stress fracture in calcaneus	Running Basketball Volleyball	Compression	Hard surface, stiff footwear.
Stress fracture in lumbar vertebrae	Weight Lifting Gymnastics Football	Compression Tension	High loads with hyperlordotic low-back posture.
Tibial plateau fractures	Skiing	Compression	Hyperextension and valgus of the knee, as in turning, with the forces applied to the inside edge of the downhill ski, and abruptly halted with heavy snow.
Stress fracture to medial malleolus	Running	Compression	Ankle sprain to outside causing compression between talus and medial malleolus or excessive pronation since the medial malleolus rotates in with tibial rotation and pronation.
Hamate fracture in the hand	Baseball Golf Tennis Racquet Sports	Compression	Relaxed grip in the swing that is stopped suddenly at the end of the swing as the club hits the ground, the bat is forcefully checked, or the racket is out of control.
Fracture of tibia	Skiing	Bending Compression Tension	Three-point bending fall in which the body weight, the boot, and the ground create a posterior bending in the tibia.
Fracture of femoral condyles	Football Skiing	Shear	Hyperextension of knee with valgus force.
Stress fracture in fibula	Running Aerobics Jumping	Tension	Jumping or deep-knee bends with a walk. Pull by soleus, tibialis posterior, peroneals, and toe flexors pulling tibia and fibula together.
Menisci tear in the knee joint	Basketball Football Jumping Volleyball Soccer	Compression Torsion	Turning on a weight-bearing limb or valgus force to the knee.
Stress fracture in metatarsal	Running	Compression	Hard surfaces, stiff footwear, high arched foot, fatigue.
Stress fracture in femoral shaft	Running Triathletes	Tension	Excessive training and mileage. Created by pull of vastus medialis or adductor brevis.

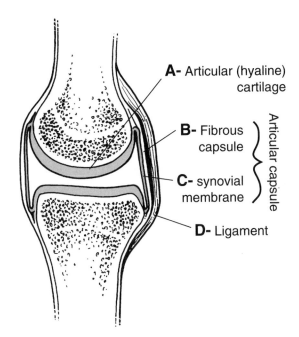

A- Articular (hyaline) cartilage

B- Fibrous capsule

C- synovial membrane

D- Ligament

Articular capsule

FIGURE 2–27. The diarthrodial joints have similar characteristics. If you study the knee, interphalangeal, elbow, or any other diarthrodial joint, you will find the same structures. These include: (A) articular or hyaline cartilage; (B) capsule; (C) synovial membrane; and (D) ligaments.

Any immobilization of the capsule alters the mechanical properties of the capsular tissue and may result in joint stiffness. Likewise, injury to the capsule usually results in the development of a thick or fibrous section that can be externally palpatable (32).

On the inner surface of the joint capsule is the synovial membrane, a loose, vascularized connective tissue that secretes synovial fluid into the joint to lubricate and provide nutrition to the joint. The fluid, having the consistency of an egg white, decreases in viscosity with an increase in shear rates. It is like catsup, hard to start but easy to move once it is going. When the joint moves slowly the fluid is highly viscous and the support is high. Conversely, when the joint moves rapidly, the fluid is elastic in its response, decreasing the friction in the joint (15).

Any injury to the joint is noticeable in both a thickening in the membrane and a change in the consistency of the fluid. The fluid fills the capsular compartment and creates pain in the joint. Physicians will drain the joint to relieve the pressure, often finding that the fluid is bloodstained.

The final structures of importance in and around the diarthrodial joint are the ligaments. A ligament connects bone to bone and is comprised of collagen, elastin, and reticulin fibers. The collagen fibers are arranged in a ligament so that it can handle both tensile loads and shear loads; however, it is best suited for tensile loading. Ligaments can be capsular, extracapsular, or intra-articular. Capsular ligaments are just thickenings in the wall of the capsule, much like the glenohumeral ligaments in the front of the shoulder capsule. Extracapsular ligaments lie outside the joint itself. The collateral ligaments found in numerous joints in the body are examples of extracapsular ligaments (i.e., fibular collateral ligament of knee). Finally, intra-articular ligaments are located inside the diarthrodial joint, such as the cruciate ligaments at the knee and the capitate ligaments within the hip joint.

Ligaments respond to loads imposed upon them by becoming stronger and stiffer over time. The strength of a ligament also diminishes rapidly with immobilization. A tensile injury to a ligament is termed a sprain and is rated 1, 2, or 3 in severity by the medical profession, depending on whether there was a partial tear of the fibers (rating=1), tear with some loss of stability (rating=2), or a complete tear with loss of joint stability (rating=3) (28).

At the end of the range of motion for every joint there is usually a ligament tightening up to terminate the motion. Since the ligaments function to stabilize, control, and limit joint motion, any injury to a ligament will influence joint motion.

Stability of the Diarthrodial Joint   The stability in a diarthrodial joint is provided by the structure, ligaments surrounding the joints, the capsule, the tendons spanning the joint, gravity, and the vacuum in the joint produced by the negative atmospheric pressure. The hip joint is one of the most stable joints in the body because it has good muscular, capsular, and ligamentous support. The structure of the hip joint has congruency between the surfaces, with a high degree of bone-to-bone contact. However, most of the stability in the hip joint is derived from the effects of gravity and the vacuum in the joint (28). The negative pressure in the joint is sufficient to hold the femur up in the joint if all other structures such as supporting ligaments and muscles are removed.

In contrast, stability of the shoulder is supplied only by the capsule and the muscles surrounding the joint. Also, the congruency of the shoulder joint is limited, with only a small proportion of the head of the humerus making contact with the glenoid cavity.

Degeneration of the Diarthrodial Joint   Injury to the structures within the diarthrodial joint can occur during high load situations or through repetitive loading over an extended period of time. The articular cartilage in the joint is especially subject to considerable wear during one's lifetime. Any trauma to or repeated wear on the joint causes a change in the articular substance to the point where there is enzymatic degradation, loss of proteoglycan, and removal of actual material by means of mechanical action. This results in diminished contact areas and erosion of the cartilage through the development of rough spots in the cartilage. The rough spots develop into fissures and eventually go deep enough so there is only subchondral bone exposed. Osteophytes or cysts form in and around the joint and this is the beginning of degenerative joint disease or osteoarthritis. The x-ray photos in FIGURE 2–28 show the areas of joint degeneration associated with osteoarthritis in the hip and vertebrae.

It is theorized that osteoarthritis develops first in the subchondral, or cancellous bone underlying the joint (33). The cartilage overlying the bone in the joint is thin; consequently, the underlying subchondral bone absorbs the shock of loading in the joint. Through repetitive load-ing or unequal loading in the joint, the subchondral bone develops micro-fractures. When the micro-fractures heal, the subchondral bone is stiffer and less able to absorb the shock, passing this role on to the cartilage. The cartilage deteriorates as a consequence of this overloading and the body lays down bone in the form of osteophytes to increase the contact area (33).

Osteoarthritis has been shown to have no relationship to hyperlaxity in the joint (34), to levels of osteoporosis (35), nor to levels of physical activity (12). However, an injured joint deteriorates at a faster rate, making it more susceptible to the development of osteoarthritis (12).

Osteoarthritis can also be created through joint immo-bilization, because the joint and the cartilage require loading and compression to exchange nutrients and wastes in the joint (36). After only 30 days of immobi-lization, the fluid is increased in the cartilage, and an early form of osteoarthritis develops. Fortunately, this process can be reversed with a return to activity (36).

Injury to other structures in the diarthrodial joint can also be very serious. An injury to the joint capsule will

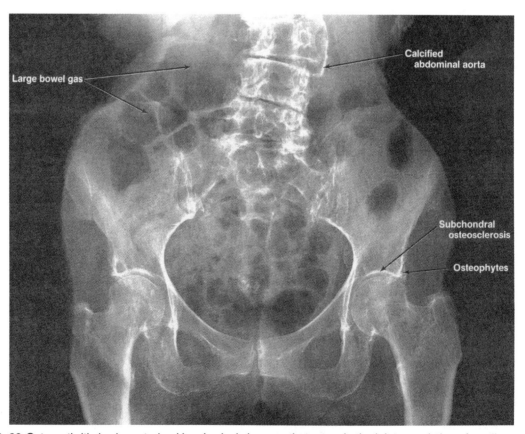

FIGURE 2–28. Osteoarthritis is characterized by physical changes that occur in the joint, consisting of cartilage erosion and formation of cysts and osteophytes. X-ray examples of osteoarthritis in the hip and the vertebrae are presented above.

result in the formation of more fibrous tissue and possible stretching of the capsule (32). Injury to the meniscus can create instability, loss of range of motion, and an increase in the synovial effusion (swelling) into the joint. Injury to the synovial membrane causes an increase in vascularity and produces a gradual fibrosis of the tissue, eventually leading to a chronic synovitis or inflammation of the membrane. Amazingly, many of these injury responses can also be reproduced through immobilization of the joint, which can produce adhesions, loss of range of motion, fibrosis, and synovitis.

Types of Diarthrodial Joints

Simple, Compound, and Complex Joints The articulating surfaces found in the different joints in the body vary in size and shape. There is typically a concave surface meeting a convex surface on the adjacent bone, termed female and male surfaces, respectively. Also, there can be more than two contact points or articulating surfaces. A joint with only two articulating surfaces is termed a simple joint, while a joint with three or more articulating surfaces is known as a compound joint. A joint with more than two articulating surfaces and with a disc or fibrocartilage is called a complex joint. An example of a simple joint is the hip or the ankle (talotibial), of a compound joint is the wrist, and of a complex joint is the knee.

Close-Packed vs Loose-Packed Positions As movement occurs through a range of motion, the actual contact area varies between the articulating surfaces. When the joint position is such that the two adjacent bones fit together in a position of best fit, and where there is maximum contact between the two surfaces, the joint is considered to be in a close-packed position. This is the position of maximum compression of the joint where the ligaments and the capsule are tense and the forces travel through the joint as if it did not exist. Examples of close-packed positions are full extension for the knee, extension of the wrist, extension of the interphalangeal joints, and maximum dorsiflexion of the foot (15). The close- and loose-packed positions of the knee joint are presented in FIGURE 2–29. Note the greater contact area in the close-packed position.

While in the close-packed position, the joint is very stable but vulnerable to injury, since the structures are taut and the joint surfaces are pressed together. The joint is especially susceptible to injury if hit by an external force, such as hitting the knee when it is fully extended.

All other joint positions are termed the loose-packed position because there is less contact area between the two

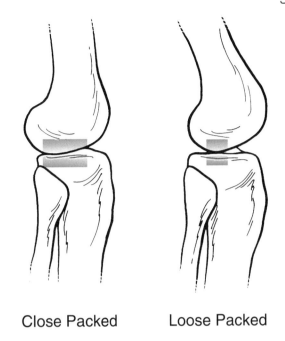

Close Packed   Loose Packed

FIGURE 2–29. In the close-packed position, there is maximum contact between the two joint surfaces and minimal mobility. In the loose-packed joint position, there is less contact between the surfaces in the joint and more mobility and movement between the two surfaces.

surfaces and the contact areas are frequently changing. There is more sliding and rolling of the bones over one another in the loose-packed position (15). This position allows for continuous movement in the joint, reducing the friction in the joint. Although the loose-packed joint position is less stable than the close-packed position, it is not as susceptible to injury because of its mobility.

There is a joint classification system that categorizes seven different types of diarthrodial joints according to the differences in articulating surfaces, the degrees of freedom allowed by the joint, and the type of movement occurring between the segments. FIGURE 2–30 offers a graphic representation of these seven joints.

The Plane or Gliding Joint The first type of joint is the plane or gliding joint, found in the foot between the tarsals and in the hand among the carpals. Movement at this type of joint does not occur about an axis and is termed non-axial since it consists of two flat surfaces that slide over each other to allow movement.

In the hand, for example, the carpals will slide over each other as the hand is moved to positions of flexion, extension, radial deviation, or ulnar deviation. Likewise, in the foot, the tarsals shift during pronation and supination, sliding over each other in the process.

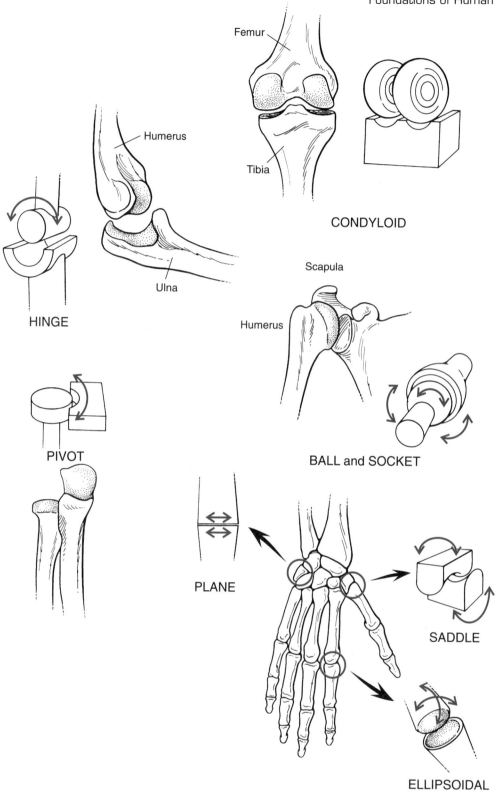

FIGURE 2–30. There are seven different types of diarthrodial joints. The non-axial joint is the plane, or gliding joint. Uniaxial joints include the hinge and pivot joints, and biaxial joints are the condylar, ellipsoid, and saddle joint. The ball-and-socket joint is the only triaxial diarthrodial joint.

The Hinge Joint   The hinge (ginglymus) joint allows movement in one plane (flexion, extension) and is termed uniaxial. Examples of the hinge joint in the body are the interphalangeal joints of the phalanges in the foot and hand, and the ulnohumeral articulation at the elbow.

The Pivot Joint   The pivot joint also allows movement in one plane (rotation; pronation, supination) and is uniaxial. Pivot joints are located at the superior and inferior radioulnar joint and the atlantoaxial articulation at the base of the skull.

The Condylar Joint   The condylar joint is a joint allowing primary movement in one plane (flexion, extension) with small amounts of movement in another plane (rotation). It is found at the knee joint and the temporomandibular joint.

The Ellipsoid Joint   The ellipsoid joint allows movement in two planes (flexion, extension; abduction, adduction) and is biaxial. Examples of this joint can be found at the radiocarpal articulation at the wrist and the metacarpophalangeal articulation in the phalanges.

The Saddle Joint   The saddle joint, only found at the carpometacarpal articulation of the thumb, allows two planes of motion (flexion, extension; abduction, adduction) with a small amount of rotation also allowed. It is similar to the ellipsoid joint in function.

The Ball-and-Socket Joint   The last type of diarthrodial joint, the ball-and-socket joint allows movement in three planes (flexion, extension; abduction, adduction; rotation) and is the most mobile of the diarthrodial joints. The hip and shoulder joints are examples of ball-and-socket joints. A summary of the major joints in the body is presented in TABLE 2–2.

Other Types of Joints

Synarthrodial or Fibrous Joints   There are other articulations that are limited in movement characteristics but, nonetheless, play an important role in stabilization of the skeletal system. Some bones are held together by fibrous articulations, such as those found in the sutures of the skull. These articulations, referred to as synarthrodial, allow little or no movement to occur between the bones and hold the bones firmly together (FIGURE 2–31).

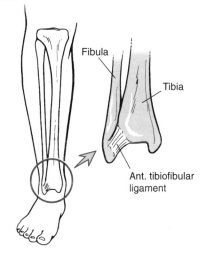

A. Synarthrodial

Distal Tibiofibular Joint

Table 2–2 Major Joints of the Body

Joint	Type	Df
Vertebrae	Amphiarthrodial	3
Hip	Ball-and-Socket	3
Shoulder	Ball-and-Socket	3
Knee	Condyloid	2
Wrist	Ellipsoid	2
Metacarpophalangeal (fingers)	Ellipsoid	2
Carpometacarpal (thumb)	Saddle	2
Elbow	Hinge	1
Radioulnar	Pivot	1
Atlantoaxial	Pivot	1
Ankle	Hinge	1
Interphalangeal	Hinge	1

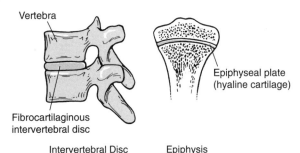

B. Amphiarthrodial

Intervertebral Disc   Epiphysis

FIGURE 2–31. (A) An example of the synarthrodial joint is the fibrous articulation at the distal tibiofibular joint. (B) The amphiarthrodial, or cartilaginous joint can be found between the vertebrae or in the epiphyseal plate of a growing bone.

Amphiarthrodial or Cartilaginous Joints   There are also cartilaginous joints, called amphiarthrodial, that are bones held together by either hyaline cartilage, such as is found at the epiphyseal plates, or by fibrocartilage, found at the pubic symphysis and the intervertebral articulations (FIGURE 2–31). The movement at these articulations is also very limited, although not to the degree of the synarthrodial joints.

Summary

The skeletal system provides a system of levers that allows a variety of movements to occur at the joints in the body, provides a support structure to the body, protects the internal structures, stores fats and minerals, and participates in blood cell formation. The types of bones that compose the skeletal system (long, short, flat, irregular, sesamoid) are shaped differently, perform different functions, and are made up of different proportions of spongy and compact bone tissue.

Bone tissue is one of the body's hardest structures because of its organic and inorganic components. Bone tissue continuously remodels through deposit and resorption, and is also very sensitive to disuse or loading. Bone tissue is deposited in response to stress on the bone and removed through resorption when not stressed. One of the ways of increasing the strength and density of bone is through a program of physical activity. Osteoporosis is a condition of the bone where bone resorption exceeds bone deposit and the bone becomes weak.

The study of the architecture of bone tissue has identified two types of bone, compact and spongy. Compact bone, found on the exterior of bone or in the shaft of the long bones, is suited to handling high levels of compression and high tensile loads produced by the muscles. Spongy bone is suited for high energy storage and facilitates stress distribution within the bone.

Bone is both anisotropic and viscoelastic in its response to loads, and will respond differently to variety in the direction of the load and to the rate at which the load is applied. When first loaded, bone responds by deforming through a change in length or shape, known as the elastic response. With continued loading, microtears occur in the bone as it yields during the plastic phase. Bone is considered to be a flexible and weak material compared to other materials such as glass or steel.

The skeletal system is subjected to a variety of loads and can handle larger compressive loads than tensile or shear loads. Commonly, bone is loaded in more than one direction such as is the case in bending where both compression and tension are applied, and in torsion loads where shear, compression, and tensile loads are all produced. Injury to bone results when the applied load exceeds the strength of the material.

The movements of the long bones occur at a synovial joint, a joint with common characteristics such as articular cartilage, a capsule, a synovial membrane, and ligaments. The synovial joint can be injured through a sprain, where the ligaments are injured. Joints are also susceptible to degeneration, characterized by breakdown in the cartilage and the bone. This is known as osteoarthritis.

The amount of motion between two segments is largely influenced by the type of synovial joint. For example, the planar joint allows simple translation between the joint surfaces, the hinge joint allows flexion and extension, the pivot joint allows rotation, the condylar joint allows flexion and extension with some rotation, the ellipsoid and the saddle joints allow flexion, extension, abduction, and adduction, and the ball-and-socket joint allows flexion, extension, abduction, adduction, and rotation. Other classifications of joints, synarthrodial and amphiarthrodial, allow little or no movement.

Review Questions

1. Describe a movement in the trunk, the upper extremity, and the lower extremity that is terminated due to a bony restriction in the joint.
2. Describe the functions of synovial fluid.
3. What are the functions of hyaline cartilage? Of menisci?
4. Compare factors stabilizing the knee joint to those stabilizing the shoulder joint.
5. Discuss the factors stabilizing the diarthrodial joint.
6. Why is bone stronger when loaded in compression as compared to tension or shear?

Additional Questions

1. What is the benefit of having spongy bone in the ends of the long bones? Why is it also beneficial *not* to have spongy bones in the shaft of the long bones?
2. What is the effect of not having a patella at the knee joint?
3. What is the benefit and the disadvantage of a close-packed joint position?
4. As a weight lifter who had included the squat exercise in your weight routine for many years, what significant changes would be apparent in the size of and the landmarks on the femur?
5. If a person weighing 700 N is standing with both feet on the ground, how much compressive force is being generated at each knee joint by body weight only (assuming 88% of the BW is above the knee)? If the person lowers 30 degrees into flexion at the knee, what is the compressive force at each knee joint? What is the shear force?
6. If the compressive stress on the tibia in running reaches a level of 3 MN $\cdot$ m^{-2} and the tensile stress on the tibia reaches values of 11 MN $\cdot$ m^{-2}, what type of injury is likely to be incurred? Provide an example of such an injury in the lower extremity.
7. What is the stiffness or modulus of elasticity of a material if the change in the strain of the material is .004 mm $\cdot$ mm^{-1} and the change in stress is 2.75 MN $\cdot$ m^{-82}.

Additional Reading

An, K.N., Himeno, S., Tsumura, H., Kawai, T., & Chao, E.Y.S.: Pressure distribution on articular surfaces; application to joint stability evaluation. Journal of Biomechanics. *23:*1013, 1991.

Blank, S.: Transverse tibial stress fractures. The American Journal of Sports Medicine. *15:*597, 1987.

Brown, T.D., Pedersen, D.R., Gray, M.L., Brand, R.A., and Rubin, C.T.: Toward an identification of mechanical parameters initiating periosteal remodeling: a combined experimental and analytic approach. Journal of Biomechanics. *23:*893, 1990.

Currey, J: The Mechanical Adaptations of Bones. Princeton, Princeton University Press, 1984.

Keller, C.S., Noyes, F.R., and Buncher, C.R.: The medical aspects of soccer injury epidemiology. The American Journal of Sports Medicine. *15:*230–237, 1987.

McAuley, E., et al.: Injuries in womens gymnastics. The American Journal of Sports Medicine. *15:*558–565, 1987.

Nambu, T., Gasser, B., Schneider, E., Bandi, W., and Perren, S.M.: Deformation of the distal femur: A contribution towards the pathogenesis of osteochondrosis dissecans in the knee joint. Journal of Biomechanics. *24:*421–433, 1991.

Nichols, J.N., and Tehranzadeh, J.: A review of tibial spine fractures in bicycle injury. The American Journal of Sports Medicine. *15:*172–174, 1987.

Nigg, B.M., and Bobbert, M.: On the potential of various approaches in load analysis to reduce the frequency of sports injuries. Journal of Biomechanics. *23:*3–12, 1990.

Parker, R.D., Berkowitz, M.S., Brahms, M.A., & Bohl, W.R.: Hook of the hamate fractures in athletes. The American Journal of Sports Medicine. *14:*517–523, 1986.

Rohl, L., Larsen, E., Linde, F., Odgaard, A., and Jorgensen, J.: Tensile and compressive properties of cancellous bone. Journal of Biomechanics. *24:*1143–1149, 1991.

Shelbourne, K.D., Fisher, D.A., Rettig, A.C., and McCarrol, J.R.: Stress fractures of the medial malleolus. The American Journal of Sports Medicine. *16:*60–63, 1988.

Thompson, N., et al.: High school football injuries: Evaluation. The American Journal of Sports Medicine. *15:*117–124, 1987.

Tursz, A., and Crost, M.: Sports-related injuries in children. The American Journal of Sports Medicine. *14:*294–299, 1986.

Viano, D.C., King, A.I., Melvin, J.W., and Weber, K.: Injury biomechanics research: An essential element in the prevention of trauma. Journal of Biomechanics. *22:*403–417, 1989.

Watson, M.D., and DiMartino, P.P.: Incidence of injuries in high school track and field athletes and its relation to performance ability. The American Journal of Sports Medicine. *15:*251–254, 1987.

Yokoe, K., and Mannoji, T.: Stress fracture of the proximal phalanx of the great toe. The American Journal of Sports Medicine. *14:*240–242, 1986.

References

1. Riegger, C.L.: Mechanical properties of bone. *In* Orthopaedic and Sports Physical Therapy. Edited by J.A. Gould and G.J. Davies. St. Louis, C.V. Mosby Co. 1985, pp. 3–49.

2. Nordin, M., and Frankel, V.H.: Biomechanics of bone. *In* Basic Biomechanics of the Musculoskeletal System. Edited by M. Nordin and V.H. Frankel. Philadelphia, Lea & Febiger, 1989, pp. 3–30.

3. Keller, T.S., and Spengler, D.M.: Regulation of bone stress and strain in the immature and mature rat femur. Journal of Biomechanics. 22:1115–1127, 1989.

4. Whalen, R.T., Carter, D.R., and Steele, C.R.: Influence of physical activity on the regulation of bone density. Journal of Biomechanics. 21:825-837, 1988.

5. Zernicke, R.F., Vailas, A.C., and Salem, G.J.: Biomechanical response of bone to weightlessness. *In* Exercise and Sport Sciences Reviews. Edited by K.B. Pandolf and J.O. Holloszy. Baltimore, Williams and Wilkins, 1990, pp. 167–192.

6. Antao, N.A.: Myositis of the hip in a professional soccer player. The American Journal of Sports Medicine. 16:82, 1988.

7. Cook, S.D., et al.: Trabecular bone density and menstrual function in women runners. The American Journal of Sports Medicine. 15:503, 1987.

8. Iskrant, A.P., and Smith, R.W.: Osteoporosis in women 45 years and related to subsequent fractures. Public Health Reports. 84:33–38, 1969.

9. Frost, H.M.: The pathomechanics of osteoporoses. Clinical Orthopaedics and Related Researc. 200:198, 1985.

10. Oyster, N., Morton, M., and Linnell, S.: Physical activity and osteoporosis in post-menopausal women. Medicine and Science in Sports and Exercise. 16:44–50, 1984.

11. Brewer, V., et al.: Role of exercise in prevention of involutional bone loss. Medicine and Science in Sports and Exercise. 15:445, 1983.

12. Lane, N.E., et al.: Running, osteoarthritis, and bone density: Initial 2-year longitudinal study. The American Journal of Medicine. 88:452–459, 1990.

13. Schaffler, M.B., and Burr, D.B.: Stiffness of compact bone: Effects of porosity and density. Journal of Biomechanics. 21:13–16, 1988.

14. Choi, K., and Goldstein, S.A.: A comparison of the fatigue behavior of human trabecular and cortical bone tissue. Journal of Biomechanics. 25:1371, 1992.

15. Soderberg, G.L.: Kinesiology: Application to Pathological Motion. Baltimore, Williams & Wilkins, 1986.

16. Shipman, P., Walker, A. and Bichell, D: *The Human Skeleton.* Cambridge, Harvard University Press, 1985.

17. Eisele, S.A.: A precise approach to anterior knee pain. The Physician and Sports Medicine. Vol 19(6): 126–130; 137–139, 1991.

18. Fine, K.M., Vegso, J.J., Sennett, B., and Torg, J.S.: Prevention of cervical spine injuries in football. The Physician and Sports Medicine. Vol 19(10): 54–64, 1991.

19. Halpbern, B.C., and Smith, A.D.: Catching the cause of low back pain. The Physician and Sports Medicine. Vol 19(6): 71–79, 1991.

20. Halpbern, B., et al.: High school football injuries: Identifying the risk factors. The American Journal of Sports Medicine. 15:316, 1987.

21. Jackson, D.L.: Stress fracture of the femur. The Physician and Sports Medicine. Vol 19(7): 1990. pp. 39–44.

22. McConkey, J.P., and Meeuwisse, W.: Tibial plateau fractures in alpine skiing. The American Journal of Sports Medicine. 16:159–164, 1988.

23. Downing, J.F., Nicholas, J.A., and Goldberg, B. (eds.): Four complex joint injuries. The Physician and Sports Medicine. Vol 19(10): 80–97, 1991.

24. Dolinar, J.: Keeping ski injuries on the down slope. The Physician and Sports Medicine. Vol 19(2): 120–123, 1990.

25. Lakes, R.S., Nakamura, S., Behiri, J.C., and Bonfield, W.: Fracture mechanics of bone with short cracks. Journal of Biomechanics. 23:967–975, 1990.

26. Matheson, G.O., et al.: Stress fractures in athletes. The American Journal of Sports Medicine. 15:46–58, 1987.

27. Mow, V.C., Proctor, C.S., and Kelly, M.A.: Biomechanics of articular cartilage. *In* Nordin, M. & Frankel, V.H. (Eds.) Basic Biomechanics of the Musculoskeletal System. Edited by M. Nordin and V.H. Frankel. Philadelphia, Lea & Febiger, 1989, pp. 31–58.

28. Hettinga, D.L.: Inflammatory response of synovial joint structures. *In* Orthopaedic and Sports Physical Therapy. Edited by J.A. Gould and G.J. Davies. St. Louis, C.V. Mosby, 1985, pp. 87–117.

29. Wallace, L.A., Mangine, R.E., and Malone, T.: The knee. *In* Gould, J.A. & Davies, G.J. (Eds.) Orthopaedic and Sports Physical Therapy. Edited by J.A. Gould and G.J. Davies. St. Louis, C.V. Mosby, 1985, pp. 342–364.

30. Henning, C.E.: Semilunar cartilage of the knee: Function and pathology. *In* Exercise and Sport Sciences Review. Edited by K.B. Pandolf. New York, Macmillan, 1988, pp. 205–214.

31 Hoffman, A.H., and Grigg, P.: Measurement of joint capsule tissue loading in the cat knee using calibrated mechanoreceptors. Journal of Biomechanics. 22:787–791, 1989.

32. Egan, J.M.: A constitutive model for the mechanical behavior of soft connective tissues. Journal of Biomechanics. 20:681–692, 1987.

33. Radin, E.L., Paul, I.L., and Rose, R.: Role of mechanical factors in the pathogenesis of primary osteoarthritis. Lancet. 1:519–522, 1972.

34. Bird, H.A.: A clinical review of the hyperlaxity of joints with particular reference to osteoarthrosis. Engineering in Medicine. 15:81, 1986.

35. Healey, J.H., Vigorita, V.J., and Lane, J.M.: The coexistence and characteristics of osteoarthritis and osteoporosis. The Journal of Bone and Joint Surgery. 67-A:586–592, 1985.

36. Navarro, A.H., and Sutton, J.D.: Osteoarthritis IX: Biomechanical factors, prevention, and nonpharmacologic management. Maryland Medical Journal. 34:591–594, 1985.

Glossary

Amphiarthrodial Joint:	A type of joint where the bones are connected by cartilage; some movement may be allowed at these joints. Also called cartilaginous joint.
Anisotropic:	Having different properties in different directions.
Apophysis:	A bony outgrowth such as a process, tubercle, or tuberosity.
Articular Cartilage:	Hyaline cartilage consisting of tough, fibrous connective tissue.
Articular Endplates:	The end of the bones, consisting of layers of hyaline cartilage, compact bone, and spongy bone.
Avulsion Fracture:	The tearing away of a part of bone when a tensile force is applied.
Ball-and-Socket Joint:	A type of diarthrodial joint which is freely movable, allowing motion through three planes.
Bending Force:	A force causing a change in the angle of the bone, offsetting in the horizontal plane. The material will bend in the region where there is no direct structural support.
Brittle:	Characterized by failure or fracture soon after the elastic limits are past; easily broken.
Cancellous Bone:	Bone tissue which is lattice-like and has a high porosity; capable of high energy storage. Also called spongy bone.
Capsular Ligament:	Ligaments contained within the wall of the capsule; thickening in the capsule wall.
Capsule:	A fibrous, connective tissue that encloses the diarthrodial joint.
Cartilaginous Joint:	A type of joint where the bones are connected by cartilage. Some movement may be allowed at these joints. Also called amphiarthrodial joint.
Close-Packed Position:	The joint position in which there is maximum contact between the two joint surfaces, and in which the ligaments are taut, forcing the two bones to act as a single unit.
Collagen:	The main protein of skin, tendon, ligament, bone, cartilage, and connective tissue.
Compact Bone:	A dense, compact tissue on the exterior of bone that provides strength and stiffness to the skeletal system. Also called cortical bone.
Complex Joint:	A joint in which two or more bones articulate, and a disc or fibrocartilage is present.
Compound Joint:	A joint in which three or more bones articulate.
Compression Force:	A force pressing the ends of the bone together, creating a shortening and widening of the structure.
Condylar Joint:	A type of diarthrodial joint that is biaxial and has one plane of movement that dominates the movement in the joint.

Cortical Bone:	A dense, compact tissue on the exterior of bone which provides strength and stiffness to the skeletal system. Also called compact bone.
Deposit:	A phase of bone remodeling during which bone is formed through osteoblastic activity.
Diarthrodial Joint:	Freely movable joint; also called synovial joint.
Ductile:	A characteristic allowing considerable deformation after the elastic region before it fails or fractures; capable of being drawn out.
Elastic:	A characteristic that allows stretch or distortion and then returns to the original shape or length; resilient.
Ellipsoid Joint:	A type of diarthrodial joint with two degrees of freedom that resembles the ball-and-socket joint.
Extracapsular Ligament:	Ligament contained outside of the joint capsule.
Fibrocartilage:	A type of cartilage having parallel, thick, collagenous bundles.
Fibrous Joint:	A type of joint where the bones are connected by fibrous material; little or no movement is allowed at these joints; also called synarthrodial joint.
Flat Bone:	A thin bone consisting of thin layers of both compact and spongy bone.
Gliding Joint:	A type of diarthrodial joint with flat surfaces that allows translation between the two bones; also called plane joint.
Hinge Joint:	A type of diarthrodial joint allowing one degree of freedom.
Intra-articular Ligament:	Ligament located inside the joint.
Irregular Bone:	A bone having a specialized shape and function.
Lever:	A simple machine that magnifies force or speed of movement.
Ligament:	A band of fibrous, collagenous tissue connecting bone or cartilage to each other; supports the joint.
Load-deformation Curve:	A plot of the relationship between the load imposed externally and the amount of deformation internally in a material.
Long Bone:	A bone longer than it is wide, having a shaft, the diaphysis, and wide ends, the metaphyses.
Loose-Packed Position:	The joint position with less than maximum contact between the two joint surfaces and in which contact areas are frequently changing.
Meniscus:	Crescent-shaped discs of fibrocartilage.
Microtrauma:	A disturbance or abnormal condition which is initially very small and cannot be seen.
Modulus of Elasticity:	The stiffness of a material determined by the slope of the curve during the elastic response phase.

Myositis Ossificans:	A condition in which bone deposits are laid down in soft tissue; ossification of muscles.
Normal Stress:	The amount of load per cross-sectional area applied perpendicular to the plane of a cross section of the loaded object.
Normal Strain:	Deformation in a material involving a change in the length of the object.
Osgood-Schlatter Disease:	A disease characterized by inflammation and formation of bone deposits at the tendon-bone junction; at the tibial tuberosity.
Osseous:	The nature or quality of bone.
Osteoarthritis:	Degenerative joint disease characterized by degeneration in the articular cartilage, osteophyte formation, and reduction in the joint space.
Osteoporosis:	A condition where the rate of bone formation is decreased, demineralization occurs, and the bone softens.
Pivot Joint:	A type of diarthrodial joint that allows movement in one plane; pronation, supination, or rotation.
Plane Joint:	A type of diarthrodial joint with flat surfaces that allows translation between the two bones; also called gliding joint.
Plastic:	A characteristic where the material deforms permanently and does not return to the original shape or length.
Porosity:	The ratio of pore space to the total volume.
Resorption:	A phase of bone remodeling where bone is lost through osteoclastic activity.
Saddle Joint:	A type of diarthrodial joint having two saddle-shaped surfaces allowing two degrees of freedom.
Sesamoid Bone:	A type of short bone embedded in a tendon or joint capsule.
Shear Force:	A force applied parallel to the surface, creating deformation internally in an angular direction.
Shear Stress:	The amount of load per cross-sectional area applied parallel to the plane of a cross section of the loaded object.
Shear Strain:	Deformation in a material involving a change in the original angle of the object.
Short Bone:	A bone having dimensions that are approximately equal.
Simple Joint:	A joint with only two articulating surfaces.
Spondylolisthesis:	A condition in which the vertebrae slip anteriorly over one another; associated with hyperlordosis in the lumbar vertebrae.
Spondylolysis:	A stress fracture of the pars interarticularis of the vertebrae, associated with hyperlordosis and excessive loading of the spine.
Spongy Bone:	Bone tissue that is lattice-like and has a high porosity; capable of high energy storage. Also called cancellous bone.
Sprain:	A joint injury where the ligament is injured.

Stiffness:	The resistance to load during deformation in a material; determined by the slope of the load-deformation curve.
Strain:	Deformation in a material measured with respect to the original length.
Strength of Material:	The load sustained before failure in a material.
Stress:	The amount of load per cross-sectional area.
Stress Fracture:	A fracture created when the loading of the skeletal system is so frequent that the bone repair process cannot keep up with the breakdown of bone tissue; also called a fatigue fracture.
Stress-Strain Curve:	A plot of the relationship between the load per cross-sectional area and the amount of deformation measured with respect to the original length.
Synarthrodial Joint:	A type of joint where the bones are connected by fibrous material; little or no movement is allowed at these joints; also called fibrous joint.
Synovial Fluid:	A liquid secreted by the synovial membrane that reduces friction in the joint; the fluid changes viscosity in response to the speed of joint movement.
Synovial Joint:	Freely movable joint; also called diarthrodial joint.
Synovial Membrane:	A loose vascularized connective tissue that lines the joint capsule.
Tension Force:	A force pulling the bone apart, creating a lengthening and narrowing of the bone.
Torsion Force:	A twisting force that creates shear stress over the entire material.
Trabecula:	Strands within spongy bone that adapt to the direction of stress on the bone.
Viscoelastic:	Material that responds with varying levels of stiffness depending on the rate and duration of the load application.
Yield Point:	The point in the loading phase at which the fibers in the material begin to give way and debond; the beginning of the plastic phase in the loading cycle.

CHAPTER 3

Muscular Considerations for Movement

Student Objectives

After reading this chapter, the student will be able to:

1. Describe the gross and microscopic anatomical structure of the muscle, and describe how a muscle contracts.
2. Explain the differences in shape and function of muscles and also differentiate between anatomic and physiologic cross section.
3. Describe the difference in the force output between the three different muscle fiber types (type I, IIa, IIb).
4. Describe the characteristics of the muscle attachment to the bone and explain the viscoelastic response of the tendon.
5. Define the four properties, the functions, and the roles of skeletal muscle.
6. Compare the differences between isometric, concentric, and eccentric muscle actions.
7. Explain how the angle of muscular attachment, the elastic component, the length-tension relationship, the force-velocity relationship, and the presence of a pre-stretch all influence tension development in the muscle.
8. Explain the physical changes that occur in the muscle as a result of strength training, and elaborate on how genetic predisposition, training specificity, intensity, and training volume will influence strength training outcomes.
9. Describe different types of resistance training and explain how training would be adjusted for the athlete or nonathlete.
10. Identify some of the major contributors to muscle injury, the location of common injuries, and means for prevention of injury to the muscle.

Muscles and gravity are the major producers of human movement. Muscles are used to hold a position, to raise or lower a body part, to slow down a fast moving segment, and to generate great speed in the body or in an object being propelled into the air. Muscles are capable of contracting rapidly and vigorously, but they tire easily and require rest after only brief periods of activity.

The tension developed by muscles applies compression to the joints enhancing their stability. However, in some joint positions, the tension generated by the muscles can act to pull the segments apart and create instability.

Muscles are used asymmetrically in most activities in which one side of the body will be using a specific set of muscles, and the other side of the body will be using opposite or different muscles. This is true in activities such as golf, bowling, baseball, walking, and racquet sports that have nonsymmetrical use of the arms, legs, and trunk. Muscles are used symmetrically in activities such as weight lifting and jumping where both sides of the body are performing the same movement using the same muscles.

Exercise programming for a young, healthy population will most likely use exercises that will push the muscular system to high levels of performance. Muscles can exert force and develop power, and many athletic training programs with the young and healthy are designed to achieve maximum performance. However, the same principles used with young and active individuals can be scaled down for use by persons of limited ability. Using the elderly as an example, it is apparent that strength decrement is one of the major factors influencing efficiency in daily living activities. The loss of strength and efficiency in the muscular system can create a variety of problems, ranging from an inability to reach overhead or open a jar lid, difficulty in going up stairs, or getting up out of a chair. Another example is the overweight individual who has difficulty walking any distance because the muscular system cannot generate sufficient power and the person fatigues easily. These two examples are really no different from the powerlifter trying to perform a maximum lift in the squat. In all three cases, the muscular system is being overloaded, with only the magnitude of the load and output varying.

All aspects of muscle structure and function related to human movement and efficiency of muscular contribution will be explored in this chapter. Since muscles are responsible for locomotion, limb movements, posture, and joint stability, a good understanding of the features and limitations of muscle action is necessary.

Gross Structure of the Muscle

Physical Organization of Muscle(s)

Muscles and muscle groups are arranged so that they may contribute individually or collectively to produce a very small movement or a very large, powerful movement. Muscles rarely act individually, but rather with other muscles in a multitude of possible roles.

To understand muscle function, the structural organization of muscle from the macroscopic external anatomy all the way down to the microscopic level of muscular action must be examined first. A good starting point is the gross anatomy and external arrangement of muscles and the microscopic view of the muscle fiber.

Groups of Muscles   Groups of muscles are contained within compartments and are defined by fascia, a sheet of fibrous tissue. The compartments divide the muscles into functional groups and it is common for muscles in a compartment to be innervated by the same nerve. Included in the thigh are three compartments: the anterior compartment containing the quadriceps femoris, the posterior compartment containing the hamstrings, and the medial compartment containing the adductors. Compartments for both the thigh and the lower leg are illustrated in FIGURE 3–1.

The compartments serve to keep the muscles organized and contained in one region, but there are times in which the compartment is not large enough to accommodate the muscle or muscle groups. In the anterior tibial region, the compartment is small and will create problems if the muscles are overdeveloped for the amount of space defined by the compartment. This is known as compartment syndrome and can be serious if the cramped compartment impinges on nerves or blood supply to the leg and foot.

Individual Muscle Organization   The anatomy of the skeletal muscle is presented in FIGURE 3–2. Each individual muscle usually has a more centralized portion where the muscle is thicker, and this is termed the belly of the muscle. Some muscles, like the biceps brachii, have very pronounced bellies while other muscles, like the wrist flexors and extensors, have bellies that are not so apparent to the observer.

Covering the outside of the muscle is another fibrous tissue, the epimysium, that plays a vital role in the transfer of muscular tension to the bone. Tensions in the mus-

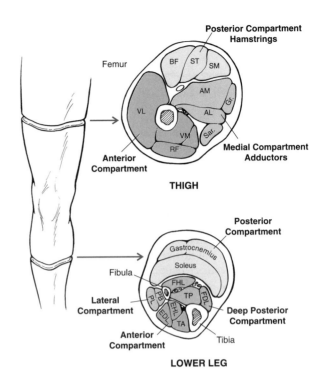

FIGURE 3–1. Muscles are grouped into compartments within each segment. Each compartment is maintained by fascial sheaths. The muscles within each compartment are functionally similar, and define groups of muscles which we would classify as extensors and flexors, etc.

cle are generated at various sites and the epimysium transfers the various tensions to the tendon, providing a smooth application of the muscular force to the bone.

Each muscle may contain thousands of muscle fibers that are carefully organized into compartments within the muscle itself. Bundles of muscle fibers are called fascicles, and each fascicle may contain as many as 200 muscle fibers.

The fascicle is covered with a dense connective sheath called the perimysium that protects the muscle fibers and creates pathways for the nerves and blood vessels. The connective tissue in the perimysium and the epimysium give muscle much of its ability to be stretched and returned to a normal resting length. The perimysium is also the focus of flexibility training because the connective tissue in the muscle can be stretched, allowing the muscle to become more elongated.

The fascicles run parallel to each other in the muscle. Each fascicle contains the long, cylindrical muscle fibers, the cells of skeletal muscles, where the force is generated in the muscle. Muscle fibers can be as large as 50 μm

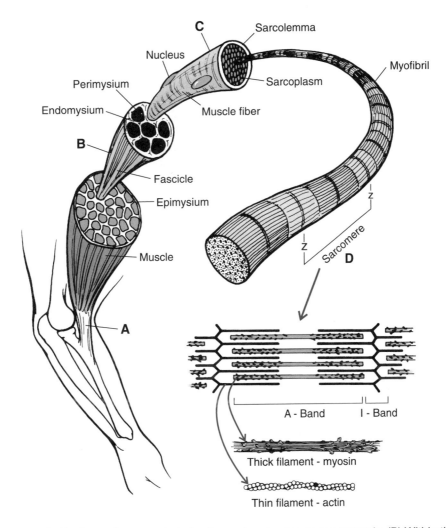

FIGURE 3–2. (A) Each individual muscle connects to the bone via a tendon or aponeurosis. (B) Within the muscle, the fibers are bundled together into fascicles. (C) Each individual fiber contains myofibril strands which run the length of the fiber. (D) The actual contractile unit is the sarcomere. Many sarcomeres are connected in series down the length of each myofibril. Muscle shortening occurs in the sarcomere as the myofilaments in the sarcomere, actin and myosin, slide towards each other.

wide and 10 centimeters long (1). Fibers also run parallel and are covered with a membrane, the endomysium. The endomysium is a very fine sheath carrying the capillaries and the nerves that nourish and innervate each muscle fiber. The vessels and the nerves usually enter in the middle of the muscle and then are distributed throughout the muscle by a path through the endomysium. The endomysium also serves as an insulator for the neurological activity within the muscle.

Directly underneath the endomysium is the sarcolemma, which is a thin plasma membrane surface that branches into the muscle. The neurologic innervation of the muscle travels through the sarcolemma and eventually reaches each individual contractile unit through chemical neurotransmission.

At the microscopic level, the fiber can be further broken down into numerous myofibrils, delicate, rod-like strands running the total length of the muscle. There can be hundreds or thousands of myofibrils in each muscle fiber, and each fiber is filled with 80% myofibrils (1). The remainder of the fiber consists of the usual organelles such as the mitochondria, the sarcoplasm, sarcoplasmic reticulum, and the T tubules. Myofibrils are 1–2 μm in diameter (about four-millionth of an inch wide) and run the length of the muscle fiber (1).

The myofibrils are cross striated due to light and dark filaments placed in an order that forms repeating patterns of bands. The dark banding is the thick protein myosin, and the light band is the thin polypeptide, actin. One unit of these bands is called a sarcomere, and this is the actual contractile unit of the muscle that develops tension as the actin filaments slide toward the middle and the myosin filaments.

Sliding Filament Theory   An explanation of the shortening of the sarcomere has been presented via the sliding filament theory (2). When calcium is released into the muscle through neurochemical stimulation, the shortening process is initiated. The sarcomere shortens as the myosin filament "walks" along the actin, forming cross-bridges between the head of the myosin and a prepared site on the actin filament. In the contracted state, the actin and myosin filaments overlap along most of their lengths (FIGURE 3–3).

The simultaneous sliding of many thousands of sarcomeres in series creates a change in length and force in the muscle (1). The amount of force that can be developed in the muscle is proportional to the number of cross-bridges formed. Through the shortening of many sarcomeres, myofibrils, and fibers, an actual movement is created by the development of tension running through the muscle and applied at both ends of the muscle to the bone.

Fiber Organization

The shape and arrangement of the fibers in the muscle will determine whether the muscle is capable of generating large amounts of force or whether the muscle has good shortening ability. In the case of the latter, the shortening ability of a muscle is reflected by both change in length and speed, depending on the movement situation. There are two basic types of fiber arrangements found in the muscle, as illustrated in FIGURE 3–4.

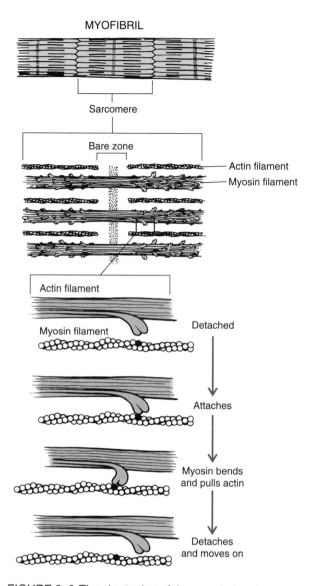

FIGURE 3–3. The shortening of the muscle has been explained by the sliding filament theory. Shortening takes place in the sarcomere as the myosin heads bind to sites on the actin filament to form a cross-bridge. The myosin head attaches and turns which moves the actin filament toward the center. It then detaches and moves on to the next actin site.

Fusiform Muscles   The fusiform fiber arrangement has parallel muscle fibers, and fascicles that run the length of the muscle. The fibers in a fusiform muscle run parallel to the line of pull of the muscle so that the fiber force is in the same direction as the musculature (3).

This spindle-shaped fiber arrangement is known for offering the potential for high amounts of shortening and

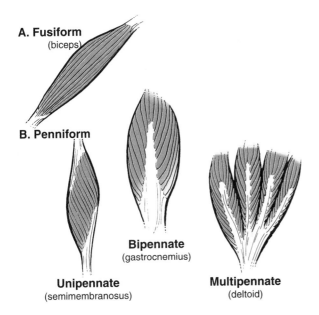

A. Fusiform
(biceps)

B. Penniform

Bipennate
(gastrocnemius)

Unipennate
(semimembranosus)

Multipennate
(deltoid)

FIGURE 3–4. (A) The fusiform muscles are spindle shaped and the fibers run so that they pull in the same direction as the whole muscle. (B) Penniform muscles have fibers which run diagonally to a tendon running through the muscle. The muscle fibers of a penniform muscle do not pull in the same direction as the whole muscle.

high velocity movements in the body (4). This is basically because the fusiform muscles are typically longer than other types of muscles and the muscle fiber length is greater than the tendon length. Fiber lengths of the fusiform muscle are shown in FIGURE 3–5.

A muscle having a greater ratio of muscle length to tendon length has the potential of shortening through greater distances. Consequently, muscles attaching to the bone with a short tendon (rectus abdominous) can move through a greater shortening distance than muscles with longer tendons (gastrocnemius) (5). Great amounts of shortening also occur because skeletal muscle can shorten up to approximately 30 to 50% of its resting length. Examples of fusiform muscles in the body are the sartorius, the biceps brachii, and the brachialis.

Penniform Muscles In the second type of fiber arrangement, penniform, the fibers run diagonally with respect to a tendon running through the muscle. The general shape of the penniform muscle is feather shaped as the fascicles are short and run at an angle. The fibers of the penniform muscle run at an angle relative to the line of pull of the muscle, so the fiber force is in a different direction than the muscle force (3). The fibers are shorter than the muscle, and the change in the individual fiber

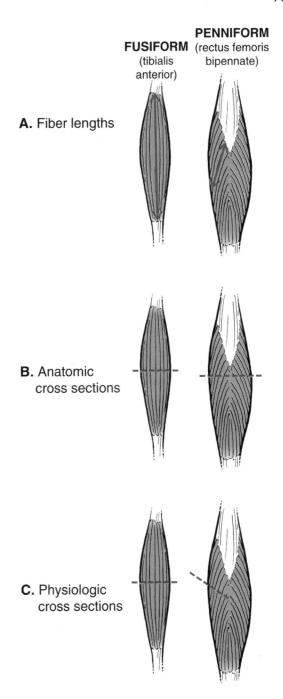

FUSIFORM
(tibialis
anterior)

PENNIFORM
(rectus femoris
bipennate)

A. Fiber lengths

B. Anatomic
cross sections

C. Physiologic
cross sections

FIGURE 3–5. (A) Fusiform muscle fiber arrangements usually have longer fiber lengths than the penniform fiber arrangements. (B) The anatomic cross sections of the fusiform and penniform fiber arrangements may or may not be similar, but the (C) physiologic cross section of the fusiform fiber arrangement is usually smaller. Thus, the fusiform muscles are typically weaker but can move through greater distances than the penniform muscle types.

length is not equal to the change in the muscle length (3). The fibers can run diagonally off of one side of the tendon, termed unipennate, off both sides of the tendon, termed bipennate, or the fiber arrangement can be a combination of the two, termed multipennate.

Since the muscle fibers are shorter and run diagonally into the tendon, the penniform fibers create slower movements, and are not capable of producing movement through a large range of motion. The trade off is a much larger physiologic cross section in the muscle that can generally produce more strength.

The physiologic cross section is the sum total of all of the cross sections of fibers in the muscle, measuring the area perpendicular to the direction of the fibers. The anatomic cross section, on the other hand, is the cross section right angles to the longitudinal axis of the muscle. Examine the anatomic and physiologic cross sections for the two types of muscles in FIGURE 3–5. In the fusiform fiber type, the physiologic and anatomic cross sections would be equal. In the penniform muscle type, the physiologic cross section is greater than the anatomic cross section, because there are a greater number of fibers in the physiologic cross section (5). Thus, a penniform muscle having the same anatomic cross section as a fusiform muscle will be capable of generating more force, because it has a greater physiologic cross section (> # of fibers). The penniform muscles are the high-force and power-producing muscles of the body. Examples of unipennate muscles are flexor pollicis longus, tibialis posterior, semimembranosus, and extensor digitorum longus. Bipennate muscles are the gastrocnemius, soleus, vastus medialis, vastus lateralis, and rectus femoris. Examples of multipennate muscles are the deltoid and the gluteus maximus.

The mechanical actions of broad pennate muscles such as the pectoralis major or trapezius, that have fibers attaching directly into bone over a large attachment site, are difficult to describe using one movement for the whole muscle (6). For example, the lower trapezius attaches to the scapula at an angle opposite of the upper trapezius; thus these sections of the same muscle are functionally independent. When the shoulder girdle is elevated and abducted as the arm is moved up in front of the body, the lower portion of the trapezius may be inactive. This presents a complicated problem when studying the function of the muscle as a whole and requires multiple lines of action and effect (6).

Fiber Type

Each muscle contains a combination of different fiber types, that are categorized as fast twitch or slow twitch.

Fiber types are an important consideration in the area of muscle metabolism and energy consumption, and muscle fiber types are thoroughly studied in exercise physiology. There are mechanical differences in the response of slow- and fast-twitch muscle fibers warranting an examination of fiber type.

Slow-Twitch Fiber Types Slow twitch or Type I, oxidative muscle fibers are found in higher quantities in the postural muscles of the body such as the upper back and the soleus (7). The fibers are red due to the high content of myoglobin in the muscle. A stained cross-section photo of the slow-twitch fiber is presented in FIGURE 3–6. These fibers have slow contraction times and are well-suited for prolonged, low-intensity work. Endurance athletes usually have a higher quantity of slow-twitch fibers.

Intermediate and Fast-Twitch Fiber Types Fast-twitch or Type II fibers are further broken down into Type IIa, oxidative-glycolytic, and Type IIb, glycolytic. The Type IIa is a red muscle known as the intermediate fast-twitch fiber because it can sustain activity for long periods, or it can contract with a burst of force and then fatigue (FIGURE 3–6). The white, Type IIb fiber provides us with rapid force production and then fatigues quickly.

Sprinters and jumpers usually have greater concentrations of fast-twitch fibers, and they are also found in higher concentrations in the phasic muscles such as the gastrocnemius. Most muscles, if not all, contain both

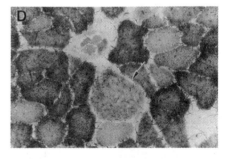

FIGURE 3–6. A histologic stained cross section of the three muscle types is presented. The arrow points to a Type IIb, fast-twitch, glycolytic fiber. A slow-twitch oxidative fiber contacts the Type IIb fiber on the underside and an intermediate fast-twitch fiber (oxidative-glycolytic) is making contact directly above and to the right of the Type IIb fiber. (From Lieber, R.L.: Skeletal Muscle Structure and Function. Baltimore, Williams and Wilkins, 1992).

fiber types. An example is the vastus lateralis, which is typically half fast-twitch and half slow-twitch fibers (7). The fiber type will influence how the muscle is trained and developed, as well as what techniques will best suit individuals with specific fiber types.

Muscle Attachment

Tendon vs Aponeurosis vs Bone A muscle attaches to the bone in one of three ways: directly into the bone, via a tendon, or via an aponeurosis. These three types of attachments are presented in FIGURE 3–7. Muscle can attach directly into the periosteum of the bone through a fusion between the epimysium and the surface of the bone, such as the attachment of the trapezius (6). Muscle can attach via a tendon that is fused with the muscle fascia, such as the hamstrings, biceps brachii, and flexor carpi radialis. Or, it can attach via a sheath of fibrous tissue known as an aponeurosis seen in the abdominals and the trunk attachment of the latissimus dorsi.

Characteristics of the Tendon The most common form of attachment, the tendon, is an inelastic bundle of collagen fibers arranged parallel to the direction of the force application of the muscle. Even though the fibers are inelastic, the tendon can respond in an elastic fashion through recoiling and the elasticity of connective tissue. Tendons can withstand high tensile forces produced by the muscles and they exhibit viscoelastic behavior in response to loading. The achilles tendon has been reported to resist tensile loads to a degree equal to or greater than steel of similar dimensions. The stress/strain response of a tendon is similar to that of the ligament.

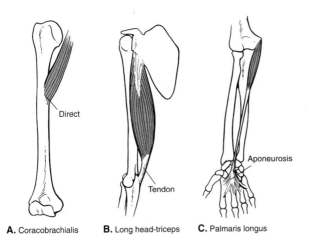

A. Coracobrachialis   **B.** Long head-triceps   **C.** Palmaris longus

FIGURE 3–7. A muscle can attach (A) directly into the bone, or indirectly via a (B) tendon or (C) aponeurosis.

Both tendons and ligaments will respond very stiffly when exposed to a high rate of loading. The differences in the strength and performance characteristics of the tendon versus muscle or bone is presented in FIGURE 3–8.

Tendon and muscle join at the myotendinous junction, where the actual myofibrils of the muscle fiber join the collagen fibers of the tendon to produce a multi-layered interface (8). The tendon connection to the bone consists of a tendon connection to fibrocartilage that joins to mineralized fibrocartilage and then into the lamellar bone. This interface blends with the periosteum and the subchondral bone (8).

Tendons and muscles work together to absorb or generate tension in the system. Tendons are arranged in series, or in line with the muscles; consequently, the tendon bears the same tension as the muscle (9). The mechanical interaction between the muscle and tendon depends on the amount of force that is being applied or generated, the speed of the muscle action, and the slack in the tendon.

If tension is generated in the muscle fibers while the tendon is slack, it will begin to recoil, or spring back to its initial length. As the slack in the tendon is taken up by the recoiling action, the time taken to stretch the tendon causes a delay in the achievement of the required level of tension in the muscle fibers (9).

Recoiling of the tendon also reduces the speed at which a muscle may shorten, which in turn increases the load a muscle can support (9). If the tendon is stiff and has no recoil, the tension will be transmitted directly to the muscle fibers, creating higher velocities and decreasing the load the muscle can support. The stiff response in a tendon allows for the development of rapid tensions in the muscle, and results in brisk, accurate movements (9).

The tendon and the muscle are very susceptible to injury if the muscle is contracting as it is being stretched. An example is the follow-through phase of a throwing action. Here, the posterior rotator cuff is stretched as it is contracting to slow the movement. Another example would be the muscle lengthening and contraction of the quadriceps femoris muscle group during the support phase of running as the body lowers through knee flexion. The tendon picks up the initial stretch of the relaxed muscle, and if the muscle contracts as it is stretched, the tension rises steeply in both muscle and tendon (9).

When tension is generated in a tendon at a slow rate, injury is more likely to occur at the tendon-bone junction than other regions. At a faster rate of tension development, the actual tendon is the more common site of failure (10). When considering the total muscle-tendon unit, the likely site of injury is the belly of the muscle or the myotendinous junction (10).

MUSCLE

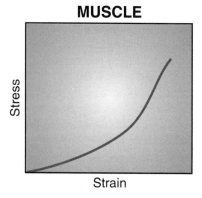

Stress

Strain

TENDON

Stress

Strain

BONE

Stress

Strain

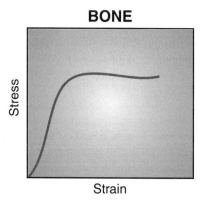

FIGURE 3–8. The stress-strain curve for muscle, tendon, and bone tissue varies. (A) Muscle is viscoelastic and thus, deforms under low load and then responds stiffly. (B) Tendon is capable of handling high loads. The end of the elastic limits of the tendon is also the ultimate strength level (no plastic phase). (C) Bone is a brittle material which responds stiffly and then undergoes minimal deformation before failure.

Many tendons travel over bony protuberances that serve to reduce some of the tension on the tendon by changing the angle of pull of the muscle and reducing the tension generated in the muscle. Examples of this can be found with the quadriceps femoris muscles and the patella, and with the tendons of the hamstrings and the gastrocnemius as they travel over condyles on the femur.

The tension developed in the tendons also produces the actual ridges and protuberances on bone. The apophyses found on a bone are developed by tension forces applied to the bone through the tendon (See Chapter 2). This is of interest to physical anthropologists because they can study skeletal remains and make sound predictions about lifestyle and occupations of a civilization by evaluating prominent ridges, size of the trochanters or tuberosities, or basic bone size of the specimen.

Origin vs Insertion   A muscle typically attaches to a bone at both ends. The attachment closest to the middle of the body, or more proximal, is termed the origin. The attachment further away from the midline or more distal is called the insertion. Traditional anatomy classes usually incorporate a study of the origins and insertions of the muscles. It is a common mistake to view the origin as the bony attachment that does not move when the muscle contracts. It is important to remember that muscles pull equally on both ends so that both attachment sites receive equal forces. The reason that both bones do not move when a muscle contracts is due to the stabilizing force of adjacent muscles or the difference in the mass of the two segments or bones to which the muscle is attached.

There are numerous examples in the body where a muscle can shift between moving one end of its attachment or the other end, depending upon the activity. One example is the psoas muscle crossing the hip, that flexes the thigh as in leg raises, or that raises the trunk, as in a curl up or sit up. The psoas example is presented in FIGURE 3–9. Another example is the gluteus medius, which moves the pelvis when the foot is on the ground and the leg when the foot is off the ground (4). The effect of tension in a muscle should be evaluated at all attachment sites, even if no movement is resulting from the force. By evaluating all attachment sites, the magnitude of the required stabilizing forces and the actual forces applied at the bony insertion can be assessed.

Functional Characteristics of the Muscle

Muscle Fiber Potential

Skeletal muscle is very resilient and can be stretched or shortened at fairly high speeds without major damage

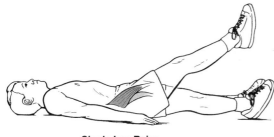

Single Leg Raise

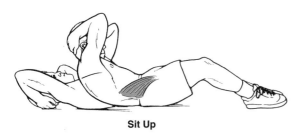

Sit Up

FIGURE 3–9. The origin of the psoas muscle is on the bodies of the last thoracic and all of the lumbar vertebrae, and the insertion is on the lesser trochanter of the femur. It is incorrect to assume that the origin remains stable in a movement. In the example above, the psoas pulls on both the vertebrae and the femur. With the trunk stabilized, the femur will move (leg raise), and with the legs stabilized, the trunk will move (sit up).

occurring in the tissue. The performance of the muscle fiber under varying load and velocity situations is determined by the four properties of the skeletal muscle tissue: irritability, contractibility, extensibility, and elasticity.

Irritability This is the ability to respond to stimulation. In a muscle, the stimulation is provided by a chemical neurotransmitter. Skeletal muscle tissue is one of the most sensitive and responsive tissues in the body. Only nerve tissue is more sensitive than skeletal muscle tissue. As an excitable tissue, skeletal muscle can be recruited quickly with significant control over what and how many muscle fibers will be stimulated for a movement.

Contractibility This is the ability of a muscle to shorten when the muscle tissue receives sufficient stimulation. Some muscles can shorten as much as 50 to 70% of their resting length. The average range is about 57% of resting length for all skeletal muscles.

The distance through which a muscle shortens is usually limited by the physical confinement of the body. For example, the sartorius can shorten more than half of its

length if the muscle is removed and stimulated in a laboratory situation, but, in the body, the shortening distance is restrained by the hip joint and positioning of the trunk and thigh.

Extensibility This is the muscle tissue's ability to lengthen out or stretch beyond the resting length. The muscle itself cannot produce the elongation—another muscle or an external force is required. Taking a joint through a passive range of motion, where an individual would push another's limb past the resting length, is a good example of elongation in the muscle tissue. The amount of extensibility in the muscle is determined by the connective tissue found in the perimysium, epimysium, and fascia surrounding and within the muscle.

Elasticity This is the ability of muscle fiber to return to its resting length once the stretch to the muscle is removed. Elasticity in the muscle is determined by the connective tissue in the muscle rather than the fibrils themselves. The properties of elasticity and extensibility are protective mechanisms in the muscle that maintain the integrity and basic length of the muscle. Elasticity is also a very critical component in facilitating output in a shortening muscle action that is preceded by a stretch. This is known as the stretch-shortening cycle, and will be discussed at length in Section III of this chapter.

Using a ligament as a comparison, it is easy to see how the property of elasticity benefits muscle tissue. The ligament, largely collagenous in nature, has little or no elasticity, and if it is stretched beyond its resting length, will not return to the original length but rather will remain extended. This can create laxity around the joint when the ligament is too long to exert much control over the joint motion. On the other hand, muscle tissue will always return to its original length. If the muscle is stretched too far, the muscle tissue will eventually tear.

Functions of Muscle

Skeletal muscles perform a variety of different functions, all of which are important to efficient performance of the human body. The three functions relating specifically to human movement are: (a) contributing to the production of skeletal movement; (b) assisting in joint stability; and (c) maintaining posture and body positioning.

Produce Movement Skeletal movement is created as muscle actions generate tensions that are transferred to the bone. The resulting movements are necessary for locomotion or other segmental manipulations.

Maintain Postures and Positions Muscle actions of a lesser magnitude are utilized to maintain postures. This muscle activity is continuous and results in small adjustments as the head is maintained in position or the body weight is balanced over the feet.

Stabilize Joints   Muscle actions also contribute significantly to stability of the joints. Muscle tensions are generated and applied across the joints via the tendons, providing stability where they cross the joint. In most joints, especially the shoulder and the knee, the muscles spanning the joint via the tendons are among the primary joint stabilizers.

Other Functions   The skeletal muscles also provide four other functions that are not directly related to human movement. First, muscles support and protect the visceral organs and protect the internal tissues from injury. Second, tension in the muscle tissue can alter and control pressures within the cavities. Third, skeletal muscle also contributes to the maintenance of body temperature by producing heat. Fourth, the muscles control the entrances and exits to the body through voluntary control over swallowing, defecation, and urination.

Role of the Muscle

In the performance of a motor skill, only a very small portion of the potential movement capability of the musculoskeletal system is used. There may be 20 to 30 degrees of freedom available to raise an arm above the head and comb the hair (11). However, many of the available movements may be undesirable in terms of the desired movement, combing the hair. To eliminate the undesirable movements and create the skill or desired movement, muscles or groups of muscles will have to play a variety of roles. To perform a motor skill at a given time, only a small percentage of the potential movement capability of the motor system will be used (11).

Prime Mover vs Assistant Mover   The muscle(s) primarily responsible for producing a given movement are called prime movers, and if more force is required, other muscles will contribute as assistant movers (4). For example, the latissimus dorsi and the pectoralis major are prime movers, as they shorten to produce extension of the arm. If more force is required in the movement of extension, the teres major and triceps brachii may assist.

Agonists and Antagonists   The various roles of selected muscles in a simple arm abduction exercise is presented in FIGURE 3–10. Muscles creating the same joint movement are termed agonists. Conversely, muscles opposing or producing the opposite joint movement are called

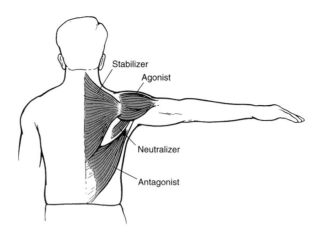

FIGURE 3–10. Muscles perform a variety of different roles when a movement is performed. In the arm abduction example above, the deltoid is the agonist since it is responsible for creating the abduction movement. The latissimus dorsi is the antagonistic muscle since it resists the abduction movement. There are also muscles stabilizing in the region so the movement can occur. In this example, the trapezius is shown stabilizing and holding the scapula in place. Lastly, there may be some neutralizing action. In this example, the teres minor may neutralize, via external rotation, any internal rotation produced by the latissimus dorsi.

antagonists. It is the antagonists that must relax to allow a movement to occur or contract concurrently with the agonists to control or slow a joint movement. Thus, when swinging the thigh forward and up, the agonists producing the movement are the hip flexors comprising the iliopsoas, rectus femoris, pectineus, sartorius, and gracilis muscles. The antagonists, or the muscles opposing the motion of hip flexion, are the hip extensors, the hamstrings, and the gluteus maximus. It is the antagonists and gravity that will slow down the movement of hip flexion and terminate the joint action.

When a muscle is playing the role of an antagonist, it is more susceptible to injury at the site of muscle attachment or in the muscle fiber itself. This is because the muscle is contracting to slow the limb, while at the same time being stretched.

Stabilizers and Neutralizers   Muscles are also used as stabilizers, acting in one segment so that a specific movement in an adjacent joint can occur. Stabilization is very important in the shoulder girdle which must be supported so that arm movements can occur smoothly and efficiently. It is also important in the pelvic girdle and hip region during gait. When one foot is on the ground in

MODEL

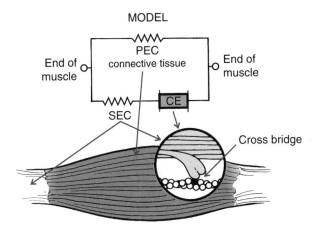

FIGURE 3–11. The three component mechanical model of the muscle consists of an active component, the contractile element (CE), and two passive components, the series elastic component (SEC) and the parallel elastic component (PEC). The CE is located at the myofibril level where cross-bridging occurs. It is believed the SEC is in the tendon, and the cross-bridging and the PEC are in the connective tissue.

walking or running, the gluteus medius contracts to maintain the stability of the pelvis so it does not drop to one side.

The last role muscles are required to play is that of synergist, or neutralizer, in which a muscle will contract to eliminate an undesired joint action of another muscle. For example, the gluteus maximus is contracted at the hip joint to produce thigh extension, but the gluteus maximus will also attempt to externally rotate the thigh. If external rotation is an undesired action, the gluteus minimus and the tensor fascia latae will contract to produce a neutralizing internal rotation action that cancels out the external rotation action of the gluteus maximus, leaving the desired extension movement.

Mechanical Components in the Muscle

Contractile, or Active, Component   Muscle contraction occurs very quickly, and the duration of the contraction depends on the contractile elements and modification of length by the elastic elements. A three-component model of the mechanical components was developed, and has been used to describe muscle actions using one contractile and two elastic components in the muscle (12,13) (FIGURE 3–11). The contractile, or active, component of the muscle is found in the myofibrils where there is cross-bridging of the actin and myosin filaments, causing a shortening of the muscle fiber (4).

Elastic, or Passive, Components   The elastic, or passive, component of muscle serves to absorb, transmit, and store energy. This passive component is further divided into two components: the series elastic component and the parallel elastic component. The series elastic component is located in the tendon (85%) and in the actin-myosin cross-bridges (15%) as the myosin filaments are rotated backwards in the stretch of a muscle (3). Since this component is in series with the contractile element, it behaves like a spring. The active component shortens and stretches the series elastic component, and the total length of the tendon-muscle complex remains constant (3). This also slows down the muscle force build up.

The parallel elastic component is found in the sarcolemma and the connective tissue surrounding the muscle that include the endomysium, perimysium, and epimysium. The parallel elastic component plays an important role when a passive muscle is being elongated. As the muscle is lengthened, the parallel elastic component offers an opposing force and prevents the contractile elements from being pulled apart by external forces (3). When the contractile elements are active, the influence of the parallel elastic component is negligible.

In a fast stretch, myosin cross-bridging contributes to the series elastic component in the muscle, while in a slow stretch, there is slipping of the cross-bridges and, consequently, no contribution from this site (3). Both the parallel and series elastic components in the muscle offer a resistive tension when stretched, and they store mechanical energy for use in a subsequent joint movement. Thus, upon receiving tension generated by the contractile force of the muscle, the elastic components stretch and store the tension which can then be transmitted to the bone.

Net Muscle Actions

Isometric Muscle Action   Muscle tension is generated against a resistance to maintain position, raise a segment or object, or to lower or control a segment. If the muscle is active and develops tension, but with no visible or external change in joint position, the muscle action is termed isometric (7). Examples of isometric muscle actions are illustrated in FIGURE 3–12. To bend over into 30 degrees of trunk flexion and hold the position, the muscle action used to hold the position would be termed isometric since there is no movement. The muscles contracting isometrically to hold the trunk in a position of flexion are the back muscles, since they are resisting gravity which wants to flex the trunk further.

To take the opposite perspective, consider the movement in which the trunk is curled up to 30 degrees and

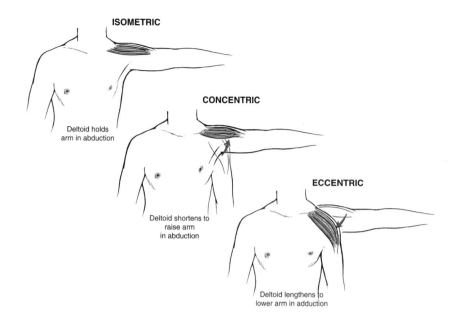

FIGURE 3–12. A muscle action is isometric when the tension generated in the muscle creates no change in joint position. A concentric muscle action occurs when the tension generated in the muscle creates a shortening of the muscle. An eccentric muscle action is generated by an external force when muscle lengthening occurs.

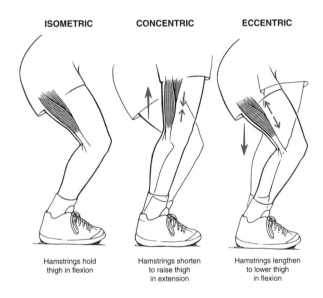

stopped. To hold this position of trunk flexion, an isometric muscle action using the trunk flexors is produced. This muscle action resists gravity which is trying to extend the trunk.

Concentric Muscle Action   If a muscle generates tension actively with visible shortening in the length of the muscle, the muscle action is termed concentric (7). In the concentrically controlled joint action, the net muscle forces producing rotation are in the same direction as the

change in joint angle, meaning that the agonists are the controlling muscles in a concentric muscle action (FIGURE 3–12). Also, the limb movement produced in a concentric muscle action is termed positive, since the joint actions are usually up against gravity, or are the initiating source of movement of a mass.

Most joint movements upward are created by a concentric muscle action. For example, flexion of the arm or forearm from the standing position will be produced by a concentric muscle action from the respective agonists or flexor muscles. Additionally, to initiate a movement of the arm across the body in a horizontal adduction movement, the horizontal adductors will initiate the movement via a concentric muscle action. Concentric muscle actions are used to generate forces against external resistances such as raising a weight, pushing off the ground, or throwing a discus.

Eccentric Muscle Action When a muscle is subjected to an external torque that is greater than the internal torque within the muscle, there is lengthening in the muscle, and the action is known as eccentric (7). The source of the external force developing the external torque that produces an eccentric muscle action is usually gravity or the muscle action of an antagonistic muscle group (1).

In the eccentric joint action, the net muscular forces producing the rotation are in the opposite direction of the change in joint angle, meaning that the antagonists are the controlling muscles in eccentric muscle action (FIGURE 3–12). Also, the limb movement produced in eccentric muscle action is termed negative, since the joint actions are usually moving down with gravity, or are controlling, rather than initiating, the movement of a mass.

Most movements downward, unless they are very fast, are controlled by an eccentric action of the antagonistic muscle groups. To reverse the example shown above, when extending the arm or forearm from flexed position, the muscle action would be eccentrically produced by the flexors, or antagonistic muscle group. Likewise, lowering into a squat position in which there is hip and knee flexion requires an eccentric movement controlled by the hip and knee extensors. Conversely, the reverse thigh and shank extension movements up against gravity would be produced concentrically by the extensors.

From these examples, the potential sites of muscular imbalances in the body can be identified, since the extensors in the trunk and the lower extremity are used to both lower and raise the segments. In the upper extremity, it is the flexors that both raise the segments concentrically and lower the segments eccentrically, thereby obtaining more use.

Eccentric actions are also used to slow down a movement. When flexing the thigh rapidly as in a kicking action, the antagonists (extensors) will eccentrically control and slow the joint action near the end of the range of motion.

Comparison of Isometric, Concentric, Eccentric Isometric, concentric, and eccentric muscle actions are not used in isolation, but rather in some combination. Typically, isometric actions are used to stabilize a body

part, and eccentric and concentric muscle actions are used sequentially to maximize energy storage and muscle performance. This natural sequence of muscle function during which an eccentric action precedes a concentric action is known as the stretch-shortening cycle.

These three muscle actions are very different in terms of their energy cost and force output. The eccentric muscle action can develop the same force output as the other two types of muscle actions, with fewer muscle fibers activated. Consequently, this muscle action is more efficient and can produce the same force output with less oxygen consumption (14) (FIGURE 3–13).

In addition, the eccentric muscle action is capable of greater force output than the isometric or concentric muscle actions (FIGURE 3–14). This occurs at the level of the sarcomere, where the force increases beyond the maximum isometric force if the myofibril is stretched and stimulated (15,16).

The concentric muscle actions generate the lowest force output of the three types of muscle actions. Force is related to the number of cross-bridges formed in the myofibril. In the isometric muscle action, the number of bridges attached remains constant. As the muscle shortens, the number of attached bridges is reduced with increased velocity (16). This reduces the level of force output generated by tension in the muscle fibers. A hypothetical torque output curve for the three muscle actions is presented in FIGURE 3–15.

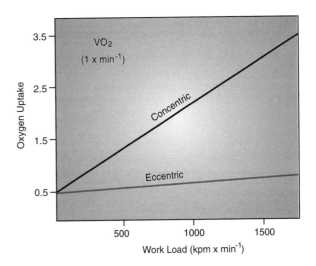

FIGURE 3–13. Asmussen (1952) illustrated that the eccentric muscle action can produce high work loads at lower oxygen uptake levels than the same loads produced with a concentric muscle action.

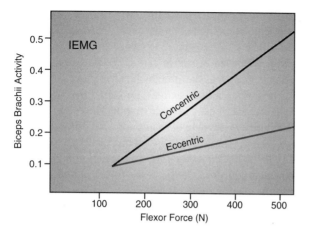

FIGURE 3–14. The activity in the biceps brachii muscle (IEMG) was higher as the same forces were generated using concentric muscle action as compared to eccentric muscle action (Komi, 1986).

An additional factor contributing to noticeable force output differences between the eccentric and concentric muscle actions is present when the muscle actions are producing movements in the vertical direction. In this case, the force output in both concentric and eccentric muscle actions is influenced by torques created by gravity. The gravitational force creates a torque that contributes to the force output in an eccentric muscle action as the muscles generate a torque that controls the lowering of the limb or body. The total force output in a lower-

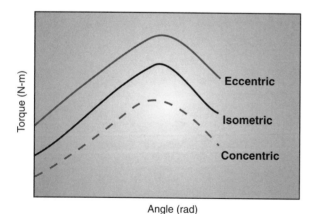

FIGURE 3–15. The eccentric muscle action is capable of generating the greatest amount of torque through a given range of motion. The isometric muscle action can generate the next highest level of torque, and the concentric muscle action generates the lowest level of torque (Enoka, 1988).

ing action is the result of both muscular torques and torques created by gravity.

The force of gravity inhibits the movement of a limb upward, and before any movement can occur, the concentric muscle action must develop a force output that is greater than the force of gravity acting on the limb or body (weight). The total force output in a raising action is predominantly muscle force. This is another reason why the concentric muscle action is more demanding than the eccentric or isometric muscle action.

This information is useful when considering exercise programs for unconditioned individuals or rehabilitation programs. Even the weakest individual may be able to perform a controlled lowering of a body part or a small weight, but may not be able to hold or raise the weight. A program that starts with eccentric exercises and then leads into isometric, followed by concentric exercises may prove to be beneficial in the progression of strength or in rehabilitation of a body part. Thus, a person unable to do a push up should start at the extended position and lower into the push up, then receiving assistance on the "up" phase until enough strength is developed for the concentric portion of the skill. Factors to consider in the use of eccentric exercises are the control of the speed at which the limb or weight is lowered, and control over the magnitude of the load imposed eccentrically, since muscle injury and muscle soreness can occur more readily with eccentric muscle action occurring under high load and high speed conditions.

Factors Influencing Muscle Force

Angle of Attachment of Muscle

The muscle supplies a certain amount of tension that is transferred via the tendon or aponeurosis to the bone. Not all of the tension or force produced by the muscle will be put to use in generating rotation of the segment. Depending on the angle of insertion of the muscle, some force will be directed to stabilizing or destabilizing the segment by pulling the bone into, or away from, the joint.

Muscular force will be primarily directed along the length of the bone and into the joint when the tendon angle is acute or lying flat on the bone. When the forearm is in the extended position, the tendon of the biceps brachii is inserting into the radius at a low angle. Initiating an arm curl from this position requires more muscle force than other positions, because most of the force generated by the biceps brachii is being directed into the elbow joint rather than into moving the segments around the joint. Fortunately, the resistance offered by the forearm weight is also minimum at the extended position, so the low muscular force available to move the segment is usually sufficient. Both the force directed along the length of the bone and that which is applied perpendicular to the bone to create joint movement can be determined by resolving the angle of the muscular force application into its respective parallel and rotary components. The parallel and rotatory components of the biceps brachii force for different attachment angles is presented in FIGURE 3–16.

It is important to recognize that even though the muscular tension may be maintained during a joint movement, the rotary component and the torque will vary depending upon the angle of insertion. Many of the neutral starting positions are "weak" positions, since most of the muscular force is directed along the length of the bone. As segments move through the midrange of the joint motion, the angle of insertion will usually increase and direct more of the muscular force into moving the segment. Consequently, when starting a weight-lifting movement from the full extended position, less weight can be lifted as compared to starting the lift with some flexion in the joint. The isometric force output of the shoulder flexors and extensors are presented for different joint positions in FIGURE 3–17.

In addition, at the end of some joint movements, the angle of insertion may move past 90 degrees, the point at which the moving force will again begin to decrease and the force along the length of the bone will act to pull the

bone away from the joint. This dislocating force is present in the elbow joint and the shoulder joint when there is a high degree of flexion in the joints (4).

Force-Time Characteristics   Somewhat related to the angle of muscular pull on the muscle is the force-time relationship in isometric muscle action. When a muscle begins to develop tension through the contractile component of the muscle, the force increases nonlinearly over time because the passive, elastic components found in the tendon and the connective tissue stretch and absorb some of the force (4). After the elastic components are stretched, the tension exerted on the bone by the muscle increases linearly over time until a maximum force is achieved.

The time to the achievement of a maximum force and the magnitude of the force produced will vary with a change in joint position. In one joint position, maximum force may be produced very quickly while in other joint positions, it may occur later in the contraction (4). This reflects the changes in tendon laxity and not changes in the tension-generating capabilities of the contractile components. If the tendon is slack, the maximum force will occur later and vice versa.

Length-Tension Relationship

The amount of force produced by a muscle is also related to the length at which the muscle is held (16). The

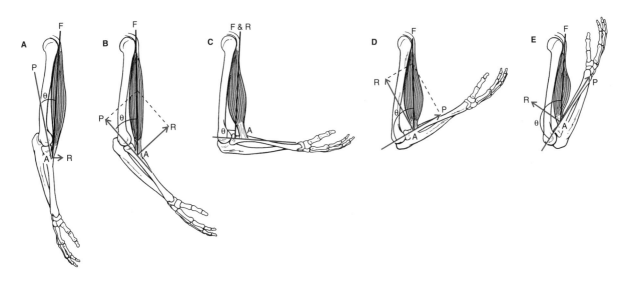

FIGURE 3–16. When muscle attachment angles are acute, the parallel component of the force (P) is highest and is stabilizing the joint. The rotatory component (R) is low (see A above). The rotatory component increases to its maximum level at a 90 degree angle of attachment (see C above). Beyond a 90 degree angle of attachment, the rotatory component diminishes and the parallel component increases to produce a dislocating force (see D and E above).

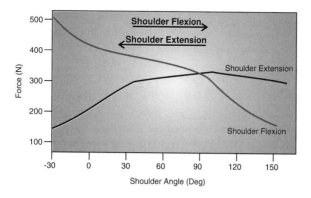

FIGURE 3–17. In this example using the shoulder joint, the isometric force output varies with the joint angle. As the shoulder joint angle increases, the shoulder extension force values increase. The reverse happens with shoulder flexion force values which decrease with an increase of the shoulder joint angle (Adapted from Kulig, Andrews, and Hay, 1984).

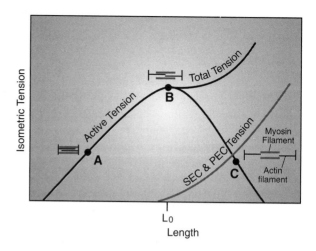

FIGURE 3–18. Muscle fibers cannot generate high tensions in the shortened state (A) because the actin and myosin filaments are doubled over. The greatest tension in the muscle fiber can be generated at a length slightly greater than resting length (B). In the elongated muscle length (C), the muscle fibers are incapable of generating tension because the cross-bridges are pulled apart. However, the total muscle tension increases because the elastic components increase their tension development.

maximum tension that can be generated in the muscle fiber will occur when a muscle is activated at a length slightly greater than resting length, somewhere between 80 to 120% of the resting length. Fortunately, the length of most muscles in the body is within this maximum force production range (16). The length-tension relationship is shown in FIGURE 3–18, demonstrating the contribution of both the active and passive components in the muscle.

Tension at Shortened Lengths The tension-developing capacity drops off when the muscle is activated at both short and elongated lengths. When a muscle has shortened up to half of its length, it is not capable of generating much more contractile tension. At the short lengths, there is less tension because the filaments have doubled over in their overlapping, creating an incomplete activation of the cross-bridges since fewer of these can be formed (16) (FIGURE 3–18). Thus, at the end of a joint movement or range of motion of a segment, the muscle is weak and incapable of generating large amounts of force due to its shortened length.

Tension at Elongated Lengths   When a muscle is lengthened and then activated, muscle fiber tension is initially greater because the cross-bridges are pulled apart after initially joining (17). This continues until the muscle length is increased slightly past the resting length. When the muscle is lengthened further and contracted, the tension generated in the muscle will drop off due to slippage of the cross-bridges, resulting in fewer cross bridges being formed (FIGURE 3–18).

Contribution of the Elastic Components The contractile component in the muscle is not the only contributor to tension in the muscle at different muscle lengths. The tension generated in a shortened muscle is shared by the series elastic component, that is predominantly tension development in the tendon. The tension in the muscle is equal to the tension in the series elastic component when the muscle contracts in a shortened length.

As the tension-developing characteristics of the active components of the muscle fibers diminish with elongation, tension in the total muscle increases due to the contribution of the passive elements in the muscle (4). The series elastic component is stretched and tension developed in the tendon and the cross-bridges as they are rotated back (3). Significant tension is also developed in the parallel elastic component as the connective tissue in the muscle offers resistance to the stretch (3). As the muscle is lengthened, passive tension is generated in these structures so that the total tension is a combination of contractile and passive components (FIGURE 3–18). At extreme muscle lengths, the tension in the muscle is almost exclusively elastic, or passive, tension.

Optimal Length for Tension The optimal muscle length for generating muscle tension is slightly greater than the resting length because the contractile components are optimally producing tension, and the passive components are storing elastic energy and adding to the total tension in the unit (18). This relationship lends support for placing the muscle on a stretch prior to using the muscle for a joint action. One of the major purposes of a wind-up or preparatory phase is to put the muscle on stretch in order to facilitate output from the muscle in the movement.

Force-Velocity Relationship (F-V)

Muscle fibers will shorten at a specific speed or velocity while concurrently developing a force used to move a segment or external load. Muscles create an active force to match the load in shortening, and the active force continuously adjusts to the speed at which the contractile system moves (16). Under low load conditions, the active force is adjusted by increasing the speed of contraction. With high loads, the muscle adjusts the active force by reducing the speed of shortening (16).

F-V Relationship in Concentric Muscle Actions In this type of muscle action, velocity is increased at the expense of a decrease in force, and vice versa. The maximum force can be generated at zero velocity, and the maximum velocity can be achieved with the lightest load. An optimal force can be created at zero velocity because a large number of cross-bridges are attached. As the velocity of the muscle shortening increases, the cycling rate of the cross-bridges increases, leaving fewer cross-bridges attached at one time (3). This equates to less force, and at high velocities, when all of the cross-bridges are cycling, the force production is negligible. This relationship is graphically presented in FIGURE 3–19.

F-V in the Muscle Fiber vs External Load The force-velocity relationship relates to the behavior of muscle fiber, and it is sometimes confusing to relate this concept to an activity such as weight lifting. As an athlete is asked to increase the load in a lift, there will most likely be a decrease in the speed of movement. Although the force-velocity relationship is still present in the muscle fiber itself, the total system is responding to the increase in the external load or weight. The muscle may still be generating the same amount of force in the fiber, but the addition of the weight slows the movement of the total system. In this case, the action velocity of the muscle is high, but the movement velocity of the high load is low (19).

Power Power, the product of force and velocity, is one of the major distinguishing features between successful and average athletes. Many sports require large power outputs with the athlete expected to move his or her body

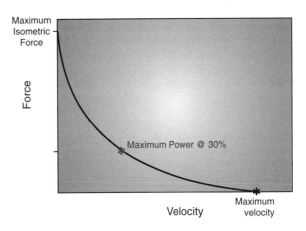

FIGURE 3–19. The force-velocity relationship in a concentric muscle action is an inverse one. The amount of tension or force-developing capabilities in the muscle decreases with an increase in velocity because of the reduced number of cross-bridges which can be maintained. Maximum tension can be generated in the isometric or zero velocity condition in which many cross-bridges can be formed. Maximum power can be generated in concentric muscle action with the velocity and force levels at 30% of maximum.

weight, or some external object, very quickly. Since velocity diminishes with the increase of load, the most power can be achieved if the athlete produces one third of his maximum force at one third of his maximum velocity (20,21). In this way, the power output is maximized even though the velocities or the forces may not be at their maximum level.

To train athletes for power, coaches must schedule high velocity activities at 30% of maximum force levels to improve power in their athletes (20). The development of power is also enhanced by the presence of fast-twitch muscle fibers that are capable of generating four times more peak power than slow-twitch fibers (20).

F-V Relationships in Eccentric Muscle Actions The force-velocity relationship in an eccentric muscle action is opposite to that seen in the shortening or concentric muscle action. Remember that an eccentric muscle action is created by an external force generated by antagonistic muscles, gravity, or some other external force. When a load greater than the maximum isometric strength value is applied to a muscle fiber, the fiber will begin to lengthen eccentrically. At the initial stages of lengthening when the load is slightly greater than the isometric maximum, the speed of lengthening and the length changes in the sarcomeres will be small (16).

If a load is as high as 50% more than the isometric maximum, the muscle will elongate at a high velocity (16). The tension increases with the speed of lengthening in the eccentric muscle action because the muscle is stretching as it is contracting (FIGURE 3–20). The eccentric force-velocity curve will end abruptly at some lengthening velocity when the muscle can no longer control the movement of the load.

Stretch-Shortening Cycle

Elastic Component Contributions If concentric muscle action, or shortening, of a muscle group is preceded by eccentric muscle action, or prestretch, the resulting concentric action is capable of generating greater force. This is because a stretch on the muscle changes the muscle's characteristics by increasing its tension through storage of potential elastic energy in the series elastic component of the muscle (22) (FIGURE 3–21). When a muscle is stretched, there is a small change in the muscle and tendon length (23), and maximum accumulation of stored energy. Thus, when a concentric muscle action follows, there is an enhanced recoil effect that adds to the force output through the muscle tendon complex (24).

A concentric muscle action beginning at the end of a prestretch will also be enhanced by the stored elastic energy in the parallel elastic component in the connective

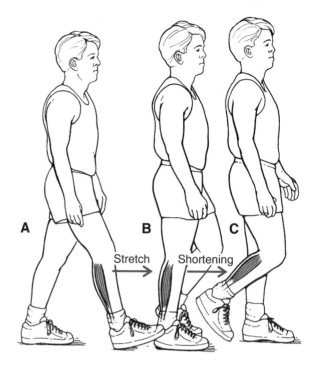

FIGURE 3–21. If a stretch of the muscle precedes a concentric muscle action, the resulting force output will be greater. The increased force output is due to contributions from stored elastic energy in the muscle, tendon, and connective tissue, and through some neural facilitation.

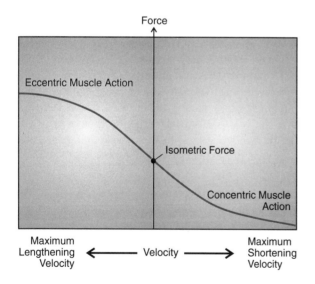

FIGURE 3–20. The relationship between force and velocity in the eccentric muscle action is opposite to that of a concentric muscle action. In the eccentric muscle action, the force increases as the velocity of the lengthening increases. The force will continue to increase until the eccentric action can no longer control lengthening of the muscle.

tissue around the muscle fibers. This contributes to a high force output at the initial portion of the concentric muscle action as these tissues return to their normal length. The contribution of the parallel elastic component drops off as the muscle continues shortening.

If the shortening contraction of the muscle occurs within a reasonable time after the stretch (0.0 to 0.9 sec), the stored energy is recovered and used. If the stretch is held too long before the shortening occurs, the stored elastic energy is lost through conversion to heat (7).

Neural Contributions The stretch preceding the concentric muscle action also initiates a stimulation of the muscle group through reflex potentiation. This activation only accounts for approximately 30% of the increase in the subsequent concentric muscle action (7). The remaining increase is attributed to stored energy. The actual process of proprioceptive activation through the reflex loop will be presented in the next chapter.

Use of the Prestretch A short-range or low-amplitude prestretch occurring over a short time period is the best technique to improve significantly the output of concentric muscle action through return of elastic energy and increased activation of the muscle (7,25). To get the greatest return of energy absorbed in the negative or eccentric action, go quickly into the stretch and don't go too far into the stretch. Also, do not pause at the end of the stretch, move immediately into the concentric muscle action. In jumping, for example, a quick counter-movement jump from the anatomical position, featuring a "drop-stop-pop" action, lowering only through 8 to 12 inches, will be much more effective than a jump from a squat position or a jump from a height that forces the limbs into more flexion (25). The influence of this type of jumping technique on the gastrocnemius activity is presented in FIGURE 3–22.

The stretch-shortening cycle is also very evident in gait when the body is subjected to impact forces and lowered due to gravity (22). The muscles lengthen eccentrically at contact, at which point the velocity of the stretch is high and large muscle forces are present at the end of the eccentric muscle action. The extensors and the plantar flexors eccentrically lengthen and absorb energy returned in the propulsive stage when the foot and lower limb push off the ground.

Fiber Type and the Prestretch   There is a difference in the manner in which slow- and fast-twitch fibers handle a prestretch. Muscles with predominantly fast-twitch fibers benefit from a very high velocity prestretch occurring over a small distance because they can store more elastic energy (7). The fast-twitch fibers can handle a fast stretch because the myosin cross-bridging occurs quickly, whereas in the slow-twitch fibers, the cross-bridging is slower (26).

In the slow-twitch fiber, the small amplitude prestretch is not advantageous since the energy cannot be stored fast enough and the cross-bridging is slower (7,26). Therefore, the slow-twitch fibers will benefit from a prestretch that is slower and advances through a greater range of motion. Some athletes with predominantly slow-twitch fibers should be encouraged to use longer prestretches of the muscle to gain the benefits of the stretch. For most athletes, however, the quick prestretch through a small range of motion is the preferred method.

Plyometrics   This principle should be incorporated into both drills and conditioning for a sport or physical activity. The use of a quick prestretch is part of a conditioning protocol known as *plyometrics*. In this protocol, the muscle is put on a rapid stretch, and a concentric muscle action is initiated at the end of the stretch. Single-leg bounding, depth jumps, and stair hopping are all examples of plyometric activities for the lower extremity. Surgical tubing or elastic bands are used to produce a rapid stretch on muscles in the upper extremity. Plyometrics will be covered in greater detail in the next chapter.

One- and Two-Jointed Muscles

As stated earlier, one cannot determine the function or the contribution of a muscle to a joint movement by just locating the attachment sites. A muscle action can move a segment at one end of its attachment or two segments at both ends of its attachment. In fact, a muscle can accelerate and create movement at all joints, whether the muscle spans the joint or not. For example, the soleus is a plantar flexor of the ankle but it can also force the knee into extension even though it does not cross the knee joint (27). This occurs in the standing posture. The soleus contracts and creates plantar flexion at the ankle joint. Since the foot is on the ground, the plantar flexion movement creates extension of the knee joint. In this manner, the soleus accelerates the knee joint twice as much as it accelerates the ankle, even though the soleus does not even span the knee (27).

Action of the Two-Joint Muscle   Most muscles cross only one joint and consequently, the dominating action of the one-joint muscle is at the joint it crosses. The two-joint muscle is a special case in which the muscle crosses two joints, creating a multitude of different joint movements that often occur in opposite sequences to each other. For example, the rectus femoris is a two-joint muscle creating both hip flexion and knee extension. Take the example of jumping up into the air. Hip extension and knee extension are used to propel upward. Does the rectus femoris, a hip flexor and a knee extensor, contribute to the extension of the knee, does it resist the movement of hip extension, or does it do both?

The action of a two-joint or biarticulate muscle depends on the position of the body and the muscle's interaction with external objects such as the ground (27). In the case of the rectus femoris, the muscle will contribute primarily to the extension of the knee because of the joint position of the hip (18). This position results in the force of the rectus femoris acting close to the hip joint, thereby limiting the action of the muscle and its effectiveness in producing hip flexion (FIGURE 3–23).

The perpendicular distance from the action line of the force of the muscle over to the hip joint is termed the *moment arm*. With an increase in the moment arm, there will be more action created at a joint if the same amount of force is applied. Thus, in the case of a two-joint muscle, the muscle will primarily act on the joint where it has the largest

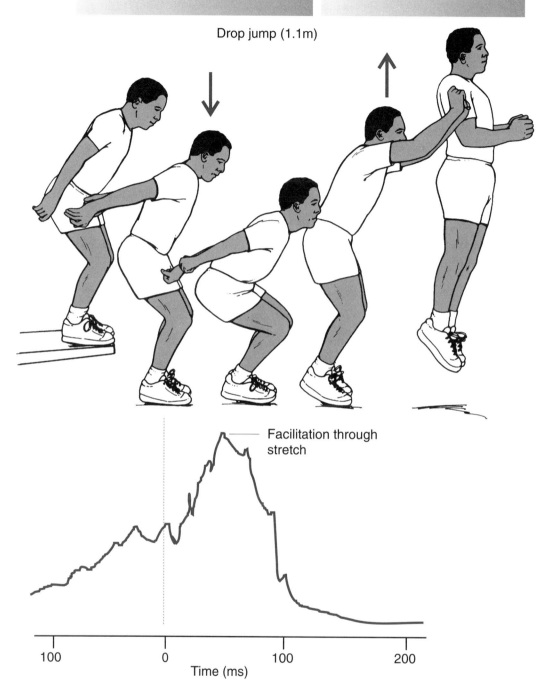

Stretch through eccentric muscle action

Concentric muscle action

Drop jump (1.1m)

Facilitation through stretch

100   0   100   200

Time (ms)

FIGURE 3–22. In trained jumpers, the prestretch is used to facilitate the neural activity of the lower extremity muscles. The neural facilitation in the gastrocnemius is shown in the example above. The neural facilitation coupled with the recoil effect of the elastic components will add to the jump if it is performed with the correct timing and amplitude (Adapted from Sale, 1992).

moment arm, or where it is further away from the joint (18). The hamstring group will primarily create hip extension rather than knee flexion because of the larger moment arm at the hip (FIGURE 3–23). The gastrocnemius will produce plantar flexion at the ankle rather than flexion at the knee joint because the moment arm is larger at the ankle.

For example, in vertical jumping, maximum height is achieved by extending the proximal joints first, and then

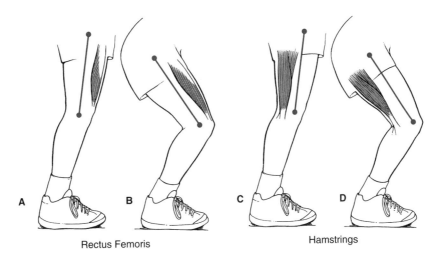

Rectus Femoris   Hamstrings

FIGURE 3–23. The rectus femoris moment arms at the hip and knee in the standing position (A) and in the squat position (B) demonstrate why this muscle is more effective as an extender of the knee joint than as a flexor at the hip. Likewise, the hamstring moment arm in the standing position (C) and in the squat position (D) demonstrate why the hamstrings are more effective as a hip extensor rather than a knee flexor.

moving distally to a point where extension (plantar flexion) occurs in the ankle joint. By the time the ankle joint is involved in the sequence, very high joint moments and extension velocities are required (28). The role of the two-joint muscle becomes very important. The biarticular gastrocnemius muscle crosses both the knee and ankle joints. Its contribution to the jumping action is influenced by the knee joint. In the jumping action, the knee joint extends and optimizes the length of the gastrocnemius (29). This keeps the contraction velocity in the gastrocnemius muscle low, even when the ankle is plantar flexing very quickly. With the velocity lowered, the gastrocnemius is able to produce more force in the jumping action (29).

The most important contribution of the two-joint muscle in the lower extremity is the reduction of the work required from the single-joint muscles. Two joint muscles initiate a mechanical coupling of the joints that allows for a rapid release of stored elastic energy in the system (30).

Two-joint muscles save energy by allowing positive work to be done at one joint and negative work to be done at the adjacent joint. Thus, while the muscles acting at the ankle are producing a concentric action and producing positive work, the knee joint muscles can be eccentrically storing elastic energy through negative work (30).

The two-joint muscle actions for walking are presented in FIGURE 3–24. Two-joint muscle strategies that work together in the walking phase are the sartorius and

rectus femoris at heel strike, the hamstrings and the gastrocnemius at midsupport, the gastrocnemius and rectus femoris at toe-off, the rectus femoris, sartorius, and hamstrings at forward swing, and the hamstrings and the gastrocnemius at foot descent (30). The relationship between these two joint muscles in the walking cycle is illustrated in FIGURE 3–24. Note at heel strike, the sartorius, a hip flexor and a knee flexor, works with the rectus femoris, a hip flexor and knee extensor. As the heel strikes, the rectus femoris performs negative work, absorbing at the knee joint as it moves into flexion. The sartorius, on the other hand, performs positive work, as the knee and the hip both flex with gravity (30).

Strengthening the Muscle

Strength is defined as the maximum amount of force produced by a muscle or muscle group at a site of attachment on the skeleton (31). Mechanically, strength is equal to maximum isometric torque that can be generated at a specific angle. However, strength is usually measured by moving the heaviest possible external load through one repetition of a specific range of motion. The movement of the load is not performed at a constant speed since joint movements are usually done at speeds varying considerably through the range of motion. There are many variables influencing strength measurement. Some of these include the muscle action (eccentric, concentric, isometric) and the speed of the limb movement

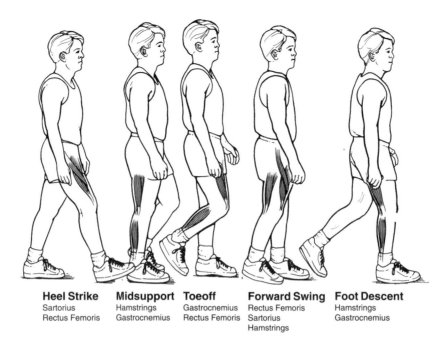

Heel Strike	**Midsupport**	**Toeoff**	**Forward Swing**	**Foot Descent**
Sartorius	Hamstrings	Gastrocnemius	Rectus Femoris	Hamstrings
Rectus Femoris	Gastrocnemius	Rectus Femoris	Sartorius	Gastrocnemius
			Hamstrings	

FIGURE 3–24. Two-joint muscles work together synergistically to optimize performance. This is illustrated above for the activity of walking.

(32). Also, length-tension, force-angle, and force-time characteristics influence strength measurements as strength varies throughout the range of motion. Strength measurements are limited by the weakest joint position.

Training of the muscle for strength focuses on developing a greater cross section in the muscle as well as developing more tension per unit of cross-section area (33). This holds true for all people, both young and old. Greater cross section, or hypertrophy, associated with weight training is due to an increase in the size of the actual muscle fibers and more capillarization to the muscle, which creates greater mean fiber area in the muscle (22,34). The size increase is attributed to increase in size of the actual myofibrils or separation of the myofibrils as shown in FIGURE 3–25. There is also some speculation that the actual muscle fibers may split (FIGURE 3–25), but this has not been experimentally substantiated in humans (34). The increase in tension per unit of cross section reflects the neural influence on the development of strength (35).

Principles of Training

Genetic Predisposition   During strength training, the magnitude of the strength gains observed will be related to numerous factors. First, there are genetic limits to the amount of strength that is obtainable, determined by distribution of fiber type and body type, or musculoskeletal

A

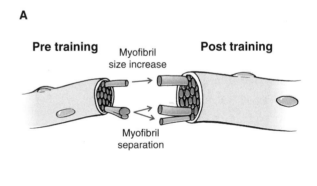

Pre training   Myofibril   Post training
                    size increase

                    Myofibril
                    separation

B   Training

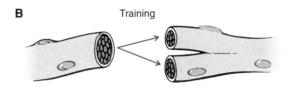

FIGURE 3–25. (A) During strength training, the muscle fibers will increase in cross section as the myofibrils become larger and separate. (B) It has been hypothesized that the fibers may also actually split; however, this has yet to be demonstrated in humans (Redrawn from McDougall, 1992).

anthropometrics (33). These cannot be altered with training. Working within these genetic limits, one can directly control the strength gains through specific attention to training principles.

Training Specificity   Training specificity, relating to the specific muscles, is also important in strength training. Only the muscles used in a specific movement pattern will be strengthened. This principle, specific adaptation to imposed demands (SAID), should direct the choice of lifts toward movement patterns related to the sport or activity in which the pattern might be used (33). This training specificity has a neurological basis, somewhat like learning a new motor skill where one is usually clumsy until the neurological patterning is established. Two sport skills, lineman drives and basketball rebounding, are presented along with lifts specific to the movement in FIGURE 3–26.

In strength training, there is a learning process that takes place in the early stages of training. This process continues on into the later stages of strength training, but it has its greatest influence at the beginning of a strength program. It is in the beginning stages that the novice lifter will demonstrate strength gains as a consequence of learning the lift rather than any noticeable increase in the physical determinants of strength, such as increase in fiber size (33,35). This is the basis for using submaximal resistance and high-repetition lifting at the beginning of a strength training program so that the lift can first be learned safely.

In addition to the specificity of the pattern of joint movement, specificity of training of the muscle also relates to the speed of training. If a muscle is trained at slow speeds, it will improve strength at slow speeds but may not be strengthened at high speeds. Thus, if power is the ultimate goal for an athlete, the strength training routine should contain movements focusing on force and velocity components to maximize and emulate power. Once a strength base is established, power is obtained with high-intensity loads and low number of repetitions (19).

Intensity   The intensity of the training routine is another important factor to monitor in the development of strength. Strength gains are directly related to the tension produced in the muscle. A muscle must be overloaded to a particular threshold level before it will respond and adapt to the training (33).

It is the amount of tension in the muscle rather than the number of repetitions that is the stimulus for strength (33). The amount of overload is usually determined as a percentage of the maximum amount of tension a muscle or muscle group can develop.

Athletes attempt to work at the highest percentage of their maximal lifting capability to increase the magnitude of their strength gains. To train regularly using a high amount of repetitions with low amounts of tension per repetition, the strength gains will be minimal because the muscle has not been overloaded beyond its threshold level. The greatest strength gains are achieved when the muscle is worked near its maximum tension before it reaches a fatigue state (2 to 6 repetitions) (33).

Rest   The quality and success of a strength development routine is also directly related to the rest provided to the muscles between sets, between days of training, and prior to competition. Rest of skeletal muscle that has been stressed through resistive training is important for the recovery and rebuilding of the muscle fiber. As the skeletal muscle fatigues, the tension development capability deteriorates and the muscle is not operating in an optimal overload situation.

Volume   It is the volume of work that a muscle performs that may be the important factor in terms of rest of the muscle. Volume of work on a muscle is the sum of the number of repetitions multiplied by the load or weight lifted (33). Volume can be computed per week, or month, or year, and should include all of the major lifts and the number of lifts. In a week, the volume of lifting for two different lifters may be the same, even though one lifter lifts 3 sets of 10 repetitions × 100 lbs for a volume of 3000 lbs and another lifter lifts 3 sets of 2 repetitions × 500 lbs, also for a volume of 3000 lbs.

At the beginning of a weight-training program, the volume is usually high. There are more sessions per week, more lifts per session, more sets per exercise, and more repetitions per set (36). The volume decreases as one progresses through the training program. This is done by lifting fewer times per week, performing fewer sets per exercise, increasing the intensity of the lifts, and performing fewer repetitions.

The yearly repetition recommendation is 20,000 lifts, which can be divided into monthly and weekly volumes as the weights are increased or decreased (36). In a month or in a week, the volume of lifting varies to offer higher- and lower-volume days and weeks.

A lifter performing heavy resistance exercises with a low number of repetitions must allow 5 to 10 minutes between sets in order for the energy systems to be replenished (33). If the rest is less than 3 minutes, a different energy system is used, resulting in lactic acid accumulation in the muscle.

FIGURE 3–26. Specific weight lifting exercises should be selected so that they reproduce some of the same movements in the sport skill. In the top example for the football lineman, the deadlift and the power clean are two lifts which include similar joint actions. Likewise, for the basketball player who uses the jumping action, the squat and heel raising exercises would be good.

Body builders use the short rest and high-intensity training to build up the size of the muscle at the expense of losing some strength gains achieved with a longer rest period. If a longer rest period is not possible, it is believed that a high repetition, low resistance form of circuit training between the high-resistance lifts may reduce the build up of lactic acid in the muscle. Body builders also exercise at loads less than those of power lifters and weight lifters (6 to 12 RM). This is the major reason for the strength differences between the weight lifter (greater strength) and the body builder (less strength).

The development of strength for performance enhancement usually follows a very detailed plan that has been outlined in the literature for numerous sports and activities. A sample weight-training cycle is presented in Table 3–1. The long-term picture usually involves some form of periodization, during which the loads are increased, and the volume of lifting is decreased over a period of months. As the athlete heads into a performance season, the lifting volume may be reduced by as much as 60%, which will actually increase the strength of the muscles. If an athlete quits lifting in preparation for a performance, strength can be maintained for at least 5 days and may be even higher after a few days of rest (33).

Strength training for the nonathlete   The principles of strength or resistance training have been discussed using the athlete as an example. It is important to recognize that these principles are applicable to rehabilitation, the elderly, children, and unconditioned individuals. Strength training is now recommended as part of one's total fitness development (37). The American College of Sports Medicine recommends at least one set of resistance training performed 2 days a week and including 8 to 12 exercises (37).

Strength training is being recognized as an effective form of exercise for the elderly. A marked strength decrement occurs with aging, and is believed to be related to reduced activity levels (38). Strength training that is maintained into the later years may counteract atrophy of bone tissue and moderate the progression of degenerative joint alterations (38). The muscle groups identified for special attention in a weight-training program for the elderly include the neck flexors, shoulder girdle muscles, abdominals, gluteals, and the knee extensors (38).

It is only the magnitude of the resistance that will vary in weight training for the athlete, elderly, young, or others. The conditioned athlete may perform a dumbbell lateral raise with a 50 lb weight in the hand, whereas an elderly person might simply raise the arm to the side, using the arm weight as the resistance. High-resistance weightlifting must be implemented with caution, especially with the young or the elderly. Excessive loading of the skeletal system through high-intensity lifting can fracture bone in the elderly, especially in the individual with osteoporosis. The epiphyseal plates in the young are also susceptible to injury under high-load conditions; thus, high-intensity programs for this age group are not recommended.

Training Modalities

Isometric Exercise   There are different ways of loading the muscle, all of which have advantages and disadvantages in terms of strength development. Isometric training loads the muscle in one joint position so the motive torque equals the resistance torque and no movement results (4,39). Individuals have demonstrated moderate strength gains using isometric exercises, and power lifters may use heavy resistance isometric training to enhance muscle size.

Isometric exercise is also used in rehabilitation and with the unconditioned individual since it is easier to perform than the concentric exercise. The major problem associated with isometric exercise is that there is minimal transfer to the real world since most of our activities involve eccentric and concentric muscle actions. Further, the isometric exercise only enhances the strength of the muscle group at the joint angle in which the muscle is stressed, thereby limiting the strength development throughout the whole range of motion.

Table 3–1

Phase	Preparation Hypertrophy	Transition Basic Strength	Competition Strength/Power	Transition (active rest) Peak/Maintain
Sets	3–10	3–5	3–5	1–3
Repetitions	8–12	4–6	2–3	1–3
Days/Wk	1–3	1–3	1–2	1
Times/Day	1–3	1–3	1–2	1
Intensity Cycle (heavy weeks/light weeks)	2–3/1	2–4/1	2–3/1	—
Intensity	Low	High	High	Very High to Low
Volume	High	Moderate to High	Low	Very Low
Int. Cycle = ratio of number of heavy training weeks to light training weeks.				

Source: NSCA 1986, 8(6), 17–24.

Isotonic Exercise   The most popular strength training modality is the isotonic form of exercise. An exercise is considered isotonic when the segment moves a specified weight through a range of motion. Although the weight of the barbell or body segment is constant, the actual load imposed on the muscle varies throughout the range of motion. In an isotonic lift, the initial load or resistance is overcome and then moved through the motion (39). The resistance cannot be heavier than the amount of motive torque developed by the weakest joint position, since the maximum load lifted is only as great as this position. Examples of isotonic modalities are the use of free weights or multijoint machines, such as universal gyms, in which the external resistance can be adjusted (FIGURE 3–27).

An isotonic movement can be produced with an eccentric or concentric muscle action. For example, the squat exercise involves eccentrically lowering a weight and concentrically raising the same weight. Even though the weight in an isotonic lift is constant, the motive torque developed by the muscle is not, due to the changes in length-tension, force-angle, and the speed of the lift. To initiate flexion of the elbow while holding a 5 lb weight, maximum tension will be generated in the flexors at the beginning of the lift to get the weight moving. Remember that this is also one of the weakest joint positions because of the angle of attachment of the muscle. Moving through the midrange of the motion requires

reduced muscular tension because the weight is moving and the musculoskeletal lever is more efficient. The resistive torque also peaks in this stage of the movement.

The isotonic lift might not adequately overload the muscle in the midrange where it is typically the strongest. This is especially magnified if the lift is performed very quickly. Perform isotonic lifts with a constant speed (no acceleration) so that the midrange is exercised. This will match the motive torque created by the muscle with the load offered by the resistance. Strength assessment using isotonic lifting is sometimes very difficult since specific joint actions are hard to isolate. Most isotonic exercises involve action or stabilization of adjacent segments.

Isokinetic Exercise   A third training modality is the isokinetic exercise, an exercise performed at a controlled velocity with varying resistance. This exercise must be performed on an isokinetic dynamometer allowing for isolation of a limb, stabilization of adjacent segments, and adjustment of the speed of movement that typically ranges from 0 degrees/sec to 600 degrees/sec (FIGURE 3–28).

When an individual applies a muscular force against the speed controlled bar of the isokinetic device, an attempt is made to push the bar at the predetermined speed. As the individual attempts to generate maximum tension at the specific speed of contraction, the tension will vary due to change in leverage and muscular attach-

FIGURE 3–27. Two forms of isotonic exercises for the upper extremity are shown above: (A) the use of free weights (bench press); and (B) the use of a machine.

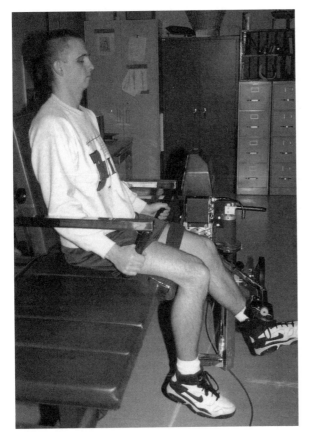

FIGURE 3–28. An example of an isokinetic exercise is presented for the movement of knee extension. The isokinetic machine shown is the Orthotron isokinetic dynamometer.

ment throughout the range of motion. Isokinetic testing has been used for quantifying strength in the laboratory or in the rehabilitation setting. There is an extensive body of literature presenting a wide array of norms for isokinetic testing of different joints, joint positions, speeds, and populations.

The velocity of the devices will significantly influence the results, so testing must be conducted at a variety of speeds or at a speed close to that which will be used in activity. This is often the major limitation of isokinetic dynamometers. For example, the isokinetic strength of the shoulder internal rotators of a baseball pitcher may be assessed at 300 degrees/second on the isokinetic dynamometer, but, in fact, the actual speed of the movement in the pitch has been shown to average 6000 degrees/second (40). Isokinetic testing does allow for a quantitative measurement of power that has been previously very difficult to measure in the field.

There are some drawbacks to using isokinetic testing and training. The movement at a constant velocity is not the type of movement typically found in sport or the activities of daily living, and the cost of most isokinetic systems and lack of mass usage make isokinetic training or testing prohibitive for many.

Close-Linked Exercise While most therapists still use isokinetic testing for assessment, many have discontinued its use for training and have gone to close-linked training, in which individuals use body weight and eccentric and concentric muscle actions. A close-linked exercise is an isotonic exercise in which the end of the chain is fixed, such as in the case of a foot or hand on the floor. An example of a close-linked exercise for the quadriceps would be a simple squat movement with the feet on the floor (FIGURE 3–29). It is believed that this form of exercise is more effective than an open-linked exercise, such as a knee extension on the isokinetic dynamometer or knee extension machine, because it uses body weight, maintains muscle relationships, and is more transferrable to normal human function.

Variable Resistive Exercise The final training modality, variable resistive exercise, or dynamic accommodating exercise, is popular as a means of developing strength in the muscle. This training modality supposedly overloads the muscle group throughout the total range of motion (39). The goal is to load the muscle group near maximum potential at each point in the range of motion.

The Nautilus system is an example of a variable resistance system in which the overload of the muscle is altered throughout the range of motion by means of a system of rotating cams that alter the mechanical advantage. (Refer to FIGURE 3–30 for an illustration of the Nautilus system.) The shape of the cams are designed to match strength curves developed throughout a range of motion. Consequently, less resistance is offered to the muscle at the beginning of a lift, in the middle of the range of motion the resistance is increased, and at the end of the range of motion, the resistance is again decreased.

One of the drawbacks with the use of this type of training is that the cams used in the equipment do not produce resistances that match all individuals or the strength curves of all individuals (41). This type of training has been shown to be no more effective than the use of free-weight isotonic lifting, and similar strength gains have been reported for both types of training (42,43,44).

Another drawback of this type of training is the loss of contribution from adjacent segments, since this exercise is usually performed with the joint isolated. For example, to work the pectoralis major muscle in the specified arm machine, the muscle receives a good workout as it is iso-

FIGURE 3–29. A close-link exercise for the quadriceps femoris muscles is shown on the left (squat) and an open-link exercise for the same muscles is shown on the right (leg extension). Close-link exercises are more effective than open link exercises.

lated in the horizontal flexion movement. But to use the horizontal flexion movement in an activity such as throwing, accompanying stabilization from the arm abductors, arm rotators, and forearm muscles is also needed. This synchrony between the stabilizers and primary muscles is lacking in many of the variable resistance exercises. A tradeoff may be in the reduction of injury with this type of lifting since proper mechanical technique is not as critical as in free weights.

Injury to Skeletal Muscle

Cause and the Site of Muscle Injury

Injury to the skeletal muscle can occur through a bout of intense exercise, exercising a muscle over a long duration, or in eccentric exercise. The actual injury to the muscle is usually a microinjury with small lesions occurring in the muscle fiber. The result of a muscle strain or microtear in the muscle is manifested by the presence of pain or muscle soreness, swelling, possible anatomical deformity, and athletic dysfunction.

Muscles at greatest risk of strain are two-joint muscles, muscles limiting the range of motion in a sport, and muscles used eccentrically (45). The two-joint muscles are at risk because they can be put on stretch at two different joints (FIGURE 3–31). To extend at the hip joint

and flex at the knee joint, the rectus femoris is put on extreme stretch and is very vulnerable for injury.

Muscles used to terminate a range of motion are at risk because they are used to eccentrically slow a limb moving very quickly. Common sites where muscles are strained as they slow a movement are the hamstrings slowing hip flexion, and the posterior rotator cuff muscles as they slow the arm in the follow-through phase of the throw (45).

While the muscle fiber itself may be the site of damage, it is believed that the source of muscle soreness immediately following exercise and strain to the system is the connective tissue. This can be in the muscle sheaths, the epimysium, perimysium, or endomysium, or it can be injury to the tendon or ligament (17). In fact, a common site of muscle strain is at the muscle-tendon junction due to the high tensions transmitted through this region. Injuries at this site are common in the gastrocnemius, pectoralis major, rectus femoris, adductor longus, triceps brachii, semimembranosus, semitendinosus, and the biceps femoris (45).

It is important to identify individuals who are at risk for muscle strain. First, the chance of injury increases with muscular fatigue as the neuromuscular system loses its ability to control the forces imposed upon the system. This commonly results in an alteration in the mechanics

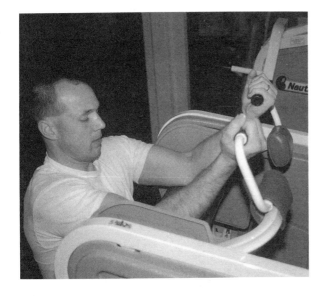

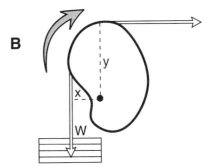

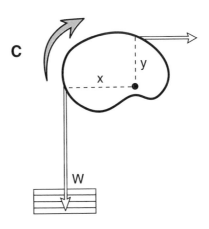

A

FIGURE 3–30. The Nautilus machine for the forearm flexors is shown in (A) above. A photograph of the actual cam is also presented. This machine uses a cam system which changes the moment arm of the weight, making it easier (B) or harder (C) to lift.

of movement and a shifting of shock absorbing load responsibilities. Practice times should be controlled and events late in the practice should not emphasize maximum load or stress conditions.

Second, an individual can also incur a muscle strain at the onset of practice if it begins with muscles that are weak from a recent usage (17). Muscles should be given ample time to recover from heavy usage. After extreme bouts of exercise, rest periods may need to be a week or

more, but normally a muscle can recover from moderate usage in a day or two (17).

Third, if trained or untrained individuals perform a unique task for the first time, they will probably experience pain, swelling, and a loss of range of motion after performing the exercise. This swelling and injury is most likely occurring in the passive elements of the muscle and will generally lessen or be reduced as the number of practices increase (17).

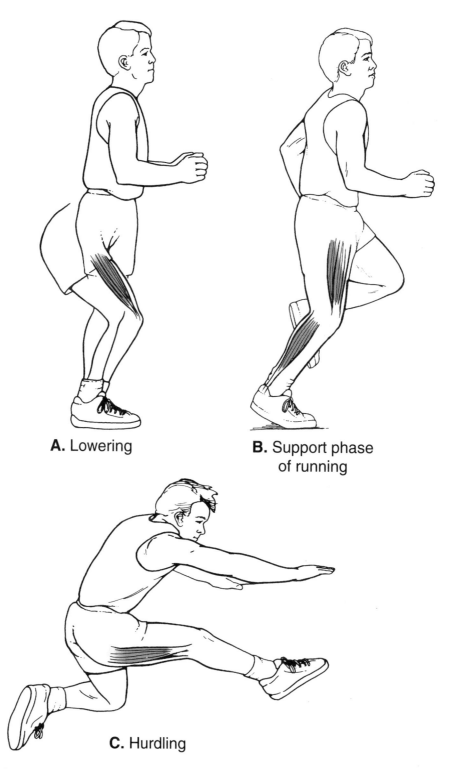

A. Lowering

B. Support phase
of running

C. Hurdling

FIGURE 3–31. Muscles that are undergoing an eccentric muscle action are at greater risk of injury. In (A), the quadriceps femoris are performing an eccentric muscle action in lowering as they control the knee flexion on the way down. In (B), the quadriceps and the gastrocnemius are eccentrically acting during the support phase of running. Two-joint muscles are also placed in injury prone positions, making them more susceptible to strain. In (C), the hamstrings are placed on extreme stretch when the hip is flexed and the knee is extended.

Last, an individual with an injury is susceptible to a recurrence of the injury or the development of an injury elsewhere in the system due to compensatory actions. For example, if the gastrocnemius is sore from a minor muscle strain, an individual may eccentrically load the lower extremity with a weak and inflexible gastrocnemius. This forces the individual to create more pronation during the support phase and run more on the balls of the feet, indirectly producing knee injuries or metatarsal fractures. With every injury there will be a functional substitution occurring elsewhere in the system, and this is where the new injury will occur.

Preventing Muscle Injury

Conditioning of the connective tissue in the muscle can greatly reduce the incidence of injury. Connective tissue responds to loading by becoming stronger. However, the rate of strengthening of connective tissue lags behind the rate of strengthening of the muscle. Therefore, base work involving low load and high repetitions should be instituted for 3 to 4 weeks at the beginning of a strength and conditioning program in order to begin the strengthening process of the connective tissue before muscle strength is increased (46).

It has been shown that different types of training influence the connective tissue in different ways. Endurance training has been shown to increase the size and tensile strength of both ligaments and tendons. Sprint training improves ligament weight and thickness, and heavy loading strengthens the muscle sheaths by stimulating the production of more collagen. When a muscle produces a maximum voluntary contraction, only 30% of the maximum tensile strength of the tendon is used (46). The remaining percentage serves as an excess to be used in very high, dynamic loading conditions. If this margin is exceeded, muscle injury occurs.

Other important considerations in preventing muscle injury are the inclusion of warm-up prior to beginning exercise routines, the development of a progressive strength program, and attention to strength and flexibility balance in the musculoskeletal system. Finally, early recognition of signs of fatigue will also help prevent injury if corrective actions are taken.

Summary

Groups of muscles are contained in fascial compartments that can be categorized by common function. The individual muscles in the group are covered by a epimysium and usually have a centralized portion called the belly. The muscle can be further divided internally into fascicles covered by the perimysium, and the fascicles contain the actual muscle fibers covered by the endomysium. Each muscle fiber contains myofibrils where the contractile unit of the muscle fiber, the sarcomere, is located. It is at the sarcomere level where cross-bridging occurs between the actin and myosin filaments, resulting in shortening of the muscle fiber.

Muscle fibers can be organized in a fusiform arrangement wherein the fibers run parallel and connect to a tendon at both ends, or in a penniform arrangement wherein the fibers run diagonally to a tendon running through the muscle. In the penniform muscle, the anatomic cross section, situated at right angles to the direction of the fibers, is less than the physiologic cross section, the sum of all of the cross sections in the fiber. In the fusiform muscle, the anatomic and physiologic cross sections are equal.

Each muscle contains different fiber types that influence the muscles ability to produce tension. Slow-twitch fiber types have slow contraction times and are well suited for prolonged, low-intensity work outs. Intermediate and fast-twitch fiber types are better suited for higher force outputs over shorter time periods.

A muscle attaches into a bone via an aponeurosis or tendon. The tendon can withstand high tensile forces and respond stiffly to high rates of loading and less stiffly at lower loading rates. The tendon recoils during muscle contraction and delays the development of tension in the muscle. This recoiling action increases the load that a muscle can support. The tendon and the muscle are more prone to injury during an eccentric muscle action.

Skeletal muscle has four properties: irritability, contractibility, extensibility, and elasticity. These properties allow muscle to respond to stimulation, shorten, lengthen beyond resting length, and return to resting length after a stretch, respectively. Given these properties, muscles can perform a variety of functions including: producing movement, maintaining postures and positions, stabilizing joints, supporting internal organs, controlling pressures in the cavities, maintaining body temperature, and controlling entrances and exits to the body. Muscles also perform various roles, such as prime mover, agonist or antagonist, and stabilizer or neutralizer.

A mechanical model of muscular contraction breaks the muscle down into active and passive components. The active component includes the contractile components found in the myofibrils and cross-bridging of the actin and myosin filaments. The passive or elastic components are located in the tendon and the cross-bridges (series-elastic), and in the sarcolemma and the connective tissue (parallel-elastic).

Muscle tension is generated to produce three different types of muscle actions: isometric, concentric, and eccentric. The isometric muscle action is used to stabilize a segment, the concentric action creates a movement, and the eccentric muscle action controls a movement. The concentric muscle actions generate the lowest force output of the three, and the eccentric muscle action generates the highest.

There are numerous factors influencing the amount of force that can be generated by a muscle, including the angle of attachment of the tendon, the laxity or stiffness in the tendon that influences the force-time relationship, the length of the muscle, the contributions of the elastic component, and the velocity of the muscle action. Greater force can be developed in a concentric muscle action if it is preceded by an eccentric muscle action, or prestretch (stretch-shortening cycle). The muscle force is increased due to facilitation via stored elastic energy and neurological facilitation. A quick, short-range prestretch is optimal for developing maximum tension in fast-twitch fibers, and a slow, larger range prestretch is beneficial for tension development in the slow-twitch fibers.

Two-joint muscles are unique since they act at two adjacent joints. Their effectiveness at one joint will depend on the positioning of the other joint, the moment arms at each joint, and the presence of muscle synergies in the movement.

The development of strength in the muscle will be influenced by genetic predisposition, training specificity, training intensity, rest of the muscle during training, and the total training volume. Training principles apply to all groups, the conditioned and the nonconditioned, and only the magnitude of the resistance need be altered. Muscles can be exercised isometrically, isotonically, isokinetically, or through some variable resistance exercise. Another important exercise consideration should be the decision on the use of open- or close-linked exercises.

Muscle injury is common and occurs most frequently in two-joint muscles and during an eccentric muscle action. To prevent muscle injury, proper training and conditioning principles should be followed.

Review Questions

1. For the following segment movements, identify the agonists and antagonists.
   a. thigh flexion
   b. thigh extension from a flexed position
   c. forearm extension against a resistance
   d. forearm flexion
   e. plantar flexion
2. Identify the eccentric muscle actions in the following
   movements.
   a. throwing a baseball (upper extremity)
   b. vertical jump (lower extremity)
   c. bench press
   d. running
   e. walking up stairs
   f. walking down stairs
3. List the advantages and disadvantages of isometric, isotonic, isokinetic, and dynamic accommodating resistive
   exercises.
4. What principles of strength training would be implemented for an athlete? An elderly female?
5. Discuss a program for power development in athletes.
6. Identify a muscle and provide an example of a movement
   for each of the following:
   a. movement of one segment to which the muscle is
      attached
   b. movement of the opposite segment attachment
   c. movement of both segment attachments
7. Working with the specificity-of-training principle,
   describe the type of training for each of the following
   activities.
   a. sprinting
   b. basketball
   c. high jumping
   d. long distance running
   e. football
   f. walking

Additional Questions

1. Speculate on why the eccentric action can sustain more
   force, why it is more efficient, and why it produces
   more injury than the concentric muscle action.
2. How does the prestretch enhance performance?
3. In each of the following movements, identify the
   sequence in the movement where injury is likely to
   occur. State reasons why.
   a. hurdling
   b. running
   c. throwing
   d. deadlift
   e. tennis forehand
   f. long jumping
   g. walking downhill
4. Describe the changes in the direction of the force
   application of the following muscle(s) as they move
   through the designated movement (force application
   with respect to the bone).
   a. muscle: biceps brachii
      movement: elbow flexion to 90 degrees of flexion
   b. muscle: quadriceps femoris
      movement: knee extension to 90 degrees of flexion
   c. muscle: deltoid
      movement: neutral arm position at side to 90
      degrees of abduction
5. The iliopsoas muscle is applying 350 N of force at an
   angle of 40 degrees. How much of the muscle force is
   directed down the vertebrae and how much of the force
   is directed perpendicular to the vertebrae?
6. How does the tendon influence the force-developing
   capabilities in the muscle?
7. For each of the following two-joint muscles, describe
   joint positions where the muscle will be inefficient due
   to loss of length in the muscle, and describe the joint
   positions that put the muscle in its fully elongated
   position.
   a. sartorius
   b. rectus femoris
   c. gastrocnemius
   d. hamstrings

Additional Reading

Armstrong, R.B.: Initial events in exercise-induced muscular injury. Medicine and Science in Sports and Exercise. *22*:429–435, 1990.

Bosco, C., Tarkka, I., and Komi, P.V.: Effect of elastic energy and myoelectrical potentiation of triceps surae during stretch-shortening cycle exercise. International Journal of Sports Medicine. *3*:137–140, 1982.

Burke, R.E.: The control of muscle force: Motor unit recruitment and firing patterns. *In* Human Muscle Power. Edited by N.L. Jones, N. McCartney, and A.J. McComas. Champaign, Ill, Human Kinetics, 1986, pp. 97–110.

Edgerton, V.R., Roy, R.R., Gregor, R.J., and Rugg, S.: Morphological basis of skeletal muscle power output. *In* Human Muscle Power. Edited by N.L. Jones, N. McCartney, and A.J. McComas. Champaign, Ill, Human Kinetics, 1986, pp. 43–64.

Enoka, R.M.: Neuromuscular Basis of Kinesiology. Champaign, Illinois, Human Kinetics, 1988.

Faulkner, J.A., Claflin, D.R., and McCully, K.K.: Power output of fast and slow fibers from human skeletal muscle. *In* Human Muscle Power. Edited by N.L. Jones, N. McCartney, and A.J. McComas. Champaign, Ill, Human Kinetics, 1986, pp. 81–94.

Fleckstein, S.J., Kirby, R.L., and MacLeod, D.A.: Effects of limited knee-flexion range on peak hip moments of force while transferring from sitting to standing. Journal of Biomechanics. *21*:915–918, 1988.

Goldstein, S.A., Armstrong, T.J., Chaffin, J.B., and Matthews, L.S.: Analysis of cumulative strain in tendons and tendon sheaths. Journal of Biomechanics. *20*:1–6, 1987.

Green, H.J.: Muscle power: Fiber type recruitment, metabolism, and fatigue. *In* Human Muscle Power. Edited by N.L. Jones, N. McCartney, and A.J. McComas. Champaign, Ill, Human Kinetics, 1986, pp. 65–80.

Grimby, L.: Single motor-unit discharge during voluntary contraction and locomotion. *In* Human Muscle Power. Edited by N.L. Jones, N. McCartney, and A.J. McComas. Champaign, Ill, Human Kinetics, 1986, pp. 111–130.

Grimby, G.: Clinical aspects of strength and power training. *In* Strength and Power in Sport. Edited by P. Komi. Boston, Blackwell Scientific Publications, 1992, pp. 338–355.

Herring, S.A.: Rehabilitation of muscle injuries. Medicine and Science in Sports and Exercise. *22*:453–456, 1990.

Kannus, P., and Yusada, K.: Value of isokinetic angle-specific torque measurements in normal and injured knees. Medicine and Science in Sports and Exercise. *24*:292–297, 1992.

Kibler, W.B.: Clinical aspects of muscle injury. Medicine and Science in Sports and Exercise. *22*:450–452, 1990.

Kirkendall, D.T.: Mechanisms of peripheral fatigue. Medicine and Science in Sports Exercise. *22*:444–449, 1990.

Mansour, J.M., and Pereira, J.M.: Quantitative functional anatomy of the lower limb with application to human gait. Journal of Biomechanics. *20*:51–58, 1987.

NSCA round table discussion on periodization. Part 2. National Strength and Conditioning Association Journal. *8*:17–24, 1986.

Otten, E.: Concepts and models of functional architecture in skeletal muscle. In K.B. Pandolf (Ed.), Exercise and Sport Sciences Reviews. Vol. 16. Edited by K.B. Pandolf. New York, MacMillan, 1988, pp. 89–137.

Rohl, L., et al.: Tensile and compressive properties of cancellous bone. Journal of Biomechanics. *24*:1143–1149, 1991.

Sargeant, A.J., et al.: Functional and structural changes after disuse of human muscle. Sport Science and Molecular Medicine. *52*:337–342, 1977.

Stein, R.B., Oguztoreli, M.N., and Capady, C.: What is optimized in muscular movements? *In* Human Muscle Power. Edited by N.L. Jones, N. McCartney, and A.J. McComas. Champaign, Ill, Human Kinetics, 1986, pp. 131–150.

Stothart, J.P.: Relationship between selected biomechanical parameters of static and dynamic muscle performance. *In* Biomechanics. Edited by S. Cerquigliani. Basel, Switzerland, Karger, 1973, pp. 210–217.

Tesch, P.A.: Training for bodybuilding. *In* Strength and Power in Sport. Edited by P. Komi. Boston, Blackwell Scientific Publications, 1992, pp. 370–380.

Wilson, G.J., Elliot, B.C., and Wood, G.A.: Stretch shorten cycle performance enhancement through flexibility training. Medicine and Science in Sports and Exercise. *24*:116–123, 1992.

References

1. Billeter, R., and Hoppeler, H.: Muscular basis of strength. *In* Strength and Power in Sport. Edited by P. Komi. Boston, Blackwell Scientific Publications, 1992, 39–63.
2. Huxley, A.F.: Muscle structure and theories of contraction. Progress in Biophysics and Biophysical Chemistry. 7:255–318, 1957.
3. Huijing, P.A.: Mechanical muscle models. *In* Strength and Power in Sport. Edited by P. Komi. Boston, Blackwell Scientific Publications, 1992, pp. 130–150.
4. Soderberg, G.L.: Kinesiology: Application to Pathological Motion. Baltimore, Williams & Wilkins, 1986.
5. Hay, J.G., and Reid, J.G.: The Anatomical and Mechanical Bases of Human Motion. Englewood Cliffs, Prentice Hall Inc, 1988.
6. Vanderhelm, F.C.T. and Veenbaas, R.: Modelling the mechanical effect of muscles with large attachment sites: Application to the shoulder mechanism. Journal of Biomechanics. 24:1151–1163, 1991.
7. Komi, P.V.: Physiological and biomechanical correlates of muscle function: Effects of muscle structure and stretch-shortening cycle on force and speed. *In* Exercise and Sport Sciences Reviews. Edited by R.L. Terjund. Lexington, Mass., Collamore Press, 1984, pp. 81–121.
8. Zernicke, R.F., and Loitz, B.J.: Exercise-related adaptations in connective tissue. *In* Strength and Power in Sport. Edited by P. Komi. Boston, Blackwell Scientific Publications, 1992, pp. 77–95.
9. Proske, U., and Morgan, D.L.: Tendon stiffness: Methods of measurement and significance for the control of movement. A review. Journal of Biomechanics. 20:75–82, 1987.
10. Stone, M.H.: Connective tissue and bone response to strength training. *In* Strength and Power in Sport. Edited by P. Komi. Boston, Blackwell Scientific Publications, 1992, pp. 279–290.
11. Kornecki, S.: Mechanism of muscular stabilization process in joints. Journal of Biomechanics. 25:235–245, 1992.
12. Hill, A.V.: Heat and shortening and the dynamic constants of muscle. Proceedings of the Royal Society of London (Biology). 126:136–195, 1938.
13. Hill, A.V.: First and Last Experiments in Muscle Mechanics. Cambridge, Cambridge University Press, 1970.
14. Asmussen, E.: Positive and negative muscular work. Acta Physiologica Scandinavica. 28:364–382, 1952.
15. Edman, K.A.P., Elzinga, G., and Noble, M.I.M.: Enhancement of mechanical performance by stretch during tetanic contractions of vertebrate skeletal muscle fibres. Journal of Physiology. 281:139–155, 1978.
16. Edman, K.A.P.: Contractile performance of skeletal muscle fibres. Strength and Power in Sport. Edited by P. Komi. Boston, Blackwell Scientific Publications, 1992, pp. 96–114.
17. Stauber, W.T.: Eccentric action of muscles: Physiology, injury, and adaptation. *In* Exercise and Sports Sciences Review. Edited by K.B. Pandolph. Baltimore, Williams & Wilkins, 1989, pp. 157–185.
18. Gowitzke, B.A.: Muscles alive in sport. *In* Biomechanics. Edited by M. Adrian and H. Deutsch. Eugene, Ore., Microform, 1984.
19. Schmidtbleicher, D.: Training for power events. *In* Strength and Power in Sport. Edited by P. Komi. Boston, Blackwell Scientific Publications, 1992, pp. 381–395.
20. Moritani, T.: Time course of adaptations during strength and power training. *In* Strength and Power in Sport. Edited by P. Komi. Boston, Blackwell Scientific Publications, 1992, pp. 226–278.
21. Perrine, J.J.: The biophysics of maximal muscle power outputs: Methods and problems of measurement. *In* Human Muscle Power. Edited by N.L. Jones, N. McCartney, and A.J. McComas. Champaign, Ill, Human Kinetics, 1986, pp. 15–26.
22. Komi, P.V.: The stretch-shortening cycle and human power output. *In* Human Muscle Power. Edited by N.L. Jones, N. McCartney, and A.J. McComas. Champaign, Ill, Human Kinetics, 1986, pp.27–40.
23. Komi, P.V.: Stretch-shortening cycle. *In* Strength and Power in Sport. Edited by P. Komi. Boston, Blackwell Scientific Publications, 1992, pp. 169–179.
24. Huijing, P.A.: Elastic potential of muscle. *In* Strength and Power in Sport. Edited by P. Komi. Boston, Blackwell Scientific Publications, 1992, pp. 151–168.
25. Asmussen, E., and Bonde-Petersen, F.: Apparent efficiency and storage of elastic energy in human muscles during exercise. Acta Physiologica Scandinavica, 92:537–545, 1974.
26. Goldspink, G.: Cellular and molecular aspects of adaptation in skeletal muscle. *In* Strength and Power in Sport. Edited by P. Komi. Boston, Blackwell Scientific Publications, 1992, pp. 211–229.
27. Zajac, F.E., and Gordon, M.E.: Determining muscle's force and action in multi-articular movement. *In* Exercise and Sports Sciences Reviews. Edited by K.B. Pandolph. Baltimore, Williams & Wilkins, 1989, pp. 187–230.
28. Van Soest, A.J., Schwaab, A.L., Bobbert, M.T., and van Ingen Schenau, G.J.: The influence of the biarticularity of the gastrocnemius muscle on vertical jumping achievement. Journal of Biomechanics, 26:1–8, 1993.
29. Bobbert, M.F. and van Ingen Schenau, G.J.: Coordination in vertical jumping. Journal of Biomechanics. 21:249–262, 1988.
30. Wells, R.P.: Mechanical energy costs of human movement: An approach to evaluating the transfer possibilities of two-joint muscles. Journal of Biomechanics. 21:955–964, 1988.
31. Kulig, K., Andrews, J.G., and Hay, J.G.: Human strength curves. *In* Exercise and Sport Sciences Reviews Edited by R.L. Terjund. Lexington, Mass., Collamore Press, 1984, pp. 417–466.

32. Knuttgen, H.G., and Komi, P.: Basic definitions for exercise. *In* Strength and Power in Sport. Edited by P. Komi. Boston, Blackwell Scientific Publications, 1992, pp. 3–6.

33. Weiss, L.W.: The obtuse nature of muscular strength: The contribution of rest to its development and expression. Journal of Applied Sport Science Research, *5*:219–227, 1991.

34. Macdougall, J.D.: Hypertrophy or hyperplasia. *In* Strength and Power in Sport. Edited by P. Komi. Boston, Blackwell Scientific Publications, 1992, pp. 230–238.

35. Sale, D.G.: Neural adaptation in strength and power training. *In* Human Muscle Power. Edited by N.L. Jones, N. McCartney, and A.J. McComas. Champaign, Ill, Human Kinetics, 1986, pp. 289–308.

36. Garhammer, J., and Takano, B.: Training for weightlifting. *In* Strength and Power in Sport. Edited by P. Komi. Boston, Blackwell Scientific Publications, 1992, pp. 357–369.

37. ACSM Position Stand.: The recommended quantity and quality of exercise for developing and maintaining cardiorespiratory and muscular fitness in healthy adults. Medicine and Science in Sports and Exercise. *22*:265–274, 1990.

38. Israel, S.: Age-related changes in strength and special groups. *In* Strength and Power in Sport. Edited by P. Komi. Boston, Blackwell Scientific Publication, 1992, pp. 319–328.

39. Ariel, G.: Resistive exercise machines. *In* Biomechanics. Edited by J. Terauds, K. Barthels, E. Kreighbaum, R. Mann, and J. Crake. Eugene, Ore., Microform, 1984.

40. Cook, E.E., Gray, V.L., Savinar-Nogue, E., and Medeiros, J.: Shoulder antagonistic strength ratios: A comparison between college-level baseball pitchers and nonpitchers. The Journal of Orthopaedic and Sports Physical Therapy. *8*:451–461, 1987.

41. Hay, J.G.: Mechanical basis of strength expression. *In* Strength and Power in Sport. Edited by P. Komi. Boston, Blackwell Scientific Publications, 1992, pp. 197–207.

42. Atha, J.: Strengthening muscle. Exercise and Sport Sciences Reviews. *9*:1–73, 1981.

43. Manning, R.J., et al.: Constant vs variable resistance training. Medicine and Science in Sports and Exercise. *22*:397–401, 1990.

44. Sylvester, L.J., Stiggins, C., McGowan, C., and Bryce, G.R.: The effect of variable resistance and free weight training programs on strength and vertical jump. National Strength and Conditioning Association Journal. *3*:30–33, 1981.

45. Garrett, W.E.: Muscle strain injuries: Clinical and basic aspects. Medicine and Science in Sports and Exercise. *22*:436–443, 1991.

46. Stone, M.H.: Muscle conditioning and muscle injuries. Medicine and Science in Sports and Exercise. *22*:457–462, 1990.

Glossary

Actin:	A protein of the myofibril, noticeable by it's light banding. Along with myosin, it is responsible for the contraction and relaxation of the muscle.
Agonist:	A muscle responsible for producing a specific movement through concentric muscle action.
Anatomic Cross Section:	The cross section at a right angle to the direction of the fibers.
Antagonist:	A muscle responsible for opposing the concentric muscle action of the agonist.
Aponeurosis:	A flattened or ribbon-like tendinous expansion from the muscle that connects into the bone.
Assistant Mover:	A muscle that assists to bring about a desired movement.
Belly:	The fleshy, centralized portion of a muscle.
Bipennate:	A feather-shaped fiber arrangement, in which the fibers run off of both sides of a tendon running through the muscle.
Close-Linked Exercise:	Exercises using eccentric and concentric muscle actions, with the feet fixed on the floor. Movements begin with segments distal to the feet (trunk and thigh) and move toward the feet, i.e., squat.
Compartment Syndrome:	A condition in which the circulation and function of the tissues within a muscle compartment become impaired due to an increase in pressure within the compartment.
Concentric:	Muscle action in which the tension developed causes a visible shortening in the length of the muscle; positive work is performed.
Contractibility:	The ability of muscle tissue to shorten when the muscle tissue receives sufficient stimulation.
Contractile Component:	The active component in the muscle where shortening takes place as cross-bridging occurs in the myofibrils.
Contraction:	The state of muscle when tension is generated across a number of actin and myosin filaments.
Cross-Bridge:	The connection and intertwining of the actin and myosin filaments of the myofibrils.
Eccentric:	Muscle action in which tension is developed in the muscle and the muscle lengthens; negative work is performed.
Endomysium:	The sheath surrounding each muscle fiber.
Elastic:	Capable of being stretched, compressed, or distorted, and then returning to the original shape.
Elasticity:	The ability of muscle tissue to return to its resting length once a stretch is removed.
Epimysium:	A dense, fibrous sheath covering an entire muscle.

Extensibility: The ability of muscle tissue to lengthen out beyond resting length.

Fascia: Sheet or band of fibrous tissue.

Fascicles: A bundle or cluster of muscle fibers.

Fast-Twitch Fibers: Large skeletal muscle fibers which are innervated by the alpha-I motor neuron and have fast contraction times. There are two subtypes of fast-twitch fibers: the low oxidative and high glycolytic (Type IIb) and the medium oxidative and high glycolytic (Type IIa).

Fibers: Elongated, cylindrical structures containing cells which comprise the contractile elements of muscle tissue.

Force-Velocity Relationship: The relationship between the tension development in the muscle and velocity of shortening or lengthening.

Fusiform: Spindle-shaped fiber arrangement in a muscle.

Inelastic: Lacking the ability to withstand compression, stretch, or distortion and return to the original shape or length.

Insertion: The more distal attachment site of the muscle.

Intensity: In weight training, the load or percentage of maximum lifting capacity lifted with each repetition.

Irritability: The capacity of muscle tissue to respond to a stimulation.

Isokinetic Exercise: An exercise in which concentric muscle action is generated to move a limb against a device that is speed controlled. Individuals attempt to develop maximum tension through the full range of motion at the specified speed of movement.

Isometric: Muscle action in which tension develops but there is no visible or external change in joint position; no external work is produced.

Isometric Exercise: An exercise that loads the muscle in one joint position.

Isotonic Exercise: An exercise in which an eccentric and/or concentric muscle action is generated to move a specified weight through a range of motion.

Length-Tension Relationship: The relationship between the length of the muscle and the tension produced by the muscle; highest tensions are developed slightly past resting length.

Myofibrils: Rod-like strands, contained within and running the length of the muscle fibers. Contains the contractile elements of the muscle.

Myosin: A thick protein of the myofibril, noticeable by its dark banding. Along with actin, it is responsible for contraction and relaxation of the muscle.

Myotendinous Junction: The site where the muscle and tendon join together, consisting of a layered interface as the myofibrils and the collagen fibers of the tendon meet.

Multipennate: A feather-shaped fiber arrangement, in which the muscle fibers run diagonally off one or both sides of a tendon running through the muscle.

Neutralizer: A muscle responsible for eliminating or cancelling out an undesired movement.

Origin:	The more proximal attachment site of the muscle.
Parallel Elastic Component:	The passive component in the muscle which develops tension with elongation; found in the connective tissue surrounding and within the muscle.
Penniform:	A feather-shaped fiber arrangement in a muscle, in which the fibers run diagonally to a tendon running through the muscle.
Perimysium:	A dense, connective tissue sheath covering the fascicles.
Physiologic Cross Section:	Area which is the sum total of all of the cross sections of fibers in the muscle; the area perpendicular to the direction of the fibers.
Plyometrics:	A training technique which uses the stretch-shortening cycle to increase athletic power.
Power:	The product of force and velocity.
Prime Mover:	The muscle that acts directly to bring about a desired movement.
Recoil:	To spring back to the original position, as seen in the elastic components in the muscle.
Sarcolemma:	A thin plasma membrane covering the muscle which branches into the muscle, carrying nerve impulses.
Sarcomere:	One contractile unit of banding on the myofibril, running Z-band to Z-band.
Series elastic component:	The passive component in the muscle that develops tension in contraction and during elongation; found in the tendon and the cross-bridges.
Sliding Filament Theory:	A theory describing muscle contraction, whereby tension is developed in the myofibrils as the head of the myosin filament attaches to a site on the actin filament.
Slow-Twitch Fibers:	Small skeletal muscle fibers innervated by the alpha-2 motoneuron, having a slow contraction time. This fiber is high oxidative and low glycolytic.
Specificity:	Training principle suggesting that specific training movements should be done in the same manner and position in which the movements are performed in the sport or activity.
Stabilizer:	A muscle responsible for stabilizing an adjacent segment.
Strength:	The maximum amount of force produced by a muscle or muscle group at a site of attachment on the skeleton; one maximal effort.
Stretch-shortening cycle:	A common sequence of joint actions where an eccentric muscle action, or pre-stretch, precedes a concentric muscle action.
Tendon:	A fibrous cord, consisting primarily of collagen, by which muscles attach to bone.
Unipennate:	A feather-shaped fiber arrangement in which the muscle fibers run diagonally off one side of the tendon.
Variable Resistive Exercise:	Exercise performed on a machine which alters the amount of resistance through the range of motion.
Volume:	In weight training, the sum of the number of repetitions multiplied by the load or weight lifted. Usually calculated over a week, month, or year.

CHAPTER 4

Neurological Considerations for Movement

After reading this chapter, the student will be able to:

1. Describe the anatomy of a motor unit including central nervous system pathways, the neuron structure, the neuromuscular junction, and the ratio of fibers to neurons that are innervated.
2. Explain the differences between the three motor unit types: Type I, Type IIa, and Type IIB.
3. Discuss the characteristics of the action potential, emphasizing how a twitch or tetany develops, as well as the influence of local graded potentials.
4. Describe the pattern of motor unit contribution to a muscle contraction through discussion of the size principle, synchronization, and rate coding of the motor unit activity.
5. Explain how electromyography (EMG) measures the action potential in the muscle, and provide some examples of the use of EMG for specific purposes.
6. Compare the myotactic, propriospinal, and supraspinal reflexes, and provide examples of each.
7. Describe the anatomy of the muscle spindle and the functional characteristics of the spindle during a stretch of the muscle or during gamma motoneuron influence.
8. Describe the anatomy of the Golgi tendon organ (GTO) and explain how the GTO responds to tension in the muscle.
9. Explain how weight training, bilateral training, or warm-up influences the neurological activity in the muscle.
10. Identify the factors influencing flexibility, and provide examples of specific stretching techniques that are successful in enhancing flexibility.
11. Describe a plyometric exercise, detailing the neurological and structural contributions to the exercise.

Human movement is controlled and monitored by the nervous system. The nature of this control is such that many muscles may need to be activated to perform a vigorous movement like sprinting, or only a few muscles may need to be activated to push a doorbell or make a phone call. The nervous system is responsible for identifying the muscles that will be activated for a particular movement, and then generating the stimulus to develop the level of force that will be required from that muscle.

Many human movements require stabilization of adjacent segments while a fine motor skill is performed. This requires a great deal of coordination on the part of the nervous system in order to stabilize such segments as the arm and forearm while very small, coordinated movements are created with the fingers, as in the act of writing.

Accuracy of movement is another task with which the nervous system is faced. The nervous system coordinates the muscles to throw a baseball with just the right amount of muscular force so that the throw is successful. Recognizing the difficulty of being accurate with a physical movement contributes to an appreciation of the complexity of neural control.

The neural network is extensive since each muscle fiber is individually serviced by a branch of the nervous system. There is information exiting the muscle and providing input to the nervous system, and there is information entering the muscle in order to initiate a muscle activity of a very specific nature and magnitude. Through this loop system, interconnected with many other loops from other muscles and with central nervous control, the nervous system is able to coordinate the activity of many muscles at once. Specific levels of force may be generated in several muscles simultaneously so that a skill such as kicking may be performed accurately and forcefully. Knowledge of the nervous system will be very helpful in the improvement of muscular output, the refinement of a skill or task, the rehabilitation of an injury, or stretching a muscle group.

General Organization of the Nervous System

The nervous system consists of two parts, the central nervous system and the peripheral nervous system, both illustrated in FIGURE 4–1. The central nervous system consists of the brain and the spinal cord, and should be viewed as the means by which human movement is initiated, controlled, and monitored.

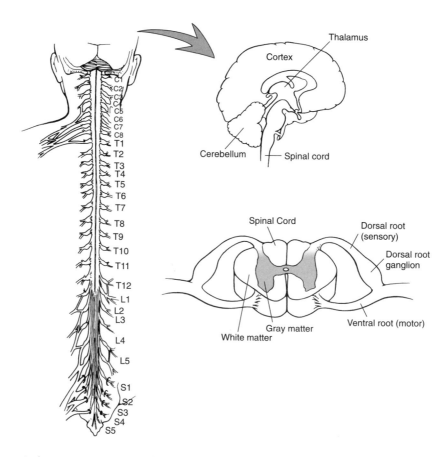

FIGURE 4–1. The central nervous system consists of the brain and the spinal cord. The peripheral nervous system consists of all of the nerves that lie outside the spinal cord. There are 31 pairs of spinal nerves that exit and enter the spinal cord at the various levels of the vertebrae. Motor information leaves the spinal cord through the ventral root (anterior) and sensory information enters the spinal cord through the dorsal root (posterior).

The peripheral nervous system consists of all the branches of nerves that lie outside the spinal cord. The peripheral nerves primarily responsible for muscular action are the spinal nerves that enter on the posterior, or dorsal side of the vertebral column, and exit on the anterior, or ventral side at each vertebral level of the spinal cord. There are eight pairs of nerves entering and exiting the cervical region, twelve pairs at the thoracic region, five at the lumbar region, five in the sacral region, and one in the coccygeal region. The pathways of the nerves are presented for the upper and lower extremities in FIGURES 4–2 and 4–3, respectively.

The nerves entering the spinal cord on the dorsal side, or back of the cord, are called sensory neurons, since they are transmitting information into the system from the muscle. The nerves exiting on the ventral side, or front of

the body, are called motoneurons, since they are carrying impulses away from the system to the muscle.

Motoneurons

Structure of the Motoneuron

The neuron is the functional unit of the nervous system carrying information to and from the nervous system. The structure of a neuron, and specifically, the motoneuron, warrants examination to understand the process of muscular contraction. Refer to FIGURE 4–4 for a close up view of the neuron and the neuromuscular junction.

The motoneuron consists of a cell body containing the nucleus of the nerve cell. The cell body, or soma, of a motoneuron is usually contained within the gray matter

Anterior View

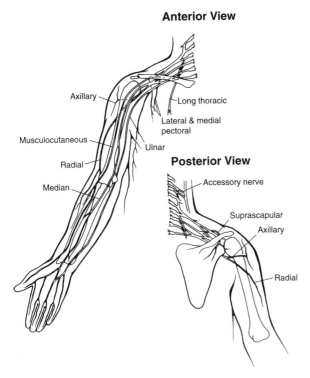

Posterior View

FIGURE 4–2. The upper extremity nerves are shown above. There are nine nerves that innervate the muscles of the upper extremity.

of the spinal cord, or is located in bundles of cell bodies found just outside the cord, termed ganglia. The cell bodies are arranged in pools spanning one to three levels of the spinal cord and innervate portions of a single muscle or selected synergists (1).

There are projections on the cell body, called dendrites, which serve as receivers and bring information into the neuron from other neurons. The dendrites are bunched together to form small bundles. A bundle will contain dendrites from other neurons and can consist of dendrites from different spinal cord levels or different neuron pools. The composition of the bundle will change as dendrites are added and subtracted. This arrangement facilitates cross-talk between neurons.

A large nerve fiber, the axon, branches out from the cell body and exits the spinal cord via the ventral root, where it is bundled together with other peripheral nerves. The axon of the motoneuron is fairly large, making it capable of transmitting nerve impulses at high velocities up to 100 m/sec. This large and fast transmitting motoneuron is also called an alpha neuron. The axon of the motoneuron is myelinated, or covered with an insulated shell. Actually, the myelination is sectioned with

Schwann's cells insulating and enveloping a specific length along the axon, followed by a gap, termed the nodes of Ranvier, and then a repeat of the insulated Schwann cell covering.

When the myelinated motoneuron approaches a muscle fiber, it breaks off into unmyelinated terminals or branches called motor endplates, which embed into fissures, or clefts, near the center of the muscle fiber. This site is called the neuromuscular junction. The neuron does not make contact with the actual muscle fiber; instead there is a small gap, termed the synaptic gap or synapse, between the terminal branch of the neuron and the muscle. This is the reason muscular contraction involves a chemical transmission, since the only way for a nerve impulse to reach the actual muscle fiber is for some type of chemical transmission to occur across the gap.

The Motor Unit

The neuron, the cell body, dendrites, axon, branches, and the muscle fibers, constitute the motor unit (FIGURE 4–5). A neuron may terminate on as many as 2000 fibers in muscles like the gluteus maximus or as few as 5 or 6 fibers, such as in the orbicularis oculi of the eye. For example, the typical ratio of neuron to the number of muscle fibers in muscles is 1:10 for the eye muscles, 1:1600 for the gastrocnemius, 1:500 for the tibialis anterior, 1:1000 for the biceps brachii, 1:300 for the dorsal interossei in the hand, and 1:96 for the lumbricales in the hand (1,2). The average number of fibers per neuron is somewhere between 100 and 200 muscle fibers (2,3). The fibers innervated by each motor unit are not bunched together and are not all in the same fascicle, but are spread out over the muscle.

When a motor unit is activated sufficiently, all the muscle fibers belonging to it will contract within a few milliseconds. This is referred to as the all-or-none principle. If a muscle has motor units with very low nerve-to-fiber ratios, there will be finer control of the movement characteristics, such as is seen around the eye or in hand movements. Many lower extremity muscles have large neuron-to-fiber ratios suitable to those functions in which large amounts of muscular output are required such as in weight bearing and ambulation.

Three different types of motor units exist, corresponding to the three fiber types discussed in the previous chapter: slow-twitch oxidative (Type I), fast-twitch oxidative (Type IIa), and fast-twitch glycolytic (Type IIb). Micrographs of the three different fiber types are presented in FIGURE 4–6. Even at this level, there is considerable variation between the three fiber types. Certain muscles, like the soleus, consist primarily of

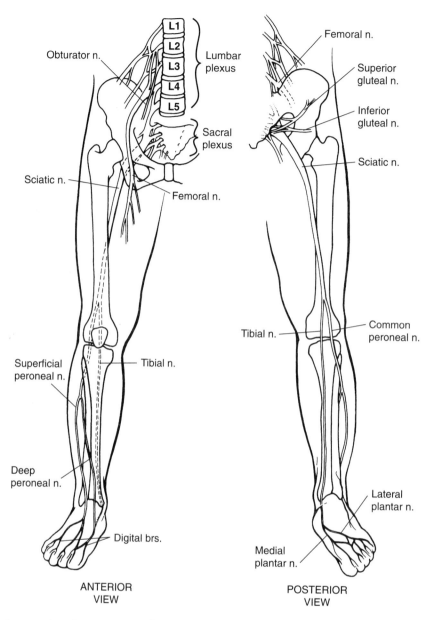

FIGURE 4–3. The lower extremity nerves are shown above. There are 12 nerves that innervate the muscles of the lower extremity.

Type I muscle fibers and motor units, whereas muscles like the vastus lateralis are approximately 50% Type I and the remainder Type II.

These fiber and motor unit types are basically genetically determined, but they can be enhanced through training. It has been shown that Type IIa muscle fibers can be converted to Type IIb fibers through specific training emphasizing power and quickness (4). It is not clear

whether a transformation can be made from Type I to Type II through specific training.

All of the muscle fibers in a motor unit are of the same type. The fast-twitch glycolytic motor units (Type IIb) are innervated by very large alpha motoneurons that conduct the impulses at very fast velocities (> 100 m/sec) creating rapid contraction times in the muscle (approximately 30 to 40/ms) (5). As a result, these large motor

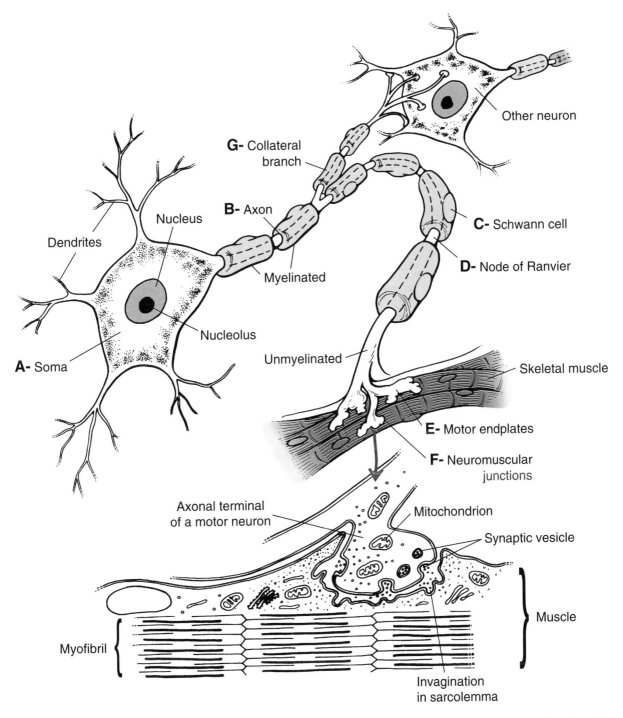

Other neuron

G- Collateral
branch

Nucleus

B- Axon

C- Schwann cell

Dendrites

Myelinated

D- Node of Ranvier

Nucleolus

Unmyelinated

A- Soma

Skeletal muscle

E- Motor endplates

F- Neuromuscular
junctions

Axonal terminal
of a motor neuron

Mitochondrion

Synaptic vesicle

Muscle

Myofibril

Invagination
in sarcolemma

FIGURE 4–4. The cell body, or (A) soma, of the neuron is located in or just outside the spinal cord. Travelling from the soma is the (B) axon, which is myelinated by (C) Schwann cells separated by gaps, the (D) nodes of Ranvier. On the ends of each axon, the branches become unmyelinated to form the (E) motor endplates that terminate at the (F) neuromuscular junction on the muscle. Neurons receive information from other neurons through (G) collateral branches.

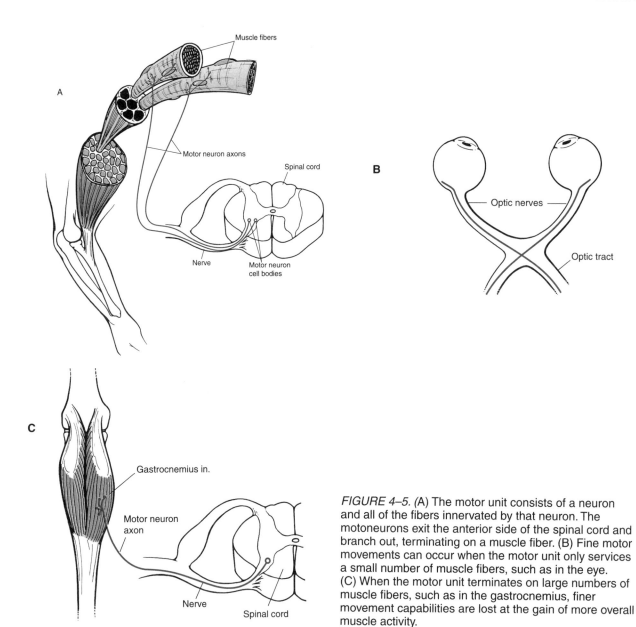

FIGURE 4–5. (A) The motor unit consists of a neuron and all of the fibers innervated by that neuron. The motoneurons exit the anterior side of the spinal cord and branch out, terminating on a muscle fiber. (B) Fine motor movements can occur when the motor unit only services a small number of muscle fibers, such as in the eye. (C) When the motor unit terminates on large numbers of muscle fibers, such as in the gastrocnemius, finer movement capabilities are lost at the gain of more overall muscle activity.

units generate muscular activity that is fast-contracting, develops high tensions, but fatigues quickly. These motor units usually have large neuron-to-fiber ratios and are found in some of the largest muscles in the body, such as the quadriceps femoris group. Utilization of these motor units would be useful in activities such as sprinting, jumping, and weight lifting.

The fast-twitch oxidative motor units (Type IIa) also have fast contraction times (approximately 30 to 50/ms), but have the advantage over fast-twitch glycolytic motor

units because they are more fatigue resistant (5). These moderately sized motor units are capable of generating moderate tensions over longer time periods. The activity from these motor units prove to be very useful in activities such as swimming and bicycling, or in job tasks seen in factories or among longshoremen.

The slow-twitch oxidative motor units (Type I) transmit the impulses slowly (approximately 80 m/s), generating slow contraction times in the muscle (> 70/ms) (5). These motor units are capable of generating very little

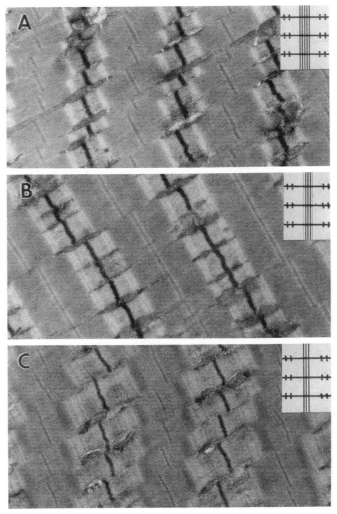

FIGURE 4–6. Micrographs allow a view of the ultrastructure of a fiber. The micrographs above demonstrate actual structural differences between the three fiber types. The Type I fibers (A) have bands of equal density; the Type IIa have bands of nonuniform density, and the Type IIb fibers have only three bands (Courtesy of J. Friden).

tension but can sustain this tension over a long time period. Consequently, the slow-twitch motor units, the smallest of the three types, are useful in maintaining postures, stabilizing joints, doing repetitive activities such as typing or gross muscular activities such as jogging.

Functional Characteristics of the Motor Unit

The nerve impulse travels through the motor unit in the form of an action potential. At rest, there is an electrical potential across the membrane of approximately 60 to 70 mV that is negative in the inside of the membrane. Upon activation by the central or peripheral nervous system, the action potential travels down through the motor unit via rapid depolarization, repolarization, to hyperpolarization.

In depolarization, the membrane potential decreases and moves closer to zero and a positive voltage of approximately 30 mV, as there is an exchange of ions across the membrane. This is followed by repolarization as the voltage moves back in the negative direction, and hyperpolarization, as the voltage becomes more negative than in the resting condition. The exchange of ions, the generation of the action potential, and the voltage change associated with the depolarization process are presented in FIGURE 4–7.

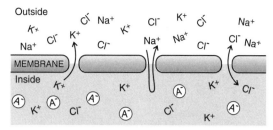

Exchange of Ions Across Membrane

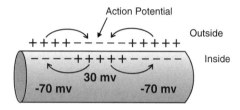

Action Potential Generated Through Change in Electrical Potential

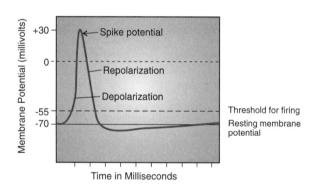

Recording of Action Potential

FIGURE 4–7. The action potential travels down through the nerve as the permeability of the nerve membrane changes, allowing an exchange of sodium (Na) and potassium (K) ions across the membrane. This creates a voltage differential that is negative on the outside of the membrane. This negative voltage, or action potential, travels down the nerve until it reaches the muscle and stimulates a muscle action potential that can be recorded.

The action potential is a propagated impulse, meaning that the amplitude of the impulse remains the same as it travels down the axon to the motor endplate. At the motor endplate, the action potential travelling down through the nerve becomes a muscle action potential travelling through the muscle. Externally, these two action potentials are not distinguishable from each other. Eventually, the muscle action potential initiates the development of the cross-bridging and shortening within the muscle sarcomere. The total process is referred to as *excitation-contraction coupling.*

Each action potential generates a twitch response in the muscle. In the twitch response, the muscle generates a peak force that drops off rapidly and returns to baseline (resting level) (FIGURE 4–8). Depending upon the motor unit and muscle fiber type, the twitch responses will vary between very fast generation of peak force in the fast-twitch fibers to very slow development of the peak force in the slow twitch fibers (1).

The muscle starts to develop outward tension and actually creates movement when multiple action potentials are generated. If action potentials are sequenced close enough together, the tensions generated by one muscle twitch will be summed with other twitches to form a tetanus, or constant tension in the muscle fiber (FIGURE 4–8). This level of tension will decline as the motor unit becomes incapable of regenerating the individual twitch responses fast enough.

The action potential in a motor unit can be facilitated or inhibited by the input it receives from the many neurons that are connecting into it within the spinal cord. As shown in FIGURE 4–9, a motor unit receives synaptic input from other neurons and from interneurons, which are small, connecting branches that can be both excitatory or inhibitory (1). The input is in the form of a local graded potential, which, unlike the action potential, does not maintain its amplitude as it travels along. Thus, the stimulus has to be sufficient to reach its destination on another neuron, and be large enough to generate a response in the neuron with which it has interfaced.

The alpha motoneuron has many collateral branches interacting with other neurons, and the number of collateral branches increases in the distal muscles (1). An interneuron receiving input from these collaterals is the Renshaw cell, an inhibitory interneuron also located in the spinal cord. The Renshaw cell is considered one of the key elements in organizing muscular response in the agonists, antagonists, and synergists when it is stimulated sufficiently by a collateral branch (6,7).

The tension or force generated by a muscle is determined by the number of motor units actively stimulated at the same time, and the frequency at which the motor units are firing. *Recruitment,* the term used to describe the order of activation of the motor units, is the prime mechanism for force production in the muscle. It usually

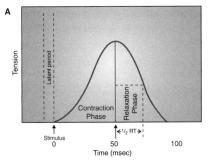

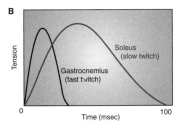

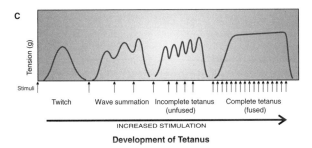

FIGURE 4–8. When a muscle receives a single excitatory stimulus, a muscle twitch is developed as shown in (A). The muscle generates a peak force during the contraction phase and then rapidly drops back down to baseline during the relaxation phase. Fiber types vary in their response to a stimulus. Slow-twitch fibers generate the peak force in twitch much more slowly than fast-twitch fibers as shown in (B). Tetany, a smooth sustained contraction, will develop as the action potentials from one twitch are summed with other twitch responses. This progression is illustrated in (C).

twitch glycolytic motor units (8). This is basically due to the fact that the small motoneurons have lower thresholds than the larger motoneurons. Thus, the small motoneurons are used over a broad tension range before the moderate or large fibers are recruited.

In walking, for example, the low threshold motor units are used for most of the gait cycle, except for some brief recruitment of the intermediate motor units during peak activation times. The high threshold, fast-twitch motor units are not usually recruited unless there is a rapid change of direction or a stumble.

In running, more motor units are recruited, with some high threshold units recruited for the peak output times in the cycle. Further, the low threshold units are recruited for

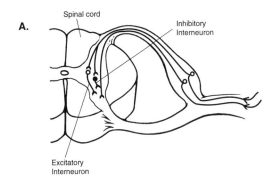

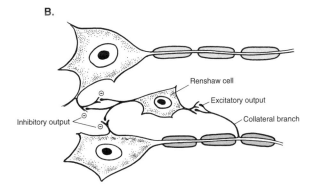

FIGURE 4–9. (A) The action potential travelling through a motor unit can be altered by input from interneurons, which are small connecting nerve branches. A local graded potential is generated that may or may not institute a change in the connecting neuron. The interneurons may produce an excitatory local graded potential, which would facilitate the action potential, or it could produce an inhibitory local graded potential sufficient to inhibit the action potential. (B) A special interneuron, the Renshaw cell receives excitatory information from a collateral branch of another neuron, stimulating an inhibitory local graded potential.

follows an orderly pattern in which pools of motor units are sequentially recruited (8). There is a functional pool of motor units for each task, whereby separate recruitment sequences can be initiated to stimulate the three different types of motor units (Type I, IIa, IIb) for the performance of different actions within the same muscle.

The sequence of motor unit recruitment usually follows the size principle, whereby the small, slow-twitch motoneurons are recruited first, followed by recruitment of the fast-twitch oxidative, and finally, the large, fast-

activities such as walking and jogging, and the fast-twitch fibers are recruited in activities such as weight lifting (8,9). Recruitment sequences for walking and for different exercise intensities are presented in FIGURE 4–10.

The motor units are recruited asynchronously, whereby the activation of a motor unit is temporally spaced but is summed with the preceding motor unit activity. If the tension is held isometrically over a prolonged period of time,

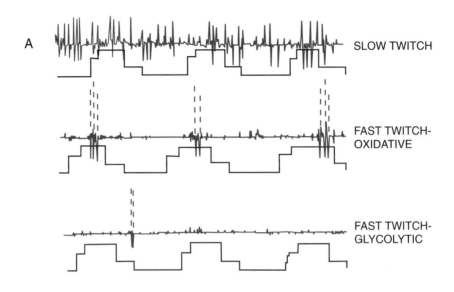

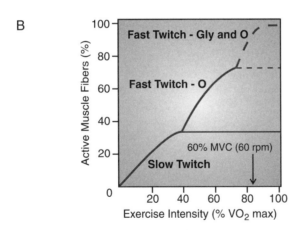

FIGURE 4–10. The order of activation of the motor units, termed recruitment, usually occurs following the size principle whereby the small slow-twitch fibers are recruited first, followed by the fast-twitch oxidative, and lastly by the fast-twitch glycolytic fibers. In (A), the muscle activity for the three muscle types is shown for three support phases in walking. Slow-twitch fibers are used for most of the gait cycle, with some recruitment of the fast-twitch fibers at peak activation times (From Grimby, L.: Single motor unit discharge during voluntary contraction and locomotion. *In* Human Muscle Power. Edited by N.L. Jones, N. McCartney, and A.J. McComas. Champaign, IL, Human Kinetics, 1986, pp. 111–129.). In (B), the recruitment pattern is similar, with slow-twitch fibers recruited for up to 40% of the exercise intensity, at which point the fast-twitch oxidative are recruited. It is not until 80% of exercise intensity is reached that the fast-twitch glycolytic fibers are recruited (From Sale, D.G.: Influence of exercise and training on motor unit activation. Exercise and Sport Science Reviews. *16*:95–151, 1987).

some of the larger motoneurons will be activated. Likewise, in vigorous, rapid movements, both small and large motoneurons will be activated.

The motor unit recruitment pattern proceeds from small to large motoneurons, slow to fast, small force to large force, and fatigue resistant to fatiguable muscles. Once a motor unit is recruited, it will remain active until the force declines, and when the force is lowered, the motor units are deactivated in reverse order of activation, with the large motoneurons going first (1). Also, the motor unit recruitment pattern is established in the muscle for a very specific movement pattern (10). If the joint position changes and a new pattern of movement is required, the recruitment pattern will change because different motor units will be involved. However, the order of recruitment from small to large will remain the same.

There is some evidence that alternative recruitment patterns may be initiated by input from the excitatory and inhibitory pathways. This is done through interneurons which alter the threshold response of the slow-and fast-twitch units. The threshold level of the fast-twitch motor unit can be lowered via excitatory interneurons.

In ballistic movements involving rapid alternating movements, there appears to be synchronous or concurrent activation of the motor unit pool whereby large motor units are recruited along with the small motoneurons. This synchronous firing has also been shown to occur as a result of weight training. It is believed that in athletic performance requiring a wide range of muscular output, the neuromuscular sequence may in fact be reversed, with the fast-twitch fibers recruited first in vigorous muscle actions (8,11).

The frequency of motor unit firing can also influence the amount of force or tension developed by the muscle. This is known as *frequency coding* or *rate coding*, and involves intermittent high-frequency bursts of action potentials or impulses ranging from 3 to 50 impulses per second (3). With increased rate coding, the rate of impulses increases in a linear fashion and only after all of the motor units are recruited (12).

In the small muscles, all of the motor units are usually recruited and activated when the external force of the muscle is at levels of only 30 to 50% of the maximum voluntary contraction level (1). Beyond this level, the force output in the muscle is increased through increases in rate coding allowing for the production of a smooth, accurate contraction.

In the large muscles, there is recruitment of motor units all through the total force range so that some muscles are still recruiting more motor units at 100% of maximum voluntary contraction. The deltoid and the biceps brachii

are examples of muscles still recruiting motor units at 80 to 100% of maximum output of the muscle (1).

The rate coding also varies with fiber type and changes with the type of movement. Examples of the rate coding of both high and low threshold fibers in two different muscle contractions is illustrated in FIGURE 4–11. In ballistic movements, the higher threshold, fast-twitch motor units fire at higher rates than the slow-twitch units. To produce rapid accelerations of the segments, the fast-twitch motor units have been shown to increase the firing rates more than the slow-twitch motor units (9). The high-threshold, fast-twitch fibers cannot be driven for any considerable length of time, but it is believed that trained athletes can drive the high threshold units longer by maintaining the firing rates. This results in the ability to produce a vigorous contraction for a limited amount of time. Eventually, the frequency of motor

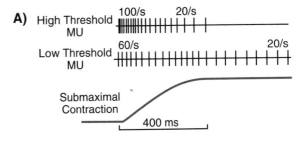

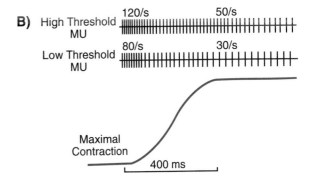

FIGURE 4–11. (A) Tension development in the muscle is influenced by the frequency at which a motor unit is activated, termed rate coding. In a submaximal muscle contract and hold, the high threshold fast-twitch fibers will increase firing rates in the ramp phase more than the low threshold units. The frequency of the motor unit firing will drop off during the hold phase, and the high threshold units will cease firing. (B) In a more vigorous contract and hold, the rate coding increases and is maintained further into the contraction by both the high and low threshold motor units (From Sale, D.G.: Influence of exercise and training on motor unit activation. Exercise and Sport Science Reviews. *16*:95–151, 1987).

unit firing will decrease during any continuous muscular contraction, whether vigorous or mild.

Measurement of Motor Unit Activity

The action potential in the muscle can be measured by recording the electrical activity in the muscle through electromyography. Instrumentation for electromyography has been available for many years. An earlier oscilloscope example of an electromyographic recording of four shoulder muscles during four different arm movements is shown in FIGURE 4–12.

The typical recording system for electromyography (EMG) consists of a set of bipolar surface or needle electrodes that are placed above, or a short distance away from the motor point, the site of the entrance of the main nerve to the muscle. An additional electrode is also placed over a bony protuberance to serve as a reference point. As the action potential in the muscle fiber travels down the muscle toward the two electrodes, the electrodes record the action potential due to depolarization and the accompanying change in voltage in the positive direction (2). When the action potential is traveling under the electrodes, the recorded voltage signal passes through zero and then moves in the negative direction as the signal travels away from the electrodes. Consequently, the EMG signal represents a sinusoidal wave, fluctuating between negative and positive. The further away the action potential is from the electrodes, the smaller the signal.

A surface electromyography records a signal that is a sum of the action potentials occurring within its measuring range. Depending on the fiber type and the velocity of contraction, some action potentials occur very quickly. Also, the propagation of the nerve impulse through depolarization may take some time before it reaches the measuring range of the electrode due to the length of the nerve fiber and the muscle (13).

EMG activity does not directly correlate with external force development in the muscle since there are internal components of the muscle, requiring tension development before any external force can be applied. In the early stages of force development in the muscle, there is a latency period, where there is EMG activity but no external force development. This is because the initial tensions developed in the muscle are used to stretch the series elastic component of the muscle first, followed by the development of tension in the contractile unit that is transferred to the bone.

This delay between the onset of the EMG signal and the beginning of tension development in the muscle is termed *electromechanical delay* (EMD). FIGURE 4–13 demonstrates EMD in a quadriceps femoris muscle performing

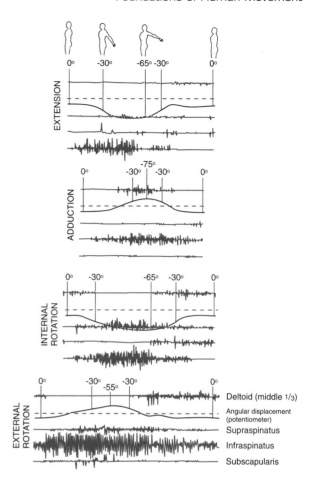

FIGURE 4–12. Electromyographical recording of a muscle(s) during a movement gives indicates whether or not the muscle is being used. In the oscilloscope example above, measurements from four muscles, the deltoid, supraspinatus, infraspinatus, and subscapularis, are recorded for four different arm movements concurrently with angular displacement data. For the movement of arm extension, the subscapularis muscle is the most active, in adduction the infraspinatus is more active, in internal rotation the subscapularis and supraspinatus are most active, and in external rotation, the infraspinatus is most active (From Close, J.D.: Functional Anatomy of the Extremities. Springfield, Ill., Charles C. Thomas Pub., 1973).

an extension movement. This delay, as long as 100 ms, is usually longer in the slow-twitch muscle fibers (14).

The EMG signal is influenced by many different factors, some of which include: the muscle fiber diameter, the electrode placement and interface with the skin, and the amount of tissue in the area. The actual activity in the muscle is also influenced by recruitment and frequency of firing of the motor units (13). Motion artifacts or interference from other noise sources such as the cable or other instrumentation can influence the EMG signal. For

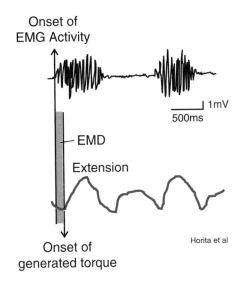

Onset of
EMG Activity

1mV
500ms

EMD

Extension

Horita et al

Onset of
generated torque

FIGURE 4–13. The electrical activity in the muscle will begin prior to the development of any external tension generation. This is known as the electromechanical delay (EMD) or latency period (Horita, T., and Ishiko, T.: Relationships between muscle lactate accumulation and surface EMG during isokinetic contractions in man. European Journal of Applied Physiology. 56:18–23, 1987).

this reason, most EMG signals are filtered to remove 8 to 20 Hz of the lower end of the signal so that these "noise" interferences can be removed.

Since the EMG signal is an alternating voltage between positive and negative, it is common to linearly rectify the EMG signal, creating an EMG record that is all positive by flipping the negative signals around. Many EMG research studies report EMG data in this fashion, allowing the visual examination of an EMG record to estimate when a muscle becomes active, how long it is active, and if there is a little or a lot of contractile activity in the muscle. The amount of contractile activity in the muscle is estimated by observing the width or amplitude of the EMG signal that only allows determination of whether a muscle is more active or less active.

To produce a quantitative number from the EMG signal, the rectified and filtered signal is integrated (IEMG), approximating the area under the EMG signal curve. The millivolt (mV) units of the raw EMG become mV∗seconds after the signal is integrated. The signal can be integrated over the total time the muscle is active, over regular intervals of time, after which the signal is set to zero and the integration repeated, or over intervals that are determined by resetting after specific voltages are achieved in the signal. This allows one to quantify the

activity in the muscle over specified time periods, or in conjunction with outside events recorded at the same time. FIGURE 4–14 demonstrates the differences in EMG signals that are raw, rectified, and integrated over a designated time period.

By analyzing the EMG signals of the various combinations of muscles, we have come to know a great deal about muscle function under various conditions. It has allowed us to identify when muscles are active in certain human movements or sport skills. For example, EMG studies have pointed out that the agonist responsible for a limb movement will typically create its muscular activity at the start of the movement and will often be silent or minimally active in the middle of the limb movement.

It has also been demonstrated that the antagonistic muscle becomes active at the end of a limb movement to create a decelerating force (1). Depending on the type of movement, there may be a co-contraction at the end of a joint movement when agonist and antagonist are simultaneously active (15). We have also learned through examination of EMG output that the muscle activity is less in an eccentric contraction, and that there is usually some co-contraction of the agonists and antagonists during most support phases (9). Additionally, rest periods in the EMG output have been identified that occur more frequently in the elite athlete. EMG analysis has allowed us to assess whether a muscle is contributing to a specific movement, for remember that even though a muscle may be anatomically placed to create a movement, it may not necessarily participate in the creation of that movement. In fact, it may actually create an opposite movement in the same segment, or a different movement in an adjacent segment. Two examples of the use of EMG data are presented in FIGURE 4–15.

Sensory Neurons

The body requires an input system to provide feedback on the condition and changing characteristics of the musculoskeletal system and other body tissues such as the skin. There are biologic sensors that collect information on such events as stretch in the muscle, heat or pressure on the muscle, tension in the muscle, or pain in the extremity. These sensors send information to the spinal cord, where the information is processed and utilized by the central nervous system in the adjustment or initiation of motor output to the muscles.

Foundations of Neural Control

When the sensory information from one of these biologic receptors brings information into the cord, triggering a predictable motor response, it is termed a *reflex*. A simple reflex arc is shown in FIGURE 4–16. There are

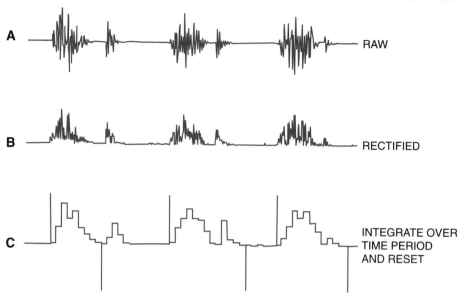

FIGURE 4-14. An electromyography signal can be processed and analyzed a number of different ways. (A) The EMG signal can be left in its original, raw form which shows the positive and negative action potentials as the signal moves toward or away from the electrodes, respectively. (B) The signal can be rectified which flips the negative signals around. This process offers an easier way of determining the activity level in the muscle. (C) The rectified signal can be integrated by measuring the area under the EMG signal curve. In the example shown, the signal is integrated over a designated time period and then reset to zero and repeated. This process eventually leads to quantification of the EMG signal.

reflexes that have sensory information entering and motor information leaving the spinal cord at the same level, creating a monosynaptic reflex arc. An example of this reflex is the stretch or myotactic reflex, stimulated by sensory neurons responding to stretch in the muscle, which in turn initiates an increase in the motor input to the same muscle (16).

Other reflexes using a simple reflex arc are the flexor reflex that initiates a quick withdrawal response after receiving sensory information indicating pain, and the cutaneous reflex, causing relaxation of a muscle after receiving stimuli on the skin in the form of massage and heat. Knowledge of the cutaneous reflex is beneficial for athletic trainers and physical therapists who use heat and massage to create relaxation in a segment or joint.

Reflexes that bring information into the spinal cord and which are processed through both sides and different levels of the spinal cord are termed *propriospinal*. An example of this type of reflex is the crossed extensor reflex that is initiated by receiving, or expecting to receive, a painful stimulus, such as stepping on something. This sensory information is processed in the spinal cord by creating a flexor and withdrawal response in the

pained limb, and an increase or excitation in the extension muscles of the other limb.

Another propriospinal reflex is the tonic neck reflex, stimulated by movements of the head that create a motor response in the arms. When the head is rotated to the left, this reflex stimulates an asymmetric response of extension of the same-side arm (left) and a flexion of the opposite arm (right). Also, when the head flexes or extends, this reflex initiates a flexion or extension of the arms, respectively.

Another type of reflex is the supraspinal reflex, which brings information into the cord and processes it in the brain. The result is a motor response. The labyrinthine righting reflex is an example of this type of reflex. This reflex is stimulated by body positioning that is leaning, upside down, or falling out of an upright posture. The response from the upper centers is to stimulate a motor response from the neck and limbs in order to maintain or move to an upright position. This reflex is a complex reflex involving many different levels of the spinal cord as well as the upper centers of the nervous system. Examples of these various reflex actions are presented in FIGURE 4-17.

A

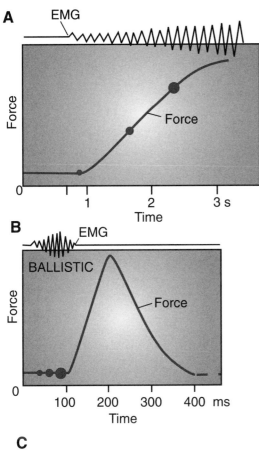

B

C

FIGURE 4–15. (A, B) Electromyography has been used to demonstrate how muscle activity is related to force development, as presented by Sale (Sale, D.G.: Neural adaptation to resistance training. Medicine and Science in Sport and Exercise. 20:S135–145, 1988). (C) EMG measurements have also provided us with information concerning a muscle's response to an endurance task at different output levels (Enoka, R.M.: Neuromechanical Basis of Kinesiology. Champaign, IL., Human Kinetics Books, p. 241.1986).

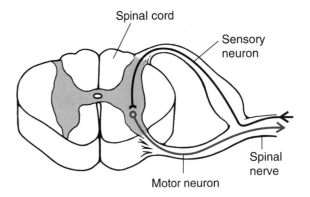

FIGURE 4–16. A simple reflex arc is shown in this illustration. Sensory information from receptors are brought into the cord where they initiate a motor response sent back out to the extremities. The stretch reflex is an example of a reflex arc which sends sensory information into the cord in response to stretch of the muscle, which in turn sends back motor stimulation to the same muscle, causing a contraction.

Sensory Receptors

The main sensory receptors for the musculoskeletal system are the proprioceptors, which transform mechanical distortion in the muscle or joint, such as any change in joint position, muscle length, or muscle tension, into nerve impulses that enter the spinal cord and stimulate a motor response (16).

The muscle spindle is a proprioceptor, found in higher abundances in the belly of the muscle, lying parallel to the muscle fibers, and actually connecting into the fascicles via connective tissue (FIGURE 4–18). The fibers of the muscle spindle are termed intrafusal as compared to muscle fibers that are termed extrafusal. The intrafusal fibers of the spindle are contained within a capsule, forming a spindle shape, hence the name muscle spindle. Some muscles such as the eye, hand, and upper back have hundreds of spindles in them, while other muscles, such as the latissimus dorsi and other shoulder muscles may have only a handful of spindles (16). Every muscle has some spindles; however, the muscle spindle is absent from some of the Type IIb fast-twitch glycolytic muscle fibers contained within some muscles.

Within each spindle capsule there may be as many as 12 intrafusal fibers that can be either of two types: nuclear bag or nuclear chain (16). Both types of fibers have non-contractile centers that contain the nuclei of the fiber, and contractile ends that can be innervated, creating shortening upon receipt of motor input. Also, both fiber types have sensory nerve fibers exiting from the equatorial

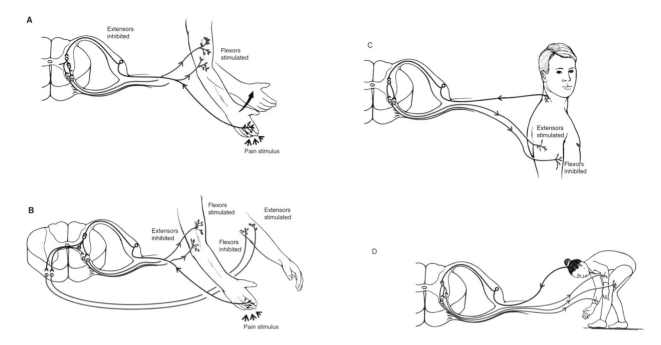

FIGURE 4–17. A reflex is a motor response developed in the central nervous system after sensory input is received. Examples of different reflex actions are shown above. (A) The flexor reflex is triggered by sensory information registering pain which facilitates a quick flexor withdrawal response away from the pain source. (B) The crossed extensor reflex is also initiated by pain and works with the flexor reflex to create flexion on the stimulated limb and extension on the contralateral limb. (C) The tonic neck reflex is stimulated by head movements and creates flexion or extension of the arms depending on the direction of the neck movement. (D) The labyrinthine righting reflex is stimulated by body positioning and causes movements of the limbs and neck in order to resume a balanced, upright posture.

region or center of the fibers, taking information into the system through the dorsal root of the spinal cord.

The nuclear bag fiber has a large clustering of nuclei in the center of the fiber. It is also thicker, and its fibers connect to the capsule and to the actual connective tissue of the muscle fiber itself. The contractile poles of the nuclear bag fiber are innervated by a gamma or fusimotor motoneuron that is smaller than the alpha motoneuron that innervates the muscle fibers. These neurons are located in the ventral horn of the spinal column and are intermingled with the alpha motoneurons. Each gamma motoneuron innervates multiple muscle spindles. The gamma motoneuron to the nuclear bag fiber is often termed the dynamic gamma motoneuron.

Exiting from the equatorial region of the nuclear bag fiber is the Type Ia or primary afferent motoneuron. There is one Type Ia afferent motoneuron per muscle spindle that sends information into the spinal cord via the dorsal horn of the spinal cord. The cell bodies, or soma, of these sensory neurons lie just outside the spinal cord, are large in diameter, and discharge in response to stretch in the muscle.

The nuclear chain fiber is smaller, with the nuclei arranged in rows in the equatorial region. The nuclear chain fiber does not connect into the actual muscle fiber and only makes connection with the spindle capsule. The ends of the nuclear chain fiber are also contractile, and are innervated by a gamma motoneuron sometimes referred to as the static gamma or static fusimotor motoneuron.

The nuclear chain fiber has two different types of sensory neurons exiting from the noncontractile portion of the fiber. In the middle of the nuclear chain fiber, the Type Ia primary afferent sensory neuron exits, taking information into the spinal cord. This sensory neuron is identical to the one leaving from the nuclear bag fiber.

From the polar ends of the nuclear chain fiber, a Type II or secondary afferent sensory neuron exits, travelling into the dorsal horn of the spinal cord. This sensory neuron is medium sized and is stimulated by stretch in the muscle, responding at a higher threshold of stretch than the Type I sensory neuron. There are generally 1 or 2 Type II sensory neurons per muscle spindle; however,

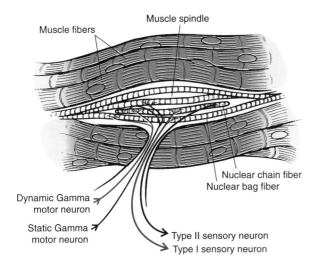

Muscle fibers
Muscle spindle
Nuclear chain fiber
Nuclear bag fiber
Dynamic Gamma motor neuron
Static Gamma motor neuron
Type II sensory neuron
Type I sensory neuron

FIGURE 4–18. The muscle spindle lies parallel with the muscle fibers. Within each spindle capsule are the actual spindle fibers which can be one of two different types: nuclear chain or nuclear bag fibers. Both of these fibers have contractile ends that are innervated by gamma motoneurons. Sensory information responding to stretch leaves the middle portion of both the chain and bag fibers through the Type Ia sensory neuron, and from the ends of the nuclear chain fibers via the Type II sensory neuron.

some muscle spindles and even some muscles (10 to 20%) have no Type II sensory neurons (16).

The muscle spindle responds to stretch of the muscle fiber because the muscle stretch also stretches and elongates the nucleated middle portion of the muscle spindle fibers, containing the Type I spiral, and the Type II flower spray sensory neuron endings. The Type I sensory neurons are more sensitive to stretch, or change in length, in the muscle because of a lower threshold, so they are the first to respond to the muscle stretch. As the muscle is stretched, the Type I generates a sensory impulse. If the muscle is stretched more rapidly, the impulses increase proportionally to the rate of stretch and fire at a higher rate. At the end of the stretch, when there is a pause in the motion, the Type I sensory impulse firing level will drop to a lower level and then fire at a constant rate. This represents a static response of muscle stretch in a fixed position.

When a stretch is imposed on the muscle, the Type I sensory neuron sends impulses into the spinal cord and connects with interneurons, generating an excitatory local-graded potential that is sent back to the muscle being stretched. If the stretch is vigorous enough, a local-graded impulse will be sent back to the same muscle with sufficient magnitude to initiate a contraction via the alpha motoneurons. This reflex arc is known as the stretch

reflex and is characterized by a quick muscular contraction following a rapid stretch of that same muscle group. The Type Ia loop is illustrated in FIGURE 4–19. It is also termed *autogenic facilitation* because of the facilitation of the alpha motoneurons of the same muscle (1). The stretch reflex primarily recruits slow-twitch muscle fibers.

The information coming into the spinal cord via the Type I sensory neuron is also sent up to the cerebellum and cerebral sensory areas to be used for feedback on muscle length and velocity. Additional connections are made in the spinal cord with inhibitory interneurons, creating a reciprocal inhibition, or relaxation of the antagonistic muscles (6). Other excitatory interneuron connections are made with the alpha motoneurons of synergistic muscles to facilitate their muscle activity along with the agonist.

When the Type II, or secondary afferent neuron is stimulated, it has a different response than the Type I sensory neuron. It produces a sensory input in response to stretch or change in length in the muscle, and is a good feedback indicator of the actual length in the muscle because its sensory impulses do not diminish when the muscle is held in a stationary position.

The Type II sensory information enters the dorsal root of the spinal cord and produces an inhibitory local-graded potential in the same muscle that was stretched. This lowers the excitability in the muscle and can cause relaxation of the muscle if the stretch is substantial enough.

The innervation of the ends of the spindle fibers by the gamma motoneuron alters the response of the muscle spindle considerably. The first important effect of gamma innervation of the spindle is that it does not allow the spindle discharge to cease when a muscle is shortened. If the muscle shortened with no gamma innervation of the ends of the spindle, the spindle activity would be silenced due to the removal of the external stretch on the muscle. The gamma motoneurons create a contraction at the ends of the spindle fibers, elongating the middle portion of the spindle in the same way an external stretch of the muscle would (16). This is called gamma bias or resetting of the muscle spindle.

The second major input from the gamma motoneuron innervation of the muscle spindle is an indirect enhancement of the motor impulses being sent to the muscle via the alpha neuron pathways. This adds to the impulses coming down through the system, alters the gain, and increases the potential for full activation via the alpha pathways. It is a main contributor to coordinating the output and patterning of the alpha motoneurons.

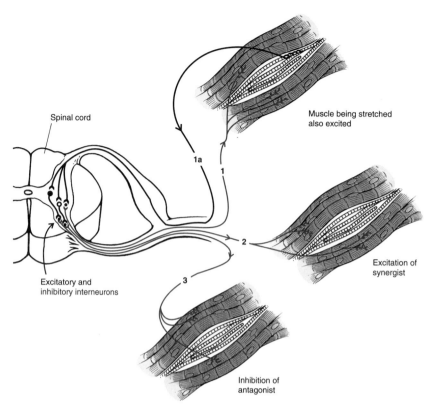

FIGURE 4–19. The Type Ia loop is initiated by a stretch of the muscle. Responding proportionally to the rate of stretch, the muscle spindle sends impulses into the spinal cord via the Type Ia sensory neuron. Within the cord, there are connections made with interneurons that produce a local graded potential which inhibits the antagonistic muscles and excites the synergists and the muscle in which the stretch occurred. This is the typical stretch reflex response, also termed autogenic facilitation.

In anticipation of lifting something heavy, the alpha and gamma motoneurons will establish a certain level of excitability in the system for accommodating the heavy resistance. If the object lifted is much lighter than anticipated, the gamma system will act to reduce the output of the Type I afferent. It will make a quick adjustment in the alpha motoneuron output to the muscle and reduce the number of motor units activated.

Finally, the gamma motoneuron is activated at a lower threshold than the alpha motoneuron, and can therefore initiate responses to postural changes by resetting the spindle and activating the alpha output (17). The afferent pathways, the gamma pathways, and the alpha pathways are all part of the gamma loop, shown in FIGURE 4–20.

Another important proprioceptor significantly influencing muscular action is the Golgi tendon organ (GTO), which monitors force or tension in the muscle. As illustrated in FIGURE 4–21, the GTO is located at the mus-

culoskeletal junction. It is a spindle-shaped collection of collagen fascicles, surrounded by a capsule that continues inside the fascicles to create compartments. The collagen fibers of the GTO are connected directly to extrafusal fibers from the muscles (16).

There are two sensory neurons exiting from a site between the collagen fascicles. When the collagen is compressed through a stretch or contraction of the muscle fibers, the Type Ib nerve endings of the GTO generate a sensory impulse proportional to the amount of deformation created in them. The response to the load, and the rate of change in the load, is linear. Several muscle fibers insert in one GTO and any tension generated in any of the muscles will generate a response in the GTO (16).

In a stretch of the muscle, the tension in the individual GTO is generated along with all other GTOs in the tendon. Consequently, the GTO response is more sensitive in a tension situation than in a stretch. This is because the

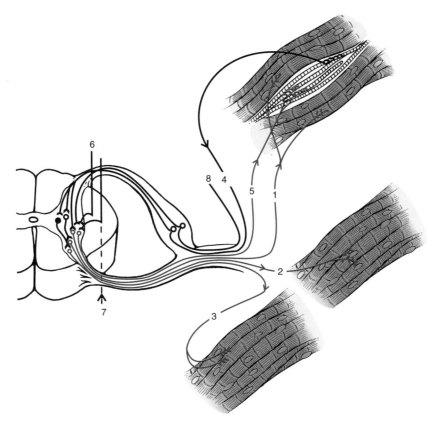

FIGURE 4–20. The Type Ia loop in which information is sent from the spindle (#4), causing inhibition (#3) and excitation of synergists and agonists (#2,#1) is facilitated by input from the gamma motoneuron (#5), which initiates a contraction of the ends of the spindle fibers, creating an "internal stretch" of the spindle fibers. The gamma motoneuron receives input via the upper centers or other interneurons in the cord (#6,7,8).

GTO measures load bearing in series with the muscle fibers, but is parallel to the tension developed in the passive elements during stretch (18). Thus, a lower threshold is present in the contraction as compared to a stretch.

The GTO generates an inhibitory local graded potential in the spinal cord known as the inverse stretch reflex. If the graded potential is sufficient, relaxation, or autogenic inhibition, will be produced in the muscle fibers connected in series with the GTO stimulated. The alpha motoneuron output to muscles undergoing a high velocity stretch or producing a high resistance output is reduced.

The GTO response is seen as a critical determinant to maximum lifting levels in weight training. It also may be responsible for uncoordinated action in the inexperienced performer by shutting down a muscle at the inappropriate time in the sequence. This is a simplistic description of the impact of GTO input. The GTO response can be inhibited in extreme circumstances, such as when a mother lifts a car off a child.

There is limited information on the sensory neuron input from the joint receptors placed in and around the synovial joints (FIGURE 4–22). One such receptor, the ruffini endings, is located in the joint capsule and responds to change in joint position and velocity of movement of the joint (19). The pacinian corpuscle is another joint receptor located in the capsule and connective tissue that responds to pressure created by the muscles, as well as to pain within the joint (19). These joint receptors, as well as other receptors in the ligaments and tendons, provide continuous input to the nervous system about the current conditions in and around the joint.

Effect of Training on Neurological Input and Output

When training the muscular system, there is a neural adaptation occurring which modifies the activation levels and patterns of the neural input to the muscle. In strength

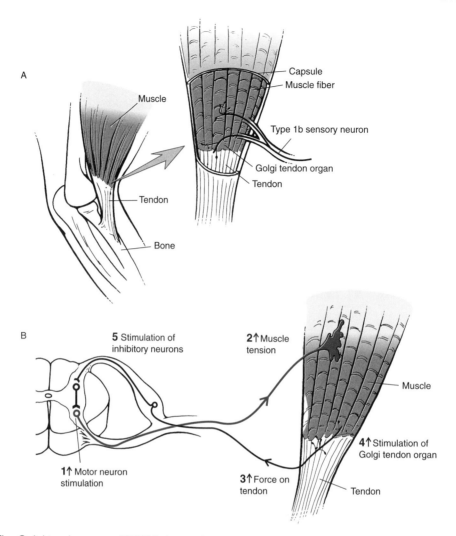

FIGURE 4–21. (A) The Golgi tendon organ (GTO) is located at the muscle-tendon junction. (B) When there is tension developed at this site, the GTO sends information into the spinal cord via Type Ib sensory neurons. The sensory input from the GTO facilitates a relaxation of the muscle via stimulation of inhibitory interneurons. This response is known as the inverse stretch reflex or autogenic inhibition.

training, for example, significant strength gains can be demonstrated after approximately 4 weeks of training. This strength gain is not due to an increase in muscle fiber size, but is rather a learning effect in which neural adaptation has occurred (20).

Stretching the Muscle

Neurological Influences

The effect of the neural adaptation is an improved muscular contraction of higher quality through coordination of motor unit activation. The neural input to the muscle, as a consequence of maximal voluntary contractions, is increased to the agonists and synergists, and there is greater inhibition of the antagonists. This neural adaptation or learning effect levels off after about 4 to 5 weeks of training, and increases in strength beyond this point are usually due to structural changes and physical increases in the cross section of the muscle. The influence of training on both the electromechanical delay and the amount of EMG activity is presented from the work of Hakkinen and Komi (21) in FIGURE 4–23.

Specificity of training is important for enhancement of neural input to the muscles. If one limb is trained at a time, greater force production is obtained, with more neural input to the muscles of that limb than if two limbs are trained at once. The loss of both force and neural

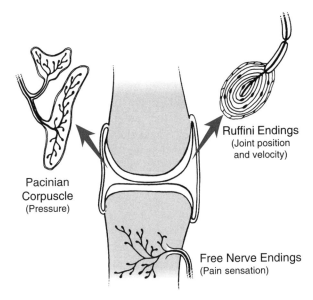

FIGURE 4–22. There are a number of other sensory receptors that send information into the central nervous system. Located in the joint capsules and connective tissue is the pacinian corpuscle, which responds to pressure, and the ruffini endings, which respond to changes in joint position. There are also free nerve endings around the joints that create the pain sensations.

input to the muscles through bilateral training of both limbs is termed bilateral deficit (3,10). In fact, training of one limb will even neurologically enhance the activity and increase the voluntary strength in the other limb.

When working with athletes who use the limbs asymmetrically, as in running or throwing, the trainer should incorporate some unilateral limb movements into the conditioning program. Those participants in sports or activities that use both limbs together, such as weightlifters, should train bilaterally.

Specificity of training also determines the fiber type that is enhanced and developed. Through resistive training, the Type II fibers can be enhanced through reduction in central inhibition and increased neural facilitation. This may serve to resist fatigue in short-term, high-intensity exercise in which the fatigue is brought on by the inability to maintain optimal nerve activation.

Even a short-term warm-up (5 to 10 minutes) preceding an event or performance will influence neural input by increasing the motor unit activity (22). Another factor enhancing the neural input into the muscle is the use of an antagonistic muscle contraction that precedes the contraction of the agonist, such as seen in preparatory movements in a skill (backswing, lowering, etc.). This diminishes the inhibitory input to the agonist and allows for more neural input and activation in the agonist contraction.

If a stretch of a muscle precedes a contraction of the same muscle, there will be some neural stimulation of the muscle via the stretch reflex arc. Athletes who must produce power, such as jumpers and sprinters, have been

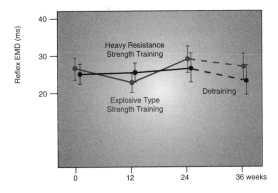

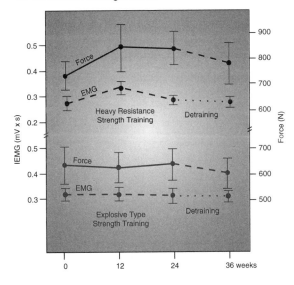

FIGURE 4–23. (A) Explosive strength training has been shown to decrease the electromechanical delay (EMD) in the muscle contraction after 12 weeks of training. However, the EMD increases again if training continues to 24 weeks, and drops off slightly with detraining. The influence of heavy resistance training on EMD is negligible. (B) The IEMG increases in the early weeks during heavy resistance training but not with explosive training. It is believed that some neural adaptation occurs in the early stages of specific types of resistance training, which facilitates an early increase in force production (From Hakkinen, K., and Komi, P.V. Training-induced changes in neuromuscular performance under voluntary and reflex conditions. European Journal of Applied Physiology. 55:147–155, 1986).

shown to have excitable systems where the reflex potentiation is high (22).

In summary, the neural input to the muscle can be enhanced through training that can increase the amount of active motor units contributing, alter the pattern of firing, and increase the reflex potentiation of the system (23). Likewise, immobilization of the muscle can create the opposite response by lowering neural input to the muscle and decreasing reflex potentiation.

Flexibility Techniques

Flexibility is an essential component of physical fitness. Increased flexibility has been shown to improve efficiency of movement, reduce the incidence of muscle strain, improve posture, and generally improve skill in certain sports. Flexibility, as it will be used in this section, is defined as the terminal range of motion of a segment. This can be obtained actively through some voluntary contraction of an agonist creating the joint movement, or passively, as when the agonist muscles are relaxed as the segment is moved through a range of motion by an external force such as another person or object (24,25).

There are many different components contributing to one's flexibility or lack thereof. First, joint structure is a determinant of flexibility, and will limit the range of motion in some of the joints in the body and produce the termination or end point of the movement. This is true in a joint such as the elbow, where the movement of extension is terminated due to bony contact between the olecranon process and fossa on the back side of the joint. A person who can hyperextend the forearm at the elbow is not one who is exceptionally flexible, but is someone who has either a deep olecranon fossa or a small olecranon process. Bony restrictions to range of motion are present in a variety of joints in the body, but this type of restriction is not the main mechanism limiting or enhancing joint flexibility.

Soft tissue around the joint is another factor contributing to flexibility. As a joint nears the ends of the range of motion, the soft tissue of one segment will be compressed by the soft tissue of the adjacent segment. This compression between adjacent tissue components will eventually contribute to the termination of the range of motion.

This means that obese individuals and individuals with greater amounts of muscle mass or hypertrophy will usually demonstrate lower levels of flexibility. However, the hypertrophied individual can obtain good flexibility in a joint by applying a larger force at the end of the range of motion, which can compress the restrictive soft tissue

to a greater degree. It is the obese individual who lacks strength who will definitely be limited in flexibility because of an inability to produce the force necessary to achieve the greater range of motion.

Ligaments restrict range of motion and flexibility by offering maximal support at the end of the range of motion. In the knee joint, for example, the ligaments of the knee terminate the extension of the leg. An individual who can hyperextend the knees is commonly called "double jointed," but actually has slightly longer ligaments that allow more joint motion before the movement is terminated.

The main factors influencing flexibility are the actual physical length of the antagonistic muscle(s) and the level of neurological innervation occurring in a muscle being stretched. Both of these factors can be influenced by specific types of flexibility training and deserve attention.

Neurological Restrictions   When a muscle is stretched, three neurological mechanisms restrict the range of motion. First, the Type Ia, primary afferent sensory neuron initiates the stretch reflex, creating increased muscular activity through alpha motoneuron innervation. This response is proportional to the rate of stretch; thus, the faster the stretch, the more the same muscle wants to contract. After the stretch is completed, the Type Ia sensory neurons will drop to a lower firing level, reducing the level of motoneuron activation or resistance in the muscle. A flexibility technique that would enhance this response is ballistic stretching, in which the segments are bounced to achieve the terminal range of motion. This type of stretching is not recommended for the improvement of flexibility because of the stimulation of the Type Ia neurons and the increase in the resistance in the muscle. However, ballistic stretching is a component of many common movements such as a preparatory wind-up in baseball or the end of the follow-through in a kick.

A better stretching technique for the improvement of range of motion would be static stretching, in which the limb is moved slightly beyond the terminal position slowly and then maintained in that position for at least 30 seconds (26). By moving the limb slowly, the response of the Type Ia sensory neuron is decreased, and by holding the position at the end, the Type Ia input has been significantly reduced, allowing minimal interference to the joint movement.

The second neurological factor in flexibility is the input of the Type II, secondary afferent sensory neuron. This sensory input will be facilitated by change in length in the muscle, producing relaxation of the muscle being stretched through the generation of a local graded potential that is also inhibitory.

The third neurological input is the Golgi tendon organ through the inverse stretch reflex. The GTO response occurs more in the active stretch when the limb is voluntarily moving into the terminal joint position. If the stretch is extreme, the GTO will initiate the inverse stretch reflex that may relax the muscle being stretched, or the muscle being contracted to produce the stretch, depending on the level of the tensions generated in each GTO.

Structural Restrictions To stretch a muscle, the primary restriction to the stretch is found in the connective tissue and tendons in and around the muscle (27,28). This includes the fascia, epimysium, perimysium, endomysium, and tendons. The actual muscle fibers do not play a significant role in the elongation of a muscle through flexibility training. To understand how the connective tissue responds to a stretch, the stress/strain characteristics of the muscle unit must be examined.

When a stretch is first imposed, the muscle creates a linear response to the load through elongation in all parts of the muscle. This is the elastic phase of the stretch, creating an elongation of the muscle in response to external stretch. If the external load is removed from the muscle during this phase of stretching, it will return to its original length within a few hours, and no residual or long-term increase in muscle length will remain. The stretching techniques working the elastic response of the muscle are very common and include short duration, repetitive joint movements. These stretches, usually preceding an activity, will produce some increase in muscle length for use in the practice or game, but will not influence any long-term improvement in flexibility.

If a muscle is put in a terminal position and maintained in the position for an extended period of time, the tissues enter the plastic region of response to the load, whereby they elongate and experience plastic deformation of the tissue (29). This plastic deformation is a long-term increase in the length of the muscle and will carry over from day to day (30). A model describing the behavior of the elastic and plastic elements acting in a stretch is presented in FIGURE 4–24.

To create increases in length due to plastic or long-term elongation, the muscle should be stretched while it is warm and the stretch should be maintained for a long period of time using a low load (31,32). Thus, to gain long-term benefits from stretching, the stretch should occur after a practice or work-out, and individual stretches should be held in the terminal joint positions for an extended period of time.

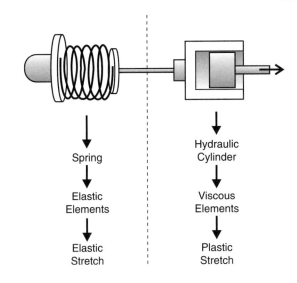

FIGURE 4–24. When a repetitive stretch of short duration is applied to the muscle, the connective tissue and the muscle respond like a spring, with a short-term elongation of the tissue, but a return to the original length after a short time period. In a long-term, sustained stretch, especially while the muscle is warm, the tissues behave more hydraulically, as a long-term deformation of the tissues takes place (From Sapega, A.A., Quedenfeld, T.C., Moyer, R.A., and Butler, R.A.: Biophysical factors in range of motion exercises. Physician and Sports Medicine. *9*:57–64, 1981).

It has been shown that the cooling of a warm muscle enhances the permanent elongation of the tissues in that muscle. It is recommended that the joint positions be held for at least 30 seconds and ideally up to a minute. However, in muscles that are inflexible and require extra attention, stretching should occur for longer time periods of 6 to 10 minutes (33). Stretching should not take place with pain to avoid experiencing any significant tissue damage.

Proprioceptive Neuromuscular Facilitation The enhancement of permanent elongation of the muscle tissue can be attained through the use of a static stretch. Another technique, proprioceptive neuromuscular facilitation (PNF), can be used to stimulate relaxation of the muscle being stretched so that the joint can be moved through a greater range of motion (34). This technique, used in rehabilitation settings, can also be put to good use with athletes or individuals who have limited flexibility in certain muscle groups, such as the hamstrings (35).

PNF incorporates various combination sequences using relaxation and contraction of the muscles being stretched. A simple PNF exercise would be to passively move an individual's limb into the terminal range of motion, have him or her contract back isometrically against the manual resistance applied by a partner, and then relax and move further into the stretch. By repeating this cycle, a significant increase in the terminal range of motion can be achieved (36). This procedure increases the range of motion because the input from the Type Ia afferent from the muscle spindle is reduced, due to the resetting of the spindle (37).

The process can be enhanced even more if a contraction of the agonist occurs at the end of the range of motion. This sets up an increase in the relaxation of the antagonist or the muscle being stretched. For example, passively move the foot into plantar flexion to stretch the dorsiflexors. Contract the dorsiflexors isometrically against a resistance applied by a partner on the top of the foot. Move the foot further into plantar flexion and then contract the plantar flexors. Both of these techniques will optimally produce the greatest increase in the range of motion. Examples of PNF exercises for the muscles of the hip and shoulder joint are presented in FIGURE 4–25.

Plyometric Training

The purpose of plyometrics is to improve the velocity of a performance. Plyometric training has been very effective in increasing power output from athletes in sports such as volleyball, basketball, high jumping, long jumping, throwing, and sprinting. Plyometrics builds on the ideas from specificity of training, whereby a muscle trained at higher velocities will improve at those velocities.

The plyometric exercise consists of rapidly stretching a muscle and immediately following with a contraction of the same muscle (38). Plyometric exercises improve power output in the muscle through facilitation of the

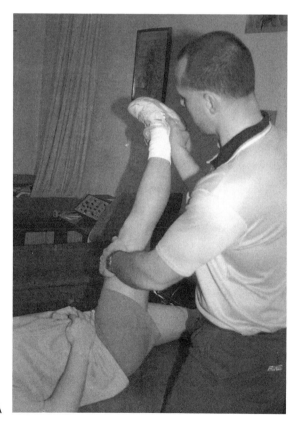
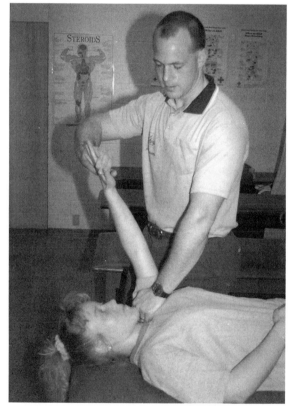

A   B

FIGURE 4–25. Proprioceptive neuromuscular facilitation (PNF) stretching techniques are effective for increasing flexibility in areas such as the hip and shoulder regions. Examples of PNF exercises for the muscles of the hip and shoulder joints are provided. (A) At the hip joint, the thigh is moved through a diagonal pattern, with manual resistance being applied at the foot and the thigh. (B) In the shoulder, the arm is moved into flexion, with manual resistance offered at the hand.

neurological input to the muscle and through increased muscle tension generated in the elastic components of the muscle. An examination of both these factors in greater detail follows.

Neurological Influences

The neurological basis for plyometrics is the input from the stretch reflex via the Type Ia sensory neuron. If the muscle is rapidly stretched, there is excitation of the alpha motoneurons, contracting that same muscle. This excitation is increased with the velocity of the stretch and is maximum at the conclusion of a rapid stretch, after which the excitation levels drop. Thus, if a muscle can be rapidly stretched and immediately contracted with no pause at the end of the stretch, there will be maximum facilitation via this reflex loop. If an individual pauses at the end of the stretch, this myoneural input will be greatly diminished. The myoelectric enhancement of the muscle being stretched accounts for approximately 25 to 30% of the increase of the force output in the plyometric, stretch-contract sequence (39).

Structural Influences

The factor accounting for the majority of the increases in output (70 to 75%) as a consequence of plyometric exercise is the restitution of elastic energy in the muscle (39). When a muscle is stretched, elastic potential energy is stored in the connective tissue and tendon, as well as in the cross-bridges as they are rotated back with the stretch (40). With a vigorous short-term stretch, there is maximal recovery of the elastic potential energy returned to the succeeding contraction of that same muscle. The net result of this short-range prestretch with a small time period between the stretch and the contraction is that larger forces can be produced for any given velocity, enhancing the power output of the system (41). Implementation of this technique suggests that a quick stretch through a limited range of motion should be followed immediately with a vigorous contraction of the same muscle.

Plyometric Examples

A plyometric exercise program includes a series of exercises imposing a rapid stretch followed by a vigorous contraction. Since the muscle is undergoing a vigorous eccentric contraction, attention should be given to the number of exercises and the load imposed through the eccentric contraction (42,43). It is suggested that plyometric exercises be done on yielding surfaces, and not more than 2 days a week. Injury rates will be higher in the use of plyometric training if these factors are not taken into account. Further, plyometric training should be used very conservatively when the participants lack strength in the muscles being trained. A strength base should be developed first. It is suggested that an individual be able to squat 60% of body weight five times in 5 seconds before beginning plyometrics (44). This is done to see if eccentric and concentric muscle actions can be reversed quickly.

Lower extremity plyometric exercises include activities such as single-leg bounds, depth jumps from various heights, stair hopping, double-leg speed hops, split jumps, bench jumps, and quick counter-movement jumping. The height from which the plyometric jump is performed is a very important consideration. Heights can range from 0.25 to 1.5 meters, and should be selected based on the fitness level of the participant. A height is too high if a quick, vigorous rebound cannot be achieved shortly after landing.

Plyometric exercises can be done 1 to 2 times per week by a conditioned athlete. A sample plyometric work-out may include 3 to 5 low intensity exercises (10 to 20 repetitions) such as jumping in place or double-leg hops; 3 to 4 moderate intensity exercises (5 to 10 repetitions) including single-leg hops, double-leg hops over a hurdle, or bounding; and 2 to 3 high-intensity exercises, including depth jumping (5 to 10 repetitions). In the beginning, the height of the box for depth jumping should be limited to avoid injury, since the amount of force to be absorbed and controlled will increase with each height increase.

Upper extremity activities can best be implemented with surgical tubing or material that can be stretched. The muscle can be pulled into a stretch by the surgical tubing, after which the muscle can contract against the resistance offered by the tubing. For example, hold surgical tubing in a diagonal position across the back and simulate a throwing motion with the right hand while holding the left hand in place. The arm will generate a movement against the surgical tube resistance and then be drawn back into a quick stretch by the tension generated in the tubing. These resistive tubes or straps can be purchased in varying resistances, offering compatibility with a variety of different strength levels.

Other forms of upper extremity plyometrics include catching a medicine ball and immediately throwing it. This puts a rapid stretch on the muscle in the catch that is followed by a concentric contraction of the same muscles in the throw. See FIGURE 4–26 for specific plyometric examples.

A

B

C

D

FIGURE 4–26. Plyometric exercises can be developed for any sport or region of the body by incorporating a stretch-contract cycle into an exercise. Examples of plyometric exercises for the lower extremity include: (A) bounding and (B) depth jumps. For the upper extremity, (C) the use of surgical tubing and (D) medicine ball throws are good exercises.

Summary

The nervous system controls and monitors human movement by transmitting and receiving signals through an extensive neural network. The central nervous system, consisting of the brain and the spinal cord, works with the peripheral nervous system via 31 pairs of spinal nerves that lie outside the spinal cord. The main signal transmitter of the nervous system is the motoneuron, which carries the impulse to the muscle.

The nerve impulse travels to the muscle as an action potential, and when it reaches the muscle, a similar action potential is developed in the muscle, which eventually initiates the shortening of the sarcomere. The actual tension in the muscle is determined by the number of motor units actively stimulated at one time.

Sensory neurons play an important role in the nervous system by providing feedback on the characteristics of the muscle or other tissues. When a sensory neuron brings information into the spinal cord and initiates a motor response, it is termed a reflex. The main sensory neurons for the musculoskeletal system are the proprioceptors. One propticeptor, the muscle spindle, brings information into the spinal cord about any change in the muscle length or velocity of a muscle stretch. Another important proprioceptor is the Golgi tendon organ (GTO), which responds to tension in the muscle.

Flexibility, an important component of fitness, is influenced by a neurological restriction to stretching that is produced by the proprioceptive input from the muscle spindle. Another area of training that uses the neurological input from the sensory neurons is plyometrics. A plyometric exercise is one that involves a rapid stretch of a muscle that is immediately followed by a contraction of the same muscle.

Review Questions

1. Why is a quick countermovement jump more effective than a deep squat jump?
2. Describe a set of plyometric exercises for the following sports:
   a. volleyball
   b. basketball
   c. shot put
   d. throwing
   e. rowing
3. Describe a progressive set of plyometric exercises for the lower extremity.
4. Describe a PNF exercise for each of the following:
   a. ankle sprain
   b. tight hamstrings
   c. shoulder joint
   d. hip joint
   e. wrist joint
5. Describe how the muscle spindle negatively influences one's attempt to increase flexibility. How should one stretch a muscle to minimize this effect?
6. Describe how the muscle spindle positively influences plyometric training. How can this effect be optimized?
7. What is the GTO influence on weight training? On coordination?

Additional Questions

1. Describe how the motor units will be recruited in the following events:
   a. long distance running
   b. sprinting
   c. weight lifting
   d. walking
   e. shot putting
   f. isometric contraction—constant load
   g. ballistic joint action
2. Describe where the precontraction stretch occurs in each of the following movements/events:
   a. running
   b. throwing
   c. jumping
   d. golfing
3. How does the interneuron control muscle function and activity? Provide an example.
4. Draw out a hypothetical EMG pattern for one of the quadriceps femoris muscles and one of the hamstring muscles for the movement of knee joint extension.
5. Describe all of the muscle and joint sensory activity occurring in each of the following movements and providing biofeedback input into the system.
   a. down phase of a squat exercise
   b. up phase of a squat exercise
   c. landing from a low height
   d. landing from a very high height

Additional Reading

Asmussen, E.: Muscle fatigue. Medicine Science and Sports. *11*:313–321, 1979.

Aura, O., and Komi, P.V.: Coupling time in stretch-shortening cycle: Influence on mechanical efficiency and elastic characteristics of leg extensor muscles. *In* Biomechanics X-A. Edited by B. Jonsson. Champaign, IL, Human Kinetics, 1987, pp. 507–511.

Beaulieu, J.E.: Developing a stretching program. The Physician and Sports Medicine. *9*(11):5–65, 1981.

Bigland-Ritchie, B.: EMG/force relations and fatigue of human voluntary contractions. *In* Exercise and Sport Science Reviews. Edited by D.I. Miller. Philadelphia, Franklin Institute Press, 1981, pp. 75–117.

Bigland-Ritchie, B., Johansson, R., Lippold, O.C.J., and Woods, J.J.: Contractile speed and EMG changes during fatigue of sustained maximal voluntary contractions. Journal of Neurophysiology. *50*(1):313–325, 1983.

Bobbert, M.F., Huijing, P.A., and van Ingen Schau, G.J.: Drop jumping. I. The influence of jumping techniques on the biomechanics of jumping. Medicine and Science in Sports and Exercise. *19*(4):332–338, 1987.

Bobbert, M.F., and van Ingen Schau, G.J. (1988).: Coordination in vertical jumping. Journal of Biomechanics. *21*(3):249–262, 1988.

Close, J.D.: Functional Anatomy of the Extremities. Springfield, Ill., Charles C. Thomas Pub., 1973.

Cracraft, J.D., and Petajan, J.H.: Effect of muscle training on the pattern of firing of single motor units. American Journal of Physical Medicine. *56*:183–194, 1977.

Edgerton, V.R.: Neuromuscular adaptation to power and endurance work. Canadian Journal of Applied Physiology. *56*:296–301, 1976.

Hagberg, M.: Muscular endurance and surface electromyogram in isometric and dynamic exercise. Journal of Applied Physiology. *51*:1–7, 1981.

Hakkinen, K., Alen, M., and Komi, P.V.: Changes in isometric force and relaxation time, electromyographic and muscle fibre characteristics of human skeletal muscle during strength training and detraining. Acta Physiologica Scandinavia. *125*:573–585, 1985.

Henriksson, J., and Bonde-Petersen, F.: Integrated electromyography of quadriceps femoris muscle at different exercise intensities. Journal of Applied Physiology. *36*:218–220, 1974.

Hickson, R.C., et al.: Potential for strength and endurance training to amplify endurance performance. Journal of Applied Physiology. *65*:2285–2290, 1988.

Horita, T., and Ishiko, T.: Relationships between muscle lactate accumulation and surface EMG during isokinetic contractions in man. European Journal of Applied Physiology. *56*:18–23, 1987.

Jorge, M., and Hull, M.L.: Analysis of EMG measurements during bicycle pedalling. Journal of Biomechanics. *19*(9):683–694, 1986.

Kranz, H., et al.: Factors determining the frequency content of the electromyogram. Journal of Applied Physiology. *55*:392–399, 1983.

LaBan, M.M.: Collagen tissue: Implications of its response to stress in vitro. Archives of Physical Medicine and Rehabilitation. *43*:461–466, 1962.

Marieb, E.N.: Human Anatomy and Physiology. Redwood City, CA, The Benjamin/Cummings Publishing Co, 1992.

Moore, M.A., and Hutton, R.S.: Electromyographic investigation of muscle stretching techniques. Medicine Science and Sports Exercise. *12*:322–329, 1980.

Ozguven, N.H., and Berme, N.: An experimental and analytical study of impact forces during human jumping. Journal of Biomechanics. *21*(2):1061–1066, 1988.

Person, R.S., and Kudina, L.P.: Discharge frequency and discharge pattern of human motor units during voluntary contraction of muscle. Electroencephalography and Clinical Neurophysiology. *32*:471–483, 1972.

Sullivan, M.K., Dejulia, J.J., and Worrell, T.W.: Effect of pelvic position and stretching method on hamstring muscle flexibility. Medicine Science and Sports Exercise. *24*:1383–1389, 1992.

Stein, R.B., Oguztoreli, M.N., and Capaday, C.: What is optimized in muscular movements? *In* Human Muscle Power. Edited by N.L. Jones, N. McCartney, and A.J. McComas. Champaign, Ill., Human Kinetics, 1986, pp. 131–150.

Tesch, P.A., et al.: Influence of lactate accumulation of EMG frequency spectrum during repeated concentric contractions. Acta Physiologica Scandinavia. *119*:61–67, 1983.

Winter, D.A.: Biomechanics of Human Movement. New York, Wiley, 1979.

References

1. Enoka, R.M.: Neuromuscular Basis of Kinesiology. Champaign, Ill., Human Kinetics, 1988.
2. Basmajian, J.V.: Muscles Alive. Their Functions Revealed by Electromyography. 4th Ed. Baltimore, Williams and Wilkins, 1978.
3. Moritani, T., and DeVries, H.A.: Neural factors versus hypertrophy in the time course of muscle strength gain. American Journal of Physical Medicine. 58(3):115–130, 1979.
4. Howald, H.: Training-induced morphological and functional changes in skeletal muscle. International Journal of Sports Medicine. 3:1–12, 1982.
5. Burke, R.E.: Motor units: Anatomy, physiology, and functional organization. In Handbook of Physiology. The Nervous System. Motor Control. Edited by J.M. Brookhart and V.B. Mountcastle. Bethesda, Md., American Physiological Society, 1981, pp. 345–422.
6. Hultborn, H., Jankowska, E., and Lindstrom, S.: Recurrent inhibition of interneurons monosynaptically activated from group Ia afferents. Journal of Physiology. 215:613–636, 1971.
7. Hultborn, H.: Convergence on interneurons in the reciprocal Ia inhibitory pathway to motoneurons. Acta Physiologica Scandinavica. 84: Suppl. 375, 1972.
8. Burke, R.E.: The control of muscle force: Motor unit recruitment and firing patterns. In Human Muscle Power. Edited by N.L. Jones, N. McCartney, and A.J. McComas. Champaign, Ill., Human Kinetics, 1986, pp. 97–109.
9. Grimby, L.: Single motor unit discharge during voluntary contraction and locomotion. In Human Muscle Power. Edited by N.L. Jones, N. McCartney, and A.J. McComas. Champaign, Ill., Human Kinetics, 1986, pp. 111–129.
10. Sale, D.G.: Influence of exercise and training on motor unit activation. Exercise and Sport Science Reviews. 16:95–151, 1987.
11. Billeter, R., and Hoppeler, H.: Muscular basis of strength. In Strength and Power in Sport. Edited by P. Komi. Boston, Blackwell Scientific Publications, 1992, pp. 39–63.
12. Bigland-Ritchie, B., et al.: Changes in motoneuron firing rates during sustained maximal voluntary contractions. Journal of Physiology. 340:335–346, 1983.
13. Stern, R.M., Ray, W.J., and Davis, C.M.: Psychophysiological Recording. New York, NY., Oxford Press, 1980.
14. Komi, P.V.: Training of muscle strength and power: Interaction of neuromotoric, hypertrophic, and mechanical factors. International Journal of Sports Medicine. 7:10–15, 1986a.
15. Bouchier, J.P., and Flieger, M.S.: Signal characteristics of EMG: Effects of ballistic forearm flexion practice. In Biomechanics IX-A. Edited by D.A. Winter, et al. Champaign, Ill., Human Kinetics, 1985, pp. 297–301.
16. Smith, J.L.: Fusimotor loop properties and involvement during voluntary movement. In Exercise and Sport Sciences Reviews. Edited by J. Keogh and R.S. Hutton. 4:297–333, 1976.
17. Soderberg, G.L.: Kinesiology: Application to Pathological Motion. Baltimore, Williams & Wilkins, 1986.
18. Jansen, J.K., and Rudford, T.: On the silent period and Golgi tendon organs of the soleus muscle of the cat. Acta Physiologica Scandinavica. 62:364–379, 1964.
19. Newton, R.A.: Joint receptor contributions to reflexive and kinesthetic responses. Physical Therapy. 62(1):23–29, 1982.
20. Sale, D.G.: Neural adaptation to resistance training. Medicine and Science in Sport and Exercise. 20:S135–145, 1988.
21. Hakkinen, K., and Komi, P.V. Training-induced changes in neuromuscular performance under voluntary and reflex conditions. European Journal of Applied Physiology. 55:147–155, 1986.
22. Koceja, D.M., and Kamen, G.: Segmental reflex organization in endurance-trained athletes and untrained subjects. Medicine and Science in Sports and Exercise. 24(2):235–241, 1992.
23. Moritani, T.: Neuromuscular adaptations during the acquisition of muscle strength, power, and motor tasks. Journal of Biomechanics. 26: 95–107, 1993.
24. Sandy, S.P., Wortmann, M., and Blanke, D.: Flexibility training: Ballistic, static, or proprioceptive neuromuscular facilitation? Archives of Physical Medicine and Rehabilitation. 6:132–138, 1982.
25. Wallin, D.V., Ekbom, V., Grahn, R., and Nordenberg, T.: Improvement of muscle flexibility. A comparison between two techniques. American Journal of Sports Medicine. 13:263–268, 1985.
26. Blanke, D.: Flexibility training: Ballistic, static, or proprioceptive neuromuscular facilitation. Archives of Physical Medicine Rehabilitation. 63:261–263, 1982.
27. Garrett, W.E., et al.: Biomechanical comparison of stimulated and nonstimulated skeletal muscle pulled to failure. American Journal of Sports Medicine. 15:448–454, 1987.
28. McHugh, M.P., Magnusson, S.P., Gleim, G.W., and Nicholas, J.A.: Viscoelastic stress relaxation in human skeletal muscle. Medicine Science and Sports Exercise. 24(12):1375–1382, 1992.
29. Taylor, D.C., Dalton, J.D., Seaber, A.V., and Garrett, W.E.. Viscoelastic properties of muscle-tendon units: The biomechanical effects of stretching. American Journal of Sports Medicine. 18:300–309, 1990.
30. Kottke, F.J., Pauley, D.L., and Ptak, R.A. The rationale for prolonged stretching for correction of shortening of connective tissue. Archives of Physical Medicine and Rehabilitation. 47:345–352, 1966.
31. Warren, C.G., Lehmann, J.F., and Koblanski, J.N.: Elongation of rat tail tendon: Effect of load and temperature. Archives of Physical Medicine and Rehabilitation. 52:465–474, 1971.

32. Warren, C.G., Lehmann, J.F., and Koblanski, J.N.: Heat and stretch procedures: An evaluation using rat tail tendon. Archives of Physical Medicine and Rehabilitation. *57:*122–126, 1976.

33. Sapega, A.A., Quedenfeld, T.C., Moyer, R.A., and Butler, R.A.: Biophysical factors in range of motion exercises. Physician and Sports Medicine. *9:*57–64, 1981.

34. Knot, M., and Voss, D.E.: Proprioceptive Neuromuscular Facilitation: Patterns and Techniques. 2nd Ed. New York, Harper and Row, 1968.

35. Osternig, L.R., Robertson, R.N., Troxel, R.K., and Hansen, P.: Differential responses to proprioceptive neuromuscular facilitation (PNF) stretch techniques. Medicine and Science in Sports and Exercise. *22:*106–111, 1990.

36. Etnyre, B.R., and Abraham, L.D.: H-reflex changes during static stretching and two variations of proprioceptive neuromuscular facilitation techniques. Electroencephalography and Clinical Neurophysiology. *63:*174–179, 1986.

37. Hardy, L., and Jones, D.: Dynamic flexibility and proprioceptive neuromuscular facilitation. Research Quarterly. *51:*625–635, 1986.

38. Bedi, J.F., Cresswell, A.G., Engle, T.J., and Nicol, S.M.: Increase in jumping height associated with maximal vertical depth jumps. Research Quarterly for Exercise and Sport. *58*(1):11–15, 1987.

39. Komi, P.V.: The stretch-shortening cycle and human power output. *In* Human Muscle Power. Edited by N.L. Jones, N. McCartney, and A.J. McComas. Champaign, Ill., Human Kinetics, 1986b, pp. 27–42.

40. Asmussen, E., and Bonde-Peterson, F.: Storage of elastic energy in skeletal muscles in man. Acta Physiologica Scandinavia. *91:*385–392, 1974.

41. Bosco, C., Ito, A., Komi, P.V., and Viitasalo, J.T.: Neuromuscular function and mechanical efficiency of human leg extensor muscles during jumping exercises. Acta Physiologica Scandinavia. *114:*543–550, 1982.

42. Chu, D., and Plummer, L.: The language of plyometrics. National Strength and Conditioning Association Journal. *6:*30–31, 1985.

43. Lundin, P.: A review of plyometric training. National Strength and Conditioning Association Journal. *7*(3):69–74, 1985.

44. Chu, D.: Plyometrics: The link between strength and speed. National Strength and Conditioning Association Journal. *5:*20–21, 1983.

Glossary

Action Potential:	An electrical current that travels through the nerve or muscle as the membrane potential changes due to the exchange of ions.
Active Range of Motion:	The degree of motion that occurs between two adjacent segments through voluntary contraction of the agonist.
All-or-None Principle:	The stimulation of a muscle fiber that will cause the action potential to either travel over the whole muscle fiber (activation threshold), or none of the muscle fiber.
Alpha Motoneuron:	An afferent neuron with a large cell body located in or near the spinal cord from which a long axon projects from the spinal cord to the muscle fibers which it innervates.
Asynchronous:	Events that do not occur at the same time. In skeletal muscle contraction, the spacing of the activation of the motor unit.
Autogenic Facilitation:	Internally generated excitation of the alpha motoneurons through stretch or some other input.
Axon:	Neuron process carrying nerve impulses away from the cell body of the neuron. The pathway through which the nerve impulse travels.
Ballistic Stretching:	Moving a limb to the terminal range of motion through rapid movements initiated by strong muscular contractions and continued by momentum.
Cell Body:	The portion of the neuron that contains the nucleus and a well-marked nucleolus. The cell body receives information through the dendrites and sends information through the axon. Also called the soma.
Central Nervous System:	The brain and the spinal cord.
Cutaneous Reflex:	Reflex that causes relaxation of the muscle upon receiving stimuli in the form of heat or massage.
Crossed Extensor Reflex:	Reflex causing extension of a flexed limb when stimulated by rapid flexion or withdrawal by the contralateral limb.
Dendrites:	Processes on the neuron that receive information and transmit information to the cell body of the neuron.
Electromechanical Delay:	Time period between the onset of electrical activity in the muscle and the first signs of tension development.
Electromyography:	The recording of the electrical activity in the muscle; recording the action potentials in a muscle or in muscle groups.
Excitation-Contraction Coupling:	Electrochemical stimulation of the muscle fiber that initiates the release of calcium and the subsequent cross-bridging between actin and myosin filaments, which leads to contraction.
Extrafusal Fiber:	Fibers outside the muscle spindle; muscle fibers.

Flexor Reflex:	Reflex initiated by a painful stimulus that causes a withdrawal or flexion of the limb away from the stimulus.
Gamma Bias:	Readjustment of the muscle spindle length by contracting the ends of the intrafusal fiber. Initiated by voluntary control such as when anticipating the receipt of a heavy weight.
Gamma Loop:	A reflex arc that works with the stretch reflex, in which descending motor pathways synapse with both alpha and gamma motoneurons of the muscle fiber and the muscle spindle.
Gamma Motoneuron:	A neuron that innervates the contractile ends of the muscle spindle.
Ganglia:	Nerve cell bodies outside the central nervous system.
Golgi Tendon Organ (GTO):	A sensory receptor located at the muscle-tendon junction that responds to tension generated during both stretch and contraction of the muscle. Initiates the inverse stretch reflex if the activation threshold is reached.
Intrafusal Fiber:	Fibers that are inside the muscle spindle.
Inverse Stretch Reflex:	Reflex initiated by high tension in the muscle, which inhibits contraction of the muscle through the GTO, causing relaxation of a vigorously contracting muscle.
Labyrinthine Righting Reflex:	Reflex stimulated by tilting or spinning of the body, which alters the fluid in the inner ear. The body responds by restoring balance by bringing the head to the neutral position or thrusting arms and legs out for balance.
Local Graded Potential:	An excitatory or inhibitory signal in the nerve or muscle that is not propagated.
Monosynaptic Reflex Arc:	The reflex arc whereby a sensory neuron is stimulated and facilitates the stimulation of a spinal motoneuron.
Motoneurons:	Neurons that carry impulses from the brain and spinal cord to the muscle receptors.
Motor Endplates:	A flattened expansion in the sarcolemma of the muscle that contains receptors to receive the expansions from the axonal terminals; also called the neuromuscular junction.
Motor Unit:	A motoneuron and all of the muscle cells it stimulates.
Muscle Spindle:	An encapsulated sensory receptor that lies parallel to muscle fibers; responds to stretch of the muscle.
Myelinated:	Nerve fibers having a myelin sheath composed of a fatty insulated lipid substance.
Myotactic Reflex:	Reflex initiated by stretching the muscle, which facilitates a contraction of the same muscle via muscle spindle stimulation; also called the stretch reflex.
Neuromuscular Junction:	Region where the motoneuron comes into close contact with the skeletal muscle; also called the motor endplate.
Neuron:	A conducting cell in the nervous system that specializes in generating and transmitting nerve impulses.

Node of Ranvier:	Gaps in the myelinated axon where the axon is enclosed only by processes of the Schwann cells.
Nuclear Bag Fiber:	An intrafusal fiber within the muscle spindle having a large clustering of nuclei in the center. The Type Ia afferent neurons exit from the middle portion of this fiber.
Nuclear Chain Fiber:	An intrafusal fiber within the muscle spindle with nuclei arranged in rows. Both the Type Ia and the Type II sensory neurons exit from this fiber.
Pacinian Corpuscle:	Sensory receptors located in the skin that are stimulated by pressure.
Passive Range of Motion:	The degree of motion that occurs between two adjacent segments through external manipulation such as gravity or manual resistance.
Peripheral Nerve System:	All nerve branches lying outside the spinal cord.
Plyometrics:	Exercises that utilize the stretch-contract sequence of muscle activity.
Primary Afferent Neuron (Type Ia):	Sensory nerve fibers from the muscle spindle that are sensitive to stretch and respond to the stretch by initiating the stretch reflex.
Proprioceptive Neuromuscular Facilitation (PNF):	Rehabilitation techniques that enhance the response from a muscle through a series of contract-relax exercises.
Proprioceptor:	A sensory receptor located in the joint, muscle, or tendon that can detect stimuli.
Propriospinal Reflex:	Reflexes processed on both sides and at different levels of the spinal cord; an example is the crossed extensor reflex.
Rate Coding:	The frequency of the discharge of the action potentials.
Reciprocal Inhibition:	Relaxation of the antagonistic muscle(s) while the agonist muscles produce a joint action.
Recruitment:	A system of motor unit activation.
Reflex:	Involuntary response to a stimuli.
Renshaw Cell:	Interneuron that receives excitatory input from collateral branches of other neurons and then produces an inhibitory effect on other neurons.
Ruffini Endings:	Sensory receptors located in the joint capsule that respond to change in joint position.
Schwann Cells:	Cells that cover the axon and produce myelination, which is numerous concentric layers of the Schwann cell plasma membrane.
Secondary Afferent Neuron (Type II):	Sensory nerve fibers from the muscle spindle that are sensitive to stretch, and that facilitate flexors and inhibit extensor activity.
Sensory Neurons:	Neurons that carry impulses from the receptors in the body into the central nervous system.
Size Principle:	The principle that describes the order of motor unit recruitment as a function of size.

Soma:

The portion of the nerve cell that contains the nucleus and the well-marked nucleolus. The soma receives information from the dendrites and sends information through the axon; also called the cell body.

Spinal Nerves:

The 31 pairs of nerves that arise from the various levels of the spinal cord.

Static Stretching:

Moving a limb to the terminal range of motion slowly and then holding the final position.

Stretch Reflex:

Reflex initiated by stretching the muscle, which facilitates a contraction of the same muscle via muscle spindle stimulation; also called the myotactic reflex.

Supraspinal Reflex:

Reflexes brought into the spinal cord but processed in the brain; example is the labyrinthine righting reflex.

Synapse:

The junction or point of close contact between two neurons, or between a neuron and a target cell.

Synchronous:

Events occurring at the same time. In muscular contraction, the concurrent activation of motor units.

Tetanus:

A smooth sustained contraction resulting from a series of excitatory inputs.

Tonic Neck Reflex:

Reflex stimulated by head movements, which stimulates flexion and extension of the limbs. The arms flex with head flexion and extend with neck extension.

Twitch:

Single, rapid contraction of a muscle in response to a single excitatory stimulus.

PART 2

Functional Anatomy

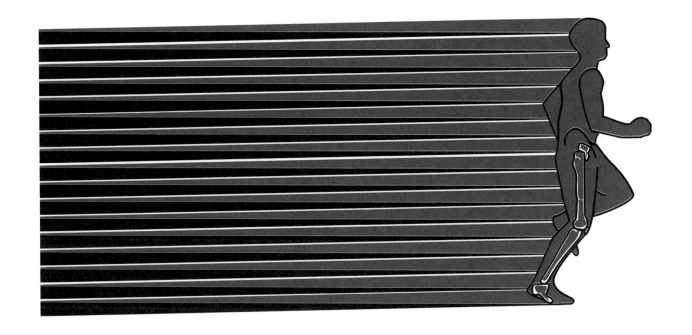

CHAPTER 5

Functional Anatomy of the Upper Extremity

Student Objectives

After reading this chapter, the student will be able to:

1. Describe the differences and similarities between the shoulder and pelvic girdles.
2. Describe the structure, support, and movements of the joints of the shoulder girdle, shoulder joint, elbow, wrist, and hand.
3. Describe the scapulohumeral rhythm in an arm elevation movement.
4. Identify the muscular actions contributing to arm and shoulder girdle movements.
5. Explain the differences in strength across the different arm movements.
6. Develop a set of strength and flexibility exercises for the shoulder girdle, arm, forearm, and hand.
7. Identify the upper extremity muscular contributions to throwing, swimming, and the golf swing.
8. List and describe some of the common injuries to the upper extremity.
9. Identify the muscles contributing to forearm, hand, and finger movements.
10. Describe some common wrist and hand positions used in precision or power.

Introduction

The upper extremity is an interesting part of the body to examine from a functional anatomy perspective because of the interplay necessary among the various joints and segments in order for smooth, efficient motion to take place. Movements of the hand are made more effective through proper hand positioning by the elbow, shoulder joint, and shoulder girdle. Also, forearm movements occur in concert with both hand and shoulder movements, and would not be half as effective if the movements occurred in isolation.

Comparison of Shoulder and Pelvic Girdles

The upper extremity is similar to the lower extremity in that both upper and lower limbs are connected to the trunk via a bony ring, or girdle. The posterior aspect of the human skeleton is presented in FIGURE 5–1 to demonstrate the two girdles; the shoulder girdle and the pelvic girdle. The upper and lower extremity girdles possess a large flat bone on the dorsal surface and they both come in close contact anteriorly via a small strut. Also, the proximal segments in the upper and lower extremities both consist of the largest bone in the extremity, the humerus and the femur, respectively. Likewise, the next adjacent segments consist of two long bones: the radius and the ulna in the forearm, and the tibia and fibula in the shank.

Finally, the hand and the foot both contain a similar number of short bones, the carpals [8] and the tarsals [7], respectively. These short bones are connected to a series of long bones in the hand and foot, the metacarpals and metatarsals, respectively, and the phalanges. The hand and the foot have a similar number of joints that are primarily of the same type.

Functionally, the arm and fingers of the upper extremity have movement capacities similar to the thigh and toes of the lower extremity. Conversely, the movement characteristics of the forearm and hand of the upper extremity vary somewhat from those of the leg and toes of the lower extremity.

There are a few major functional differences between the upper and lower extremity that need to be pointed out before taking a closer look at their individual structures and functions. The two girdles, shoulder and pelvic, connecting the upper and lower extremity to the trunk, connect in different ways, creating one of the main functional differences between upper and lower extremities. The

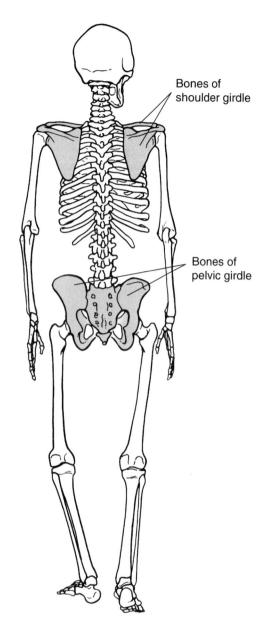

Bones of
shoulder girdle

Bones of
pelvic girdle

FIGURE 5–1. The upper and lower extremity are similar in many respects. They both have girdles which serve as the attachment site for the limbs. Both the shoulder and pelvic girdles protect, serve as attachment sites for muscles, and functionally participate in arm and thigh movements by moving to accommodate the shoulder and hip joints, respectively.

upper extremity connects to the trunk via the sternum, and the shoulder girdle forms an incomplete ring since the scapulae do not make contact with each other in the back. On the other hand, the lower extremity connects to the trunk via the sacrum, and a complete ring is formed by the pelvic girdle since both sides of the pelvis are con-

nected to each other anteriorly and posteriorly (FIGURE 5–1). This has major functional consequences, allowing independent motion of the right and left upper extremity and dependent motion of the right and left lower extremity. Thus, a movement of the right arm will have minimal influence on the function of the left arm while a movement of the left leg will have a direct effect on the function of the right limb.

The upper and lower extremities also have different functional roles to play, with the lower extremity involved primarily with weight bearing, ambulation, posture, and most gross motor activities. The upper extremity participates in activities that require skills in manipulation, dexterity, striking, catching, and fine motor abilities. Let's begin our examination of the most mobile of the extremities, the upper limb.

The Shoulder Complex

The shoulder complex has many articulations, each one contributing to the movement of the arm through coordinated joint actions. The movement of the scapula is the best way of observing movements being generated at two of the articulations: the sternoclavicular and acromioclavicular joints. Arm movements take place at the glenohumeral joint. While it is possible to create a small amount of movement at any one of these articulations in isolation of movement at the other joints, usually movement is generated at all three of these joints concomitantly as the arm is raised, lowered, or if any other significant arm action is produced (1).

Anatomical and Functional Characteristics of the Joints

The only point of skeletal attachment of the upper extremity to the trunk occurs at the sternoclavicular joint. It is here where the clavicle is joined to the manubrium of the sternum. A close-up view of the sternoclavicular joint is shown in FIGURE 5–2. This joint is a gliding, synovial joint that has a fibrocartilaginous disc in the joint (2). The joint is reinforced by three ligaments: the interclavicular, the costoclavicular, and the sternoclavicular, of which the costoclavicular is the main supporter of the joint (3). The joint is also reinforced and supported by muscles, such as the short, powerful subclavius. Additionally, a strong joint capsule contributes to making the joint resilient to dislocation or disruption.

Movements of the clavicle at the sternoclavicular joint occur in three directions, giving it three degrees of freedom. The clavicle can move up and down in the movements of elevation and depression, respectively. This

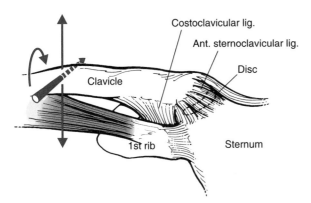

FIGURE 5–2. The sternoclavicular joint is a sturdy, well-reinforced joint that allows movement to occur in three different planes. Movements include elevation and depression, protraction and retraction, and rotation.

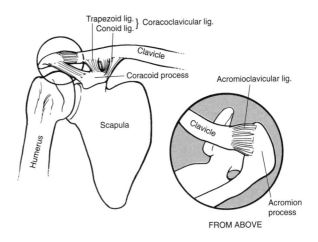

FIGURE 5–3. The acromioclavicular joint is a small joint that allows movement of the scapula on the clavicle. It is vulnerable to injury when forces come up through the arm and the shoulder joint.

movement takes place between the clavicle and the meniscus in the sternoclavicular joint and has a range of motion of approximately 30 to 40 degrees (2).

The clavicle can also move anteriorly and posteriorly via the movements termed protraction and retraction, respectively. This movement occurs between the sternum and the meniscus in the joint through a range of motion of approximately 30 degrees. Finally, the clavicle can also rotate anteriorly and posteriorly along its long axis through approximately 40 to 50 degrees (2).

The clavicle is connected to the scapula at its distal end via the acromioclavicular (AC) joint. This is a small, gliding, synovial joint that is not present in all individuals. It frequently has a fibrocartilaginous disc like the sternoclavicular joint (3). It is at this joint that most of the movements of the scapula on the clavicle occur. The acromioclavicular joint is illustrated in FIGURE 5–3.

The AC joint lies over the top of the humeral head and can serve as a bony restriction to arm movements above the head. The joint is reinforced with a very dense capsule and a set of acromioclavicular ligaments lying above and below the joint. Close to the AC joint is the important coracoclavicular ligament, which assists by serving as an axis of rotation during scapular movements.

The movement of the scapula at the acromioclavicular joint can occur in three different directions, as shown in FIGURE 5–4. The scapula can move anteriorly and posteriorly about a vertical axis, and these movements are known as protraction or abduction, and retraction or adduction, respectively. Protraction and retraction occur as the acromion process moves on the meniscus in the joint, and as the scapula rotates about the medial coracoclavicular ligament, the conoid. There can be anywhere

from 30 to 50 degrees of protraction and retraction of the scapula (3).

The second scapular movement is swinging out and back in the frontal plane, termed upward and downward rotation. This movement occurs as the clavicle moves on the meniscus in the joint and as the scapula rotates about the lateral coracoclavicular ligament, the trapezoid portion. This movement can occur through a range of approximately 60 degrees (2).

The third and final movement potential, or degree of freedom, is the scapular movement up and down, termed elevation and depression. This movement occurs at the acromioclavicular joint and is not assisted by rotations occurring about the coracoclavicular ligament. The range of motion at the acromioclavicular joint for elevation and depression is approximately 30 degrees (2,3).

The scapula movements are also dependent on the movement and position of the clavicle. The movements at the sternoclavicular joint are opposite to the movements at the acromioclavicular joint for elevation, depression, protraction, and retraction. For example, as elevation occurs at the acromioclavicular joint, depression occurs at the sternoclavicular joint, and vice versa. This is not true for rotation, since the clavicle will rotate in the same direction along its length. The clavicle does rotate in different directions to accommodate the movements of the scapula, rotating anteriorly with protraction and elevation, and posteriorly with retraction and depression.

The scapula interfaces with the thorax via the scapulothoracic joint. This is not the typical articulation, connecting bone to bone, and is called a *physiological* joint

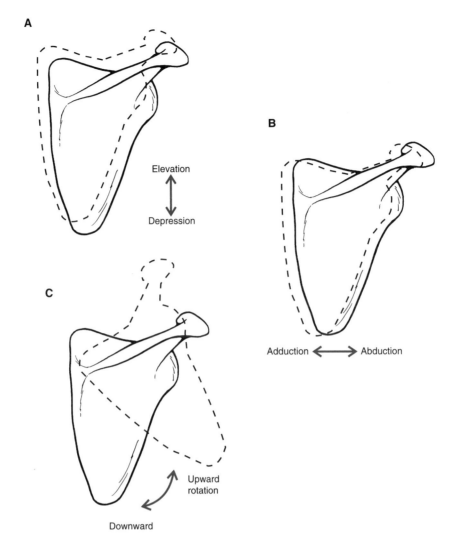

A

Elevation

Depression

B

Adduction ← → Abduction

C

Upward
rotation

Downward

FIGURE 5–4. Scapular movements take place in three different directions. (A) Elevation and depression of the scapula occur with a shoulder shrug or when the arm raises. (B) Abduction (protraction) and adduction (retraction) occur when the scapulae are drawn away or toward the vertebrae, respectively, or when the arm is brought in front or behind the body, respectively. (C) The scapula also rotates upward and downward as the arm raises and lowers, respectively.

(2). The scapula actually rests on two muscles, the serratus anterior and the subscapularis, which are both connected into the scapula and move across each other as the scapula moves. Underneath these two muscles lies the thorax.

The scapula moves across the thorax as a consequence of actions at the acromioclavicular and the sternoclavicular joints, and the total range of motion for the scapulothoracic articulation is approximately 60 degrees of motion for 180 degrees of arm abduction or flexion.

Approximately 65% of this range of motion is a result of motion occurring at the sternoclavicular joint, and 35% of the motion occurs as a result of acromioclavicular joint motion (2).

The final articulation in the shoulder complex is the shoulder joint, or the glenohumeral joint, illustrated in FIGURE 5–5. This is a synovial, ball-and-socket joint, offering the greatest range of motion and movement potential of any joint in the body. The reason for the laxity and the excessive range of motion allowed in the joint

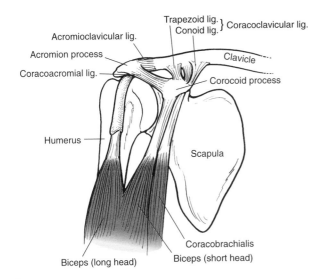

FIGURE 5–5. The glenohumeral joint is a ball-and-socket joint that relies on the muscles, ligaments, and other soft tissue for support. It is one of the most mobile joints in the body.

is the structural make-up, a lax joint capsule, and limited ligamentous support.

The joint contains a small, shallow socket, the glenoid fossa, which is only one fourth the size of the humeral head that must fit into it. The joint cavity is deepened by a rim of fibrocartilage, the glenoid labrum, that receives supplementary reinforcement from the surrounding ligaments and tendons. The labrum varies from individual to individual and is even absent in some cases (4). The glenoid labrum increases the contact area to 75%. The joint capsule has approximately twice the volume of the humeral head, allowing the arm to be raised through a considerable range of motion (5). Because there is minimal contact between the glenoid fossa and the head of the humerus, the shoulder joint largely depends on the ligamentous and muscular structures for stability. On the anterior side of the joint, support is provided by the capsule, the glenoid labrum, the glenohumeral ligaments, three reinforcements in the capsule, the coracohumeral ligament, and fibers of the subscapularis and the pectoralis major that blend into the joint capsule (5). Both the coracohumeral and the middle glenohumeral ligament support and hold up the relaxed arm. They also offer functional support through the movements of abduction, external rotation, and extension (3). Posteriorly, the joint is reinforced by the capsule, the glenoid labrum, and fibers from the teres minor and infraspinatus that blend into the capsule. Location and support

action of ligaments of the shoulder girdle and shoulder joint are included in Appendix A for your review.

The superior aspect of the shoulder joint is often termed the *impingement area*. The support on the superior portion of the shoulder joint is offered by the capsule, the glenoid labrum, the coracohumeral ligament, and muscular support and capsular reinforcement is given by the supraspinatus and the long head of the biceps brachii. Above the supraspinatus muscle lies the subacromial bursae and the coracoacromial ligament, forming an arch underneath the acromioclavicular joint. This area and a typical impingement position is presented in FIGURE 5–6.

A bursa is a fluid-filled sac found at strategic sites around the synovial joints that reduces the friction in the joint. The supraspinatus muscle and the bursae in this area are compressed as the arm raises above the head and can be irritated if the compression is of sufficient magnitude or duration. The inferior portion of the shoulder joint is minimally reinforced by the capsule and the long head of the triceps brachii.

The range of motion of the arm at the shoulder joint is considerable because of the aforementioned structural reasons (FIGURE 5–7). The arm can move through approximately 180 degrees of flexion to approximately 60 degrees of hyperextension in the sagittal plane (2,6). The amount of flexion can be very limited if the shoulder joint is also externally rotated, and with the joint in maximal external rotation, the arm can only be flexed through 30 degrees (6). Also, during passive flexion and extension movements, there is accompanying anterior and pos-

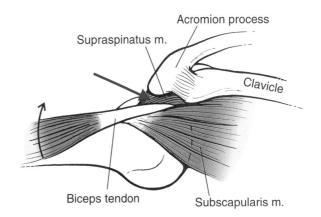

FIGURE 5–6. The area in the shoulder joint termed the *impingement area* contains structures that can be damaged with repeated overuse. The actual impingement occurs in the abducted position with the arm in a rotated position.

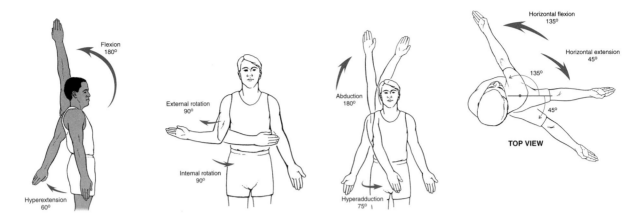

FIGURE 5–7. There is considerable range of motion at the shoulder joint. The arm can move through 180 degrees of flexion or abduction, 60 degrees of hyperextension, 75 degrees of hyperadduction, 90 degrees of internal and external rotation, 135 degrees of horizontal flexion, and 45 degrees of horizontal extension.

terior translation, respectively, of the head of the humerus on the glenoid (7).

The arm can also abduct through 180 degrees. The abduction movement can be limited by the amount of internal rotation occurring simultaneously with abduction. If the joint is maximally rotated internally, the arm can only produce about 60 degrees of abduction (6). As the arm adducts down to the neutral position, it can continue past the neutral position for approximately 75 degrees of hyperadduction across the body.

The arm can rotate internally and externally 90 degrees each, for a total of 180 degrees of rotation (5). Rotation is limited by abduction of the arm. In the neutral position, the arm can rotate through the full 180 degrees, but in 90 degrees of abduction, the arm can only rotate through 90 degrees (6). Finally, the arm can move across the body in an elevated position for 135 degrees of horizontal flexion or adduction, and 45 degrees of horizontal extension or abduction (2).

Extreme range of motion is required in many different activities, such as throwing, tennis, swimming, and gymnastics (1). It has been demonstrated that some athletes who commonly produce movements using extreme joint positions can have a loose shoulder joint, where the humeral head can actually lose contact with the glenoid cavity in the terminal or extreme positions. At these extreme joint positions, the joint capsule and the ligaments surrounding the shoulder joint are strained. With continued motion at the terminal joint positions, the humeral head can be forced over the glenoid labrum rim into a dislocated or subluxated position, from which it will usually relocate back into the socket (1).

With the arm in the neutral position, the ligaments and many of the supporting muscles are loose. If the arm is externally rotated in this position, the capsule is tightened (8). Internal rotation in this position has no effect on tightening the capsule.

The inferior portion of the joint capsule is loose, allowing the arm to both abduct and externally rotate through a significant range of motion. As the arm moves into abduction through 45 degrees, the subscapularis and the lower glenohumeral ligament begin to tighten and support the joint (9). The joint can be stabilized even more with the addition of more external rotation. This stabilization continues up to 90 degrees of abduction with the activity level of the subscapularis diminishing.

Other muscles such as the supraspinatus, infraspinatus, and the teres minor also provide stability up through 90 degrees of abduction by compressing the humeral head in the socket. Their contribution to the stability in the joint diminishes above 90 degrees of abduction. However, in extreme abduction with external rotation, the shoulder joint is in a close-packed position with the ligaments tight around the joint, offering some stability (3).

Combined Movement Characteristics of the Shoulder Complex

The movement potential of each joint has been examined in the previous section. This section will examine the movement of the shoulder complex as a whole, sometimes referred to as the scapulohumeral rhythm. Any time the arm is raised in flexion or abduction, there are accompanying scapular and clavicular movements. A posterior

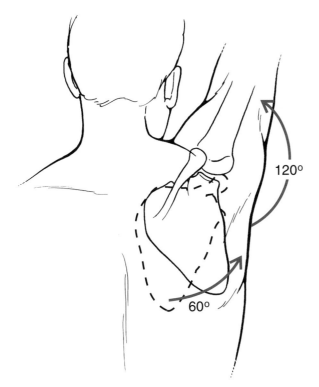

FIGURE 5–8. The movement of the arm is accompanied by movements of the shoulder girdle. The working relationship between the two is known as the scapulohumeral rhythm. The arm can move through only 30 degrees of abduction and 45 to 60 degrees of flexion with minimal scapular movements. Past these points, the scapula movements occur concomitantly with the arm movements. For 180 degrees of flexion or abduction, approximately 120 degrees of motion occurs in the glenohumeral joint and 60 degrees of motion occurs as a result of scapular movement on the thorax.

view of the relationship between the arm and scapular movements is shown in FIGURE 5–8.

In the first 30 degrees of abduction, or the first 45 to 60 degrees of flexion, the scapula moves either toward or away from the vertebral column to seek a position of stability on the thorax (3). After stabilization has been achieved, the scapula moves laterally, anteriorly, and superiorly in the movements described as upward rotation, protraction or abduction, and elevation. The clavicle also rotates posteriorly, elevates, and protracts as the arm is moving through flexion or abduction (10).

In the early stages of abduction or flexion, the movements are primarily glenohumeral except for the stabilizing motions of the scapula. Past 30 degrees of abduction, or 45 to 60 degrees of flexion, the ratio of glenohumeral to scapular movements becomes 5:4 so that there are 5 degrees of humeral movement for every 4 degrees of

scapular movement on the thorax (3,11). For the total range of motion through 180 degrees of abduction or flexion, the glenohumeral to scapula ratio is 2:1; thus, the 180 degree range of motion is produced by 120 degrees of glenohumeral motion and 60 degrees of scapular motion (5). The contributing joint actions to the scapular motion are 20 degrees produced at the acromioclavicular joint, 40 degrees produced at the sternoclavicular joint, and 40 degrees of posterior clavicular rotation (10).

As the arm abducts up to 90 degrees, the greater tuberosity on the humeral head approaches the coracoacromial arch, compression of the soft tissue begins to limit further abduction, and the tuberosity makes contact with the acromion process (10). If the arm is externally rotated, 30 degrees more abduction can occur as the greater tuberosity is moved out from under the arch. Abduction is limited even more and can only occur through 60 degrees with arm internal rotation, since the greater tuberosity is held under the arch (10). Also, full abduction cannot be achieved without some extension of the upper trunk to assist the movement.

Muscular Actions

The insertion, action, and nerve supply for each individual muscle of the shoulder joint and shoulder girdle is outlined in Appendix B. Please refer to the figures and charts if you are unfamiliar with these muscles. Special interactions between the muscles will be presented in this section.

The muscles contributing to the movements of shoulder abduction and flexion are similar. The deltoid generates about one half of the muscular force for elevation of the arm in abduction or flexion. The contribution of the deltoid increases as abduction increases, and the muscle is most active through 90 to 180 degrees (12). However, the deltoid has been shown to be most resilient to fatigue in the range of motion through 45 to 90 degrees of abduction (2), making this range of motion more popular for arm raising exercises.

When the arm elevates, the rotator cuff (teres minor, subscapularis, infraspinatus, supraspinatus) also plays an important role since the deltoid cannot abduct or flex the arm without stabilization of the humeral head (2). The rotator cuff as a group is also capable of generating a flexion or abduction arm movement, with about 50% of the force normally generated in these movements (5).

In the early stages of arm flexion or abduction, the teres minor works with the deltoid to depress the humeral head and stabilize it so that the arm can be raised by the deltoid (3). The muscle force of the teres minor is equal and opposite to that of the deltoid, forming a force couple. The subscapularis and the infraspinatus join a little later in flexion or abduction to assist with the humeral head stabilization (13,14). The latissimus dorsi also contracts eccentrically to assist with the stabilization of the head and increases in activity as the angle increases (14). The interaction between the deltoid and the rotator cuff in abduction and flexion is shown in FIGURE 5–9. The downward and inward force of the rotator cuff allows the deltoid to elevate the arm.

Above 90 degrees of flexion or abduction, the rotator cuff force decreases, leaving the shoulder joint more vulnerable to injury (5). However, one of the rotator cuff muscles, the supraspinatus, remains a major contributor above 90 degrees of flexion or abduction. In the upper range of motion, the deltoid begins to pull the humeral head down and out of the joint cavity, creating a subluxating force (3). In order to move through 90 to 180 degrees of flexion or abduction, there must be external rotation in the joint. If the humerus externally rotates 20 degrees or more, the biceps brachii can also abduct the arm (5).

When the arm is abducted or flexed, the shoulder girdle must protract, or abduct, elevate, and upwardly rotate with clavicular rotation posteriorly to maintain the gle-

noid fossa in the optimal position. As shown in FIGURE 5–10, the serratus anterior and the trapezius work as a force couple to create the lateral, superior, and rotation movements of the scapula (5). These muscle actions take place after the deltoid and the teres minor have initiated the elevation movements of the arm and continue up through 180 degrees, with the greatest muscular activity through 90 to 180 degrees (12). The serratus anterior is also responsible for holding the scapula to the thorax wall and preventing any winging of the vertebral border.

If the arm is slowly lowered, producing the movements of adduction or extension of the arm with accompanying retraction, depression, and downward rotation of the shoulder girdle with forward clavicular rotation, the muscle actions are eccentric, and are therefore being controlled by the muscles previously described in the arm abduction and flexion section. However, if the arm is forcefully lowered, or if it is lowered against an external resistance such as a weight machine, the muscle action will be concentric.

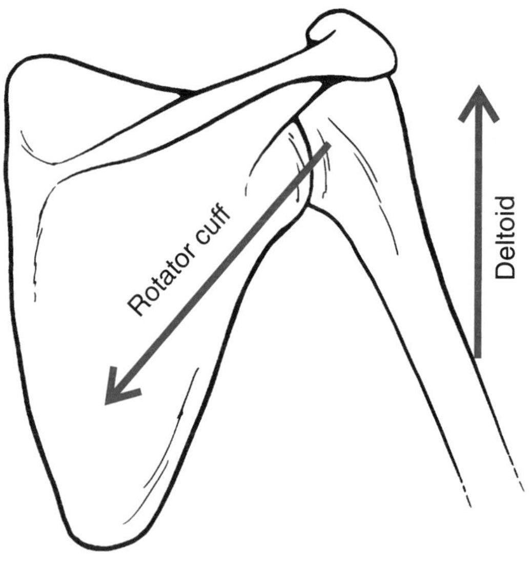

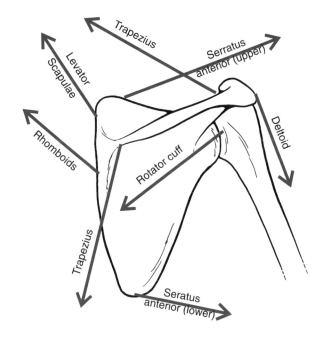

FIGURE 5–9. For efficient flexion or abduction of the arm, the deltoid and the rotator cuff work together. In the early stages of abduction and flexion up through 90 degrees, the rotator cuff applies a force to the humeral head which keeps the head depressed and stabilized in the joint while the deltoid applies a force to elevate the arm. This relationship changes in the upper range of motion as the rotator cuff activity drops off.

FIGURE 5–10. The direction of pull of various shoulder girdle muscles, the deltoid, and the rotator cuff for the resting arm position. Notice the line of pull of the trapezius and the serratus anterior, which work together to produce abduction, elevation, and upward rotation of the scapula necessary in arm flexion or abduction. Likewise, notice the pull of the levator scapulae and the rhomboid, which will also assist in elevation of the scapula.

In a concentric adduction or extension against an external resistance, as in swimming, the muscles responsible for creating these joint actions are the latissimus dorsi, teres major, and the sternal portion of the pectoralis major. The teres major is active only against a resistance, whereas the latissimus dorsi has been shown to be active in these movements even when there is no resistance offered (15).

As the arm is adducted or extended, the shoulder girdle retracts, depresses, and downwardly rotates with forward clavicular rotation. The rhomboid downwardly rotates the scapula and works with the teres major and the latissimus dorsi in a force couple to control the arm and scapular motions during lowering. Other muscles actively contributing during the movement of the scapula back to the resting position while working against a resistance are the pectoralis minor, which depresses and downwardly rotates the scapula, and the middle and lower portions of the trapezius, which contribute to the retraction of the scapula with the rhomboid. These muscular interactions are illustrated in FIGURE 5–11.

Two other movements of the arm, very important to many sport skills and the efficient movement of the arm above 90 degrees (measured from arm at the side), are internal and external rotation. An example of both external and internal rotation in a throwing action is shown in

FIGURE 5–12. External rotation is an important component of the preparatory or cocking phase and internal rotation is important in the force application and follow-through phase.

External rotation is necessary when the arm is above 90 degrees, and is produced by the infraspinatus and the teres minor (3). The activity of both of these muscles increases with the external rotation in the joint (16). Since the infraspinatus is also an important muscle in humeral head stabilization, it fatigues early in elevated arm activities.

Internal rotation is produced primarily by the subscapularis, latissimus dorsi, teres major, and portions of the pectoralis major. The teres major is an active contributor to internal rotation only when the movement is produced against a resistance. The muscles contributing to the internal rotation joint movement are capable of generating a large amount of force, yet the internal rotation actions in most upper-extremity factions never require or use very much internal rotation force (3).

The shoulder girdle movements accompanying internal and external rotation depend on the position of the arm. In an elevated arm position, the shoulder girdle movements described in conjunction with abduction and flexion would be necessary. Rotation produced with the arm in the neutral position would require minimal shoul-

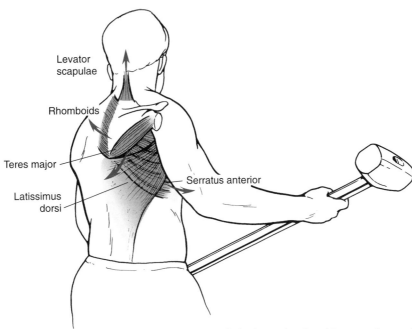

FIGURE 5–11. Lowering the arm against a resistance utilizes the latissimus dorsi and teres major working as a force couple with the rhomboid. Other muscles that contribute to the lowering action are the pectoralis major, pectoralis minor, levator scapulae, and serratus anterior.

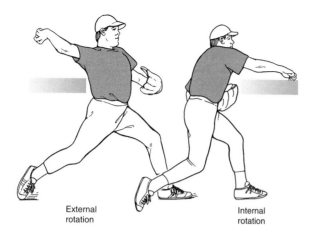

External
rotation

Internal
rotation

FIGURE 5–12. Shoulder joint rotation is an important contributor to the overhand throw. In the preparatory or cocking phase, the arm externally rotates to increase the range of motion and the distance over which the ball will travel. Internal rotation is an active contributor to the force application phase, and the movement continues through the follow-through phase as the arm is slowed down.

der girdle assistance. It is also in this position that the full range of rotation through 180 degrees can be obtained, because as the arm is raised, muscles used to rotate the humerus are also being used to stabilize the humeral head, and the humeral head will be restrained in rotation in the upper range of motion. Specifically, internal rotation is very difficult in the elevated arm positions since the tissue under the acromion process is very compressed with the greater tuberosity pushed up against it (12).

Two final joint actions that are actually combinations of the elevated arm positions are horizontal flexion or adduction, and horizontal extension or abduction. Since the arm is elevated, the same muscles described earlier for abduction and flexion will also be contributing to these movements of the arm across the body.

Muscles contributing more significantly to the horizontal flexion joint movement are the pectoralis major and the anterior head of the deltoid. This movement brings the arms across the body in the elevated position and is important in power movements of upper extremity skills. The movement of horizontal extension where the arm is brought back in the elevated position is produced primarily by the infraspinatus, teres minor, and posterior head of the deltoid. This joint action is common in the backswing and preparatory actions in upper extremity skills (2).

Strength of the Shoulder Muscles

The shoulder muscles can generate the greatest strength output in the adduction movement, in which there are mus-

cle fibers from the latissimus dorsi, teres major, and the pectoralis major contributing to the movement. The adduction strength of the shoulder muscles is twice that of the opposite abduction movement, even though the abduction movement and muscle group is used more frequently in activities of daily living or sport (2).

The movement capable of generating the next highest level of strength after the adductors is the extension movement, which uses the same muscles that contribute to arm adduction. The extension action is slightly stronger than its opposite movement, flexion. Following flexion, the next strongest joint action is abduction, illustrating the fact that the shoulder joint actions are capable of generating more output in the lowering phase using the adductors and extensors, as compared to the raising phase when the flexors and abductors are used.

The weakest joint actions in the shoulder are the rotation joint movements, with external rotation the weaker of the two. The strength output of the rotators is influenced by the arm position, and the greatest internal rotation strength can be obtained with the arm in the neutral position, while the greatest external rotation strength can be obtained with the shoulder in a position of 90 degrees of flexion. However, with the arm elevated to 45 degrees, both internal and external rotation strength outputs are greater in 45 degrees of abduction than 45 degrees of flexion (17). The external rotation joint movement is important in the upper 90 degrees of arm elevation, providing stability to the joint. The internal rotation joint movement creates instability in the joint, especially in the upper elevation levels, as it compresses the soft tissue in the joint.

The muscle activity in the shoulder complex generates high forces in the shoulder joint itself. The rotator cuff muscle group as a whole, capable of generating a force 9.6 times the weight of the limb, generates maximum forces at 60 degrees of abduction (2). Since each arm weighs approximately 7% of body weight, the rotator cuff generates a force in the shoulder joint approximately 70% of body weight. At 90 degrees of abduction, the deltoid generates a force averaging 8 to 9 times the weight of the limb, creating a force in the shoulder joint ranging from 40 to 50% of body weight (2). In fact, the forces in the shoulder joint at 90 degrees of abduction have been shown to be close to 90% of body weight. These forces can be cut in half if the forearm is flexed to 90 degrees at the elbow.

Conditioning

The shoulder muscles are easy to both stretch and strengthen due to the mobility in the joint. The examples include both active and passive stretches. In the active stretches, the individual moves the limb to the terminal position without any external force, while in passive stretches, an external force is imposed, usually a weight or force created manually by the other limb.

In the strengthening categories, both manual resistance and weight lifting illustrations offer suggestions for the poorly conditioned individual or rehabilitation setting as well as for any individual with access to a weight-training facility. A manual resistance exercise requires another individual to apply an external force while the subject moves through a range of motion. The benefit of manual resistance exercising is that the external force applied by a partner can be readily adjusted to the level the subject can handle. It is a good technique for individuals who lack strength in specific regions. Another technique for a weak muscle group is to begin exercises using body parts such as an arm or leg as the resistance.

The muscles acting on the shoulder joint and shoulder girdle usually work in combination, making it difficult to isolate a specific muscle in an exercise. Examples of stretching, manual resistance, and weight training for the shoulder abductors and flexors are presented in FIGURE 5–13. Examples for shoulder adductors and extensors are shown in FIGURE 5–14. A rowing action is good for the trapezius, levator scapula, and rhomboid, and the push-up is good for the serratus anterior and pectoralis minor (18).

An important muscle group to emphasize in a stretching or strengthening routine of the shoulder complex is the rotator cuff, since these muscles stabilize the shoulder joint and perform a wide variety of shoulder movements. Be sure and include some exercises for this muscle group. Examples of rotator cuff exercises are presented in FIGURE 5–15. The final example is for the shoulder girdle, and includes exercises for the abductors, adductors, and elevators (FIGURE 5–16).

Some resistance exercises may cause irritation to the shoulder joint and should be avoided by individuals with specific injuries to that area. Any lateral dumbbell raise using the deltoid may cause impingement in the coracoacromial area. This is magnified if the shoulder is internally rotated. A solution for those wishing to avoid impingement or who have injuries in this area would be to externally rotate the arm and then perform the lateral raise (10). It is important to recognize that when an adjustment like this is made, the muscle activity and the forces generated internally will also change. External rotation during a lateral raise will alter the activity of the deltoid and will facilitate activity in the internal rotators.

Exercises like the bench press and push-ups should be avoided by individuals having instability in the posterior portion of the shoulder joint caused by adduction and internal rotation. Likewise, stress on the anterior portion of the capsule is produced by the pullover exercise that moves from an extreme flexed, abducted, and externally rotated position. Other exercises to be avoided by individuals with anterior capsule problems are behind the neck pull-downs, horizontal flexion/extension movements, and rowing exercises. The risks in these three exercises can be minimized if no external rotation or even some internal rotation is maintained in the joint while performing these actions. The external rotation position produces strain on the anterior portion of the shoulder (10). In an exercise such as the squat, which utilizes the lower extremity musculature, the position of the shoulder in external rotation may even prove to be harmful because of the strain on the anterior capsule created by weights held in external rotation. Attempts should be made to minimize this joint action by balancing a portion of the weight on the trapezius, or to use alternative exercises such as the dead lift.

Finally, if an individual is experiencing problems with the rotator cuff musculature, heavy lifting in the abduction movement should be minimized or avoided. This is due to the fact that the rotator cuff muscles must generate a large amount of force during this abduction action to support the shoulder joint and complement the activity of the deltoid. Heavy weight lifting above the head should be avoided to reduce strain on the rotator cuff muscles (19).

Contribution of Shoulder Musculature to Sport Skills or Movements

In order to appreciate fully the contribution of a muscle or muscle group to an activity, the activity or movement of interest must be evaluated and studied. This will provide an understanding of the functional aspect of the movement, ideas for training and conditioning of the appropriate musculature, and a better comprehension of injury sites and mechanisms. Selected activities will be presented in this section. These are examples of a functional anatomy description of a movement and are gathered primarily from electromyographic research collected over the years. It is important to note that these examples do not include all of the muscles that may be active in these activities, but only the major contributing muscles.

The first example is a daily living activity (ADL) of just pushing up out of a wheelchair or chair (FIGURE 5–17). This activity places a tremendous load on the upper extremity muscles as full body weight is supported in the transfer (20). Three different push-up techniques are presented. First, if you simply push up out of a chair or

FLEXIBILITY

A

B

MANUAL RESISTANCE

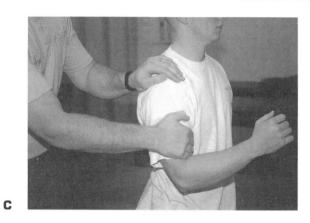

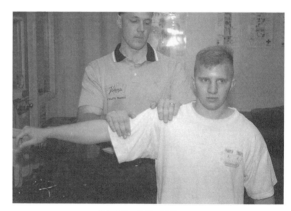

C

D

ARM ELEVATION

WEIGHT TRAINING

E

F

FIGURE 5–13. The photos illustrate stretching, manual resistance, and weight training examples for shoulder flexion. (A, C, E) and abduction (B, D, F). The flexors and abductors can be stretched as the arm is brought horizontally across the chest. Manual resistance in flexion and abduction can be applied to the humerus as the arm is brought forward and up, respectively. A good weight training exercise for the flexors is the bench press, and for the abductors, a lateral fly.

FLEXIBILITY

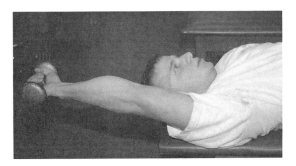

A **B**

MANUAL RESISTANCE

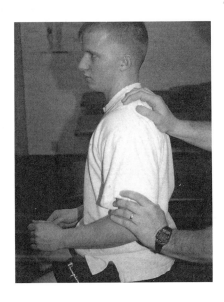

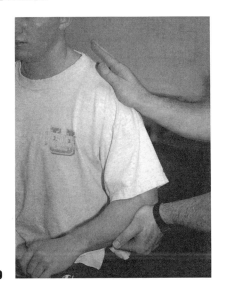

C **D**

WEIGHT TRAINING

E **F**

FIGURE 5–14. Flexibility exercises for the extensors and adductors include (A) a passive stretch imposed by the other arm or (B) a stretch imposed by a lightweight dumbbell. Manual resistance can be applied to the posterior humerus as the arm is pulled back into hyperextension (C) or at the elbow as the arm is pulled down into adduction (D). Two good weight training exercises for the extensors and adductors are the (E) lat pull down and (F) dips.

FLEXIBILITY

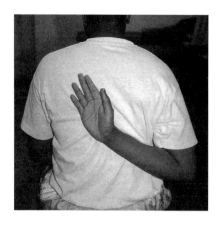

A

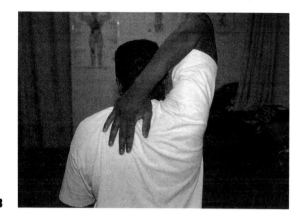

B

MANUAL RESISTANCE

C

D

FIGURE 5–15. Exercises for the rotator cuff usually involve some shoulder rotation exercises. The "scratch test" is a good flexibility exercise and stretches the rotator cuff when the scratch is made behind the back (A) or over the head (B). The rotator cuff can be strengthened with a manual resistance (C) applied to the lower forearm as the arm rotates externally. Another good manual resistance exercise is resistance applied to the humerus (D) as the arm is being raised in abduction.

wheelchair, the muscle primarily used is the triceps brachii, followed by the pectoralis major, with some minimal contribution from the latissimus dorsi. If you alter the technique and push up from a long sitting position with the elbows flexed to 90 degrees, the triceps brachii will still be extensively used (21). In this position, the activity of the pectoralis major will drop off and the activity of the latissimus dorsi will increase, reversing their order of contribution as compared to the chair-rising activity just presented. Finally, if a push up from a long sitting position is done with the elbow maximally flexed and the arm abducted, the muscular output from all three of these muscles must be maximum, making this the most difficult of the three push-up techniques presented (21).

Let's look at a second activity, freestyle swimming, which incorporates many shoulder movements and uses shoulder muscles as the primary source of force generation in the activity. The activity and the muscles contributing to freestyle swimming are presented in FIGURE 5–18.

The freestyle swimming technique can be broken down into two basic phases: pull-through and recovery. In the pull-through phase, the propelling forces are generated by moving the arm through the water. Both internal rotation and adduction are the primary movements in this phase, generated predominantly by the latissimus dorsi, teres major, and the pectoralis major (22,23). These muscles are active throughout the pull-through phase with activity maximal at the mid-pull position. The pull-

WEIGHT TRAINING

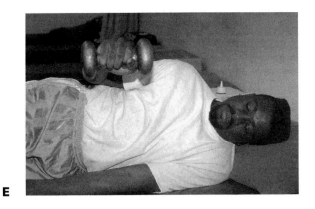

E

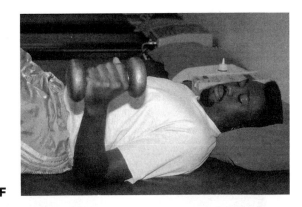

F

G

FIGURE 5–15. (continued) The arm should be rotated internally 20–30 degrees in this exercise. High resistance exercises for the rotator cuff include dumbbell rotation exercises in external rotation (E), internal rotation (F), or arm rotation exercises with dumbbells from the abducted position (G).

through phase is terminated at full adduction and internal rotation of the arm.

The recovery phase is initiated when the arm begins to abduct and externally rotate to bring it out of the water and prepare for another reentry. The main muscles active in the recovery phase are the supraspinatus and the infraspinatus which serve to abduct and externally rotate the humerus, the middle deltoid which also abducts, and the serratus anterior, which is very active in the hand lift as it rotates the scapula. The role of the serratus anterior is viewed as being crucial since it is operating at near-maximal levels during the swimming stroke. It is believed that impingement in the swimmer's shoulder may be a result of serratus anterior fatigue which produces scapular rotation that does not coincide, and is out of synchrony with humeral abduction, thereby creating the impingement in the subacromial area (22,23).

The third activity for evaluation in terms of shoulder musculature is throwing, an overhead pattern common to many different sports, including football, baseball, track and field, and others. This pattern is shown in FIGURE 5–19. Throwing places a lot of strain on the shoulder

joint and requires a great deal of upper extremity muscular action to control and contribute to the throwing movement, even though the lower extremity is a major contributor to the power generation in a throw.

The throwing action described in this section will be a pitch in baseball, and the description will be provided from the perspective of a right-hand thrower. Throwing, or pitching, can be divided into three basic phases. First, during the cocking or preparatory phase, the front leg strides forward and the hand and ball are moved as far back behind the body as possible. This is accomplished through abduction of the arm to 90 degrees, external rotation of the arm, scapula retraction or adduction, and forearm flexion. Also, the body is propelled forward through extension of the lower extremity and the trunk rotates to move the hand even further behind the body. This phase ends when the arm reaches 90 degrees of abduction with maximal external rotation (22,24).

The muscular activity in the cocking phase is minimal in the early stages but becomes very substantial at the end when the arm is maximally externally rotated and the trunk is beginning left rotation as it moves for-

FLEXIBILITY

MANUAL RESISTANCE

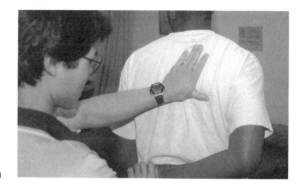

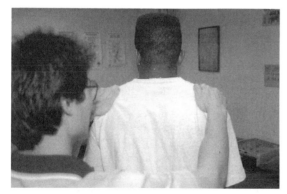

SHOULDER GIRDLE

FIGURE 5–16. The abductors of the shoulder girdle can be stretched in a horizontal extension partner stretch (A) and the adductors can also be passively stretched as the arm is pulled across the body (B). Manual resistance exercises for the shoulder girdle require special attention to isolating a specific movement. The adductors can be strengthened by stabilizing the upper back and applying resistance to the posterior distal humerus as it is horizontally extended (C). With pressure on the scapula, the adductors can also be exercised with pressure placed on the hand as the arm is pulled back (D). The elevators can be isolated in a manual resistance exercise with pressure placed on top of the shoulder in a shrug action (E).

ward. In the early part of the cocking phase, the deltoid and the supraspinatus are active in producing the abduction of the arm. The infraspinatus and the teres minor are also active, assisting with abduction and initiating the external rotation action, and the subscapularis is minimally active to assist during the shoulder abduction. The trapezius and the rhomboid retract or adduct the scapula during the cocking phase, although their level of activity is minimal (22).

During late cocking, when the front stride foot hits the ground and the shoulder is nearing maximal external rotation, there are other muscles beginning to contribute

WEIGHT TRAINING

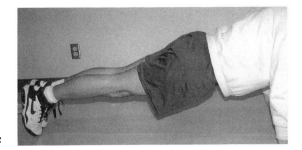

F

G

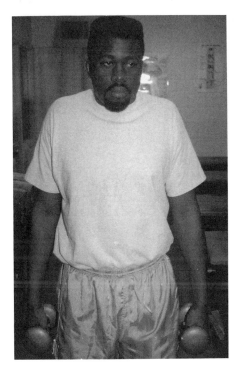

H

FIGURE 5–16. (continued) Weight-training exercises for the shoulder girdle include a push up for the abductors (F), bent-over row for the adductors (G), and dumbbell shrugs for the elevators (H).

to the movement. The latissimus dorsi and the pectoralis major muscles demonstrate a rapid increase in activity as they begin to slow the backward arm movement and initiate movement forward. The activity in these two muscles increases as tension is developed through stretching of the muscle fibers. This is an example of the stretch-shortening cycle discussed in a previous chapter. Likewise, the triceps brachii activity also increases to slow the forearm flexion, and the serratus anterior activity is increased to control the scapular movements. The biceps brachii also reaches its peak activity as elbow flexion reaches it maximum level, producing reciprocal muscle activity within the triceps brachii (25).

Muscles previously active in the early portion of the cocking phase also change in their level of output as the arm nears the completion of this phase. Teres minor and infraspinatus activity increase at the end of cocking to generate maximal external rotation. The activity of the supraspinatus increases as it maintains the abduction late into the cocking phase. The subscapularis activity also has been shown to increase up to maximum levels in preparation for the acceleration of the arm. The deltoid muscle is the only muscle whose activity diminishes late into the cocking phase (22).

At the end of the cocking phase, the external rotation motion is terminated by the anterior capsule and ligaments, and the subscapularis, pectoralis major, triceps brachii, teres major, and the latissimus dorsi muscles. Consequently, in this phase of the throwing action, the anterior capsule and ligaments and the tissue of the speci-

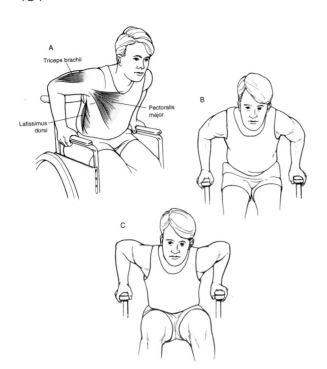

Push-up Actions
Types of Raising/Push-up Activities

Muscles	Chair Raise	Long Sit Elbow Flex	Long Sit Elbow Flex + ABD
Latissimus dorsi	*	**	***
Pectoralis major	**	*	***
Triceps brachii	***	***	***

* = low activity; ** = moderate activity; *** = high activity

FIGURE 5–17. (A) If you push up out of a chair, the muscle primarily used is the triceps brachii with some contribution from the pectoralis major and minimal assistance from the latissimus dorsi. (B) From the long sitting position with the elbows flexed, the triceps is still active but there is more contribution from the latissimus dorsi. (C) If the arms are abducted, all three muscles are very active in the raising movement.

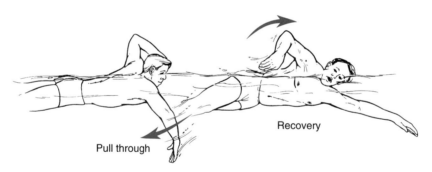

Swimming Muscle Actions
Phases

Muscles	Pull Through	Recovery
Deltoid	*	***
Infraspinatus		***
Latissimus dorsi	***	
Pectoralis major	***	
Serratus anterior		***
Supraspinatus		***
Teres major	***	

* = low activity; ** = moderate activity; *** = high activity

FIGURE 5–18. The shoulder muscles are significant contributors to the swimming action, generating the major percentage of the muscular power. In the pull-through phase of swimming the latissimus dorsi and the pectoralis major are very active. In the recovery phase as the arm is lifted out of the water, the deltoid, infraspinatus, serratus anterior, and the supraspinatus are active.

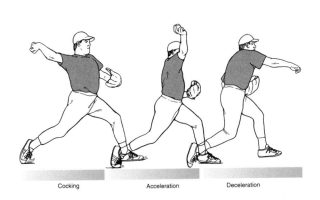

| | Throwing Muscle Actions | | |
| | Phases | | |
Muscles	Cocking	Acceleration	Deceleration
Biceps brachii	**	*	***
Brachialis	**	*	***
Deltoid	****	**	***
Infraspinatus	***	**	**
Latissimus dorsi	**	***	*
Pectoralis major	**	***	
Rhomboid	*	*	***
Serratus anterior	**	***	
Subscapularis	***	***	
Supraspinatus	***	**	***
Teres minor	***	**	**
Trapezius	*	*	***
Triceps brachii	*	***	

* = low activity; ** = moderate activity; *** = high activity

FIGURE 5–19. Although throwing uses the lower extremity to develop a significant portion of the power, upper extremity musculature also contributes to the activity. In the cocking phase, the arm is drawn back behind the body. In this phase, numerous muscles are active, with the infraspinatus, subscapularis, teres minor, and supraspinatus the most active. In the acceleration phase, the latissimus dorsi, pectoralis major, serratus anterior, subscapularis, and triceps brachii are active. In the follow-through or deceleration phase, there is substantial activity in the biceps brachii, brachialis, supraspinatus, and trapezius.

fied muscles are at greatest risk of injury (22,24). Examples of injuries developing in this phase are tendinitis of the insertion of the subscapularis and muscle strain of the pectoralis major, teres major, or latissimus dorsi.

The second phase of the throw is termed the acceleration phase and is an explosive action characterized by the initiation of elbow extension, arm internal rotation with maintenance of the 90 degrees of abduction, scapula protraction or abduction, and some horizontal flexion as the arm is brought around to the front. The muscles most active in the acceleration phase are those picking up activity late in the cocking phase, including: the subscapularis, latissimus dorsi, teres major, and pectoralis major, generating the horizontal flexion and the internal rotation movements; the serratus anterior, which pulls the scapula forward into protraction or abduction; and the triceps brachii, initiating and controlling the extension of the forearm. Muscles minimally active during the acceleration phase are the biceps brachii, the trapezius, the infraspinatus, and the teres minor (22,25). Sites of irritation and strain in this phase of the throw are found at the sites of the muscular attachment and in the subacromial area. This area is subjected to compression during the abduction, internal rotation movements in this phase.

The last phase of throwing is the follow-through or the deceleration phase, when the arm travels across the body and eventually stops over the opposite knee joint. This phase begins after ball release and, in the early phases,

after maximum internal rotation is achieved in the joint, there is a very quick muscular action resulting in external rotation and horizontal flexion of the arm. Following this into the later stages of the follow-through is trunk rotation, and the replication of the shoulder and scapular movements seen in the cocking phase. This includes an increase in the activity of the deltoid as it attempts to slow the horizontally flexed arm, the latissimus dorsi as it creates further internal rotation, the trapezius which creates slowing of the scapula, and the supraspinatus to maintain the arm abduction and continue to produce internal rotation (22,25). In this phase of throwing, the posterior capsule and corresponding muscles are at risk for injury since they are being rapidly stretched.

The last activity, the golf swing, presents a more complicated picture of shoulder muscle function because the left and right arms must work in concert, producing opposite movements and using opposing muscles to produce the action. The golf swing can be divided into four phases: the takeaway, forward swing, acceleration, and follow-through. These phases and their muscular actions are shown in FIGURE 5–20.

In the takeaway phase for a right-handed golfer, the club is brought up and back behind the body as the left arm comes across the body and the right arm abducts minimally. The shoulder muscular activity in this phase is minimal except for moderate subscapularis activity on the left arm to produce internal rotation, and marked

Golf Muscle Actions

Muscles	Takeaway	Forward Swing	Acceleration	Follow Through
		Phases		
L. Deltoid	*	*	*	*
R. Deltoid	*	*	*	*
L. Infraspinatus			**	
R. Infraspinatus			*	
L. Latissimus dorsi		**	***	**
R. Latissimus dorsi		**	***	
L. Pectoralis major			***	**
R. Pectoralis major		***	****	**
L. Subscapularis	**	**	***	***
R. Subscapularis		**	***	
L. Supraspinatus				
R. Supraspinatus	**	*		*

* = low activity; ** = moderate activity; *** = high activity

FIGURE 5–20. The upper extremity muscle activity in the take-away and forward swing phase of the golf swing is moderate, with only the pectoralis major registering high activity in the forward swing phase. However, in the acceleration phase, the muscular activity is high in the latissimus dorsi, pectoralis major, and the subscapularis on both sides of the body. Finally, high activity is seen in the subscapularis in the follow-through.

activity from the supraspinatus on the right side to abduct the arm (22).

In the next phase, the forward swing, movement of the club is initiated in the forward direction by moderate activity from the latissimus dorsi and subscapularis on the left side. On the right side, there is accompanying high activity from the pectoralis major, moderate activity from the latissimus dorsi and subscapularis, and minimal activity from the supraspinatus and the deltoid. This phase brings the club around to shoulder level through continued internal rotation of the left arm and the initiation of internal rotation of the right arm with some adduction.

The acceleration phase begins at approximately shoulder level and continues until contact is made with the golf ball. On the left side, there is substantial muscular activity in the pectoralis major, latissimus dorsi, and subscapularis as the arm is extended and maintained in inter-

nal rotation. On the right side, there is even more activity from these same three muscles as the arm is brought vigorously downward (22).

Once the ball is hit, the follow-through phase begins. Here, there is continued movement of the arm and club across the body to the left. This movement must be decelerated and slowed. In the follow-through phase, there is some minimal activity from the supraspinatus and the deltoid on the right side as the arm moves up and across the body. The left side has high activity in the subscapularis, and moderate activity in the pectoralis major, latissimus dorsi, and the infraspinatus as the upward movement of the arm is curtailed and slowed (22). It is here in the follow-through phase that considerable strain can be placed on the posterior portion of the right shoulder and the anterior portion of the left shoulder as the rapid deceleration occurs.

Injury Potential in the Shoulder Complex

The shoulder complex is subject to a wide variety of injuries that can be incurred through some trauma, usually in the form of contact with an external object such as the ground or another individual, or through repetitive joint actions, creating inflammatory sites in and around the joints or muscular attachments. When examining the various types of injuries to an anatomical region, pay particular attention to the cause of the injury so that you may work on injury reduction through recognition and elimination of factors predisposing an individual to injury.

The injuries occurring to the shoulder girdle are primarily a result of impacts received during falls or contact with an external object. The sternoclavicular joint can sprain or dislocate anteriorly if an individual falls on the top of the shoulder in the area of the middle deltoid. An individual with a sprain to this joint would experience pain in horizontal extension movements of the shoulder such as are found in the golf swing or the backstroke in swimming (26).

Anterior subluxations of this joint in adolescents have also occurred spontaneously during throwing because they have more mobility in this joint as compared to adults. A posterior dislocation or subluxation of the sternoclavicular joint can be quite serious since the trachea, esophagus, and numerous veins and arteries lie below this structure. This injury occurs as a consequence of force to the sternal end of the clavicle, and the individual may experience symptoms such as choking, shortness of breath, or difficulty in swallowing (26). Overall, the sternoclavicular joint is well reinforced with ligaments, and fortunately, the incidence of injury in the form of sprains, subluxations, and dislocations are not common.

The clavicle is frequently a site of injury by direct trauma received through contact in football or some other sport. The most common injury is a fracture to the middle third of the clavicle, received by falling on the shoulder or outstretched arm, or receiving a blow on the shoulder that sends a force down the shaft of the clavicle. Other less common fractures occur to the medial clavicle as a result of direct trauma, or the lateral end of the clavicle as a result of direct trauma to the tip of the shoulder (26). Clavicular fractures in adolescents heal very quickly and effectively, whereas in adults, the healing and repair process is not as efficient or effective. This is related to the differences in the level of skeletal maturation since in the adolescent, new bone is being formed at a much faster rate than in the mature individual.

Injuries to the acromioclavicular joint can cause a considerable amount of disruption to shoulder movements. Again, if you fall on the point of the shoulder, the acromioclavicular joint can subluxate or dislocate. This can also occur if you fall on your elbow or an outstretched arm. This joint is also frequently subjected to overuse injuries in sports using the overhand pattern such as throwing, tennis, and swimming. Other sports that repeatedly load the joint in the overhead position, such as weight lifting or wrestling, may also cause the overuse syndrome to occur. The consequence of overuse of the joint is capsule injury, ectopic calcification in the joint, and possible degeneration of the cartilage (26).

The scapula is rarely placed in a position where it receives sufficient force to create an injury. However, if an athlete or an individual falls on his/her upper back, it is possible to fracture the scapula and bruise the musculature so that abduction of the arm is quite painful. Another site of fracture on the scapula is the corocoid process, which can be fractured with separation of the acromioclavicular joint. Throwers can also acquire a bursitis at the inferomedial border of the scapula, creating pain as the scapula moves through the cocking and acceleration phases in the throw. The pain is relieved in the follow-through phase.

A traction force created during activities such as weight lifting (bench press, push-ups), lifting above the head, playing tennis, or carrying a backpack can produce trauma to the brachial nerve plexus. If the long thoracic nerve is impinged, there can be isolated paralysis of the serratus anterior, causing winging of the scapula and a decreased ability to abduct and flex at the shoulder joint (26).

The shoulder joint is commonly injured either through direct trauma or repeated overuse. Dislocation or subluxation in the glenohumeral joint is frequent due to the lack of bony restraint and the dependence on soft tissue for restraint and support of the joint. The glenoid fossa faces anterolaterally, creating more stability in the posterior joint than the anterior. Thus the most common direction of dislocation is anteriorly, with the anterior-inferior dislocation accounting for 95% of all dislocates (27).

The usual cause of the dislocation is contact or some force applied to the arm when it is in an abducted and externally rotated, overhead position. This drives the humeral head anteriorly, and it may tear the capsule or the glenoid labrum. The rate of recurrence of the dislocation depends on the age of the individual, and the magnitude of the force producing the dislocation (28). The recurrence rate for the general population is 33 to 50%, increasing to 66 to 90% if the dislocation occurs in an individual less than 20 years of age (29). In fact, the younger the age at the first dislocation, the more likely a recurrent dislocation will follow. Also, if a low amount of force created the dislocation, it is more likely that a recurrent dislocation will occur.

Recurrent dislocations also depend on the amount of initial damage and whether the glenoid labrum was also damaged (30). A tear to the glenoid labrum is similar to tearing the meniscus in the knee and results in clicking and pain in the overhead position (1). An anterior dislocation also makes it difficult to internally rotate the arm so that the opposite shoulder cannot be touched with the hand on the injured side.

Posterior dislocation of the shoulder is rare (2%) and usually is associated with a force applied with the arm adducted and internally rotated, and the hand below shoulder level (1). The clinical signs of a posterior dislocation are an inability to abduct and externally rotate the arm.

Soft tissue injuries around the shoulder joint are numerous and are associated most often with overhead motions of the arm such as are seen in throwing, swimming, and racket sports. The rotator cuff muscles, active in controlling the humeral head and motion during the overarm pattern, are very susceptible to injury.

In an upper extremity throwing pattern, when the arm is in the preparatory phase with the shoulder abducted and externally rotated, the anterior capsule and, specifically, the subscapularis muscle are susceptible to strain or tendinitis at the insertion on the lesser tuberosity (31). In late cocking and early acceleration phase, the posterior portion of the capsule and posterior labrum are susceptible to injury as the anterior shoulder is tightened, driving the head of the humerus backward (7). In the follow-through phase, when the arm is brought horizontally across the body at a very high speed, the posterior rotator cuff, infraspinatus, and teres minor, are very susceptible to muscle strain or tendinitis on the greater tuberosity insertion site as they work to decelerate the arm (32).

The most common mechanism of injury to the rotator cuff occurs when the greater tuberosity pushes up against the underside of the acromion process. This is known as the impingement syndrome and occurs during the acceleration phase of the overhand pattern when the arm is internally rotating while still maintained in the abducted position. This condition occurs in the range of 70 to 120 degrees of flexion or abduction and is more common in such activities as the tennis serve, throwing, butterfly stroke in swimming, and the overhand crawl in swimming (5). If a pitcher opens up too soon in the pitch or if an athlete maintains the shoulder joint in an internally rotated position, impingement is more likely to occur. It is also commonly injured in wheelchair athletes or in individuals transferring from the wheelchair to a bed or chair (33,34). The supraspinatus muscle, lying in the subacromial space, is compressed and can be torn with impingement, and with time, calcific deposits can be laid down in the muscle or tendon. This irritation can occur with any overhead activity, creating a painful arc of arm motion through 60 to 120 degrees of abduction or flexion (3).

Another injury that is a consequence of impingement is subacromial bursitis, an irritation of the bursae lying above the supraspinatus muscle and underneath the acromion process (5). This develops in wheelchair propulsion due to high pressure in the joint and abnormal distribution of stress in the subacromial area (33). Finally, the tendon of the long head of the biceps brachii can also become irritated when the arm is forcefully abducted and rotated. Bicipital tendinitis develops as the biceps tendon is subluxated or irritated within the bicipital groove. In throwing, the arm externally rotates to 160 degrees in the cocking phase and the elbow moves through 50 degrees of motion. Since the biceps brachii acts on the shoulder and also is responsible for decelerating the elbow in the final 30 degrees of extension, it is often maximally stressed (35). In a rapid throw, the long head of the biceps brachii may also be responsible for tearing the anterosuperior portion of the glenoid labrum. Irritation to the biceps tendon is manifested in a painful arc syndrome similar to that of the rotator cuff injury.

In summary, the shoulder complex provides us with the greatest mobility of any region in the body, but as a consequence of this great mobility, it is an unstable area subjected to numerous injuries. In spite of its high probability of injury, successful rehabilitation following surgery is quite common. It is important to maintain the strength and flexibility of the musculature surrounding the shoulder complex since there is considerable dependence on the musculature and soft tissue for support and stabilization.

The Elbow and Radioulnar Joints

The role of forearm movement, generated at the elbow or radioulnar joints, is to assist the shoulder in the application of force and in controlling the placement of the hand in space. The combination of shoulder and elbow/radioulnar joint movements affords us the capacity to place our hand in hundreds of thousands of different positions, allowing us tremendous versatility. Whether you are working above your head, shaking someone's hand, writing a note, or tying your shoes, hand position is important and is generated by the working relationship between the shoulder complex and the forearm.

Anatomical and Functional Characteristics of the Joints

The elbow joint is considered a very stable joint, with structural integrity, good ligamentous support, and good muscular support of the joint. In the elbow joint region, there are three joints allowing motion between the three bones of the arm and forearm (radius, ulna, humerus). Movement between the forearm and the arm takes place at the ulnarhumeral and radiohumeral articulations, while movements between the radius and the ulna take place at the radioulnar articulations (3). The ulnarhumeral, radiohumeral, and proximal radioulnar articulations are shown in FIGURE 5–21.

The ulnarhumeral joint is the articulation between the ulna and the humerus and is the major contributing joint to flexion and extension actions of the forearm. The joint is the union between the spool-like trochlea on the distal end of the humerus and the trochlear notch on the ulna. On the front of the ulna is the coronoid process, which blocks up against the coronoid fossa of the humerus and limits flexion in the terminal range of motion. Likewise, on the posterior side of the ulna is the olecranon process, which impacts the olecranon fossa on the humerus and terminates extension. An individual who can hyperextend at the elbow joint may have a smaller olecranon process or a larger olecranon fossa, thereby allowing more extension before blocking occurs.

The trochlear notch of the ulna fits snugly around the trochlea, offering good structural stability. The trochlea has articular cartilage over the anterior, inferior, and posterior surfaces, and is asymmetrical with an oblique posterior projection (36). In the extended position, this asymmetrical trochlea creates an angulation of the ulna laterally, creating a valgus position. This is termed the *carrying angle* and ranges from 10 to 15 degrees in males and 20 to 25 degrees in females (3,36). An illustration of how the carrying angle is measured is shown in FIGURE 5–22. As the forearm flexes, this valgus position is reduced and may even result in a varus position with full flexion.

The second joint participating in flexion and extension of the forearm is the radiohumeral articulation. At the distal end of the humerus is the articulating surface

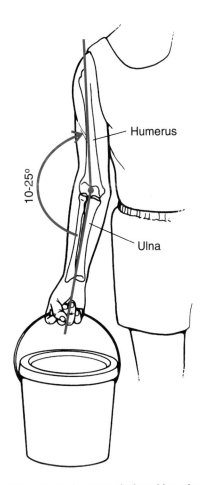

FIGURE 5–22. In the extended position, the ulna and the humerus form an angle because of asymmetry in the trochlea. This angle is called the *carrying angle*, and ranges from 10–25 degrees.

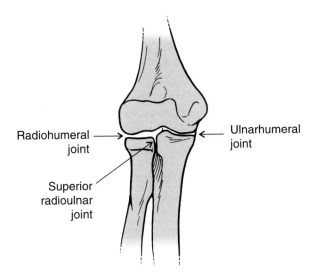

FIGURE 5–21. There are three articulations in the elbow joint region that are significant in terms of movement. Flexion and extension at the elbow joint occur at the ulnarhumeral and radiohumeral articulations. Pronation and supination occur at the superior and inferior radioulnar articulations (superior shown above).

for this joint, the capitullum, which is spheroidal in shape and covered with cartilage on the anterior and inferior surfaces. The top of the round radial head butts up against the capitullum, allowing for radial movement around the humerus during flexion and extension. The capitullum acts as a buttress for lateral compression and other rotational forces absorbed during throwing or other rapid forearm movements.

The third articulation, the radioulnar joint establishes the movement characteristics between the radius and the ulna in pronation and supination. There are actually two radioulnar articulations, the superior in the elbow joint region and the inferior in the wrist. Also, midway down between the radius and the ulna is another fibrous connection between the radius and the ulna, recognized by some as a third radioulnar articulation.

The superior or proximal radioulnar joint consists of the articulation between the radial head and the radial fossa on the side of the ulna. The radial head rotates in a fibrous, osseus ring and can turn both clockwise or counterclockwise, creating a movement of the radius relative to the ulna (37). In the neutral position, the radius and ulna lie next to each other, but in full pronation, the radius has crossed over the ulna diagonally. As the radius crosses over in pronation, the distal end of the ulna moves laterally, and vice versa during supination.

An interosseus membrane connects the radius and ulna, running the length of the two bones. This fascia increases the area for muscular attachment and ensures that the radius and ulna maintain a specific relationship to each other. It also transmits forces received distally from the radius to the ulna. The membrane is taut in the semiprone position (37).

Two final structural components in the elbow region worth examining are the medial and lateral epicondyles, prominent landmarks on the medial and lateral sides of the elbow. These extensions of the humerus serve as a site of muscular attachment for many of the hand muscles and are also sites of injury from repetitive overuse. Both epicondyles serve as good anatomical landmarks for the elbow joint.

The elbow joint is supported on the medial and lateral sides by collateral ligaments. The medial, or ulnar, collateral ligament connects the ulna to the humerus and offers support and resistance to valgus stresses imposed upon the elbow joint. Support in the valgus direction is very important in the elbow joint since most forces are directed medially creating a valgus force. Consequently, the ulnar collateral is taut in all joint positions. There is also a set of collateral ligaments on the lateral side of the

joint, termed the lateral or radial collateral ligaments. Since varus stresses are rare, these ligaments are not as significant in supporting the joint (2).

A ligament important for the function and support of the radius is the annular ligament, which wraps around the head of the radius, attaching to the side of the ulna. This ligament holds the radius up into the elbow joint while still allowing it to turn in the movements of pronation and supination. The insertion and stabilizing actions of ligaments supporting the elbow joint region are reviewed in Appendix A.

The three joints of the elbow complex do not all reach a close-packed position at the same point in the range of motion. Close-packed position for the radiohumeral is achieved when the forearm is flexed 80 degrees and is in the semiprone position (37). The fully extended position is the close-packed position, or position of maximum joint surface contact and ligamentous support, for the ulnarhumeral joint. Thus, when the ulnarhumeral articulation is most stable in the extended position, the radiohumeral articulation is loose-packed and least stable. The proximal radioulnar joint is in its close-packed position in the semiprone position, complementing the close-packed position of the radiohumeral (37).

The range of motion at the elbow in flexion and extension is approximately 145 degrees of active flexion, 160 degrees of passive flexion, and 5 to 10 degrees of hyperextension (37). The extension movement is limited by the joint capsule and the flexors. It is also terminally restrained by bone-on-bone impact with the olecranon process, as described in an earlier section.

Flexion at the joint is limited by the soft tissue, the posterior capsule, the extensor muscles, and the bone-on-bone contact of the coronoid process with its respective fossa. A significant amount of hypertrophy or fatty tissue will limit the range of motion in flexion considerably. Approximately 100 to 140 degrees of flexion and extension is required for most daily activities through the range of motion from 30 degrees of flexion to 130 degrees of flexion (2,3,38). For example, to reach the back of the head and comb the hair, 140 degrees of flexion is required, while only 15 degrees of flexion is required to tie a shoe (38).

The range of motion for pronation is approximately 70 degrees, and is limited by the ligaments, the joint capsule, and soft tissue compressing as the bones cross. Range of motion for supination is 85 degrees and is limited by ligaments, the capsule, and the pronator muscles. Approximately 50 degrees of pronation and 50 degrees of supination are required to perform most of our daily activities (2).

Muscular Actions

There are 24 muscles crossing the elbow joint, some of which act on the elbow joint exclusively and others that act at the wrist and finger joints (39). The majority of these muscles are capable of producing as many as three different movements at the elbow, wrist, or phalangeal joints, but one movement is dominant, and is the movement with which the muscle or muscle group is associated. The location, action, and nerve supply of the muscles acting at the elbow joint can be found in Appendix B.

The elbow flexors (biceps brachii, brachialis, brachioradialis, pronator teres, extensor carpi radialis) become more effective as flexion progresses because their mechanical advantage increases with an increase in the moment arm (39). The strongest flexor of the group is the brachialis, the "workhorse" of elbow flexion, because it is the only pure flexor and produces the greatest amount of work in comparison to the other muscles (3). The brachialis output is not influenced by pronation or supination of the forearm, and reaches its maximum output at approximately 120 degrees of flexion (40). It is active in all positions, at all speeds, and with or without resistance.

The biceps brachii has a long and short head that attaches to the superior aspect of the shoulder joint and on the corocoid process of the scapula, making both heads a two-joint muscle. At the other end of the biceps brachii, the muscle inserts into the radial tuberosity of the radius, which can be moved with supination and pronation of the forearm. Consequently, the contribution of the biceps brachii to forearm flexion will be dependent on both the position of the arm and the position of the forearm in pronation and supination.

The biceps brachii is most effective as a flexor in the supine forearm position when the attachment to the radius is not twisted under (2,3). The influence of pronation on the tendon of the biceps brachii is illustrated in FIGURE 5–23. The activity drops off in the semiprone position where it becomes active only against resistance, while in the prone position, the biceps brachii minimally contributes even against resistance. It has also been shown to make the most contribution to the flexion action during the middle 90 degrees of flexion, from 30 to 120 degrees of flexion (3,41).

The biceps brachii contribution can be increased if the arm is extended or hyperextended at the shoulder joint, where the insertion of the long head of the biceps brachii is put on stretch. The output from the biceps brachii is optimized at approximately 120 degrees of forearm flex-

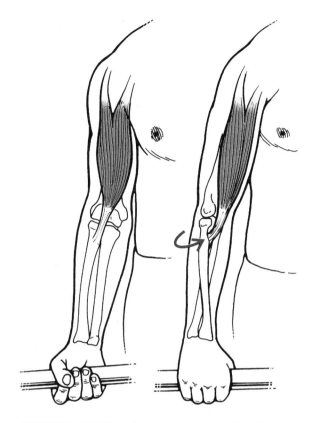

FIGURE 5–23. When the forearm is pronated, the attachment of the biceps brachii to the radius is twisted under. This position interferes with the flexion-producing action of the biceps brachii. The biceps brachii is more efficient in producing flexion when the forearm is supinated and the tendon is not twisted under the radius.

ion and can be maximized further with the forearm supinated and the arm extending (40).

The brachioradialis is a muscle with small volume and very long fibers. It is a very efficient muscle, used more with rapid flexion movements or against resistance (2). The brachioradialis has been shown to produce the most activity at 120 degrees of flexion with the forearm in the supine position (40). The brachioradialis does not increase its activity when the arm is semiprone or prone.

In the extensor muscle group is the powerful triceps brachii, the strongest muscle of all of the forearm muscles, with great strength potential and work capacity due to muscle volume and the mechanical advantage it gains from its large moment arm (39). The triceps brachii has three portions: the long head, medial head, and lateral head. Of these three, only the long head crosses the shoulder joint, making it dependent partially on shoulder position for its effectiveness. It is the least active of the triceps, but it can be increasingly more involved with shoulder flexion as its insertion on the shoulder is stretched.

The medial head of the triceps brachii is considered the "workhorse" of the extension movement because it is active in all positions, at all speeds, and against maximal or minimal resistance. The lateral head of the triceps brachii, although the strongest, remains relatively inactive unless movement occurs against a resistance (3). The output of the triceps brachii is not influenced by forearm positions of pronation and supination.

A third movement of the forearm occurring at the radioulnar articulations is pronation, produced by the pronator quadratus and pronator teres. The activity and the overall contribution of the pronator quadratus is considerably greater than that of the pronator teres. The pronator quadratus is more active regardless of forearm position, whether the activity is slow or fast, and whether working against a resistance or not. The pronator teres is called upon to become more active when the pronation action becomes rapid or against a high load (3). The pronator teres is most active at 60 degrees of forearm flexion (40).

The final movement of the forearm, supination, is produced by the supinator muscle and the biceps brachii under certain circumstances. The supinator is the only muscle contributing to a slow unresisted supination action in all forearm positions (2). The biceps brachii can supinate during rapid or rested movements when the elbow is flexed. The flexion action of the biceps brachii is neutralized by actions from the triceps brachii, allowing contribution to the supination action. At 90 degrees of flexion, the biceps brachii becomes a very effective supinator (3).

The muscles surrounding the elbow joint are positioned to also create forces in the mediolateral direction. The muscles on the lateral side of the joint, the anconeus, brachioradialis, extensor carpi radialis, extensor digitorum communis, and the extensor carpi ulnaris create a valgus force. The muscles on the medial side of the joint, the pronator teres, flexor carpi radialis, flexor digitorum superficialis, and flexor carpi ulnaris, create varus forces (39). These muscles should be considered and included in a conditioning program when stability at the elbow joint is required.

Strength of the Forearm Muscles

The flexor muscle group is almost twice as strong as the extensors in all positions, making us better pullers than pushers (3). The joint forces created by a maximum isometric flexion in an extended position is equal to approximately two times body weight (2).

The semiprone position is the position in which maximum strength in flexion can be developed, followed by the supine, and lastly, the pronated position (3,42). The semiprone position is most commonly used in daily activities. Semiprone flexion exercises should be included in a conditioning routine to take advantage of the strong position of the forearm.

Extension strength is greatest from a position of 90 degrees of flexion (2). This is a common forearm position for daily living activities and for power positions of upper-extremity sport skills. Finally, pronation/supination strength is greatest from the semiprone position, with the torque dropping off considerably at higher or lower angles (3).

The load carrying capacity of the elbow joint is considerable. In a push-up, the peak axial forces on the elbow joint average 45% of body weight (43). These forces depend on hand position, with the force reduced to 42.7% of body weight when the hands were placed further apart from the normal position (43).

Conditioning

The effectiveness of exercises used to strengthen or stretch will depend on the various positions of the arm and the forearm. In stretching the muscles, the only positions putting any form of stretch on the flexors and extensors must incorporate some hyperextension and flexion at the shoulder joints. Stretching these muscles while the arm is in the neutral position is almost impossible because of the bony restrictions to the range of motion.

Strengthening the flexors and extensors can be enhanced through improvement of the length-tension relationship in the muscle by placing the arm in a position of hyperextension when working the flexors, and in flexion when working the extensors. The use of stretch-contract in a strength exercise will facilitate the activity and output from the muscles.

The position of the forearm is also important in forearm strengthening activities. The forearm position in which the flexors and extensors are the strongest is the semiprone position. For the flexors specifically, the biceps brachii can be brought more or less into the exercise by supinating or pronating, respectively. There are numerous exercises available for both the flexors and extensors; examples are provided in FIGURES 5–24 and 5–25.

The pronators and supinators offer a greater challenge in the prescription of strength or resistive exercises (FIGURE 5–26). Stretching these muscle groups presents no

FLEXIBILITY

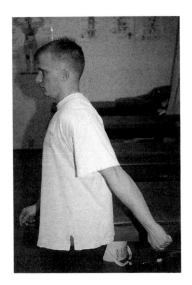

A

MANUAL RESISTANCE

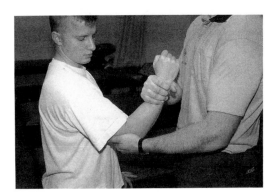

B

WEIGHT TRAINING

C

E

D

FIGURE 5–24. Only the biceps brachii, which crosses the shoulder joint, can be stretched to any significant degree. A stretch can be imposed by hyperextending the arm with the elbow extended (A). A good manual resistance exercise is one in which the elbow is stabilized and resistance is applied at the wrist region as the forearm is pulled into flexion (B). The flexors are an easy group to weight train using exercises like the upright row (C), dumbbell curls (D), and the pull-up (E).

FLEXIBILITY

MANUAL RESISTANCE

A

B

C

WEIGHT TRAINING

D

E

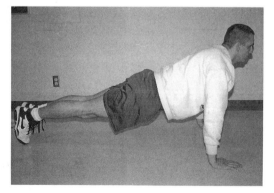

F

FIGURE 5–25. The triceps brachii can be stretched by pulling the arm up, behind the head with the elbow flexed (A). Manual resistance can be used to strengthen the triceps brachii by stabilizing the elbow and applying resistance at the wrist as the individual extends (B) or the same exercise can be performed from the supine position (C). Weight-training exercises for the forearm extensors are common and include the french curl (D), the triceps press (E), and the push-up (F).

FLEXIBILITY

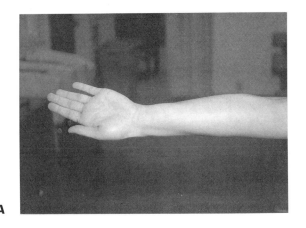

A

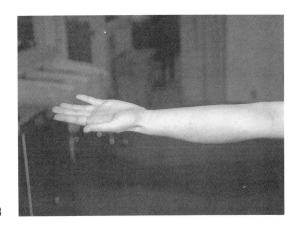

B

MANUAL RESISTANCE

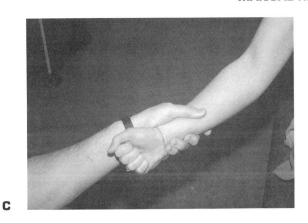

C

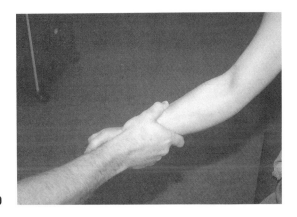

D

WEIGHT RESISTANCE

E

FIGURE 5–26. The pronators and supinators are not usually included in a stretching or strengthening routine, but there is much to be gained by including them. To stretch the supinators, turn the hand from the supine, anatomical position all the way around until the palm is facing posteriorly (A). To stretch the pronators, move the forearm into the maximum supination position which will be one close to anatomical position (B). The supinators and pronators can be strengthened with manual resistance by applying resistance on the outside of the forearm during supination (C) or on the inside of the forearm during pronation (D). The elbow should be flexed to reduce contribution from the shoulder rotators. Higher resistance exercises can be performed for both supination and pronation using a dumbbell (E).

problem, since a maximal supination position will adequately stretch the pronation musculature, and vice versa. Also, low resistance exercises can be implemented by applying a force in a turning action to a door knob or some other immovable object. However, high resistance exercises implore the use of creativity since there are no standardized sets of exercises for these muscles.

Contribution to Sports Skills or Movements

The forearm musculature, specifically the forearm extensors, are very important sources of force for raising up out of a chair. As shown in FIGURE 5–17, raising up or wheelchair activities require a great amount of extension at the elbow joint, thus the triceps brachii becomes a very important muscle to strengthen in the elderly or in wheelchair-bound individuals (21).

In throwing, the forearm flexors and extensors are active throughout the total throwing action (FIGURE 5–19). In the cocking phase, the biceps brachii and the brachialis are active as the forearm flexes and the arm is abducted. The activity of the triceps brachii begins at the end of the cocking phase, when the arm is in maximum external rotation and the elbow is maximally flexed. There is a co-contraction of the biceps brachii and the triceps brachii at this time. Additionally, the forearm is pronated to 90 degrees at the end of the cocking phase via the pronator teres and pronator quadratus (25).

In the acceleration phase of throwing, the activity of the triceps brachii increases significantly as the forearm is rapidly extended, and the activity continues until a high forearm extension velocity is achieved. The activity of both the biceps brachii and the brachialis drops off rapidly in the acceleration phase (25).

Just after release, the negative acceleration phase begins and is marked by a rapid drop off in the activity of the triceps brachii. There is a very rapid increase in the activity of the biceps brachii and the brachialis in the follow-through phase as these muscles attempt to reduce the tensile loads on the rapidly extending forearm. It is here in this portion of the throwing action that the biceps brachii is most susceptible to strain (35). The overall activity of the biceps brachii and the brachialis increases with corresponding increases in speed of the throw.

Injury Potential in the Forearm

The elbow joint is subjected to injuries caused by the absorption of a high force, such as in falling, but the majority of the injuries around the elbow joint are a consequence of participating in repetitive activities such as throwing. The high-impact injuries will be presented first, followed by the more common overuse injuries.

One of the injuries occurring as a consequence of absorbing a high force is dislocation, usually occurring in sports such as gymnastics, football, or wrestling, when the athlete falls on an outstretched arm, forcing a posterior dislocation (44). With the dislocation, there can be a fracture that commonly occurs in the medial epicondyle. Other areas that may fracture with a fall include the olecranon process, the radial head, or the shaft of either or both radius and ulna. Additionally, spiral fractures of the humerus can be incurred through a fall or through high forces absorbed in a throwing action.

Direct blows to any muscle can culminate in a condition known as myositis ossificans, where the body deposits ectopic bone in the muscle in response to the severe bruising and repeated stress to the muscle tissue. Although most common in the quadriceps femoris in the thigh, the brachioradialis muscle in the lower arm is the second most common area of the body to be susceptible to repeated blows and thus this condition (44).

A high muscular force can create a rupture of the long head of the biceps brachii. Commonly seen in the adult, the joint movements facilitating this injury are arm hyperextension, forearm extension, and forearm pronation, and if these three movements occur concomitantly, the strain on the biceps brachii may be tremendous. Finally, falling on the elbow can irritate the olecranon bursa, creating olecranon bursitis, looks very disabling due to the swelling, but is actually minimally painful and quite functional (37).

The overuse injuries occurring around the elbow are usually associated with throwing or some overhead pattern such as the tennis serve. In throwing, there are stringent demands placed upon the medial side of the elbow joint. Through the high-velocity actions of the throw, there are large tensile forces developed on the medial side of the elbow joint, compressive forces developed on the lateral side of the joint, and shear forces generated on the posterior side of the joint. A maximal valgus force is applied to the medial side of the elbow during the latter part of the cocking phase and through the initial portion of the acceleration phase, and is responsible for creating the medial tension syndrome or pitchers elbow (44). This excessive valgus force is responsible for sprain or rupture of the ulnar collateral ligaments, medial epicondylitis, tendinitis of the forearm or wrist flexors, avulsion fractures to the medial epicondyle, and osteochondritis dissecans to the capitellum or olecranon (3,44).

Medial epicondylitis is an irritation of the insertion site of the wrist flexors attached to the medial epicondyle. They are stressed with the valgus force accompanied by wrist actions. The osteochondritis dissecans, a lesion in the bone and articular cartilage, commonly occurs on the capitellum due to the compression during the valgus position that forces the radial head up against the capitellum. During the valgus overload, coupled with forearm extension, the olecranon process can be wedged up against the fossa creating an additional site for osteochondritis dissecans and breakdown in the bone. Additionally, the olecranon is subject to high tensile forces and can experience a traction apophysitis, or bony outgrowth, similar to that seen with the patellar ligament of the quadriceps femoris group (44).

The lateral overuse injuries to the elbow usually occur as a consequence of overuse of the wrist extensors at their attachment site on the lateral epicondyle. The overuse of the wrist extensors occurs as they eccentrically slow down or resist any flexion movement at the wrist. Lateral epicondylitis, or tennis elbow, is associated with force overload due to improper technique or use of heavy rackets. If the backhand stroke in tennis is executed with the elbow leading or if the performer hits the ball consistently off-center, the wrist extensors and the lateral epicondyle will become irritated (45). Also, a large grip or tight strings may increase the load on the epicondyle by the extensors.

The Wrist and Fingers

The hand is primarily used for manipulation activities requiring very fine movements incorporating a wide variety of hand and finger postures. Consequently, there is much interplay between the wrist joint positions and efficiency of finger actions. The hand region has many stable, yet very mobile segments, with very complex muscle and joint actions.

Anatomical and Functional Characteristics of the Joints

Beginning with the most proximal joints of the hand and working to the tips of the fingers offers the best perspective on the interaction between segments and joints in the hand. All of the joints of the hand are illustrated in FIGURE 5–27. To review the anatomical structures in the wrist and hand, refer to Appendix A for ligamentous structure and Appendix B for muscular location, action, and innervation. The wrist joint, or radiocarpal joint, is the articulation where movement of the whole hand occurs. The radiocarpal involves the broad distal end of the radius and two carpals, the scaphoid and the lunate. There is also minimal contact and involvement with the triquetrum. This ellipsoid joint allows movement in two planes: flexion/extension and radial/ulnar flexion.

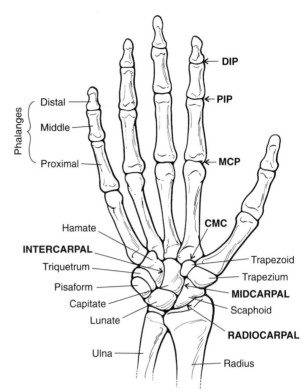

FIGURE 5–27. The wrist and hand can perform both precision and power movements because of numerous joints controlled by a large number of muscles. Most of the muscles originate in the forearm and enter the hand as tendons.

Adjacent to the radiocarpal joint, but not participating in any wrist movements is the distal radioulnar articulation. The ulna makes no actual contact with the carpals and is separated by a fibrocartilage disk. This arrangement is important so that the ulna can glide on the disk in pronation and supination, while at the same time not influencing wrist or carpal movements.

In order to understand wrist joint function, the structure and function at the joints between the carpals must be examined. There are two rows of carpals, the proximal row, containing the three carpals participating in wrist joint function (lunate, scaphoid, triquetrum) and the little pisiform bone that sits on the medial side of the hand serving as a site of muscular attachment. In the distal row there are also four carpals, the trapezium interfacing with the thumb at the saddle joint, the trapezoid, the large capitate, and the hamate.

The articulation between the two rows of carpals is called the *midcarpal joint*, and the articulation between a pair of carpals is known as an *intercarpal joint*. All of these joints are gliding joints in which translatory movements are produced concomitantly with wrist movements, although the proximal row is more mobile than the distal row (46). There is a concave transverse arch running across the carpals, forming the floor and walls of the carpal tunnel through which the tendons of the flexors and the median nerve travel.

The scaphoid may be one of the most important carpals since it supports the weight of the arm, transmits forces received from the hand to the bones of the forearm, and is also a key participant in wrist joint actions. The scaphoid supports the weight of the arm and transmits forces when the hand is fixed and the forearm weight is applied to the hand. Because the scaphoid interjects into the distal row of carpals, it sometimes will move with the proximal row, and other times with the distal row.

When the hand flexes at the wrist joint, the movement is initiated at the midcarpal joint, accounting for 60% of the total range of flexion motion (3). Forty percent of the wrist flexion is attributable to movement of the scaphoid and lunate on the radius. The total range of motion for flexion is 70 to 90 degrees, and it is believed that only 10 to 15 degrees of wrist flexion is needed for most daily activities involving the hand (2). Wrist flexion range of motion will be reduced if flexion is performed with the fingers flexed because of the resistance offered by the finger extensors.

Wrist extension is also initiated at the midcarpal joint, where the capitate moves quickly and becomes close-packed with the scaphoid, drawing the scaphoid in with the movements of the second row of carpals. This reverses the role of the midcarpal and radiocarpal joints to the movement, with over 60% of the movement now being produced at the radiocarpal joint and over 30% at the midcarpal joint (3). This switch is attributed to the fact that the scaphoid moves with the proximal row of carpals in the flexion movement and with the distal row of carpals in extension. The range of motion for extension is 70 to 80 degrees, with approximately 35 degrees of extension needed for daily activities (2,46). The range of motion of wrist extension will be reduced if the extension is performed with the fingers extended.

The hand can also move laterally in radial and ulnar flexion or deviation. These movements are created as the proximal row of carpals glide over the distal row. In the radial flexion movement, the proximal row moves toward the ulna and the distal row moves toward the radius. The opposite occurs for ulnar flexion. The range of motion for radial flexion is 15 to 20 degrees and for ulnar flexion, 30 to 40 degrees (2).

The close-packed position for the wrist, in which maximal support is offered, is in a position of hyperex-

tension. The close-packed position for the midcarpal joint is a position of radial flexion (3). Both of these positions should be considered when selecting positions that would maximize stability in the hand. For example, in racket sports, the wrist will be more stable in a slightly hyperextended position. Also, when one falls on the hand with the arm outstretched and the wrist hyperextended, the wrist, and specifically, the scaphoid carpal bone are especially susceptible to injury because it is in the close-packed position.

Moving distally, the next articulation is the carpometacarpal (CMC) joint, which connects the carpals with each of the five fingers via the metacarpals. Each metacarpal and phalange is also called a ray and is numbered from the thumb to the little finger, with the thumb being the first ray and the little finger the fifth. The CMC articulation is the joint providing the most movement for the thumb and the least movement for the fingers, and thus, will be discussed separately.

For the four fingers, the CMC joint offers very little movement, being a gliding joint that moves directionally with the carpals. The movement is very restricted at the second and third CMC but increases to allow as much as 10 to 30 degrees of flexion and extension at the CMC of the ring and little finger (2). There is also a concave transverse arch across the metacarpals of the fingers similar to that of the carpals. This arch facilitates the gripping potential of the hand.

The CMC joint of the first ray, or thumb, is a saddle joint consisting of the articulation between the trapezium and the first metacarpal. It provides the thumb with a major portion of its range of motion, allowing for 50 to 80 degrees of flexion and extension, 40 to 80 degrees of abduction and adduction, and 10 to 15 degrees of rotation (3,46). Since the thumb is oriented sideways to the hand, it can be moved through a much more diversified range of motion than the big toe that attaches to face the same direction as the toes. The thumb sits at an angle of 60 to 80 degrees to the arch of the hand and has a wide range of functional movements (47).

The thumb can touch each of the fingers in the movement of opposition, and is very important in all gripping and prehension tasks. Opposition can take place through a range of motion of approximately 90 degrees. Without the thumb, and specifically the movements allowed at the CMC, the function of the hand would be very limited.

The metacarpals connect with the phalanges to form the metacarpophalangeal joints (MCP). Again, the function of the MCP of the four fingers differs from that of the thumb, so they will be presented separately. The MCP of the four fingers is a condyloid joint allowing movement in two planes: flexion/extension and abduction/adduction. The joint is well reinforced on the dorsal side by the dorsal hood of the fingers, on the palmar side by the palmar plates that span the joint, and on the sides by the collateral ligaments or deep transverse ligaments.

The fingers can flex through 70 to 90 degrees of flexion, with most flexion allowed in the little finger and the least allowed in the index finger (3). Flexion, which determines grip strength, can be more effective and produces more force when the wrist joint is held in 20 to 30 degrees of hyperextension, a position that increases the length of the finger flexors.

Extension of the fingers at the MCP can take place through about 25 degrees of motion. The extension motion can be limited by the position of the wrist, and would be limited with the wrist hyperextended and enhanced with the wrist flexed.

The fingers spread in abduction and are brought back together in adduction at the MCP joint. Approximately 20 degrees of abduction and adduction are allowed (46). Abduction is extremely limited if the fingers are flexed, since the collateral ligaments become very tight and disallow the movement. Thus, fingers can be abducted when extended and in the process of catching an object, and then cannot be abducted or adducted when flexed around the object, assuring a firm grip.

The MCP joint for the thumb is a hinge joint allowing motion in only one plane. The joint is reinforced with collateral ligaments and the palmar plates, but is not connected with the other fingers via the deep transverse ligaments. Approximately 30 to 90 degrees of flexion and 15 degrees of extension can take place at this joint (46).

The most distal joints in the upper extremity link are the interphalangeal articulations (IP). Each finger has two IP joints, the proximal interphalangeal (PIP) and the distal interphalangeal joints (DIP). The thumb has one IP joint and consequently has only two sections or phalanges, the proximal and distal phalanges, whereas the fingers have three phalanges, the proximal, middle, and distal phalanges. The IP joints are hinge joints allowing for movement in one plane only (flexion/extension), and they are reinforced on the lateral sides of the joints by collateral ligaments, disallowing movement outside of flexion and extension. The range of motion in flexion of the fingers is 110 degrees at the PIP, and 90 degrees at the DIP and the IP of the thumb (2, 46).

Like the MCP joint, the flexion strength at these joints determines grip strength and can be enhanced with the wrist hyperextended 20 degrees, and would be impaired if the wrist were flexed. Various finger positions can be obtained through antagonistic and synergistic actions

from other muscles so that all fingers can flex or extend at the same time, there can be extension of the MCP with flexion of the IP, and vice versa. There is usually no hyperextension allowed at the IP joints unless an individual has ligaments that are longer and allow extension to occur because of joint laxity.

Muscular Actions

Most of the muscles acting at the wrist and finger joints originate outside the hand in the region of the elbow joint and are termed *extrinsic* muscles (See Appendix B). These muscles enter the hand as tendons that can be quite long, as in the case of some finger tendons that eventually terminate on the distal tip of a finger. The tendons are held in place on the dorsal and palmar wrist area by extensor and flexor retinaculums, bands of fibrous tissue running transversely across the distal forearm and wrist. During wrist and finger movements, the tendons move through considerable distances, but are still maintained by the retinacula.

In addition to the muscles originating in the forearm, there are intrinsic muscles, originating within the hand, that create movements at the metacarpophalangeal and interphalangeal joints. The four intrinsic muscles of the thumb form the fleshy region in the palm known as the thenar eminence, and three intrinsic muscles of the little finger form the smaller hypothenar eminence, the fleshy ridge on the little finger side of the palm.

The wrist flexors (flexor carpi ulnaris, flexor carpi radialis, palmaris longus) are all fusiform muscles originating in the vicinity of the medial epicondyle on the humerus and running about halfway down the forearm before becoming a tendon. The flexor carpi radialis and flexor carpi ulnaris contribute the most to wrist flexion, since the palmaris longus is so variable and may be as small as a tendon, or even absent in 13% of the population (3). The strongest flexor of the group, the flexor carpi ulnaris, gains some of its power by encasing the pisiform bone and using it as a sesamoid bone to increase mechanical advantage and reduce the overall tension on the tendon. Since most activities require the use of a small amount of wrist flexion, attention should always be given to the conditioning of this muscle group.

The wrist extensors (extensor carpi ulnaris, extensor carpi radialis longus, extensor carpi radialis brevis) originate in the vicinity of the lateral epicondyle, and the muscles become tendons about a third of the way down the forearm. The wrist extensors also act and create movements at the elbow joint, thus elbow joint position is important for extensor function. The extensor carpi radialis longus and extensor carpi radialis brevis create flexion at the elbow joint, and thus can be enhanced as a wrist extensor with extension at the elbow. The extensor carpi ulnaris creates extension at the elbow and would be enhanced as a wrist extensor in elbow flexion. Also, wrist extension is an important action accompanying and supporting the gripping action using finger flexion; thus the wrist extensors are active with this activity.

Both the wrist flexors and extensors pair up to produce the movements of ulnar and radial flexion. Ulnar flexion is produced by the ulnaris wrist muscles consisting of the flexor carpi ulnaris and the extensor carpi ulnaris. Likewise, radial flexion is produced by the flexor carpi radialis, extensor carpi radialis longus, and the extensor carpi radialis brevis. The radial flexion joint movement, although only capable of half the range of motion as ulnar flexion, is important in many racket sports since it creates the close-packed position, stabilizing the hand (46).

Finger flexion of rays 2 through 5 is performed primarily by the flexor digitorum profundus and flexor digitorum superficialis, the extrinsic muscles originating in the vicinity of the medial epicondyle. The flexor digitorum profundus cannot independently flex each finger; thus, flexion at the middle, ring, and little fingers will usually occur together since the flexor tendons all arise from a common tendon and muscle. However, the index finger can independently flex because of the separation of the flexor digitorum profundus muscle and tendon for this digit.

The flexor digitorum superficialis is capable of flexing each finger independently; thus, the fingers can be independently flexed at the PIP but not at the DIP. Flexion of the little finger is also assisted by one of the intrinsic muscles, the flexor digiti minimi brevis. Flexion of the fingers at the metacarpophalangeal articulation is produced by two sets of intrinsic muscles lying in the palm and between the metacarpals, the lumbricales, and the interossei. These muscles also produce extension at the interphalangeal joints since they attach into the fibrous extensor hood running the length of the dorsal surface of the fingers. Consequently, to achieve full flexion of the MP, PIP, and DIP joints, the long finger flexors must override the extension component of the lumbricales and interossei (3). This is easier if tension is taken off the extensors by creating some wrist extension.

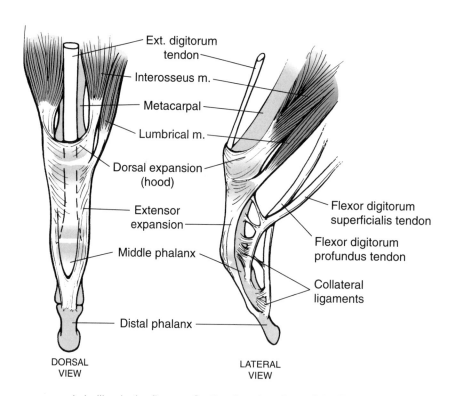

FIGURE 5–28. There are no muscle bellies in the fingers. On the dorsal surface of the fingers is the extensor expansion and the extensor hood into which the finger extensors attach. Tendons of the finger flexors travel the ventral surface of the fingers. The fingers flex and extend as tension is generated in the tendons via muscular activity in the upper forearm.

The extension of fingers 2 through 5 is created primarily by the extensor digitorum, originating from the lateral epicondyle, entering the hand as four tendon slips that branch off at the metacarpophalangeal articulation. The tendons create a main slip that inserts into the extensor hood and two collateral slips that connect into adjacent fingers. The extensor hood, formed by the tendon of the extensor digitorum and fibrous connective tissue, wraps around the dorsal surface of the phalanges and runs the total length of the finger to the distal phalanx. The structures in the finger are shown in FIGURE 5-28.

Since the lumbricales and interossei connect into this hood, they also assist with extension of the PIP and DIP joints, and their actions are facilitated as the extensor digitorum contracts, applying tension to the extensor hood and stretching these muscles (3,46).

Abduction of fingers two, three, and four is performed by the dorsal interossei, consisting of four intrinsic muscles lying between the metacarpals. They connect to the lateral sides of digits two and four and to both sides of the middle finger, digit three. The little finger, digit five, is abducted by one of its intrinsic muscles, the abductor digiti minimi brevis.

The three palmar interossei, lying on the medial side of digits two, four, and five, pull the fingers back into adduction. The middle finger is adducted by the dorsal interossei, which is connected to both sides of the middle finger. Abduction and adduction movements are very necessary for grasping, catching, and gripping objects. Abduction is severely limited when the fingers are flexed due to the tightening of the collateral ligament, and the limited length-tension relationship in the interossei, which are also flexors of the metacarpophalangeal joint (3,46).

The thumb has eight muscles controlling and generating an expansive array of movements. The muscles of the thumb are presented in Appendix B. The thumb movement of opposition is the most important because it provides the opportunity to pinch, grasp, or grip an object by bringing the thumb across to any of the fingers. Although all of the hypothenar muscles contribute to opposition, the main muscle responsible for initiating the movement is the opponens pollicis. The little finger is also assisted in opposition by the opponens digiti minimi.

Strength of the Hand and Fingers

Strength in the hand is usually associated with grip strength and there are many different ways or purposes to a grasp or grip of an object. A firm grip requiring maximum output will utilize the extrinsic muscles, whereas fine movements such as a pinch will use more of the intrinsic muscles to finely tune the movements.

The strength of a grip can be enhanced by the position of the wrist. Placing the wrist in ulnar flexion increases the strength output of the PIP and DIP flexors the most, followed by wrist hyperextension, and lastly wrist flexion. Thus, a grip can be strengthened if the wrist is placed in a position of slight ulnar flexion and hyperextension. Likewise, the grip can be loosened if the wrist is put in a flexion position. Grip strength at 40 degrees of wrist hyperextension is more than three times greater than that of grip strength measured in 40 degrees of wrist flexion (2). The strength of the grip may increase with specific wrist positioning, but the incidence of strain or impingement on structures around the wrist also increases. The neutral position of the wrist is the safest position and reduces strain on the wrist structures.

The strongest muscles in the hand region, capable of the greatest work capacity, are, in order from high to low, the flexor digitorum profundus, the flexor carpi ulnaris, the extensor digitorum, the flexor pollicis longus, the extensor carpi ulnaris, and the extensor carpi radialis longus (2). Two muscles that are weak and capable of little work capacity are the palmaris longus and the extensor pollicis longus.

Conditioning

There are three main reasons why people condition the hand region. First, the fingers can be strengthened to enhance the grip strength in athletes who participate in racket sports, individuals who work with implements, or who lack the ability to grasp or grip objects.

Secondly, the muscles acting at the wrist joint are usually strengthened and stretched to facilitate a wrist position for racket sports or to enhance wrist action in a throwing or striking event such as volleyball.

The final reason for conditioning the hand region is injury prevention or reduction. The tension developed in the hand and finger flexors and extensors places considerable strain on the medial and lateral aspect of the elbow joint. Some of this strain can be reduced through stretching and strengthening exercises.

Overall, the conditioning of the hand region is simple and can be done in a very limited environment, such as an office, with minimal equipment. Examples of some flexibility and resistance exercises for the wrist flexors and extensors and the fingers are presented in FIGURES 5–29 and 5–30, respectively. Wrist curls and tennis-ball-gripping exercises are the most popular for this region.

Contribution to Sport Skills or Movements

At the wrist joint, there can be a dynamic contribution to a skill or movement, or there can be stabilization of the

FLEXIBILITY

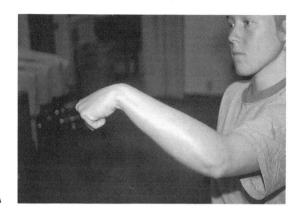

A

B

MANUAL RESISTANCE

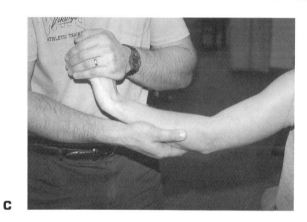

C

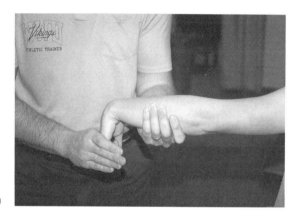

D

WEIGHT TRAINING

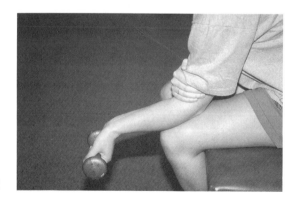

E

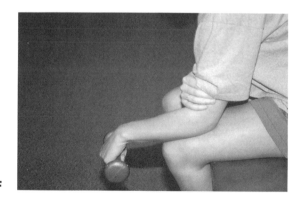

F

FIGURE 5–29. Stretch the wrist extensors and flexors by placing the hand in extreme flexion (A) or hyperextension (B), respectively. The extensors and flexors can be strengthened with manual exercises by applying pressure on the top of the hand during extension (C), or holding the hand as it moves into flexion (D). The flexors can be further strengthened with wrist curls (E) and the extensors with the use of the reverse wrist curl (F).

FLEXIBILITY

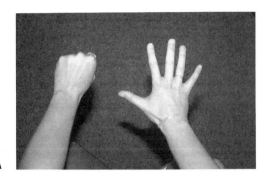

A

MANUAL RESISTANCE

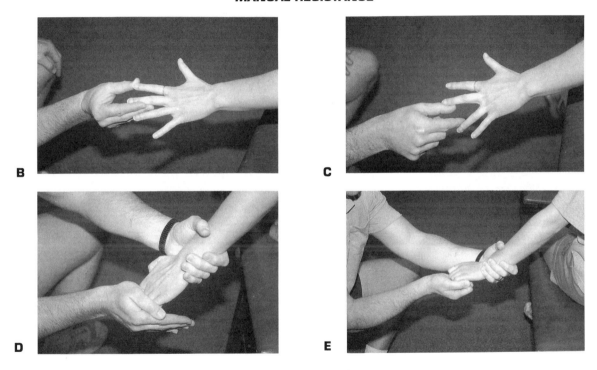

B

C

D

E

WEIGHT TRAINING

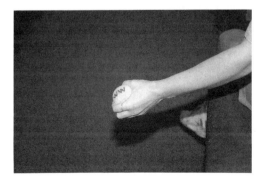

F

FIGURE 5–30. The finger muscles can be stretched by opening and closing the fingers (A), strengthened by applying a resistance on the fingers as they abduct (B), or adduct (C), or over the top as they extend (D), and underneath as the fingers attempt to form a fist (E), squeezing a tennis ball is also good (F).

joint in a fixed position. In golf, for example, the wrist actively flexes radially via the flexor carpi radialis and extensor carpi radialis muscles in the preparatory phase, and ulnar flexes via the opposite muscle groups in the power phase. These same joint actions would be seen in the baseball swing, and are significant contributors to the success of both events.

Wrist extension draws the hand back, and wrist flexion snaps the wrist through in activities such as serving and spiking in volleyball, dribbling in basketball, and throwing a baseball. Even though the speed of the flexion and extension movement may be determined by contributions from adjacent joints, strengthening of the flexors and extensors would enhance the force production.

Commonly, the wrist is maintained in position so that an efficient force application can occur. In tennis and racket sports, for example, the wrist is either held in the neutral position or in a slightly radially flexed position. By holding the wrist stationary, the force applied to the ball by the racket will not be lost through movements occurring at the wrist. This position is maintained by both the wrist flexors and extensors.

Another example of maintaining a wrist position is in the volleyball underhand pass, in which the wrist is maintained in an ulnar flexed position. This opens up a broader area for contact and locks the elbows so they maintain an extended position upon contact. Finally, the wrist must be maintained in a stable, static position to achieve maximal performance from the fingers. Thus, while playing a piano or typing, the wrist must be maintained in the optimal position for finger usage, which is usually slight hyperextension via the wrist extensors. All of these wrist positions, while allowing for stability or creating more efficient muscle action, also create additional stress on the joint and may result in an overuse injury to the region.

In the hand, grip strength and the type of grip are the most important contributions to sport skills or daily activities and involve maintaining a static position of the fingers and the thumb. In a grip, the fingers flex to wrap around an object. If a power grip is needed, the fingers will flex more, with the most powerful grip being the fist position, where there is flexion at all three finger joints, the MP, PIP, and DIP. If a fine precision grip is required, there may be only limited flexion at the PIP and DIP, and only one or two fingers may be involved, such as in pinching or in writing (2). Examples of both power and precision grips are shown in FIGURE 5–31.

It is really the thumb that determines whether a fine precision position or power position is generated. If the thumb remains in the plane of the hand in an adducted position and the fingers flex around an object, a power position is created. An example of this would be the grip used in the javelin throw and in the golf swing. This power position still allows for some precision, important in directing the club or javelin.

Power can be enhanced in the grip by producing a fist with the thumb wrapped over the fully flexed fingers. With this grip, there is minimal, if any, precision, thus, accuracy is lost. In addition, most power positions of the fingers are accompanied by an ulnar flexion and extension at the wrist joint.

In activities requiring precise actions, the thumb is positioned more perpendicular to the hand and moved into opposition. There is also accompanying limited flexion at the fingers. An example of this type of position would be in pitching, writing, and pinching. In a pinch or prehensile grip, more force can be generated if the pulp of the thumb is placed against the pulps of the index and long fingers. This pinch is 40% stronger than the pinch grip with the tips of the thumb and fingers (48).

Finally, the dynamic actions of the fingers present in a wide variety of activities, such as writing, tying shoes, or playing a musical instrument, require a very coordinated action between flexors, extensors, abductors, and adductors of all the fingers. An analysis of the muscular actions contributing to an activity such as playing a piece on a piano would be extensive and is beyond the scope of this text.

Injury Potential in the Hand and Fingers

There are many injuries that can occur to the hand as a result of absorbing a blunt force, as in impact with a ball, the ground, or another object. Common injuries of this type in the wrist region usually are associated with a fall on the hand, forcing the wrist into an extreme position of flexion or extension, with extreme hyperextension being the most common. This can result in a sprain of the wrist ligaments, a strain of the wrist muscles, a fracture of the scaphoid (70%) or other carpals (30%), a fracture of the distal radius, or a dislocation between the carpals and the wrist or other carpals (49).

The end of the radius is one of the most frequently fractured areas of the body since it is less dense and the force from a fall is absorbed by the radius. A common fracture of the radius, the Colles' fracture, is a diagonal fracture that forces the radius into more radial flexion and shortens it. These injuries are associated more with activities such as hockey, fencing, football, rugby, skiing, soccer, bicycling, parachuting, mountain climbing, and hang-gliding, in which the chance of a blunt macrotrauma is higher than other activities.

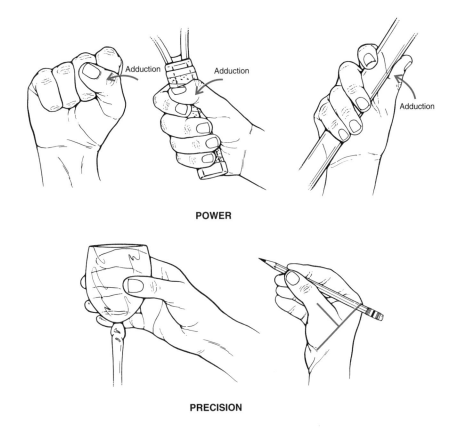

POWER

PRECISION

FIGURE 5–31. If power is needed in grip, the fingers will flex at all three finger joints to form a fist. Also, if the thumb adducts the grip will be more powerful. A precision grip, on the other hand, usually involves slight flexion at a small number of finger joints with an accompanying thumb position that is perpendicular to the hand.

Examples of injuries to the fingers and the thumb as a result of blunt impact are fractures, dislocations, and tendon avulsions. The thumb can be injured by jamming it or forcing it back into extension, causing severe strain of the thenar muscles and strain of the ligaments surrounding the MCP joint. A Bennet's fracture is a common fracture to the thumb, occurring at the base of the first metacarpal as a result of the jamming or forced thumb positioning. Thumb injuries are common in skiing due to jamming created by the ski pole (50). Thumb injuries are also common in biking (51).

The fingers are also frequently fractured or dislocated by a ball hitting the tip of the finger and forcing it into extreme flexion or extension, or by some other impact with the ground or another object. Fractures are more common in the proximal phalanx and rarely occur in the middle phalanx. High-impact collisions with the hand, such as are seen in boxing or the martial arts result in more fractures or dislocations of the ring and little fingers since they are least supported in the fist position.

The finger flexor or extensor mechanisms can be disrupted with a blow forcing the finger into extreme positions. *Mallet finger* is an avulsion injury to the extensor tendon at the distal phalanx due to forced flexion, and results in the loss of the ability to extend the finger. *Boutonniere deformity*, caused by the avulsion or stretching of the middle branch of the extensor mechanism creates a stiff and immobile PIP articulation (3). Avulsion of the finger flexors is called *jersey finger* and is caused by forced hyperextension of the distal phalanx. The finger flexors can also develop nodules on them, creating a *trigger finger*, resulting in snapping during flexion and extension of the fingers. These finger and thumb injuries are also commonly associated with the sports and activities listed above, due to the incidence of impact occurring to the hand region.

There are also some overuse injuries associated with repetitive use of the hand in sports, work, or other activities. Tenosynovitis of the radial flexors and thumb muscles is common in activities such as canoeing, rowing, rodeo activities, tennis, and fencing. Tennis, golf, throwing, javelin, hockey and other racket sports, in which the wrist flexors and extensors are used to stabilize the wrist or create a repetitive wrist action, are susceptible to tendinitis of the wrist muscles inserting into the medial and lateral epicondyles. Medial or lateral epicondylitis may also result from this overuse. Basically, medial epicondylitis is associated with overuse of the wrist flexors, while lateral epicondylitis is associated with overuse of the wrist extensors.

A disabling overuse injury occurring to the hand is *carpal tunnel syndrome*. Next to low back injuries, carpal tunnel syndrome is one of the most frequent work injuries reported by the medical profession. The floor and sides of the carpal tunnel are formed by the carpals, and the top is formed by a transverse ligament. Travelling through this tunnel are all of the wrist flexor tendons and the median nerve (FIGURE 5–32). Through repetitive actions at the wrist, usually involving repeated wrist flexion, the wrist flexor tendons may be inflamed to the point

where there is pressure and constriction put on the median nerve. The median nerve innervates the radial side of the hand, specifically the thenar muscles of the thumb. Impingement of this nerve can cause pain, atrophy of the thenar muscles, and tingling sensations in the radial side of the hand.

To eliminate this condition, the source of the irritation must be removed by examining the workplace environment, a wrist stabilizing device can be applied to reduce the magnitude of the flexor forces, or a surgical release can be administered. It is recommended that the wrist be maintained in a neutral position while performing tasks in the workplace in order to avoid carpal tunnel syndrome.

Ulnar nerve injuries can also result in loss of function to the ulnar side of the hand, specifically the ring and little finger. Damage to this nerve can occur as a result of trauma to the elbow or shoulder region. Ulnar neuropathy is associated with activities such as cycling (52).

Summary

The upper extremity is much more mobile than the lower extremity, even though they have structural simi-

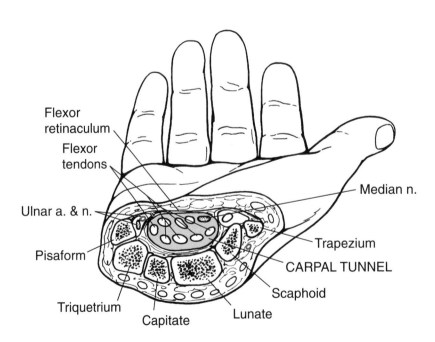

FIGURE 5–32. The floor and sides of the carpal tunnel are formed by the carpals and the top of the tunnel is covered by ligament and the flexor retinaculum. Within the tunnel are wrist flexor tendons and the median nerve. Through overuse of the wrist flexors, the median nerve can be impinged causing carpal tunnel syndrome.

larities. There are similarities in the connection into girdles, the number of segments, and the decreasing size of the bones toward the distal end of the extremities.

The shoulder complex consists of the sternoclavicular joint, the acromioclavicular joint, and the glenohumeral joint. The sternoclavicular joint is very stable, and allows the clavicle to move in elevation and depression, protraction and retraction, and rotation. The acromioclavicular joint is a small joint that allows the scapula to protract and retract, elevate and depress, and upwardly and downwardly rotate. The glenohumeral joint provides movement of the humerus through flexion and extension, abduction and adduction, medial and lateral rotation, and combination movements of horizontal abduction and adduction, and circumduction. A final articulation, the scapulathoracic joint is termed a *physiological joint* because of the lack of connection between two bones. It is here that the scapula moves on the thorax.

The movements of the arm at the shoulder joint are considerable. The arm can move through 180 degrees of abduction, flexion, and rotation because of the interplay between movements occurring at all of the articulations. The timing of the movements between the movements of the arm, scapula, and clavicle is termed the scapulohumeral rhythm. Through 180 degrees of elevation (flexion or abduction), there is approximately 2:1 degrees of humeral movement to scapular movements.

The muscles creating movement of the shoulder and shoulder girdle are also very important for maintaining stability in the region. In abduction and flexion, for example, the deltoid produces about 50% of the muscular force for the movement, but it requires assistance from the rotator cuff (teres minor, subscapularis, infraspinatus, supraspinatus) to stabilize the head of the humerus so that elevation can occur. Also, the shoulder girdle muscles contribute as the serratus anterior and the trapezius assist to stabilize the scapula and produce accompanying movements of elevation, upward rotation, and protraction.

To extend the arm against a resistance, the latissimus dorsi, teres major, and pectoralis major act on the humerus and are joined by the rhomboid and the pectoralis minor, which retract, depress, and downwardly rotate the scapula. Similar muscular contributions are made by the infraspinatus and teres minor in external rotation of the humerus and the subscapularis, latissimus dorsi, teres major, and pectoralis major in internal rotation.

The shoulder muscles can generate considerable amounts of force in the adduction and extension movements in which the muscle mass is the greatest. The next strongest movement is flexion and the weakest move-

ments are abduction and rotation. The muscles surrounding the shoulder joint are capable of generating high forces in the range of 8 to 9 times the weight of the limb.

Conditioning of the shoulder joint muscles is easy because of the mobility of the joint. Numerous strength and flexibility exercises are used to isolate specific muscle groups or to replicate an upper extremity pattern used in a skill. There should be some special exercise considerations for individuals with shoulder injury that exclude any exercise creating impingement in the joint.

Shoulder joint muscles are very important contributors to specific sport skills or movements. In the push-up, for example, the pectoralis major, latissimus dorsi, and triceps brachii are important contributors. In swimming, the latissimus dorsi, teres major, pectoralis major, supraspinatus, infraspinatus, middle deltoid, and serratus anterior make important contributions. In throwing, the deltoid, supraspinatus, infraspinatus, teres minor, subscapularis, trapezius, rhomboid, latissimus dorsi, pectoralis major, teres major, and deltoid all contribute.

Injury to the shoulder joint complex can be acute in the case of dislocations of the sternoclavicular or glenohumeral joints and fractures of the clavicle or humerus, or the injuries can be chronic like bursitis or tendinitis. Common injuries associated with impingement of the shoulder joint are subacromial bursitis, bicipital tendinitis, and tears in the supraspinatus muscle.

The elbow and the radioulnar joints assist the shoulder in applying force and putting the hand in a proper position. The joints making up the elbow joint are the ulnarhumeral and radiohumeral, where flexion and extension occur, and the superior radioulnar, where pronation and supination of the forearm occur. The region is well supported by ligaments and the interosseus membrane running between the radius and the ulna. The joint structures allow approximately 145 to 160 degrees of flexion and 70 to 85 degrees of pronation and supination.

There are 24 muscles spanning the elbow joint and these can be further classified into flexors (biceps brachii, brachioradialis, brachialis, pronator teres, extensor carpi radialis), extensors (triceps brachii, anconeus), pronators (pronator quadratus, pronator teres), and supinators (biceps brachii, supinator). The flexor muscle group is considerably stronger than the extensors. Maximum flexion strength can be developed from the semiprone forearm position. Extension strength is maximum from a flexion position of 90 degrees, and pronation and supination strength is also maximum from the semiprone position.

The forearm muscles are used in a variety of different skills and movements. For example, the triceps brachii is an important contributor to rising from a chair, wheelchair activities, and throwing. Likewise, the biceps brachii and the pronator muscles are important in various phases of throwing.

The elbow and forearm are vulnerable to injury as a result of falling or of some repetitive overuse. In absorbing high forces, the elbow can dislocate or fracture, or muscles can rupture. Through overuse, injuries such as medial or lateral tension syndrome can produce epicondylitis, tendinitis, or avulsion fractures.

The wrist and hand consist of very complex structures that work together to provide fine movements used in a variety of daily activities. The main joints of the hand consist of the radiocarpal joint, the inferior radioulnar joint, the midcarpal and intercarpal joints, the carpometacarpal joints, the metacarpophalangeal joints, and the interphalangeal joints. The hand is capable of moving through 70 to 90 degrees of wrist flexion, 70 to 80 degrees of extension, 15 to 20 degrees of radial flexion, and 30 to 40 degrees of ulnar flexion. The fingers can flex through 70 to 110 degrees of flexion depending on the actual joint of interest (MCP or IP), 20 to 30 degrees of hyperextension, and 20 degrees of abduction. The thumb has special structural and functional characteristics that are related to the role of the carpometacarpal joint.

The extrinsic muscles acting on the hand enter the region as tendons. The muscles work in groups to produce wrist flexion (flexor carpi ulnaris, flexor carpi radialis, palmaris longus), extension (extensor carpi ulnaris, extensor carpi radialis longus, extensor carpi radialis brevis), ulnar flexion (flexor carpi ulnaris, extensor carpi ulnaris), and radial flexion (flexor carpi radialis, extensor carpi radialis longus, extensor carpi radialis brevis). Finger flexion is produced by the flexor digitorum profundus and flexor digitorum superficialis, and extension is produced primarily by the extensor digitorum. The fingers are abducted by the dorsal interossei and adducted by the palmar interossei.

Strength in the fingers is important in activities or sports in which a firm grip is essential. The strength of the grip can be enhanced by placing the thumb in a position parallel with the fingers (fist position). Where precision is needed the thumb should be placed perpendicular to the fingers. The muscles of the hand can be exercised via a series of exercises that incorporates various wrist and finger positions.

The fingers and hand are frequently injured because of their vulnerability, especially when performing activities such as catching balls. Sprains, strains, fractures, and dislocations are common results of injuries sustained by the fingers or hands in the absorption of an external force. There are also common injuries in the hand that are associated with overuse, and these include medial or lateral tendinitis or epicondylitis, and carpal tunnel syndrome.

Review Questions

1. List the range of motion for movements at the shoulder, elbow, wrist, and phalangeal joints. Estimate the degrees of motion at each joint that would be necessary to produce the following movements:
   a. combing your hair
   b. rowing
   c. forehand drive in tennis
   d. jump shot in basketball
   e. scratching the middle of your back (overhead)
2. Summarize the upper extremity muscular contributions to the following:
   a. pull-up
   b. push-up
   c. baseball batting
   d. backstroke in swimming
   e. opening a door
3. Describe a set of flexibility exercises for the following muscle(s):
   a. arm flexors and abductors
   b. arm extensors and adductors
   c. rotator cuff
   d. forearm flexors and extensors
   e. forearm pronators and supinators
   f. hand flexors and extensors
   g. finger flexors
4. Design both a low resistance and high resistance exercise routine for the following muscle(s):
   a. arm flexors and abductors
   b. arm extensors and adductors
   c. rotator cuff
   d. forearm flexors and extensors

   e. forearm pronators and supinators
   f. hand flexors and extensors
   g. finger flexors
5. For each of the following movements or skills, describe the upper extremity muscles or joint structures that are susceptible to injury, state why they are susceptible, and at what phase in the activity they are susceptible to the injury.
   a. volleyball
   b. bench press
   c. butterfly stroke in swimming
   d. backhand in tennis
   e. golf swing
   f. arm abduction
   g. arm horizontal flexion
   h. parallel bar dips
6. Compare the differences in musculature used in a wide-grip bench press versus a narrow-grip press.
7. How does forearm flexion strength vary with pronation and supination positioning? What forearm position would you select for exercising this muscle group? Why?
8. Analyze the dumbbell lateral raise and compare it with a pull-over exercise in terms of shoulder muscles used.
9. Analyze the listed exercises and offer reasons why you think each exercise is a good or bad exercise for the upper extremity.
   a. push-up
   b. pull-up
   c. jumping jacks
   d. arm circles
   e. upright row

Additional Questions

1. Select an activity using finger movements and describe how the movement can be made more efficient with positioning of the wrist, elbow, and radioulnar joints.
2. Describe the scapulohumeral rhythm that would take place in both raising and lowering the arm.
3. Compare biceps brachii action in supination of the forearm in the extended forearm position with the forearm in flexed position. Explain.
4. Vigorously flex the forearm, then check the movement

suddenly before completing the full range of movement. Palpate the triceps brachii. Does it contract during any part of the movement? Explain.
5. In what movement does the extensor carpi ulnaris and flexor carpi ulnaris act as mutual neutralizers?
6. Perform a push-up. How does the muscular activity change in the lowering phase of the push-up compared to the raising phase?

Additional Reading

An, K.N., et al.: Tendon excursion and moment arm of index finger muscles. Journal of Biomechanics. *16*:419–426, 1983.

Andrews, J.R., and Whiteside, J.A.: Common elbow problems in the athlete. Journal of Sports Physical Therapy. *17*:289–295, 1993.

Baker, C.L., Uribe, J.W., and Whitman, C.: Arthroscopic evaluation of acute initial anterior shoulder dislocations. The American Journal of Sports Medicine. *18*:25–28, 1990.

Barmakian, J.T.: Anatomy of the joints of the thumb. Hand Clinics. *8*:683–691, 1992.

Basmajian, J.V.: Recent advances in the functional anatomy of the upper limb. American Journal of Physical Medicine. *18*:165–177, 1969.

Basset, R.W., Browne, A.O., Morrey, B.F., and An, K.N.: Glenohumeral muscle force and moment mechanics in a position of shoulder instability. Journal of Biomechanics. *23*:405–412, 1990.

Blakely, R.L., and Palmer, M.L.: Analysis of shoulder rotation accompanying a proprioceptive neuromuscular facilitation approach. Physical Therapy. *66*:1224–1227, 1986.

Bowers, K.D.: Treatment of acromioclavicular sprains in athletes. The Physician and Sports Medicine. *11*:79–89, 1979.

Boyd, H.B., and Sisk, D.: Recurrent posterior dislocation of the shoulder. The Journal of Bone and Joint Surgery. *54-A*:779–786, 1972.

Brewer, B.J.: Aging of the rotator cuff. The American Journal of Sports Medicine. *7*:102–110, 1972.

Brown, L.P., et al.: Upper extremity range of motion and isokinetic strength of the internal and external shoulder rotators in major league baseball players. The American Journal of Sports Medicine. *16*:577–585, 1988.

Brunet, M.E., Haddad, R.J. Jr., and Porshe, E.B.: Rotator cuff impingement syndrome in sports. The Physician and Sports Medicine. *10*:86–94, 1982.

Bryan, R.S., and Dobyns, J.H.: Fractures of the carpal bones other than the lunate and navicular. Clinical Orthopaedics and Related Research. *149*:107–111, 1979.

Carmichael, S.W., and Hart, D.L.: Anatomy of the shoulder joint. The Journal of Orthopaedic and Sports Physical Therapy. *6*:225–228, 1985.

Carson, W.G.: Rehabilitation of the throwing shoulder. Clinics in Sports Medicine. *8*:657–689, 1989.

Coleman, A.E.: In-season strength training in major league baseball players. The Physician and Sports Medicine. *10*:125–132, 1982.

Constant, C.R.: Injuries to the elbow. Journal of Royal College of Edinburgh. *35(Supp)*:S31–32, 1990.

Conwell, E.: Injuries to the wrist. Clinical Symposia. *22*:3–30, 1970.

Cook, E.E., Gray, V.L., Savinar-Nogue, E., and Medeiros, J.: Shoulder antagonistic strength ratios: A comparison between college-level baseball pitchers and nonpitchers. The Journal of Orthopaedic and Sports Physical Therapy. *8*:451–461, 1987.

Craig, S.M.: Anatomy of the joints of the fingers. Hand Clinics. *8*:693–700, 1992.

Danzig, L., Resnick, D., and Greenway, G.: Evaluation of unstable shoulders by computed tomography. The American Journal of Sports Medicine. *10*:138–141, 1982.

Dobyns, J.H., Franklin, H.S., and Linsheid, R.L.: Sports stress syndromes of the hand and wrist. The American Journal of Sports Medicine. *6*:236–253, 1978.

Donatelli, R., and Greenfield, B.: Case study: Rehabilitation of a stiff and painful shoulder: A biomechanical approach. The Journal of Orthopaedic and Sports Physical Therapy. *9*:118–126, 1987.

Ellenbecker, T.S., and Derscheid, G.L.: Rehabilitation of overuse injuries of the shoulder. Clinics in Sports Medicine. *8*:583–604, 1989.

Ellenbecker, T.S., Davies, G.L., and Rowinsky, M.J.: Concentric versus eccentric isokinetic strengthening of the rotator cuff. The American Journal of Sports Medicine. *16*:64–69, 1988.

Engle, R.P., and Canner, G.C.: Posterior shoulder instability: Approach to rehabilitation. The Journal of Orthopaedic and Sports Physical Therapy. *10(12)*:488–494, 1989.

Ferrari, D.A.: Capsular ligaments of the shoulder. The American Journal of Sports Medicine. *18*:20–24, 1990.

Fisk, G.R.: An overview of the injuries of the wrist. Clinical Orthopaedics and Related Research. *149*:137–143, 1980.

Gerdle, B., Eriksson, N.E., and Hagberg, C.: Changes in the surface electromyogram during increasing isometric shoulder forward flexions. European Journal of Applied Physiology. *57*:404–408, 1988.

Gerdle, B., Elert, J., and Henriksson-Larsen, K.: Muscular fatigue during repeated isokinetic shoulder forward flexions in young females. European Journal of Applied Sciences. *58*:666–673, 1989.

Gerdle, B., Eriksson, N.E., Brundin, L., and Edstrom, M.: Surface EMG recordings during maximum static shoulder forward flexion in different positions. European Journal of Applied Physiology. *57*:415–419, 1988.

Gowan, D., et al.: A comparative electromyographic analysis of the shoulder during pitching. The American Journal of Sports Medicine. *15*:486–490, 1987.

Gregg, J.R., et al.: Serratus anterior paralysis in the young athlete. Journal of Bone and Joint Surgery. *61-A*:825–832, 1979.

Hagberg, M.: Electromyographic signs of shoulder muscular fatigue in two elevated arm positions. American Journal of Physical Medicine. *60*:111–121, 1981.

Hagberg, C., and Hagberg, M.: Surface EMG amplitude and frequency dependence on exerted force for the upper trapezius muscle: A comparison between right and left sides. European Journal of Applied Physiology. *58*:641–645, 1989.

Hart, D.L., and Carmichael, S.W.: Biomechanics of the shoulder. The Journal of Orthopaedic and Sports Physical Therapy. *16*:229–278, 1985.

Howell, S.M., Imoberseg, A.M., Seger, D.H., and Marone, P.J.: Clarification of the role of the supraspinatus muscle in shoulder function. The Journal of Bone and Joint Surgery. *68(3)*:398–404, 1985.

Imaeda, T., An, K., Cooney, W. P., and Linscheid, R.: Anatomy of trapeziometacarpal ligament. The Journal of Hand Surgery. *18A*:226–231, 1993.

Jobe, F.W., and Bradley, J.P.: The diagnosis and nonoperative treatment of shoulder injuries in athletes. Office Practice of Sports Medicine. *8*:419–433, 1989.

Jobe, F.W., and Moynes, D.R.: Delineation of diagnostic criteria and a rehabilitation program for rotator cuff injuries. The American Journal of Sports Medicine. *10*:336–339, 1982.

Jobe, F.W., Tibone, J.E., Perry, J., and Moynes, D.: An EMG analysis of the shoulder in throwing and pitching. The American Journal of Sports Medicine. *11*:3–5, 1983.

Jones, L., and Hunter, I.W.: Changes in pinch force with bidirectional load forces. Journal of Motor Behavior. *24*:157–164, 1992.

Kauer, J.M.: Functional anatomy of the wrist. Clinical Orthopaedics and Related Research. *149*:9–19, 1980.

Knudsen, D.V.: Hand forces and impact effectiveness in the tennis forehand. Journal of Human Movement Studies. *17*:1–7, 1989.

Knudsen, D.V: Factors affecting force loading on the hand in the tennis forehand. The Journal of Sports Medicine and Physical Fitness. *31*:527–531, 1991.

Knudsen, D.V., and White, S.C.: Forces on the hand in the tennis forehand drive: Application of force sensing resistors. International Journal of Sport Biomechanics. *5*:324–331, 1989.

Kuhlman, J.R., et al.: Isokinetic and isometric measurement of strength of external rotation and abduction of the shoulder. The Journal of Bone and Joint Surgery. *74-A*:1320–1333, 1992.

Landerjerit, B., and Maton, B.: In vivo muscular force analysis during the isometric flexion of a monkey's elbow. Journal of Biomechanics. *21*:577–584, 1988.

Martin, P.E., and Heise, G.D: Archery bow grip force distribution: Relationships with performance and fatigue. International Journal of Sport Biomechanics. *8*:305–319, 1992.

Minamikawa, Y., et al.: Stability and constraint of the proximal interphalangeal joint. The Journal of Hand Surgery. *18A*:198–204, 1993.

Moynes, D.R.: Prevention of injury to the shoulder through exercises and therapy. Clinics in Sports Medicine. *2*:413–422, 1983.

Neer, C.S., and Foster, C.R.: Inferior capsular shift for involuntary inferior and multi-directional instability of the shoulder. The Journal of Bone and Joint Surgery. *62-A*:897–907, 1980.

Nemeth, G., Kronberg, M., and Brostrom, L. Electromyogram (EMG) recordings from the sub-scapularis muscle: Description of a technique. Journal of Orthopaedic Research. *8*:151–153, 1990.

Nicholson, G.G.: Rehabilitation of common shoulder injuries. Clinics in Sports Medicine. *8*:657–689, 1989.

Nicholson, G.G.: The effects of passive joint mobilization on pain and hypomobility associated with adhesive capsulitis of the shoulder. The Journal of Orthopaedic and Sports Physical Therapy. *6*:238–246, 1985.

Nielsen, A.J.: Case study: Myofascial pain of the posterior shoulder relieved by spray and stretch. The Journal of Orthopaedic and Sports Physical Therapy. *3*:21–26, 1981.

Nikolaou, P.K., et al.: Biomechanical and historical evaluation of muscle after controlled strain injury. The American Journal of Sports Medicine. *15*:9–14, 1987.

O'Driscoll, S.W., Morrey, B. F., Korinek, S., and An., K.: Elbow subluxation and dislocation. Clinical Orthopaedics and Related Research. *280*:186–197, 1992.

Otis, J.C., et al.: Torque production in the shoulder of the normal young adult male. The American Journal of Sports Medicine. *18*:119–123, 1990.

Pappas, A.: Elbow problems associated with baseball during childhood and adolescence. American Journal of Sports Medicine. *11*:30–41, 1982.

Paulos, L.E., and Franklin, J.L.: Arthroscopic shoulder decompression development and application. The American Journal of Sports Medicine. *18*:235–244, 1990.

Pearl, M.L., Perry, J., Torburn, L., and Gordon, L.H.: An electromyographic analysis of the shoulder during cones and planes of arm motion. Clinical Orthopaedics and Related Research. *284*:116–127, 1992.

Peat, M.: Functional anatomy of the shoulder complex. Physical Therapy. *66*:1855–1865, 1986.

Perry, J.: Normal upper extremity kinesiology. Physical Therapy. *58*:265–269, 1978.

Regan, W.D., Korinek, S.L., Morrey, B.F., and An, K.: Biomechanical study of ligaments around the elbow joint. Clinical Orthopaedics and Related Research. *271*:170–179, 1991.

Richards, R.R., Gordon, R., and Beaton, D.: Measurement of wrist, metacarpophalangeal joint, and thumb extension strength in a normal population. The Journal of Hand Surgery. *18A*:253–261, 1993.

Sarrafian, S.K., Melamed, J.L., and Goshgarian, G.M.: Study of wrist motion in flexion and extension. Clinical Orthopaedics and Related Research. *126*:153–159, 1977.

Schenkman, M., and De Cartaya, V.R.: Kinesiology of the shoulder complex. The Journal of Orthopaedic and Sports Physical Therapy. *8*:438–450, 1987.

Silfverskiold, J., and Waters, R.L.: Shoulder pain and functional disability in spinal cord injury patients. Clinical Orthopaedics and Related Research. *272*:141–145, 1991.

Soderberg, G.J., and Blaschak, M.J.: Shoulder internal and external rotation peak torque production. The Journal of Orthopaedic and Sports Physical Therapy. *8*:518–524, 1987.

Stroyan, M., and Wilk, K.E.: The functional anatomy of the elbow complex. Journal of Sport Physical Therapy. *17*:279–288, 1993.

Tank, R., and Halbach, J.: Physical therapy evaluation of the shoulder complex in athletes. The Journal of Orthopaedic and Sports Physical Therapy. *3*:108–119, 1982.

Taylor, D.C., Dalton, J.D., Seaber, A.V., and Garret, W.E.: Viscoelastic properties of muscle tendon units. The American Journal of Sports Medicine. *18*:300–309, 1990.

Thein, L.A.: Impingement syndrome and its conservative management. Journal of Sports Physical Therapy. *11*:183–191, 1989.

Tibone, J.E., et al.: Surgical treatment of tears of the rotator cuff in athletes. The Journal of Bone and Joint Surgery. *68-A*:887–891, 1986.

Tolbert, J.R., Blair, W.F., Andrews, J.G., and Crowninshield, R.D.: Kinetics of normal and prosthetic wrists. Journal of Biomechanics. *18*:887–897, 1985.

Van Woensel, W., and Arwert, H.: Effects of external load and abduction angle on EMG level of shoulder muscles during isometric action. Electromyographic Clinical Neurophysiology. *33*:185–191, 1993.

Veeger, H.E.J., et al.: Inertia and muscle contraction parameters for musculoskeletal modeling of the shoulder mechanism. Journal of Biomechanics. *24*:615–629, 1990.

Volz, R.G., Lieb, M., and Benjamin, J.: Biomechanics of the wrist. Clinical Orthopaedics and Related Research. *149*:112–117, 1980.

Wadsworth, C.T.: Clinical anatomy and mechanics of the wrist and hand. Journal of Orthopaedics and Sports Physical Therapy. *4*:206–216, 1983.

Walmsley, R.P., and Szybbo, C.: A comparative study of the torque generated by the shoulder internal and external rotator muscles in different positions and at varying speeds. The Journal of Orthopaedic and Sports Physical Therapy. *9*:217–222, 1987.

Wilson, F.: Valgus extension overload in the pitching elbow. American Journal of Sports Medicine. *11*:123–132, 1983.

Youm, Y., McMurty, R.Y., Flatt, A.E., and Gillespie, T.E.: Kinematics of the wrist. The Journal of Bone and Joint Surgery. *60-A*:423–432, 1978.

Youm, Y, Gillespie, T.E., Flatt, A.E., and Sprague, B.L.: Kinematic investigation of normal MCP joint. Journal of Biomechanics. *11*:109–118, 1978.

Zemel, N.P.: Metacarpophalangeal joint injuries in the fingers. Hand Clinics. *8*:745–754, 1992.

References

1. Zarins, B., and Rowe, R.: Current concepts in the diagnosis and treatment of shoulder instability in athletes. Medicine and Science in Sports and Exercise. *16*:444–448, 1984.

2. Nordin, M., and Frankel, V.H.: Basic Biomechanics of the Musculoskeletal System. Philadelphia, Lea & Febiger, 1989.

3. Soderberg, G.L.: Kinesiology: Application to Pathological Motion. Baltimore, Williams & Wilkins, 1986.

4. Prodromos, C.C., Ferry, J.A., Schiller, A.L., and Zarins, B.: Histological studies of the glenoid labrum from fetal life to old age. The Journal of Bone and Joint Surgery. *72-A*:1344–1348, 1990.

5. Halbach, J.W., and Tank, R.T.: The shoulder. *In* Orthopaedic and Sports Physical Therapy. Edited by J.A. Gould and G.J. Davies. St. Louis, C.V. Mosby, 1985, pp. 497–517.

6. Blakely, R.L., and Palmer, M.L.: Analysis of rotation accompanying shoulder flexion. Physical Therapy. *64*:1214–1216, 1984.

7. Harryman, D.T., et al.: Translation of the humeral head on the glenoid with passive glenohumeral motion. The Journal of Bone and Joint Surgery. *72-A*:1334–1343, 1990.

8. Heinrichs, K.I.: Shoulder anatomy, biomechanics and rehabilitation considerations for the whitewater slalom athlete. National Strength and Conditioning Association Journal. *13*:26–35, 1991.

9. Turkel, S.J., Panio, M.W., Marshall, J.L., and Girgis, F.G.: Stabilizing mechanisms preventing anterior dislocation of the glenohumeral joint. The Journal of Bone and Joint Surgery. *63(8)*:1208–1217, 1981.

10. Einhorn, A.R.: Shoulder rehabilitation: Equipment modifications. The Journal of Orthopaedic and Sports Physical Therapy. *6*:247–253, 1985.

11. Poppen, N.K., and Walker, P.S.: Normal and abnormal motion of the shoulder. The Journal of Bone and Joint Surgery. *58-A*:195–200, 1976.

12. Peat, M., and Graham, R.E.: Electromyographic analysis of soft tissue lesions affecting shoulder function. American Journal of Physical Medicine. *56*:223–240, 1977.

13. Blackburn, T.A., McLeod, W.D., White, B., and Wofford, L.: EMG analysis of posterior rotator cuff exercises. Athletic Training. *25(1)*:40–45, 1990.

14. Kronberg, M., Nemeth, G., and Brostrom, L.: Muscle activity and coordination in the normal shoulder. Clinical Orthopaedics and Related Research. *257*:76–85, 1990.

15. Broome, H.L., and Basmajian, J.V.: The function of the teres major muscle: An electromyographic study. Anatomical Record. *170(3)*:309–310, 1970.

16. Jiang, C.C., et al.: Muscle excursion measurements and moment arm determinations of rotator cuff muscles. Biomechanics in Sport. *13*:41–44, 1987.

17. Hageman, P.A., Mason, D.K., Rulund, K.W., and Humpal, S.A.: Effects of position and speed on eccentric and concentric isokinetic testing of the shoulder rotators. The Journal of Orthopaedic and Sports Physical Therapy. *11*:64–69, 1989.

18. Moseley, J.B., et al.: EMG analysis of the scapular muscles during a shoulder rehabilitation program. The American Journal of Sports Medicine. *20*:128–134, 1992.

19. McCann, P.D., Wootten, M.E., Kadaba, M.P., and Bigliani, L.U.: A kinematic and electromyographic study of shoulder rehabilitation exercises. Clinical Orthopaedics and Related Research. *288*:179–188, 1993.

20. Gellman, H., Sie, I., and Waters, R.L.: Late complications of the weight-bearing upper extremity in the paraplegic patient. Clinical Orthopaedics and Related Research. *233*:132–135, 1988.

21. Anderson, D.S., Jackson, M.F., Kropf, D.S., and Soderberg, G.L.: Electromyographic analysis of selected muscles during sitting push-ups. Physical Therapy. *64*:24–28, 1984.

22. Moynes, D.R., Perry, J., Antonelli, D.J., and Jobe, F.W.: Electromyography and motion analysis of the upper extremity in sports. Physical Therapy. *66*:1905–1910, 1986.

23. Nuber, G.W., et al.: Fine wire electromyography analysis of muscles of the shoulder during swimming. The American Journal of Sports Medicine. *14*:7–11, 1986.

24. Fleisig, G.S., Dillman, C.J., and Andrews, J.R.: A biomechanical description of the shoulder joint during pitching. Sports Medicine Update. *6*:10–24, 1991.

25. Jobe, F.W., Moynes, D.R., Tibone, J.E., and Perry, J.: An EMG analysis of the shoulder in pitching. The American Journal of Sports Medicine. *12*:218–220, 1984.

26. Whiteside, J.A. and Andrews, J.R.: On-the-field evaluation of common athletic injuries: Part VI: Evaluation of the shoulder girdle. Sports Medicine Update. *7*:24–28, 1992.

27. Nitz, A.J.: Physical therapy management of the shoulder. Physical Therapy. *66*:1912–1919, 1986.

28. Henry, J.H., and Genung, J.A.: Natural history of glenohumeral dislocation-revisited. The American Journal of Sports Medicine. *10*:135–137, 1982.

29. Smith, R.L., and Brunolli, J.: Shoulder kinesthesia after anterior glenohumeral joint dislocation. The Journal of Orthopaedic and Sports Physical Therapy. *11*:507–513, 1990.

30. Pappas, A.M., Goss, T.P., and Kleinman, P.K.: Symptomatic shoulder instability due to lesions of the glenoid labrum. The American Journal of Sports Medicine. *11*:279–288, 1983.

31. Simon, E.R., and Hill, J.A.: Rotator cuff injuries: An update. The Journal of Orthopaedic and Sports Physical Therapy. *10(10)*:394–398, 1989.

32. Duda, M.: Prevention and treatment of throwing arm injuries. The Physician and Sports Medicine. *13*:181–186, 1985.

33. Bayley, J.C., Cochran, T.P., and Sledge, C.B.: The weight-bearing shoulder. The Journal of Bone and Joint Surgery. *69-A*:676–678, 1987.

34. Burnham, R.S., et al.: Shoulder pain in wheelchair athletes. The American Journal of Sports Medicine. *21*:238–242, 1993.

35. Andrews, J.R., Carson, W.G., and McLeod, W.D.: Glenoid labrum tears related to the long head of the biceps. The American Journal of Sports Medicine. *13*:337–341, 1985.

36. Yocum, L.A.: The diagnosis and nonoperative treatment of elbow problems in the athlete. Office Practice of Sports Medicine. *8*:439–437, 1989.

37. Bowling, R.W., and Rockar, P.: The elbow complex. *In* Orthopaedics and Sports Physical Therapy. Edited by J. Gould and G.J. Davies. St. Louis, C.V. Mosby, 1985, pp. 476–496.

38. Morrey, B.F., Askew, L.J., An, K.N., and Chao, E.Y.: A biomechanical study of normal and functional elbow motion. The Journal of Bone and Joint Surgery. *63-A*:872–877, 1981.

39. An, K.N., et al.: Muscles across the elbow joint: A biomechanical analysis. Journal of Biomechanics. *14*:659–669, 1981.

40. Stewart, O.J., Peat, M., and Yaworski, R.T.: Influence of resistance, speed of movement, and forearm position on recruitment of the elbow flexors. American Journal of Physical Medicine. *60(4)*:165–179, 1981.

41. Van Zuylen, E.J., and Van Velzen, A.: A biomechanical model for flexion torques of human arm muscles as a function of elbow angle. Journal of Biomechanics. *21*:183–189, 1988.

42. Ober, A.G.: An electromyographic analysis of elbow flexors during sub-maximal concentric contractions. Research Quarterly for Exercise and Sport. *59*:139–143, 1988.

43. An, K., Chao, E.Y.S., Donkers, M.J., and Morrey, B.F.: Intersegmental elbow joint load during pushup. Biomedical Scientific Instrumentation. *28*:69–74, 1992.

44. Ireland, M.L., and Andrews, J.R.: Shoulder and elbow injuries in the young athlete. Clinics in Sports Medicine. *7*:473–494, 1988.

45. Kulund, D.N., Rockwell, D.A., and Brubaker, C.E.: The long-term effects of playing tennis. The Physician and Sports Medicine. *7*:87–91, 1979.

46. Wadsworth, C.T.: The wrist and hand. *In* Orthopaedic and Sports Physical Therapy. Edited by J.A. Gould and G.J. Davies. St. Louis, C.V. Mosby, 1985.

47. Imaeda, T., An, K., and Cooney, W.P.: Functional anatomy and biomechanics of the thumb. Hand Clinics. *8*:9–15, 1992.

48. Jones, L.A.: The assessment of hand function: A critical review of techniques. The Journal of Hand Surgery. *14A*:221–228, 1989.

49. Mayfield, J.K.: Mechanism of carpal injuries. Clinical Orthopaedics and Related Research. *149*:45–54, 1980.

50. Wadsworth, L.T.: How to manage skier's thumb. The Physician and Sports Medicine. *20*:69–78, 1992.

51. Shea, K.G., Shumsky, I.B., and Shea, O.F.: Shifting into wrist pain. The Physician and Sports Medicine. *19*:59–63, 1991.

52. Munnings, F.: Cyclist's palsy. The Physician and Sports Medicine. *19*:113–119, 1991.

Glossary

Abduction:	Sideward movement away from the midline or sagittal plane.
Abductor Digiti Minimi Brevis:	Muscle inserting on the pisiform bone and the fifth proximal phalanx; abducts the little finger.
Abductor Pollicis Brevis:	Muscle inserting on the scaphoid, trapezium and the first proximal phalanx; abducts the thumb.
Abductor Pollicis Longus:	Muscle inserting on the radius and the base of the first metacarpal; abducts the thumb.
Acromioclavicular Joint:	Articulation between the acromion process of the scapula and the lateral end of the clavicle.
Adduction:	Sideward movement toward the midline or sagittal plane; return movement from abduction.
Adductor Pollicis:	Muscle inserting on the capitate, second and third metacarpal, and base of first phalanx; adducts the thumb.
Anconeus:	Muscle inserting on the lateral epicondyle and the olecranon process; extends the forearm.
Annular Ligament:	Ligament inserting on the anterior and posterior margins of the radial notch; supports the head of the radius.
Bennet's Fracture:	Longitudinal fracture of the base of the first metacarpal.
Biceps Brachii:	Muscle inserting on the supraglenoid tubercle, coracoid process and radial tuberosity; flexes the arm and forearm and supinates the forearm.
Bicipital Tendinitis:	Inflammation of the tendon of the biceps brachii.
Boutonniere Deformity:	A stiff proximal interphalangeal articulation caused by injury to the finger extensor mechanism.
Brachial Plexus:	A network of vessels and nerves located in the neck and axilla regions; with nerves from the ventral branches of the last four cervical spinal nerves and the first thoracic nerves.
Brachialis:	Muscle inserting on the lower humerus and the coronoid process; flexes the forearm.
Brachioradialis:	Muscle inserting on the lateral supracondylar ridge and the styloid process; flexes the forearm.
Bursae:	A fibrous, fluid-filled sac located between bones and tendons or other structures in order to reduce friction during movement.
Bursitis:	Inflammation of the bursae.
Capitulum:	Eminence on the distal end of the lateral epicondyle of the humerus which articulates with the head of the radius at the elbow joint.

Carpal Tunnel Syndrome:	Pressure and constriction of the median nerve caused by repetitive actions at the wrist.
Carpometacarpal Joint:	Articulation between the carpals and the metacarpals in the hand.
Carrying Angle:	Angle between the ulna and the humerus with the elbow extended; 10–25 degrees.
Clavicle:	An s-shaped long bone articulating with the scapula and the sternum.
Collateral Ligament:	Ligament(s) located on the sides of a joint.
Conoid Ligament:	A branch of the coracoclavicular ligament running from the coracoid process to the clavicle; prevents anterior and posterior scapular and upward and downward clavicular movements.
Coracoacromial Ligament:	Ligament running from coracoid process to acromion process which forms an arch over the shoulder.
Coracohumeral Ligament:	Ligament from coracoid process to the greater and lesser tuberosity on the humerus; supports the weight of the arm and stabilizes the humeral head while checking external rotation of the arm.
Corocoid Process:	A curved process arising from the upper neck of the scapula; overhangs the shoulder joint.
Coronoid Fossa:	Cavity in the humerus that receives the coronoid process of the ulna during elbow flexion.
Coronoid Process:	Wide eminence on proximal end of ulna; forms anterior portion of trochlear fossa.
Degeneration:	Deterioration of tissue; a chemical change in the body tissue; change of tissue to a less functionally active form.
Deltoid:	Muscle inserting on the clavicle, acromion process, spine of the scapula and the deltoid tubercle on the humerus; produces arm abduction, internal rotation, horizontal flexion and extension.
Depression:	Movement of the segment downward (scapula, clavicle); return of the elevation movement.
Dislocation:	Bone displacement, which causes separation of the bony surfaces in a joint.
Ectopic Bone:	Bone formation that is displaced and located away from the normal sites.
Ectopic Calcification:	The hardening of organic tissue through deposit of calcium salts in areas located away from the normal sites.
Elevation:	Movement of the segment upward, e.g., scapula, clavicle.
Epicondylitis:	Inflammation of the epicondyle or tissues connecting into the epicondyle, e.g., medial or lateral epicondylitis.
Extensor Carpi Radialis Brevis:	Muscle inserting on the lateral epicondyle and the base of the third metacarpal; produces forearm flexion, hand extension, and radial flexion.

Extensor Carpi Radialis Longus: Muscle inserting on the lateral supracondylar ridge and the base of the second metacarpal; produces forearm flexion, hand extension, and radial flexion.

Extensor Carpi Ulnaris: Muscle inserting on the lateral epicondyle and the base of the fifth metacarpal; produces forearm extension, hand extension, and ulnar flexion.

Extensor Digitorum: Muscle inserting on the lateral epicondyle and the dorsal hood of phalanges 1–4; extends the fingers and the hand.

Extensor Indicis: Muscle inserting on the ulna, the interosseous membrane, and the dorsal hood of the second phalanx; extends the index finger.

Extensor Pollicis Brevis: Muscle inserting on the radius, ulna, and the base of the proximal phalanx of the thumb; extends the thumb.

Extensor Pollicis Longus: Muscle inserting on the ulna, interosseous membrane, and distal phalanx of the thumb; extends the thumb.

Flexor Carpi Radialis: Muscle inserting on the medial epicondyle and base of second and third metacarpal; produces hand flexion and radial flexion.

Flexor Carpi Ulnaris: Muscle inserting on the medial epicondyle, pisiform, hamate, and base of the fifth metacarpal; produces hand flexion and ulnar flexion.

Flexor Digiti Minimi Brevis: Muscle inserting on the hamate and the proximal phalanx of the little finger; flexes the little finger.

Flexor Digitorum Profundus: Muscle inserting on the ulna and the base of the distal phalanx of fingers 2–5; produces finger and hand flexion.

Flexor Digitorum Superficialis: Muscle inserting on the medial epicondyle and the middle phalanx of fingers 2–5; produces finger and hand flexion.

Flexor Pollicis Brevis: Muscle inserting on the trapezium, trapezoid, capitate, and proximal phalanx of the thumb; produces thumb flexion.

Flexor Pollicis Longus: Muscle inserting on the radius and the interosseus membrane and the base of the distal phalanx of the thumb; produces thumb flexion.

Force Couple: Two forces that are equal in magnitude and, acting in opposite directions, produce rotation about an axis.

Fracture: A break in a bone.

Glenohumeral Joint: The articulation between the head of the humerus and the glenoid fossa on the scapula.

Glenohumeral Ligament: Ligament connecting the edge of the glenoid to the humeral head and offering support in the movements of external rotation and abduction; resists anterior dislocation of the arm.

Glenoid Labrum: Ring of fibrocartilage around the rim of the glenoid fossa which deepens the socket in the shoulder and hip joints.

Glenoid Fossa: A depression in the lateral superior scapula that forms the socket for the shoulder joint.

Horizontal Extension (Abduction): Movement of an elevated segment (arm, leg) away from the body in the posterior direction.

Horizontal Flexion (Adduction): Movement of an elevated segment (arm, leg) toward the body in the anterior direction.

Hypothenar Eminence: The ridge on the palm on the ulnar side created by the presence of intrinsic muscles acting on the little finger.

Impingement Syndrome: Irritation of structures above the shoulder joint due to repeated compression as the greater tuberosity is pushed up against the underside of the acromion process.

Infraspinatus: Muscle inserting on the infraspinous fossa and the greater tubercle; produces external rotation and horizontal extension of the arm.

Intercarpal Joint: Articulation between the carpal bones.

Interossei: Muscle inserting on the sides of the metacarpals and the proximal phalanx of the fingers; produces finger abduction and adduction (dorsal and palmar).

Interosseus Membrane: A thin layer of tissue running between two bones (radius and ulna, tibia and fibula).

Interphalangeal Joint: Articulation between the phalanx of the fingers and toes.

Jersey Finger: Avulsion of a finger flexor tendon through forced hyperextension.

Lateral Epicondyle: Projection from the lateral side of the distal end of the humerus giving attachment to the hand and finger extensors.

Latissimus Dorsi: Muscle inserting on the spinous processes of the lower thoracic and lumbar vertebrae, lower ribs, iliac crest, inferior angle of the scapula and the intertubercular groove on the humerus; internally rotates, adducts and extends the arm.

Levator Scapula: Muscle inserting on the transverse processes of the cervical vertebrae and the superior angle of the scapula; elevates the scapula.

Lumbricales: Muscles inserting on the tendon of the flexor digitorum profundus and the dorsal hood of fingers 2–5; flexes and extends the fingers.

Mallet Finger: Avulsion injury to the finger extensor tendons at the distal phalanx, produced by a forced flexion.

Medial Epicondyle: Projection from the medial side of the distal end of the humerus giving attachment to the hand and finger flexors.

Medial Tension Syndrome: Also termed pitcher's elbow, a medial pain brought on by excessive valgus forces, which may cause ligament sprain, medial epicondylitis, tendinitis, or avulsion fractures to the medial epicondyle.

Metacarpophalangeal Joint: Articulation between the metacarpals and the phalanges in the hand.

Midcarpal Joint: Articulation between the proximal and distal row of carpals in the hand.

Olecranon Bursitis: Irritation of the olecranon bursae caused commonly by falling on the elbow.

Olecranon Fossa:	A depression on the posterior, distal humerus; creates a lodging space for the olecranon process of the ulna in forearm extension.
Olecranon Process:	Projection on the proximal, posterior ulna; fits into the olecranon fossa during forearm extension.
Opponens Digiti Minimi:	Muscle inserting on the hamate and the fifth metacarpal; produces opposition of the little finger.
Opponens Pollicis:	Muscle inserting on the trapezium and the first metacarpal; produces opposition of the thumb.
Osteochondritis Dissecans:	Inflammation of bone and cartilage resulting in splitting of pieces of cartilage into the joint (shoulder, hip).
Palmaris Longus:	Muscle inserting on the medial epicondyle and the palmar aponeurosis; produces flexion of the hand.
Pectoralis Major:	Muscle inserting on the clavicle, sternum, ribs 1–6, greater tubercle of the humerus and intertubercular groove; produces internal rotation, horizontal flexion, flexion, and extension of the arm.
Pectoralis Minor:	Muscle inserting on ribs 3–5 and the corocoid process; produces shoulder girdle depression, downward rotation, and protraction.
Pitcher's Elbow:	Also termed medial tension syndrome, a medial pain brought on by excessive valgus forces which may cause ligament sprain, medial epicondylitis, tendinitis, or avulsion fractures to the medial epicondyle.
Power Grip:	A powerful hand position produced by flexing the fingers maximally around the object at all three finger joints and the thumb adducted in the same plane as the hand.
Precision Grip:	A fine-movement hand position produced by positioning the fingers in a minimal amount of flexion with the thumb perpendicular to the hand.
Pronation:	The inward rotation of a body segment (forearm).
Pronator Quadratus:	Muscle inserting on the ulna and radius; produces pronation of the forearm.
Pronator Teres:	Muscle inserting on the medial epicondyle, coronoid process, and radius; produces forearm pronation and flexion.
Protraction:	Also called abduction, the movement of the scapula forward and away from the vertebral column.
Radiocarpal Joint:	Articulation between the radius and the carpals (scaphoid and lunate).
Radiohumeral Joint:	Articulation between the radius and the humerus.
Radioulnar Joint:	Articulation between the radius and the ulna (superior and inferior).
Retinaculum:	Fibrous band that contains tendons or other structures.
Retraction:	Also called adduction, the movement of the scapula backward and toward the vertebral column.

Rhomboid:	Muscle inserting on the spinous processes of vertebrae C-7, T1–5 and the medial border of the scapula; retracts and elevates the shoulder girdle.
Rotation:	Movement of a segment about an axis running longitudinally through it.
Rotator Cuff:	Four muscles surrounding the shoulder joint consisting of the infraspinatus, supraspinatus, teres minor, and subscapularis.
Rupture:	An injury in which the tissue is torn or disrupted in a forcible manner.
Scapulohumeral Rhythm:	The movement relationship between the humerus and the scapula during arm raising movements; the humerus moves two degrees for every one degree of scapular movement through 180 degrees of arm flexion or abduction.
Scapula:	A flat, triangular bone on the upper, posterior thorax.
Scapulothoracic:	A physiologic joint between the scapula and the thorax.
Serratus Anterior:	A muscle inserting on ribs 1–8 and the ventral, medial surface of the scapula; protracts, elevates, and upwardly rotates the scapula.
Shoulder Girdle:	An incomplete bony ring in the upper extremity formed by the two scapulae and clavicles.
Sprain:	An injury to a ligament surrounding a joint; rupture of fibers of a ligament.
Sternoclavicular Joint:	The articulation between the sternum and the clavicle.
Strain:	Injury to the muscle, tendon, or muscle-tendon junction due to overstretching or excessive tension applied to the muscle; tearing and rupture of the muscle or tendon fibers.
Subacromial Bursae:	The bursae between the acromion process and the insertion of the supraspinatus muscle.
Subacromial Bursitis:	Inflammation of the subacromial bursae, common to impingement syndrome.
Subluxation:	An incomplete or partial dislocation between two joint surfaces in the joint.
Subscapularis:	Muscle inserting on the ventral surface of the scapula and the lesser tubercle of the humerus; produces internal rotation of the arm.
Supination:	The outward rotation of a body segment (forearm).
Supinator:	Muscle inserting on the lateral epicondyle of the humerus and the lateral side of the radius; supinates the forearm.
Supraspinatus:	Muscle inserting on the supraspinatus fossa and the lesser tubercle of the humerus; abducts and flexes the arm.
Tendinitis:	Inflammation of a tendon.
Tenosynovitis:	Inflammation of the sheath surrounding a tendon.
Teres Major:	Muscle inserting on the dorsal surface of the scapula and the lesser tubercle of the humerus; produces internal rotation, adduction, and extension of the arm.
Teres Minor:	Muscle inserting on the lateral, dorsal surface of the scapula and the greater tubercle of the humerus; produces external rotation and horizontal extension of the arm.

Thenar Eminence:	Ridge or mound on the radial side of the palm formed by the intrinsic muscles acting on the thumb.
Traction Apophysitis:	Inflammation of the apophysis (process, tuberosity) created by a pulling force of tendons.
Trapezius:	Muscle inserting on the occipital bone, ligamentum nuchae, spinous processes of C-7, T1–12, and the acromion process; produces upward rotation, elevation, and retraction of the shoulder girdle, and arm abduction.
Triceps Brachii:	Muscle inserting on the infraglenoid tubercle, middle and posterior shaft of the humerus, and the olecranon process; extends the forearm and the arm.
Trigger Finger:	Snapping during flexion and extension of the fingers created by nodules on the tendons.
Trochlea:	Medial portion of the distal end of the humerus which articulates with the trochlear notch of the ulna.
Trochlear Notch:	A deep groove in the proximal end of the ulna which articulates with the trochlea of the humerus.
Ulnarhumeral Joint:	The articulation between the ulna and the humerus; commonly called the elbow joint.
Winged Scapula:	Tilting out of the vertebral border of the scapula created due to serratus anterior weakness or insufficiency.

CHAPTER 6

Functional Anatomy of the Lower Extremity

I. The Pelvic and Hip Complex
  A. Anatomical and Functional Characteristics of the
     Joints
     1. pubic symphysis
        a) ligaments
     2. sacroiliac joint
        a) ligaments   b) movements
     3. hip joint
        a) ligaments   b) movements
  B. Combined Movements of the Pelvis and Thigh
  C. Muscular Actions
     1. thigh flexion and extension
     2. thigh abduction and adduction
     3. thigh external rotation and internal rotation
  D. Strength and Force of the Hip Joint
  E. Conditioning
  F. Contribution to Sport Skills or Movements
     1. walking
     2. running
     3. stair ascent and descent
     4. cycling
  G. Injury Potential in the Pelvic and the Hip Complex

II. The Knee Joint
  A. Anatomical and Functional Characteristics of the
     Joint
     1. tibiofemoral joint
        a) menisci b) ligaments c) movements
     2. patellofemoral joint
        a) ligaments   b) Q angle   c) movements
     3. tibiofibular joint
        a) movements
     4. screw home mechanism
  B. Muscular Actions
     1. leg extension and flexion
     2. leg internal and external rotation
  C. Strength and Forces at the Knee Joint
  D. Conditioning
  E. Contribution to Sport Skills or Movements
     1. walking
     2. running

     3. stair ascent and descent
     4. lifting
     5. cycling
  F. Injury Potential in the Knee Joint

III. The Ankle and Foot
  A. Anatomical and Functional Characteristics of the
     Joints
     1. talocrural joint
        a) ligaments   b) movements
     2. subtalar joint
        a) ligaments   b) movements
     3. midtarsal joint
        a) ligaments   b) movements
     4. tarsometatarsal joints
        a) ligaments   b) movements
     5. metatarsophalangeal joints
        a) movements
     6. interphalangeal joints
        a) movements
     7. arches of the foot
        a) medial longitudinal   b) lateral longitudinal
        c) transverse
     8. foot types
     9. alignment abnormalities
  B. Muscle Actions
     1. foot plantar flexion and dorsiflexion
     2. foot supination
        a) calcaneal inversion b) forefoot adduction
     3. foot pronation
        a) calcaneal eversion   b) forefoot abduction
  C. Strength and Forces of the Ankle and Foot
  D. Conditioning
  E. Contribution to Sport Skills or Movements
     1. walking
     2. running
     3. stair ascent and descent
     4. cycling
  F. Injury Potential in the Ankle and Foot

IV. Summary

After reading this chapter, the student will be able to:

1. Describe the structure, support, and movements of the hip, knee, ankle, and subtalar joints.
2. Identify the pelvic and sacral movements accompanying the movements of the thigh and the trunk.
3. Explain how changes in the femoral neck angles will influence hip position.
4. Identify the muscular actions contributing to movements at the hip and knee joints.
5. Compare the movements of the thigh in terms of flexibility and strength.
6. Identify the lower extremity muscular contributions to walking, running, stair climbing, cycling, and lifting.
7. Describe how alterations in the alignment in the lower extremity influences function at the knee, hip, ankle, and foot.
8. Discuss various loads that must be absorbed and transmitted by the hip, knee, ankle, and foot in daily activities.
9. Identify the muscular contributions to specific movements at the knee, ankle, and foot.
10. List and describe some of the common injuries to the hip, knee, ankle, and foot.
11. Develop a set of strength and flexibility exercises for the hip, knee, and ankle joints.
12. Differentiate between the movements of pronation and supination during both weight bearing and non-weight bearing.
13. Discuss the structure and function of the arches of the foot.

The lower extremity is subjected to high forces, generated via repetitive contacts the foot makes with the ground, while at the same time being responsible for supporting the central mass and the upper extremities. Both lower limbs are connected to each other and to the trunk by the pelvic girdle. This establishes a link between both extremities and the trunk that must always be considered when examining movements and the muscular contributions to movements in the lower extremity.

Movement in any part of the lower extremity, pelvis, or trunk will influence every other aspect of the lower extremity, thus, a foot position or movement can influence the position or movement at the knee or hip of either limb, or a pelvic position can influence actions throughout the whole lower extremity. It is important to evaluate movement and actions throughout both limbs, the pelvis, and the trunk, rather than focus on a single joint, in order to understand lower extremity function for the purpose of rehabilitation, sport performance, or exercise prescription.

For example, in a simple kicking action, it is not just the kicking limb that is critical to the success of the skill. The contralateral limb plays a very important role in stabilization and support of body weight, the pelvis establishes the correct positioning for the lower extremity, and trunk positioning determines the efficiency of the lower extremity musculature. Likewise, in evaluating a limp in walking, attention should not be focused exclusively on the limb in which the limp occurs, since it may have been created by something happening in the other extremity.

The Pelvic and Hip Complex

The pelvic girdle, including the hip joint, plays a very integral role in supporting the weight of the body, while at the same time offering mobility by increasing the range of motion in the lower extremity. The pelvic girdle serves as a site of muscular attachment for 28 trunk and thigh muscles, none of which are positioned to act solely on the pelvic girdle (1). Like the shoulder girdle, the pelvis must be oriented to place the hip joint in a favorable position for lower extremity movement, therefore, concomitant movement of the pelvic girdle and the thigh at the hip joint are necessary for efficient joint actions.

The pelvic girdle and hip joints are part of a closed kinetic chain system whereby forces travel up through the hip and the pelvis into the trunk, or down from the trunk through the pelvis and the hip to the knee and foot and the ground. Finally, pelvic girdle and hip joint positioning contribute significantly to the maintenance of balance and standing posture by employing continuous muscular action to fine tune and ensure equilibrium.

The pelvic region is one area of the body where there are noticeable differences between the sexes in the general population. As illustrated in FIGURE 6–1, females have pelvic girdles that are lighter, thinner, and wider than their male counterparts (2). The female pelvis flares out more laterally in the front, and the sacrum is also wider in the back, creating a broader pelvic cavity than males. This skeletal difference will be referred to later in this chapter since it has a direct influence on muscular function in and around the hip joint.

Anatomical and Functional Characteristics of the Joints

The bony attachment of the lower extremity to the trunk occurs via the pelvic girdle (FIGURE 6–2). The pelvic girdle has right and left coxal bones, with each consisting of a fibrous union between the top ilium bone, the posterior and inferior ischium bone, and the anterior, inferior pubic bone.

The right and left sides connect anteriorly at the pubic symphysis, a cartilaginous joint, having fibrocartilage connecting the two bones together. This joint is firmly supported by a pubic ligament that runs along the anterior, posterior, and superior sides of the joint. The movement at this joint is very limited, maintaining a firm connection between right and left coxal bones.

The pelvis is connected to the trunk at the sacroiliac joint, a strong, synovial joint containing a soft, fibrocartilage and powerful ligamentous support (FIGURE 6–2). The articulating surface on the sacrum faces posteriorly and laterally and articulates with the ilium, which faces forward and medially (3).

The sacroiliac joint transmits the weight of the body to the hip and is subject to loads from the lumbar region or from the ground. It is also an energy absorber of shear forces during gait (1). There are three sets of ligaments supporting the left and right sacroiliac joints, and these ligaments are the strongest in the body (See Appendix A).

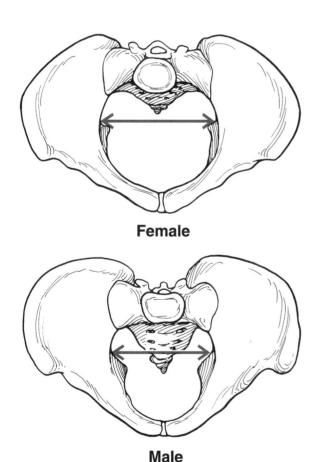

Female

Male

FIGURE 6–1. The pelvis of a female is lighter, thinner, and wider than that of a male. The female pelvis also flares out in the front and has a wider sacrum in the back.

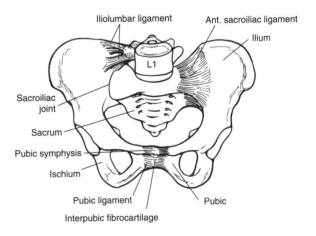

FIGURE 6–2. The pelvic girdle supports the weight of the body, serves as an attachment site for numerous muscles, contributes to the efficient movements of the lower extremity, and helps maintain balance and equilibrium. The girdle consists of two coxal bones, each created through the fibrous union of the ilium, ischium, and pubic bones. The right and left coxal bones are joined anteriorly at the pubic symphysis, and connect posteriorly via the sacrum and the two sacroiliac joints.

Even though the sacroiliac joint is well reinforced by very strong ligaments, there is movement occurring at the joint. The amount of movement allowed at the joint varies considerably between individuals and sexes. Males have thicker and stronger sacroiliac ligaments and consequently, do not have mobile sacroiliac joints. In fact, 3 out of 10 males have fused sacroiliac joints (3).

In females, the sacroiliac joint is more mobile due to more laxity in the ligaments supporting the joint. This laxity increases with monthly cycles of the hormones and the joint is extremely lax and mobile during pregnancy (4).

Another reason why the sacroiliac joint is more stable in males is related to positioning differences in the center of gravity. In the standing position, body weight forces the sacrum down, tightening the posterior ligaments and forcing the sacrum and ilium together, providing stability

to the joint. This position is the close-packed position for the sacroiliac joint (1).

In females, the location of the center of gravity is in the same plane as the sacrum, and in males, the center of gravity is more anterior. This means that in males, there will be more of a load placed on the sacroiliac joint, which in turn creates a tighter and more stable joint (3).

Movements at the sacroiliac joint can best be described by sacral movements. The movement of the sacrum accompanying each specific trunk movement is presented in FIGURE 6–3. The triangular-shaped sacrum is actually five fused vertebrae that will move with the pelvis and trunk. The top of the sacrum, the widest part, is the base of the sacrum, and when this base moves anteriorly it is termed *sacral flexion* (1). This movement occurs with extension of the trunk or with flexion of the thigh.

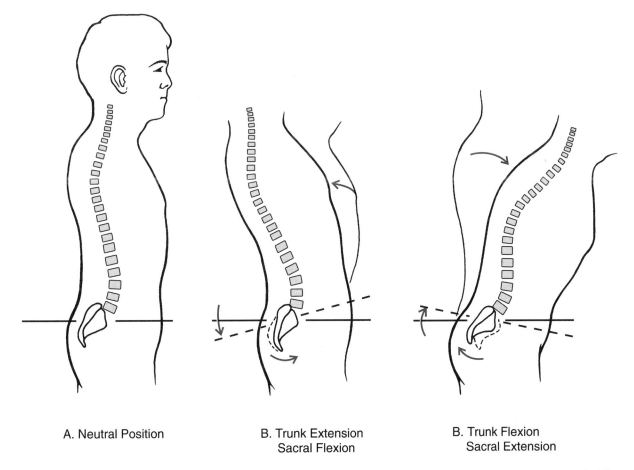

A. Neutral Position

B. Trunk Extension
Sacral Flexion

B. Trunk Flexion
Sacral Extension

FIGURE 6–3. (A) In the neutral position, the sacrum is placed in the close-packed position by the force of gravity. The sacrum responds to movements of both the thigh and the trunk. (B) When the trunk extends or the thigh flexes, the sacrum will flex. Flexion of the sacrum occurs when the wide base of the sacrum moves anteriorly. (C) During trunk flexion or thigh extension, the sacrum extends as the base moves posteriorly. The sacrum will also rotate to the right or left with lateral flexion of the trunk (not shown above).

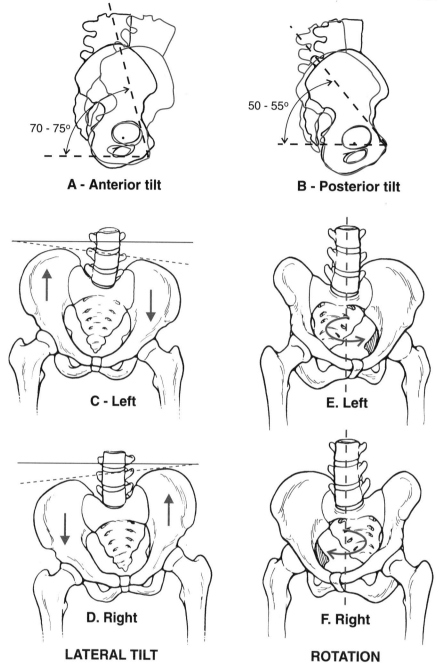

A - Anterior tilt

70 - 75°

B - Posterior tilt

50 - 55°

C - Left

D. Right

E. Left

F. Right

LATERAL TILT   **ROTATION**

FIGURE 6–4. The pelvis moves in six different directions in response to a trunk or thigh movement. (A) Anterior tilt of the pelvis accompanies trunk flexion or thigh extension. (B) Posterior tilt accompanies trunk extension or thigh flexion. (C) Left and (D) right lateral tilt accompany weight bearing on the right and left limbs, respectively, or lateral movements of the thigh or trunk. (E) Left and (F) right rotation accompany left and right rotation of the trunk, respectively, or unilateral leg movement.

The sacrum extends as the base moves posteriorly, with trunk flexion or thigh extension. The sacrum also rotates along an axis running diagonally across the bone, and right rotation is designated if the anterior surface of the sacrum faces to the right. This sacral torsion is produced by the piriformis muscle in a side bending exercise of the trunk (1).

In addition to the movement between the sacrum and the ilium, there is also movement of the pelvic girdle as a whole. These movements, shown in FIGURE 6–4, accompany trunk and thigh movements to facilitate positioning of the hip joint and the lumbar vertebrae. Although muscles will facilitate the movements of the pelvis, there is no one set of muscles acting on the pelvis specifically; thus, pelvic movements occur as a consequence of movements of the thigh or the lumbar vertebrae.

Movements of the pelvis are described by monitoring the ilium, and specifically the anterior, superior, and inferior iliac spines on the front of the ilium. Anterior tilt of the pelvis occurs when the trunk flexes or the thighs extend and is defined as a forward tilting and downward movement of the pelvis. This anterior tilt can be created by protruding the abdomen and creating a swayback position in the low back.

Posterior tilt is created through trunk extension, flattening of the low back, or thigh flexion, and it occurs as the pelvis moves posteriorly. The pelvis can also tilt laterally, and will naturally try to move through a right lateral tilt when weight is supported by the left limb. This movement is controlled by muscles so that it is not pronounced unless the controlling muscles are weak. Thus, right and left lateral tilt will occur with weight bearing and any lateral movement of the thigh or trunk. Finally, the pelvic girdle will rotate to the left and right as unilateral leg movements take place. As the right limb is swung forward in a walk, run, or kick, the pelvis rotates to the left.

The final joint in the pelvic girdle complex is the hip joint, which can be generally characterized as a very stable, yet mobile joint. The hip joint is a 3 degree of freedom (dF), ball-and-socket joint, consisting of the articulation between the acetabulum on the pelvis and the head of the femur. The structure of the hip joint is illustrated in FIGURE 6–5.

The acetabulum is the concave surface of the ball and socket, facing anteriorly, laterally, and inferiorly (5,6). Interestingly, the three bones forming the pelvis, the ilium, ischium, and pubis, make their fibrous connections with each other in the acetabular cavity. The cavity is lined with articular cartilage that is thicker at the edge and thickest on the top part of the cavity (5,7). There is no cartilage on the underside of the acetabulum.

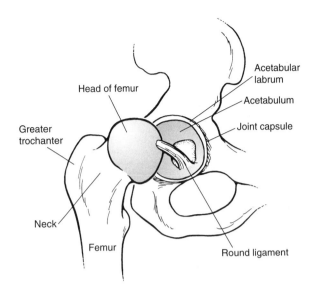

FIGURE 6–5. The hip joint is a very stable joint that also has considerable mobility in three different directions. The joint is formed by the concave surface of the acetabulum on the pelvis and the large head of the femur.

Similar to the shoulder, there is a rim of fibrocartilage that encircles the acetabulum, the acetabular labrum, that serves to deepen the socket and increase stability (8). The spherical head of the femur fits snugly into the acetabular cavity. Both the femoral head and the acetabulum have large amounts of spongy, trabecular bone that facilitate the distribution of the forces absorbed by the hip joint (5). The head is also lined with articular cartilage, and is thicker in the middle central portions of the head where most of the load is supported. The cartilage on the head thins out at the edges where the acetabular cartilage is thick (5). Approximately 70% of the head of the femur articulates with the acetabulum, in comparison to 20 to 25% for the head of the humerus with the glenoid cavity.

Surrounding the whole hip joint is a loose but strong capsule that is reinforced by ligaments and the tendon of the psoas muscle. The capsule is more dense in the front and top of the joint where the stresses are the greatest, and is quite thin on the back side and on the bottom of the joint (9).

There are three ligaments blending with the capsule and receiving nourishment from the joint. The insertions and actions of these ligaments can be found in Appendix A. The iliofemoral ligament, or Y-ligament, is strong, and supports the anterior hip joint in the standing posture and resists the movements of extension, internal rotation, and some external rotation (8). This ligament is capable of supporting most of the body weight. Also, the movement of hyperextension may be so limited by this ligament that it may not actually occur in the hip joint itself, but rather as a consequence of anterior pelvic tilt.

The second ligament on the front of the hip joint, the pubofemoral ligament, primarily resists abduction with some resistance to external rotation. The final ligament on the outside of the joint is the ischiofemoral ligament, located on the posterior capsule where it resists adduction and internal rotation (8). None of the ligaments surrounding the hip joint resist during the movement of flexion and all of them are loose during flexion. This makes flexion the movement with the largest range of motion.

The femur is held away from the hip joint and the pelvis by the femoral neck. The neck is formed by cancellous trabecular bone with a thin corticol layer for strength. The corticol layer is reinforced on the lower surface of the neck where great strength is required in response to high tension forces. Also, the medial portion of the femoral neck is the portion responsible for withstanding the ground reaction forces. The lateral portion of the neck resists compression forces created by the muscles (5).

The femoral neck joins up with the shaft of the femur, which slants medially down to the knee. The shaft is very narrow in the middle where it is reinforced with the thickest layer of corticol bone. Also, the shaft bows anteriorly to offer the optimal structure for sustaining and supporting high forces (9).

The femoral neck is positioned at a specific angle in both the frontal and transverse planes in order to facilitate a congruent articulation within the hip joint and to hold the femur away from the body. The angle of inclination is the angle of the femoral neck in the frontal plane, which is approximately 125 degrees with respect to the femoral shaft (9) (FIGURE 6–6). This angle is larger at birth by almost 20 to 25 degrees and it lowers as the person matures and assumes weight bearing positions. It is also believed that the angle continues to lower into later adult years by approximately 5 degrees.

The range of the angle of inclination is usually within 90 to 135 degrees (5). The angle of inclination is important because it determines the effectiveness of the hip abductors, the length of the limb, and the forces imposed on the hip joint. If the angle of inclination is greater than 125

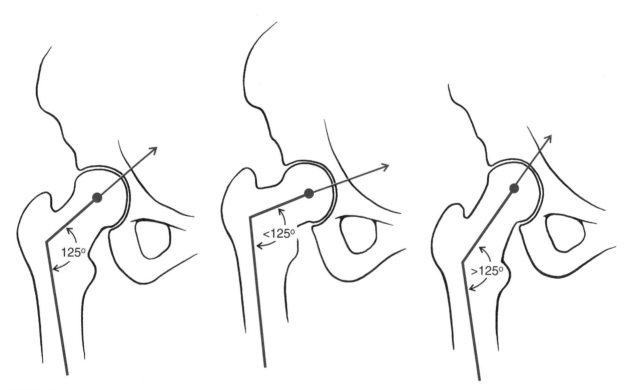

FIGURE 6–6. The angle of inclination of the neck of the femur is approximately 125 degrees. If the angle is less than 125 degrees, it is termed *coxa vara*. When the angle is reduced, the limb is shortened, the abductors are more effective, there is less load on the femoral head, but more load on the femoral neck. When the neck angle is greater than 125 degrees, it is termed *coxa valga*. This positioning lengthens the limb, reduces the effectiveness of the abductors, increases the load on the femoral head, and decreases the load on the neck.

degrees, it is termed coxa valga. This increase in the angle of inclination lengthens the limb, reduces the effectiveness of the hip abductors, increases the load on the femoral head, and decreases the stress on the femoral neck (9).

Coxa vara, in which the angle of inclination is less than 125 degrees, shortens the limb, increases the effectiveness of the hip abductors, decreases load on the femoral head, and increases stress on the femoral neck. This varus position gives the hip abductors a mechanical advantage, needed to counteract the forces produced by body weight. The result is a reduction in the load imposed upon the hip joint and the reduction in the amount of muscular force needed to counteract the force of body weight (9).

The angle of the femoral neck in the transverse plane is termed the *angle of anteversion* (FIGURE 6–7). Normally the femoral neck is rotated anteriorly 12 to 14 degrees with respect to the femur (8). Anteversion in the hip increases the mechanical advantage of the gluteus maximus, making it more effective as an external rotator (6).

If there is excessive anteversion in the hip joint in which it rotates beyond 14 degrees to the anterior side,

the head of the femur is uncovered, and a person must assume an internally rotated posture or gait to keep the head in. The toeing-in accompanying anteversion is illustrated in FIGURE 6–7. Other accompanying lower extremity adjustments to the excessive anteversion include an increase in the Q-angle, patellar problems, increase in leg lengths, more pronation at the subtalar joint, and an increase in lumbar curvature (5,9).

If the angle of anteversion is reversed so that it moves posteriorly, it is termed *retroversion*, which creates an externally rotated gait, a supinated foot, and a decrease in the Q-angle (9). See FIGURE 6–7 for an illustration.

The hip joint is a very stable joint, even though the acetabulum is not deep enough to cover all of the femoral head. The acetabular labrum deepens the socket to increase stability, and the joint is in a close-packed position in full extension when the lower body is stabilized on the pelvis. The joint is stabilized by gravity during stance when body weight presses the femoral head against the acetabulum (9). There is also a difference in atmospheric pressure in the hip joint, creating a vacuum and a suction of the femur up into the joint. Even if all of

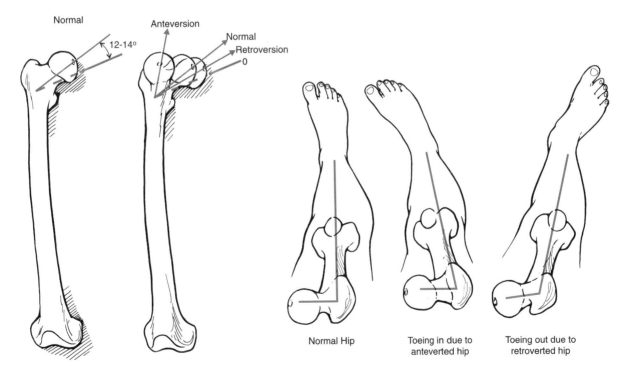

FIGURE 6–7. The angle of the femoral neck in the frontal plane is called the angle of anteversion. The normal angle is approximately 12 to 14 degrees to the anterior side. If this angle increases, a toe-in position is created in the extremity. If the angle of anteversion is reversed so the femoral neck moves posteriorly, it is termed retroversion. Retroversion causes a toeing-out of the extremity.

the ligaments and muscles are removed from around the hip joint, the femur would still remain up in the socket (10).

The hip joint also has strong ligaments and muscular support in all directions to support and maintain stability in the joint. The position where there is maximum congruence between the femoral head is a position of 90 degrees of flexion with a small amount of rotation and abduction. This is a very stable and comfortable hip joint position and is common in sitting. A position of instability for the hip joint is in flexion and adduction where the legs are crossed (10).

The hip joint allows the thigh to move through a wide range of motion in three directions (FIGURE 6–8). The thigh can move through 70 to 140 degrees of flexion and 4 to 15 degrees of hyperextension in the sagittal plane (5,11). These measurements are made with respect to a fixed axis and will vary considerably if measured with respect to the pelvis (12).

A wide range of hip flexion and extension is used in daily activities. Roughly 80 to 100 degrees of flexion and extension is needed to lower into or raise out of a chair, respectively (13). Sixty-three degrees of flexion is used to climb a stair, while only 24 to 30 degrees of hip flexion is required in the descent of the same stair (9,13). These values would change with a corresponding increase or decrease in the rise height of the stairs.

In the walking gait, the maximum amount of hip flexion is 35 to 40 degrees, achieved during late swing prior to heel strike (5). Full extension is required at the hip in gait, and occurs just as the heel lifts off the ground. If thigh extension is limited or impaired, there will be compensatory joint actions at the knee or in the lumbar vertebrae to accommodate the lack of hip extension.

Flexion of the thigh occurs freely with the knees flexed but is severely limited by the hamstrings if the flexion occurs with the knee extension (10). The hyperextension movement is limited by the anterior capsule, the strong hip flexors, and the iliofemoral ligament. These three structures impose a major restriction on the hyperextension movement.

The thigh can abduct through approximately 30 degrees, and can adduct 25 degrees beyond the anatomical position (5). Most activities require 20 degrees of abduction and adduction (5).

In walking, there is roughly 12 degrees, total, of abduction and adduction (9). Maximum abduction of the thigh occurs in swing just after toe-off, while maximum adduction is present through most of the support phase (5). To bend down and pick something up from the squat position, or tie a shoe, requires 18 to 20 degrees of abduc-

tion (9). The abduction movement is limited by the adductors and the adduction movement is limited by the tensor fascia latae.

Finally, the thigh can internally rotate through 70 degrees, and externally rotate through 90 degrees from the anatomical position (5). The range of motion for rotation at the hip can be enhanced by the position of the thigh. Both internal and external rotation range of motions can be increased by flexing the thigh (10).

In gait, the thigh externally rotates through 8 to 10 degrees in the swing phase, and internally rotates 4 to 6 degrees, beginning just before heel strike and lasting into late stance (5). To bend down and pick something up requires 10 to 15 degrees of external rotation (9). Both the internal and external rotation joint actions are limited by their antagonistic muscle group and the ligaments of the hip joint.

Combined Movements of the Pelvis and Thigh

There is no movement in the pelvic girdle similar to the scapulohumeral rhythm seen in the upper extremity. This is because the right and left lower limbs do not move independently as do the upper limbs, and because the pelvis moves concomitantly with the lumbar vertebrae, or with the thigh. Also, the thigh movements are not as limited by the acetabulum as the humerus is by the scapula.

Although movements of the thigh can occur without pelvic movement, the pelvis and the thigh will commonly move together unless the trunk restrains the pelvic activity. For example, if the pelvis is tilted forward from a relaxed standing posture, flexion of both thighs, and hyperextension or swayback in the lumbar spine will occur. Conversely, the thighs will hyperextend and the lumbar vertebrae will flex or flatten out as a result of a posterior tilt.

If the trunk is stabilized by being positioned on the ground in a prone or supine position, or in a hanging position, the pelvis will be stabilized and a different combination of movements will occur. Flexion of the thigh in the supine position, such as in a double straight-leg raise, does not produce anterior tilt of the pelvis if the trunk muscles can stabilize the pelvis. If not, the hip flexors will produce hyperextension of the lumbar vertebrae that will in turn create an anterior tilt of the pelvis. Thigh flexion with the knee flexed and the feet flat on the ground, such as in the case of a curl up, will produce a posterior tilt of the pelvis. Likewise, flexing both limbs in the hanging position will produce a posterior tilt of the pelvis since it is being stabilized by the trunk.

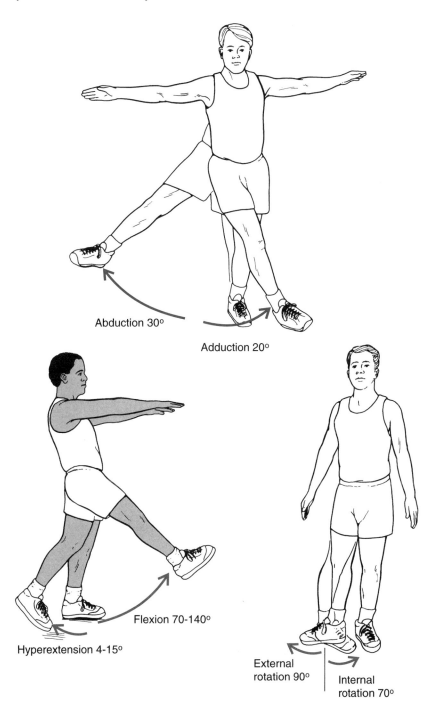

FIGURE 6–8. The thigh can move through a wide range of motion in three different directions. The thigh moves through 70 to 140 degrees of flexion, 4–15 degrees of hyperextension, 30 degrees of abduction, 20 degrees of adduction, 90 degrees of external rotation, and 70 degrees of internal rotation.

Hyperextension of both thighs from the prone face-down position will produce anterior tilt of the pelvis due, again, to the trunk stabilization. Finally, swinging both legs into extension from the supported, hanging position will also result in an anterior tilt of the pelvis.

Since many lower extremity movements occur unilaterally, the combined movements of the pelvis and thigh are different from those movements just described for bilateral movement of both limbs. For example, if one thigh flexes forward and the other limb is supporting the body weight such as in walking or kicking, the pelvis will rotate to the opposite side of the thigh flexing, posteriorly tilt, and laterally flex down to the support limb. With the thigh back behind in hyperextension and the opposite limb supporting the body weight, such as in toe-off in the gait cycle, or in the preparatory phase of a kick, the pelvis will rotate in the swing-leg direction, anteriorly tilt, but still laterally tilt to the support limb side (8). Many of these pelvic movements are counteracted and dampened by arm swinging and trunk rotation movements that absorb some of the pelvic activity.

Muscular Actions

The thigh movement of flexion is used in walking and running to bring the leg forward and through. It is also an important movement in climbing stairs or walking uphill, and is forcefully used in the activity of kicking. Little emphasis is placed on training the joint for the movement of flexion, since most consider flexion at the hip to play a minor role in activities. However, hip flexion is a very important joint movement for sprinters, hurdlers, high jumpers, and others who must develop quick leg action. Elite athletes in these activities usually have proportionally stronger hip flexors and abdominals than do less skilled athletes. Recently, there has been more attention given to the training of the hip flexors in long distance runners as well, because it has been shown that fatigue in the hip flexors during running may alter gait mechanics and lead to injuries that may have been avoided through better conditioning of this muscle group.

The strongest hip flexor is the iliopsoas muscle, a combination of three different muscles, the psoas major, psoas minor, and the iliacus (9). This is a two-joint muscle acting on both the lumbar spine of the trunk and the thigh. If the trunk is stabilized, the iliopsoas will produce flexion at the hip joint that is slightly facilitated with the thigh abducted and externally rotated. If the thigh is fixed, the iliopsoas will produce hyperextension of the lumbar vertebrae and flexion of the trunk.

The iliopsoas becomes more active in the midrange of the flexion movement. For example, in the double-leg raise, the initial part of the leg raise action will be facilitated by the abdominals and the iliopsoas, with the iliopsoas activity increasing after the movement is initiated in the middle range of motion. Likewise, in a curl-up or sit-up, the abdominals will contribute significantly through the first 45 degrees of trunk flexion, and then the iliopsoas will be active through the mid-range. This does not seem to change even if one moves from a straight-leg sit up to a bent-knee curl up, although the iliopsoas activity will be increased if the feet are held by a partner.

Iliopsoas activity during a sit-up or curl-up should be controlled by preceding the activity with a posterior pelvic tilt so that a swayback or lumbar hyperextension position is avoided. The loss of iliopsoas muscle function will only slightly impair thigh flexion, since some of the flexion can be initiated by the abdominals and other flexors. The impairment does increase with increasing angles of flexion, demonstrating the importance of the iliopsoas in the middle range of motion.

The rectus femoris is another hip flexor whose contribution depends on knee joint positioning, since it acts as an extensor of the knee joint as well. It is called the "kicking muscle" because it is in maximal position for output at the hip during the preparatory phase of the kick when the thigh is drawn back into hyperextension and the leg is flexed at the knee. This position puts the rectus femoris on stretch and into an optimal length-tension relationship for the succeeding joint action in which the rectus femoris makes a powerful contribution to both hip flexion and knee extension during the kick. During the kicking action, the rectus femoris is very susceptible to injury and avulsion at its insertion site, the anterior inferior spine on the ilium. Loss of function of the rectus femoris will diminish thigh flexion strength as much as 17% (14).

There are three other secondary flexors of the thigh, the sartorius, the pectineus, and the tensor fascia latae (See Appendix B). The sartorius is a two-joint muscle crossing the knee joint to the medial side. It is a weak fusiform muscle producing abduction and external rotation in addition to the flexion action, creating a common dance position with the thigh raised and turned out. Knee extension will put this muscle on stretch.

The pectineus is one of the upper groin muscles. It is primarily an adductor of the thigh except in walking, when it actively contributes to thigh flexion. It is accompanied by the tensor fascia latae, which is more of an internal rotator, except in walking, when it aids thigh flexion. The tensor fascia latae is considered a two-joint muscle since it attaches to the fibrous band of fascia, the iliotibial band, running down the lateral thigh and attaching across the knee joint on the lateral aspect of the tibia.

Thus, this muscle will be stretched by a position of knee extension.

During the thigh flexion movement, the pelvis is pulled anteriorly by these muscles unless stabilized and counteracted by the trunk. The iliopsoas muscle will pull the pelvis anteriorly and so will the tensor fascia latae. If either of these muscles are tight, there may be pelvic torsion, pelvic instability, or a functional short leg created.

Extension of the thigh is important in the support of the body weight in stance as it maintains and controls the hip joint actions in response to the gravitational pull downward. Thigh extension also assists in propelling the body up and forward in walking, running, or jumping by producing hip joint actions that counteract gravity. The extensors attach into the pelvis and, consequently, play a major role in stabilizing the pelvis in the anterior and posterior directions.

The muscles contributing under all conditions of extension at the hip joint are the hamstrings. The two medial hamstrings, the semimembranosus and the semitendinosus, are not as active as the lateral hamstring, the biceps femoris, considered the "workhorse" of extension at the hip.

Since all of the hamstrings cross the knee joint, producing both flexion and rotation of the leg, their effectiveness as hip extensors will depend on positioning at the knee joint. With the knee joint extended, the hamstrings are put on stretch for optimal action at the hip. The hamstring output also increases with increasing amounts of thigh flexion; however, the hamstrings can be lengthened to a position of muscle strain if the leg is extended with the thigh in maximal flexion.

The hamstrings also control the pelvis by pulling down on the ischial tuberosity, creating a posterior tilt of the pelvis, and are responsible for maintaining upright posture in this manner. Tightness in the hamstrings can create significant postural problems by flattening the low back and producing a continuous posterior tilt of the pelvis.

In level walking or in low-output hip extension activities, the hamstrings are the predominant muscles contributing to the movement. With loss of function in the hamstrings, there would be a significant impairment in the extension movement at the hip.

If the resistance in the extension movement is increased, or if a more vigorous hip extension is needed, the gluteus maximus is recruited as a major contributor to the joint action (8). This occurs in running up hills, climbing stairs, raising out of a deep squat, sprinting, or rising from a chair. It also occurs in an optimal length-tension position, with thigh hyperextension and external rotation (8).

The gluteus maximus appears to dominate the pelvis during gait rather than contribute significantly to the generation of extension forces. Since the thigh is almost extended during the walking cycle, the function of the gluteus maximus is more of trunk extension and posterior tilt of the pelvis. At foot strike when the trunk flexes, the gluteus maximus will prevent the trunk from pitching forward. Since the gluteus maximus also externally rotates the thigh, an internal rotation position will place the muscle on stretch. Loss of function of the gluteus maximus muscle will not result in a significant impairment in the extension strength of the thigh, since the hamstrings dominate the extension strength production (14).

Finally, since the flexors and extensors control the pelvis in the anterior and posterior direction, it is important that they be balanced in both strength and flexibility so that the pelvis is not drawn forward or backward as a result of one group being stronger or less flexible.

The abduction of the thigh is an important movement in many dance and gymnastics skills, but abduction and the abduction muscles are more important in their role as stabilizers of the pelvis and thigh during gait. The abductors can raise the thigh laterally in the frontal plane or, if the foot is on the ground, they can move the pelvis on the femur in the frontal plane. When abduction occurs, such as in the splits on the ground, both hip joints will displace the same number of degrees in abduction, even though only one limb may have moved. The relative angle between the thigh and the trunk will be the same in both hip joints in abduction because of the pelvic shift in response to abduction initiated in one hip joint.

The main abductor of the thigh at the hip joint is the gluteus medius. This multipennate muscle contracts during the stance in a walk, run, or jump to stabilize the pelvis so that it does not drop to the nonstance limb. The effectiveness of the gluteus medius muscle is determined by its mechanical advantage. It is more effective if the angle of inclination of the femoral neck is less than 125 degrees, taking the insertion further away from the hip joint, and it is also more effective for the same reason in the wider pelvis (6). As the mechanical advantage of the gluteus medius is increased, the stability of the pelvis in gait will also improve.

The gluteus minimus, tensor fascia latae, and the piriformis also contribute to abduction of the thigh, with the gluteus minimus being the most active of the three. A 50% reduction in the function of the abductors will result in a slight-to-moderate impairment in abduction function (14). If the abductors are weak, there will be an excessive tilt in the frontal plane, with a higher pelvis on the weaker side (15). Additionally, the shear forces across the sacroiliac joint will greatly increase, and the individual will walk with greater side-to-side sway.

The adductor muscle group works to bring the thigh across the body such as seen commonly in dance, soccer, gymnastics, and swimming. The adductors, like the abductors, also work to maintain the pelvic position during gait. The adductors as a group constitute a large muscle mass, with all of the muscles originating on the pubic bone and running down the inner thigh. Although the adductors are important in specific activities, it has been shown that a 70% reduction in the function of the thigh adductors will result in only a slight or moderate impairment in hip function (14).

On the medial side of the thigh is the gracilis, on the anterior side of the thigh is the adductor longus, in the middle is the adductor brevis, and on the back side of the inner thigh is the adductor magnus. Up high in the groin is the pectineus, previously discussed briefly in its role as hip flexor. The adductors are active during the swing phase of gait as they work to bring the limb through (8).

The adductors work with the abductors to balance the pelvis. The abductors on one side of the pelvis will work with the adductors on the opposite side to maintain pelvic positioning and prevent tilting. The abductors and the adductors must be balanced in strength and flexibility so that the pelvis can be balanced side to side. FIGURE 6–9 illustrates how pelvic tilt can be created through abduction and adduction imbalances. If the abductors overpower the adductors through contracture or a strength imbalance, the pelvis will tilt to the side of the strong, contracted abductor. Adductor contracture or strength imbalances will produce a similar effect in the opposite direction.

The external rotation of the thigh is important in preparation for power production in the lower extremity as it follows the trunk during rotation. The muscles primarily responsible for external rotation are the gluteus maximus, the obturator externus, and the quadriceps femoris. The obturator internus, inferior and superior gemellus, and the piriformis contribute to external rotation when the thigh is extended. Since most of these muscles attach to the anterior face of the pelvis, they also exert considerable control over the pelvis and the sacrum.

Internal rotation of the thigh is basically a weak movement because it is a secondary movement for all of the muscles contracting to produce the joint action. The two muscles most involved in creating the movement of internal rotation are the gluteus medius and the gluteus minimus. Internal rotation is also aided by contractions of the gracilis, the adductor longus, adductor magnus, the

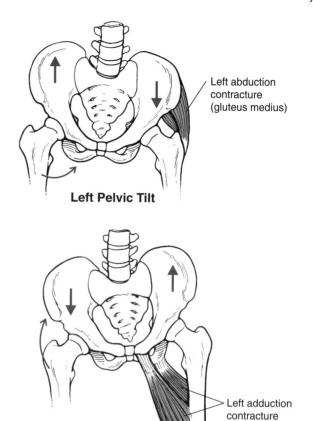

Left Pelvic Tilt

Left abduction contracture (gluteus medius)

Right Pelvic Tilt

Left adduction contracture (adductor longus & medius)

FIGURE 6–9. The abductors and adductors work in pairs to maintain pelvic height and levelness. If an abductor or adductor muscle group is stronger than the contralateral group, the pelvis will tilt to the strong side. As shown above, this will also happen with contracture of the muscle group.

tensor fascia latae, the semimembranosus, and the semitendinosus.

Strength and Force of the Hip Joint

The hip muscles can generate the greatest strength output in the movement of extension, with the most massive muscle in the body, the gluteus maximus, combining with the hamstrings to produce this movement. Extension strength is maximum with the hip flexed to 90 degrees and diminishes by about half as the hip flexion angle approaches the zero, neutral position (8). Extension strength also depends on the knee position, since the hamstrings cross the knee joint. The hamstring contribution to hip extension strength is enhanced with the knees extended (10).

There are many muscles contributing to hip flexion strength, but many of the muscles do so secondarily to other main roles. Hip flexion strength is primarily generated with the powerful iliopsoas muscle, though the strength of this muscle diminishes with trunk flexion. Additionally, the flexion strength of the thigh can be enhanced if there is flexion at the knee joint, increasing the contribution of the rectus femoris to the flexion strength generation.

Abduction strength is maximal from the neutral position and diminishes more than half at 25 degrees of abduction (8). This reduction is associated with decreases in muscle length, in spite of the fact that the ability of the gluteus medius to abduct the leg improves as a consequence of improving the direction of the pull of the muscle. The strength output of the abduction movement can also be increased if it is performed with the thigh flexed (8).

The potential for the development of adduction strength is substantial since the muscles contributing to the movement are massive when combined as a group. However, the adduction movement is not the primary contributor to most movements or sport activities and, consequently, is minimally loaded or strengthened through activity. Adduction strength values are greater from a position of slight abduction as a stretch is placed on the muscle group.

The strength of the external rotators is 60% greater than the internal rotators, except in a position of hip flexion when the internal rotators are slightly stronger (8). The strength output of both the internal and external rotators is greater in the seated position than in the supine position.

Forces generated by the muscles combine with those created by the body weight to produce very substantial loads on the hip joint. Standing on two limbs loads the hip joint with a force equivalent to 30% of body weight (8). This force is generated primarily by the body weight above the hip joint and is shared by right and left joints.

When a person stands on one limb, the force imposed on the hip joint increases significantly to approximately 2.5 to 3 times body weight (8,9). This is as a result of the increase in the amount of body weight, previously shared with the other limb, and a vigorous muscular contraction of the abductors, accounting for the majority of the increase in the force. This force generation at the hip joint is illustrated in FIGURE 6–10, showing the effect of gravity downward, the muscular force of the abductors upward to control the pelvis, and the corresponding forces applied to the hip joint. With increased muscular activity, the loads become even higher.

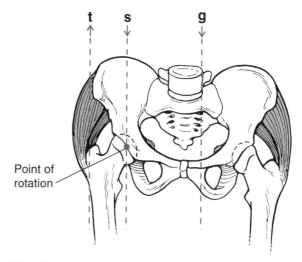

FIGURE 6–10. In stance, the force of gravity pulls the pelvis down as shown with the arrow *g* above. The rotation of the pelvis downward is counteracted with the creation of an abductor force *t* to counteract the effect of gravity. The result is an increase of the force acting on the hip joint *s*.

In stair climbing, forces can reach levels of 3 times body weight; in walking, the forces range from 4 to 7 times body weight; and in running, the forces can be as high as 10 times body weight (5,8,9). Fortunately, the hip joint can withstand 12 to 15 times body weight before fracture or breakdown in the osseus component will occur (9).

Conditioning

The muscles surrounding the hip joint receive some form of conditioning during walking, raising or lowering from a chair, or performing other common daily activities, such as climbing stairs. The hip musculature should be balanced so that the extensors do not overpower the flexors and the abductors are balanced by the adductors. This will ensure that there is sufficient control over the pelvis.

Also, since the hip joint muscles are used in all support activities, it is better to design exercises using a closed kinetic chain, which has the forces coming up through the system from the foot, such as would be the case in a squat. An example of an open kinetic chain exercise would be one using a machine, in which the muscle group would move the limb through a prescribed arc of motion. Finally, there are many two-joint muscles acting at the hip joint; thus, careful attention should be paid to adjacent joint positioning in order to maximize a stretch or strengthening exercise.

The flexors are best exercised in the supine or hanging position so that the thigh can be raised against gravity. The hip flexors are minimally used in a lowering activity such as a squat, when there is flexion of the thigh, because the extensors control the movement eccentrically. Since the hip flexors attach on the trunk and across the knee joint, their contribution to the flexion action can be enhanced with the trunk extended. Flexion at the knee will also enhance thigh flexion. It is easy to stretch the flexors with both trunk and thigh placed in hyperextension. The rectus femoris can be placed in a very strenuous stretch with thigh hyperextension and maximal knee flexion. Examples of both stretching and strengthening of the flexor muscle group have been provided in FIGURE 6–11.

A sampling of conditioning exercises for the extensors is presented in FIGURE 6–12. The success of conditioning the extensors will depend on trunk and knee joint positioning as well. The greater the amount of knee flexion, the less the hamstrings will contribute to the extension action, requiring more contribution from the gluteus maximus. For example, in a quarter squat activity with the extensors used eccentrically to lower the body and concentrically to raise the body, the hamstrings are the most active contributors. However, in a deep squat, with the amount of knee flexion increased to 90 degrees and beyond, the gluteus maximus will be used more, since the hamstrings are incapacitated due to reduced length.

Trunk positioning is also important and the activity of the hamstrings is enhanced with trunk flexion, since it increases the length of both the hamstrings and the gluteus maximus. The extensors are best exercised using a standing, weight-supported position, since they are used in this position in most cases and are one of the propulsive muscle groups in the lower extremity.

The extensors can be stretched to maximum levels with hip flexion accompanied by full extension at the knee. The stretch on the gluteus maximus can be increased with thigh internal rotation and adduction.

The abductors and adductors are difficult to condition because they influence balance and pelvic position so significantly. Examples of some exercises are presented in FIGURE 6–13. In a standing position, the thigh can be abducted against gravity, but it will cause the pelvis to shift dramatically so that the person loses balance. The adductors present even more of a problem, since it is very difficult to place the adductors in a position where they work against gravity since the abductors are responsible for lowering the limb back to the side after abducting it. Consequently, the supine position is best for both strengthening and stretching the abductors and adduc-

tors. Resistance can be offered manually or through some machine with external resistance to the movement.

The abductors and adductors can be exercised from the side-lying position so that they can work against gravity, but this position requires stabilization of the pelvis and low back before it can be effective. It is hard to independently exercise the abductors or adductors on one side without working the other side as well since both sides will be affected equally due to the action of the pelvis. For example, 20 degrees of abduction at the right hip joint will result in 20 degrees of abduction at the left hip joint because of the pelvic tilt accompanying the movement.

The rotators of the thigh are the most challenging in terms of conditioning because it is so difficult to apply a resistance to the rotation action. Some examples are presented in FIGURE 6–14. The seated position is recommended for strengthening the rotators since the rotators are strong in this position, and resistance to the rotation can be applied to the leg easily either with surgical tubing or manually. Since the internal rotators lose effectiveness in the extended supine position, they definitely should be exercised in the seated orientation. Both muscle groups can be stretched in the same way they are strengthened, using the opposite joint action for the stretch.

Contribution to Sport Skills or Movements

During walking, the muscles around the pelvis and the hip joint contribute minimally to the actual propulsion in the walk and are more involved with control of the pelvis (16). The muscular contribution of the lower extremity muscles active in walking are summarized in FIGURE 6–15. During the support phase of walking, there are two additional phases, braking and propulsive. The braking phase begins with heel strike and terminates with midsupport as the body is directly over the foot.

At heel strike in the braking phase, there is moderate activity in the gluteus medius of the weight bearing limb that keeps the pelvis balanced against the weight of the trunk. The abduction muscle force balances the trunk and the swing-leg about the supporting hip (17). This activity continues until midsupport (9). Similar activity is seen in the gluteus minimus, which contracts to assist the gluteus medius in its pelvis stabilization. The adductors also work concurrently with both of these muscles to control the limb during support.

The hamstrings contract at heel strike to control both hip flexion and trunk flexion eccentrically. Shortly after heel strike, the activity of the hamstrings drops off. The gluteus maximus is also active at heel strike to assist with the movement of the body over the leg. Finally, the

FLEXIBILITY

MANUAL RESISTANCE

WEIGHT TRAINING

FIGURE 6–11. Examples of flexibility, manual resistance, and weight-training exercises for the flexors are presented. The hip flexors can be stretched by pulling the leg toward the buttock (A) or by performing a lunge movement (B). Manual resistance can be applied on top of the thigh as the individual pulls upward (C). Weight training exercises for the hip flexors include hanging leg lifts (D) or flexion on a hip machine (E).

FLEXIBILITY

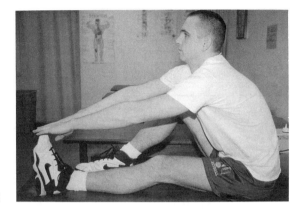

FIGURE 6–12. Thigh extensors are stretched maximally with the hip flexed and the knee extended. Examples of flexibility exercises are the passive straight-leg raise (A), a toe-touch from a seated position (B), and a toe-touch with the contralateral leg stabilized on the floor (C).

tensor fascia latae is active from heel strike to mid-support to assist with frontal plane control of the pelvis (9).

The propulsive phase of support begins with the mid-support position and ends with the toe leaving the ground. Late in the propulsive phase, the thigh begins to abduct, and there is increased activity in the gluteus medius with support contributions from the adductors of the opposite limb. Both the gluteus maximus and the hamstrings are also active at the end of the propulsive phase to make their contribution to the propulsion in the walk (9). As the toe begins to leave the ground, the thigh begins to internally rotate and flex, and the activity of the tensor fascia latae increases.

At the beginning of the swing phase, the limb must be swung forward rapidly. This movement is initiated by a vigorous contraction of the iliopsoas, sartorius, and the tensor fascia latae. The thigh adducts about mid-swing phase and internally rotates just after toe-off. The adduc-

tors are active at the beginning of the swing phase and continue into the stance phase. At the end of the swing phase, the limb is decelerated by activity from the hamstrings and the gluteus maximus (9).

If one graduated to a brisk race walking technique, the muscular activity would change. Since the race walk creates an exaggerated gait pattern in the frontal plane in which there is a lateral tilt of the pelvis away from the straight support limb, more activity from the abductors is required (18). This technique of exaggerated pelvic shift is necessary because it moderates the rise of the center of gravity created by the straight leg and increases the ability to produce a greater force on the ground. There is also greater hip flexion activity halfway into the swing phase, and more rapid extension by the extensors to propel the body forward quickly (18).

In running, the motions at the hip joint occur through a greater range as compared to walking, except for the

MANUAL RESISTANCE

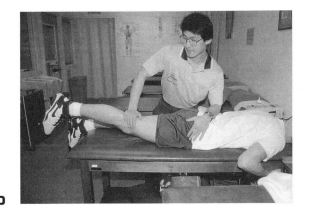
D

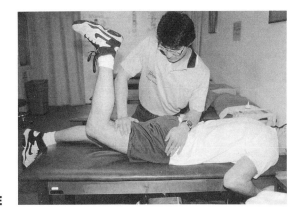
E

WEIGHT TRAINING

F

G

FIGURE 6–12. (continued) Manual resistance exercises are done from the prone position and include applying resistance to a straight leg that is extending (D). The gluteus maximus can be somewhat isolated if the same exercise is performed with the knee flexed (E). Weight-training exercises for the extensors include the leg curl (F) or the squat (G).

movement of hyperextension, which is greater in walking due to the increased time in stance. The muscular activity in running is similar to that seen in walking (FIGURE 6–16). In running, there are between 800 and 2000 foot contacts with the ground per mile, and 2 to 3 times the body weight is absorbed up through the extremity by the foot, leg, thigh, pelvis, and spine (19).

At heel strike, the force acting at the hip joint is approximately 4 times body weight due to the absorption of the force coming from the ground, the support of the body weight, and the muscular contraction of the abductors. This force increases to approximately 7 times body weight just before toe-off, due again to increases in the activity of the abductors (5).

During the swing phase, when there are no external forces acting on the joint, the load on the joint is reduced to 1 times body weight. This load is generated by con-

traction of the extensors. All of these forces are less in women because of their wider pelvises, making the abductors more effective so they do not have to generate as high a force output (5).

In the support phase of running, the gluteus medius and the tensor fascia latae are active just prior to contact and in the initial braking phase of support. These muscles control the pelvis on the ground to keep it from tilting to the opposite side. As the speed of the run increases, the activity of the gluteus medius and gluteus minimus will decrease slightly (20).

Also active in the initial portion of the support phase are the gluteus maximus and the hamstrings, which eccentrically control the limb in flexion. The hamstrings become more active in the support phase as the speed increases, whereas the gluteus maximus becomes less active at this point (20).

FLEXIBILITY

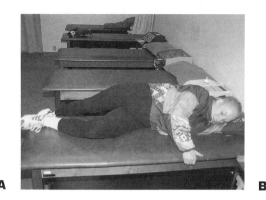

MANUAL RESISTANCE

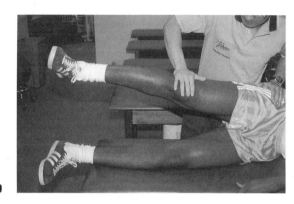
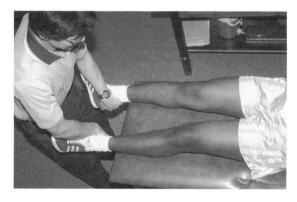

FIGURE 6–13. Thigh abductor and adductor muscles can be best exercised from the side-lying or supine position. Flexibility exercises include side-lying with the leg hanging over the table to stretch the abductors (A), a cross-over sit to stretch the abductors (B), and the adductor stretch with the feet together (C). Manual resistance exercise of the abductors can be administered from the side-lying position with resistance applied to the thigh as the limb is abducted (D). Resistance can be applied to the inside of the ankles to resist the adduction movement (E).

During the propulsive phase of running, the hamstrings are very active during the extension of the thigh. The gluteus maximus also contributes to extension during late stance, while also generating external rotation until toe-off.

Once the foot leaves the ground to begin the swing phase, the limb is brought forward by the iliopsoas and rectus femoris, slowing the thigh in hyperextension, and moving the thigh forward into flexion. This muscle is the most important muscle for forward propulsion of the body since it accounts for the large range of motion in the lower extremity. It initiates the flexion movement so vigorously that the iliopsoas action also contributes to the knee extension action. The iliopsoas is active for more than 50% of the swing phase in running (20). In the early part of the swing phase, there is activity in the adductors, which, as in walking, is working with the abductors to control the pelvis.

At the end of the swing phase, there is a great amount of muscular activity in the gluteus maximus and the hamstrings as they begin to decelerate the rapidly flexing thigh. As the speed of the run increases, the activity of the gluteus maximus increases as it assumes more of the responsibility for slowing the thigh down in preparation for foot contact in descent. Also, in the later portion of the swing phase, the abductors will become active again as they lower the thigh down eccentrically to produce adduction.

The pelvic activity in both running and walking is controlled by the arms, the trunk position, and the abductor muscles. Pelvic rotation to the left occurs as the right thigh swings forward. This rotation is countered by arm

WEIGHT TRAINING

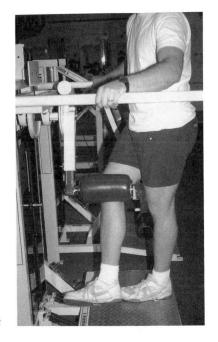

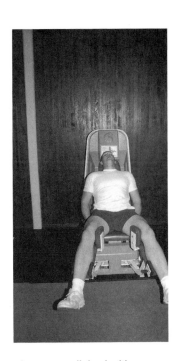

F   **G**   **H**

FIGURE 6–13. (continued) Weight-training exercises for the abductors and adductors can be accomplished with special hip machines (F), cables (G), or Nautilus hip machines (H).

swing of the contralateral arm in the opposite direction. If the arm swing is excessive or insufficient in its control of the leg swing, the pelvis will rotate too much, placing stress on muscular attachments on the iliac crest.

Pelvic activity in right and left lateral tilt is controlled by the action of the abductors. As the pelvis tilts down to the unsupported limb, the abductors of the support limb stabilize the pelvis so it does not tilt to the opposite side. If the abductors cannot control the pelvic action in the frontal plane, a Trendelenburg gait will be present in which the pelvis will drop to the unsupported side. An illustration of the trendelenburg gait is presented in FIGURE 6–17. Excessive movements in the frontal plane generates shear forces at the pubic symphysis (8).

The pelvis moves very little in anterior and posterior tilt as the leg flexes and extends. However, position of the trunk can greatly influence anterior and posterior tilt of the pelvis during running and walking. In uphill running or walking, the trunk is flexed, creating an anterior tilt of the pelvis. The opposite occurs in downhill walking or running where the trunk hyperextends and creates a posterior tilt of the pelvis.

In another activity, stair ascent and descent, there are similar patterns to those just described in walking and running; however, the hip muscles generally contribute less than the muscles acting on the knee and ankle joints. A review of the muscles contributing to ascent and descent is presented in FIGURE 6–18. Going up a stair, termed ascent, is first initiated with a leg pull via vigorous contraction of the iliopsoas, which pulls the leg up against gravity to the next stair (21). The rectus femoris becomes active in this phase as it assists in the thigh flexion and eccentrically slows the knee flexion.

Next, foot placement occurs as the limb is lowered down to the next step. At this point, there is activity in the hamstrings, primarily working to slow down the extension at the knee joint (21). As the foot makes contact with the next step, there is weight acceptance involving some activity in the extensors of the thigh.

The next phase is pull-up, in which extension occurs to bring the body up to the step. The majority of the extension is generated at the knee joint, and there is minimal contribution from the hip, other than contraction by the gluteus medius to pull the trunk up over the limb (21). Finally, in the forward propelling stage in which the limb pushes up to the next step, there is minimal activity at the hip with the ankle joint generating the most force.

Going down the stairs, or descent, requires minimal hip muscular activity. In the leg pull phase, the hip flexors are active, followed by hamstring activity in the foot

FLEXIBILITY

MANUAL RESISTANCE

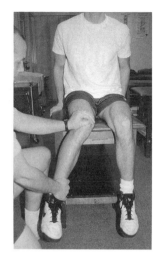

A

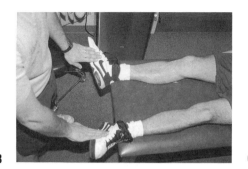

B

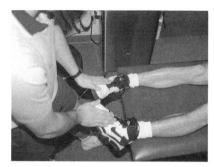

C

WEIGHT TRAINING

D

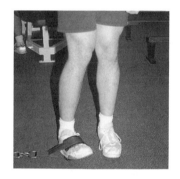

E

FIGURE 6–14. The rotators can be easily stretched by using the opposite movement; however, strengthening these muscles requires creativity and attention to stabilization of adjacent body parts. To stretch the rotators, the individual should be placed in a seated position and the leg pulled out to stretch the external rotators and pushed in to stretch the internal rotators (A). Manual resistance to rotation can be applied to the inside of the foot as the individual internally rotates (B) or to the outside of the foot during external rotation (C). Weight-training exercises for the rotators include a squat with the toes pointing in (D) or cable exercises (E).

placement phase when the limb is lowered to the ground (21). As the limb makes contact with the next step in weight acceptance, the hip is minimally involved as most of the weight is eccentrically absorbed at the knee and ankle joints. The muscles acting at the knee joint are primarily responsible for generating the forces in the forward propelling phase.

In the final phase of support, the controlled lowering phase, the body is eccentrically lowered onto the step primarily through activity at the knee joint. There is a minimum amount of hip extensor activity at the end of this phase.

In stair climbing, it is the higher step leg producing the greatest effort for both ascent and descent. The knee joint is more active in ascent than descent, and the hip joint activity is small in both, being almost negligible in the descent phase (21).

A final activity for analysis is cycling, which can be divided into a cycle of 360 degrees with 0 degrees being the top of the cycle and 180 degrees being the bottom of the cycle. As shown in FIGURE 6–19, the muscles of the thigh can contribute significantly to the generation of force in cycling. The muscles active in the top half of the cycle, from 270 to 360 degrees and from 0 to 90 degrees, are the sartorius and tensor fascia latae. These two muscles are active through 270 to 360 to pull the thigh up. They work with the gracilis through the initial 90 degrees of the cycle to assist with the flexion action (22).

Both the gluteus medius and the gluteus maximus become active after about 30 degrees into the cycle and continue in through approximately 150 degrees as the thigh is extended. As the activity from the gluteus medius and maximus begins to decline, the activity of the hamstrings increases and continues from approximately 130

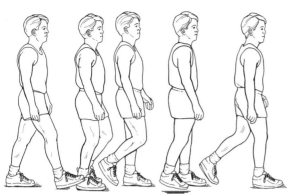

Foot strike Midsupport Toe off Forward swing Deceleration

	Walking Phases				
		Support		Swing	
				Forward	
Muscles	Footstrike	Midsupport	Toe-off	Swing	Deceleration
Dorsiflexors	***	**	**	**	**
Intrinsic Foot Muscles		***			
Gluteus Maximus	*	**	***		*
Gluteus Medius	**	***	**	*	
Gluteus Minimus	**	***	**	*	
Hamstrings	***	**	**	*	**
Iliopsoas				***	
Plantar Flexors		*	**		
Quadriceps	*	***	**		*
Sartorius				**	*
Tensor Fascia Latae	*	**	*	***	
Thigh Adductors	**	**	*	**	*

* = low activity
** = moderate activity
*** = high activity

FIGURE 6–15. The contribution of the lower extremity muscles to the walking gait are presented with indications of level of activity. During foot strike, there is a high level of activity in the dorsiflexors and hamstrings; in midsupport, the gluteus minimus, gluteus medius, and the quadriceps femoris group are most active; at toe-off, the intrinsic foot muscles and the gluteus maximus are most active; in the forward swing, the iliopsoas and the tensor fascia latae are active; at the end of the swing phase, the activity extensors is low to moderate (Adapted from Mann, 1986).

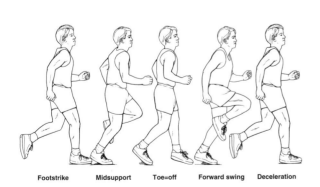

Footstrike   Midsupport   Toe=off   Forward swing   Deceleration

	Running Phases				
		Support		Swing	
				Forward	
Muscles	Footstrike	Midsupport	Toe-off	Swing	Deceleration
Dorsiflexors	*	**	**	**	**
Intrinsic Foot Muscles		***			
Gluteus Maximus	**	**	***		*
Gluteus Medius	**	***	**	*	
Gluteus Minimus	**	***	**	*	
Hamstrings	***	**	***	*	**
Iliopsoas				***	
Plantar Flexors	**	*	**		
Quadriceps	**	***	***		*
Sartorius				**	*
Tensor Fascia Latae	**	**	*	***	
Thigh Adductors	**	**	*	**	*

* = low activity
** = moderate activity
*** = high activity

FIGURE 6–16. In running, there is high level of muscular activity in the hamstrings, gluteus minimus, gluteus maximus, the quadriceps femoris group, and the intrinsic muscles of the foot, during the support phase of the activity. During the swing phase, there is substantial activity in the iliopsoas and the tensor fascia latae (Adapted from Mann, 1986).

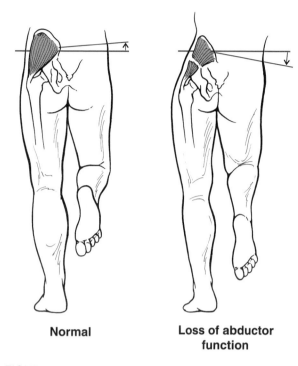

Normal   **Loss of abductor function**

FIGURE 6–17. If the hip abductors lack the strength to control the pelvis in the frontal plane (lateral tilt), a Trendelenburg gait will result. This gait is characterized by a drop to the unsupported side during walking or running.

degrees to 250 degrees into the cycle as they work both the hip and knee (22).

Before leaving this section, it is necessary to mention the contribution of the hip joint to movements commonly seen in dance and gymnastics. There are excessive hip joint movements required in these activities, often with 180 degrees of turnout required at the hip joint to perform specific skills. Turnout at the hip through this range of motion stresses the rotators and the adductors. Inability to generate maximal external rotation of the hip joint often leads to overuse injury of one form or another in the lower extremity as other adjustments are made to accommodate the lack of motion at the hip joint.

Injury Potential in the Pelvic and Hip Complex

Injuries to the pelvis and hip joint account for a small percentage of injuries incurred in the lower extremity. In fact, overuse injuries to this area account for only 5% of the total for the whole body (23). This may be attributable to the strong ligamentous support, significant muscular support, and solid structural characteristics of the region.

Injuries to the pelvis primarily occur in response to some abnormal function that excessively loads areas of the pelvis. Iliac apophysitis is an example of such an injury in which excessive arm swing in gait causes exces-

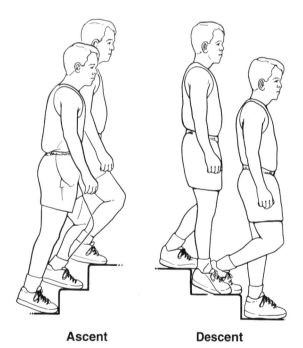

Ascent   **Descent**

Muscles	Stair Climbing Phases	
	Ascent	Descent
Dorsiflexors	*	*
Gluteus Medius	**	*
Hamstrings	*	*
Illiopsoas	**	*
Plantar Flexors	**	**
Quadriceps	***	**

* = low activity
** = moderate activity
*** = high activity

FIGURE 6–18. In stair ascent, there is significant contribution from the quadriceps, with assistance from the plantar flexors and the iliopsoas. In descent, the same muscles control the movement eccentrically. For stair climbing as a whole, there is less contribution from the hip muscles than in walking or running (Adapted from McFayden, et al., 1988).

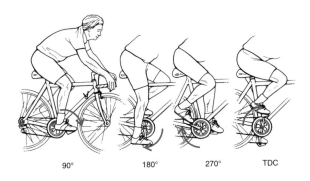

Muscles	Cycling Phases			
	TDC-90	90–180	180–270	270-TDC
Biceps Femoris	*	***	*	*
Gastrocnemius	**	***	**	*
Gluteus Maximus	***	**	*	*
Gracilis	**	*	*	
Rectus Femoris	**	**	*	***
Sartorius	*	*	*	*
Semimembranosus	**	***	**	*
Tensor Fascia Latae	*	*	*	*
Tibialis Anterior	**	**	**	***
Vastus Lateralis	***	**	*	**
Vastus Medialis	***	**	*	**

* = low activity
** = moderate activity

FIGURE 6–19. In the first 90 degrees of the cycling stroke, from top center to a crank position parallel to the ground, there is high activity in the quadriceps femoris group and the gluteus maximus. From 90 to 180 degrees, the gastrocnemius and hamstrings are more active. In the upswing of the cycle, from 180 to 270 degrees, there is some activity from the hamstrings, the dorsiflexors, and the gastrocnemius. In the last phase of the cycle, 270–0 or 360 (top center), there is high activity in the rectus femoris and the tibialis anterior (Adapted from Houtz, et al., 1959; Jorge, et al., 1986).

sive rotation of the pelvis, creating stress on the attachment site of the gluteus medius and tensor fascia latae on the iliac crest (23). Apophysitis is an inflammation of an apophysis, or bony outgrowth, and can also develop into a stress fracture.

Other sites in the pelvis subjected to apophysitis or stress fracture is the anterior superior iliac spine where the sartorius attaches (24). At the anterior inferior iliac spine, the rectus femoris can produce the same type of injury. The iliac crest can also experience an avulsion fracture from a strong pull by the abdominal attachments (24).

A stress fracture in the pubic rami can be produced by strong contractions from the adductors, often associated with over striding in a run (25). Finally, the hamstrings can exert enough force to create an avulsion fracture on the ischial tuberosity. All of these injuries are more common in activities such as sprinting, jumping, soccer, football, basketball, or figure skating in which sudden bursts of motion are required (24).

The sacrum and the sacroiliac joint can dysfunction as a result of injury or poor posture. Posturally, if one assumes a round shoulder, forward head posture, the center of gravity of the body is moved forward. This increase in the curvature of the lumbar spine produces a ligamentous laxity in the dorsal sacroiliac ligaments and stress on the anterior ligaments (26). Also, any skeletal asymmetry such as a short leg will produce a ligament laxity in the sacroiliac joint (1).

With excessive mobility, there are large forces transferred to the sacroiliac joint, producing an inflammation of the joint known as sacroiliitis. Inflammation of the joint may occur in an activity such as long jumping, in which the landing is absorbed with the leg extended at the knee and the hip flexed, or in extreme flexion of the trunk, or flexion of the trunk combined with lateral flexion (1). The sacroiliac joint also becomes very mobile in pregnant women, making them more susceptible to sacroiliac sprain (4).

The functional position of the sacrum and the pelvis is also important for maintaining an injury free lower extremity. A functional short leg can be created by posterior rotation of the same side ilium, anterior ilium rotation of the opposite side, superior ilium movement on the same side, forward or backward sacral torsion to the same side, or sacral flexion of the opposite side (1). A functional short leg requires adjustments all down the limb, creating stress at the sacroiliac joint, the knee, and the foot.

The hip joint, although not susceptible to high rates of injury, does incur some age-related conditions that must be considered when working with children or older adults. In children 3 to 12 years old, the condition known as Legg-Calve-Perthes disease may appear (9). This condition, also called coxa plana, is a condition in which the femoral head experiences a degeneration and damage to the proximal femoral epiphysis. This disorder strikes males four times more frequently than females, and usually occurs to one limb. It is caused by trauma to the joint, synovitis or inflammation to the capsule, or some vascular condition that limits blood supply to the area.

Slipped capital femoral epiphysitis is another disorder, striking children between the ages of 10 and 17 years old. It is usually caused by some traumatic event that forces the femoral neck into external rotation. This tilts the femoral head back and medially and tilts the growth plate forward and vertical, producing a nagging pain on the front of the thigh. An individual with this disorder will also walk with an externally rotated gait and have limited internal rotation with the thigh flexed and abducted (9). An example of where slippage may occur is the case of the baseball player who is running the bases and rounds third base with the left foot fixed in internal rotation while the trunk and pelvis rotate in the opposite direction.

The final major childhood disorder to the hip joint is congenital hip dislocation, a disorder striking females more than males (9). Usually this condition is diagnosed early as the infant assumes weight on the lower extremity. The hip joint subluxates or dislocates with no reason. The thigh cannot abduct, the limb shortens, and a limp is usually present. Fortunately, this condition is easily corrected with an abduction orthotic.

An age-related disorder of the hip joint seen commonly in the elderly is osteoarthritis, a degeneration of the joint cartilage and the underlying subchondral bone, a narrowing of the joint space, and the growth of osteophytes in and around the joint. This affliction strikes millions of elderly people, creating a significant amount of pain and discomfort during weight support and gait activities. To reduce the pain in the joint, individuals often assume a position of flexion, adduction, and external rotation, whichever is the position of least tension for the hip.

Over 60% of all injuries to the hip occur in the soft tissue (15). Of these injuries, 62% will occur in running, 62% are associated with a varum alignment in the lower extremity, and 30% are associated with a leg-length discrepancy (15). These types of injuries are usually muscle strains, tendinitis of the muscle insertions, or bursitis.

The most common soft tissue injury to the hip region is gluteus medius tendinitis, occurring more frequently in women as a result of excessive pull by the gluteus medius during running (15). A hamstring strain is also common and is seen in activities such as hurdling in which the lower limb is placed in a position of maximum hip flexion and knee extension. It can also occur with speed or hill running, or in individuals performing with poor flexibility or conditioning in this muscle group.

Iliopsoas strain can occur in activities such as sprinting, in which a rapid forceful flexion taxes the muscle, or the muscle is used to eccentrically slow a rapid extension at the hip. The adductors are often strained in an activity such as soccer in which the lower extremity is rapidly abducted and externally rotated in preparation for contact with the ball. Strain to the rectus femoris can occur in a rapid forceful flexion of the thigh such as is seen in sprinting, or in a vigorous hyperextension of the thigh such as in the preparatory phase of a kick.

A piriformis strain caused by excessive external rotation and abduction when the thigh is flexed creates pain in the adduction, flexion, and internal rotation movement of the thigh. A piriformis syndrome can develop, which is an impingement of the sciatic nerve aggravated by internal and external rotation movement of the thigh during walking (15). The syndrome can also be created by a functional short leg that lengthens the piriformis and then stretches it as the pelvis drops to the shorter leg. The irritation of the sciatic nerve causes pain in the buttock area, which can travel down the outside of the leg.

Other soft tissue injuries to the hip region are seen in the bursae. The most common of these is greater trochanteric bursitis, which is caused by hyperadduction of the thigh. This could be produced by running with more of a leg crossover in each stride, imbalance between the abductors and adductors, running on banked surfaces, having a leg length difference, or remaining on the outside of the foot during the support phase of a walk or run (19,23). It is especially prevalent in runners with a wide pelvis, a large Q-angle, and an imbalance between the abductors and adductors (19,23).

Since the right hip adductors work with the left hip abductors, and vice versa, any imbalance creates an asymmetrical posture. For example, a weak right abductor would create a lateral pelvic tilt, with the right side high and the left side low. This creates stress on the lateral hip, setting up the conditions for the bursitis. Pain on the outside of the hip is accentuated with trochanteric bursitis when the legs are crossed.

Ischial bursitis can develop with prolonged sitting and is aggravated by walking, stair climbing, and flexion of the thigh. Finally, iliopectineal bursitis may develop in reaction to a tight iliopsoas muscle or osteoarthritis of the hip (9).

Two remaining soft tissue injuries seen in dancers is lateral hip pain created by iliotibial band syndrome and snapping hip syndrome. The strain to the iliotibial band is created because dancers warm up with the hip abducted and externally rotated. They have very few flexion and extension routines in both warm-up and dance routines. The stress to the iliotibial band occurs with thigh adduction and internal rotation, extremely limited in dancers due to technique (27). Iliotibial band syndrome can also be caused by excess tension in the tensor fascia latae in abducting the hip in single-stance weight bearing. The snapping hip is also common, producing a clicking sound as the hip capsule moves or the iliopsoas tendon snaps over a bony surface.

The bony or osseus injuries to the hip are usually a result of a strong muscular contraction creating an avulsion fracture. The abductors can create an avulsion fracture on the greater trochanter, and the iliopsoas can pull hard enough to produce an avulsion fracture at the lesser trochanter (9). Stress fractures can also appear in the femoral neck with the causes not yet determined. It is believed that these stress fractures may be related to some type of vascular necrosis in which the blood supply is limited, or to some hormonal deficiency that reduces the bone density in the neck (15). The stress fracture at this site produces a pain in the groin area.

The Knee Joint

The knee joint is a condyloid joint with 2 df, in which flexion and extension occur much like flexion and extension at the elbow joint, but where the flexion movement is accompanied by a small but significant amount of rotation (28). The joint is vulnerable when it comes to injury because of the mechanical demands placed upon it and the reliance on soft tissue for support.

The knee joint supports the weight of the body and transmits forces from the ground while allowing a great deal of movement between the femur and the tibia. In the extended position, the knee joint is stable due to the vertical alignment, the congruency of the joint surfaces, and the effect of gravity on the joint. In any flexed position, the knee joint is mobile and requires special stabilization from the powerful capsule, ligaments, and muscles surrounding the joint (28).

The ligaments surrounding the knee support the joint passively, as they are loaded in tension only. The muscles support the joint actively and are also loaded in tension, and bone offers support and resistance to compressive loads (29). Functional stability of the joint is derived from the passive restraint of the ligaments, the joint geometry, the active muscles, and the compressive forces pushing the bones together.

Anatomical and Functional Characteristics of the Joint

There are three articulations in the region known as the knee joint: the tibiofemoral joint, the patellofemoral joint, and the superior tibiofibular joint (30). The structure of the knee joint is illustrated in FIGURE 6–20.

The tibiofemoral joint, commonly referred to as the *actual knee joint*, is the articulation between the two longest and strongest bones in the body, the femur and the tibia. At the end of the femur are two large convex surfaces, the medial and lateral condyles, separated by the intercondylar notch in the back and the patellar, or trochlear groove in the front (28) (FIGURE 6–20).

It is important to review the anatomical characteristics of these two condyles because their differences and the corresponding differences on the tibia account for the presence of rotation in the knee joint. The lateral condyle is flatter, it has a larger surface area, it is more prominent anteriorly to hold the patella in place, and it is basically aligned with the femur (30). The medial condyle projects more longitudinally and medially, it is longer in the anteroposterior direction, it angles away from the femur in the rear, and it is aligned with the tibia (30). Above the condyles on both sides are the epicondyles, which are the sites of capsule, ligament, and muscular attachment.

The condyles rest on the tibial plateau, a medial and lateral surface separated by a ridge of bone termed the intercondylar eminence. This ridge of bone serves as an attachment site for ligaments, centers the joint, and stabilizes the bones in weight bearing (30). The medial surface of the plateau is oval in shape, longer in the anteroposterior direction, and concave to accept the convex condyle of the femur. The lateral tibial plateau is circular and slightly convex (30). Consequently, the medial tibia and femur fit fairly snugly together, but the lateral tibia and femur do not fit together because both surfaces are convex (28). This structural difference is one of the determinants of rotation as the lateral condyle has more excursion with flexion and extension at the knee.

Two separate, fibrocartilage menisci lie between the tibia and the femur. As shown in FIGURE 6–21, the lateral meniscus is oval-shaped with attachments at the anterior and posterior horns via coronary ligaments (30,31). It also receives attachments from the quadriceps femoris in the front, and the popliteus muscle and the posterior cruciate ligament in the rear. The lateral meniscus occupies a larger percentage of the area in the lateral compartment than the medial meniscus in the medial compartment. Also, the lateral meniscus is more mobile,

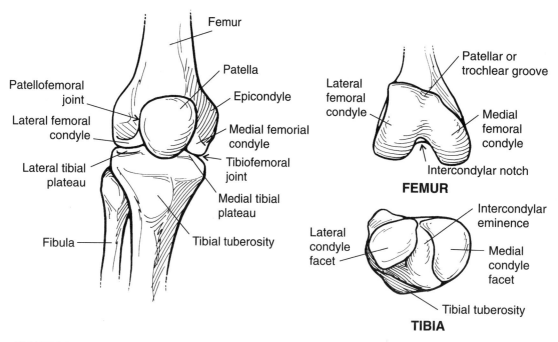

FIGURE 6–20. The knee joint is a condyloid joint allowing movement in two directions, flexion and extension, and rotation. Movements at the knee joint are complex due to the structural differences in the medial and lateral compartments of the joint.

capable of moving over twice the distance of the medial meniscus in the anteroposterior direction (30).

The medial meniscus is more crescent-shaped with a wide base of attachment on both the anterior and posterior horns via the coronary ligaments (FIGURE 6–21). It is connected to the quadriceps femoris and the anterior cruciate ligament in the front, the tibial collateral ligament on the side, and the semimembranosus muscle to the back of the side (30). The menisci are connected to each other at the anterior horns by a transverse ligament. The menisci are avascular, having no blood supply to the inner portion of the fibrocartilage, thus if a tear occurs, healing is almost impossible. There is some blood supply to the fibrocartilage on the outer portion of the menisci, making healing of the structure possible.

The menisci are important structures in the knee joint, participating in many functions. The menisci enhance stability in the joint by deepening the contact surface on the tibia. They participate in shock absorption by transmitting half of the weight-bearing load in full extension and a significant portion of the load in flexion (32). In flexion, the lateral meniscus carries the greater portion of the load.

By absorbing some of the load, the menisci protect the underlying articular cartilage and subchondral bone. The menisci transmit the load across the surface of the joint, reducing the load per unit area on the tibiofemoral contact sites (31). The contact area in the joint is reduced by two times when the menisci are absent. This increases the pressure and the susceptibility to injury (33).

During low-load situations, the contact is primarily on the menisci, and in high-load situations, the contact area increases with 70% of the load still on the menisci (31). The lateral meniscus carries a significantly greater percentage of the load.

The menisci also enhance lubrication of the joint. By acting as a space-filling mechanism, more fluid is dispersed to the surface of the tibia and the femur. It has been demonstrated that a 20% increase in friction within the joint will occur with the removal of the meniscus (32).

Finally, the menisci limit motion between the tibia and femur. In flexion and extension, the menisci move with the femoral condyles. As the leg flexes, the menisci move posteriorly because of the rolling of the femur and muscular action of the popliteus and semimembranosus muscles (32). At the end of the flexion movement, the menisci fill up the posterior portion of the joint, acting as a space-filling buffer. The reverse movement occurs in extension, as the quadriceps femoris and the patella assist in moving the menisci forward on the surface. Add-

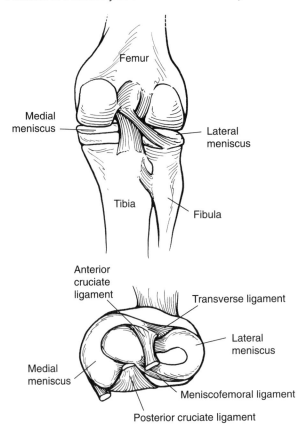

FIGURE 6–21. Two fibrocartilage menisci are located in the lateral and medial compartments of the knee. The medial meniscus is crescent-shaped and the lateral meniscus is oval-shaped to match the surfaces of the tibial plateau and the differences in the shape of the femoral condyles. Both menisci serve important roles in the knee joint by offering shock absorption, stability, lubrication, and by increasing the contact area between the tibia and the femur.

itionally, the menisci follow the tibia during rotation movements.

The tibiofemoral joint is supported by four main ligaments, two collaterals and two cruciates, which assist in maintaining the relative position of the tibia and femur so that contact is appropriate and at the right time. See Appendix A for insertions, actions, and illustration of these ligaments. They are the passive load-carrying structures of the joint and serve as a back up to the muscles (29).

On the sides of the joint are the collateral ligaments. The medial collateral ligament supports the knee against any valgus force and offers some resistance to both internal and external rotation (34). It is taut in extension and reduces in length by approximately 17% in full flexion (35). The medial collateral ligament offers 78% of the total valgus restraint at 25 degrees of knee flexion (36).

The lateral collateral ligament, thinner and rounder than the medial ligament, offers the main resistance to varus forces acting at the knee. It is also taut in extension and reduces its length by approximately 25% in full flexion (35). The lateral collateral ligament offers 69% of the varus restraint at 25 degrees of knee flexion (36). The lateral collateral ligament is not affected by rotation at the knee joint (35).

In full extension, the collateral ligaments are assisted by the tightening of the posteromedial and posterolateral capsules, making the extended position the most stable. There is a deep layer of ligaments under the collaterals that have been considered a branch of the collaterals and termed *capsular ligaments* (37).

The cruciate ligaments are intrinsic, lying inside the joint in the intercondylar space, and control both anteroposterior and rotational motion in the joint. The anterior cruciate ligament provides the primary restraint for the anterior movement of the tibia relative to the femur. It accounts for 85% of the total restraint in this direction (36). The anterior cruciate is 40% longer than its counterpart, the posterior cruciate. It elongates by about 7% as the knee moves from extension to 90 degrees of flexion, and then maintains the same length up through maximum flexion (35).

If the joint is internally rotated, the insertion of the anterior cruciate ligament moves forward, elongating the ligament slightly more. With the joint externally rotated, the ligament does not elongate up through 90 degrees of knee flexion, but elongates up to 10% from 90 degrees to full flexion (35). Different parts of the anterior cruciate ligament are taut in different positions, with the anterior fibers tight in extension, the middle fibers tight in internal rotation, and the posterior fibers tight in flexion. The ligament as a whole is considered to be taut in the extended position (FIGURE 6–22).

The posterior cruciate ligament offers the primary restraint to posterior movement of the tibia on the femur, accounting for 95% of the total resistance to this movement (36). The ligament decreases in length and slackens by 10% at 30 degrees of knee flexion, and then maintains that length (35). It increases in length by about 5% with internal rotation of the joint up to 60 degrees of flexion, and then decreases in length by 5 to 10% as flexion continues. The posterior cruciate is not affected by external rotation in the joint, maintaining a fairly constant length. It is maximally strained through 45 to 60 degrees of flexion (35) (FIGURE 6–22). Like the anterior cruciate, the fibers of the posterior cruciate participate in different functions. The posterior fibers are taut in extension, the anterior fibers taut in mid-flexion, and the posterior fibers in full flexion.

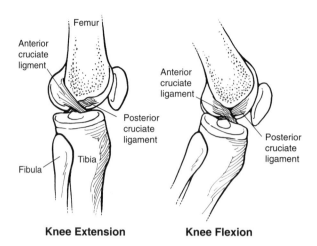

Knee Extension   **Knee Flexion**

FIGURE 6–22. The anterior cruciate ligament provides anterior restraint of the movement of the tibia relative to the femur. It is taut in knee extension. The posterior cruciate ligament offers restraint to posterior movement of the tibia relative to the femur. It is taut in flexion. Both sets of cruciates limit rotation at the knee joint.

Both the cruciates stabilize, limit rotation, and cause the sliding of the condyles over the tibia in flexion. In a standing posture, with the tibial shaft vertical, the femur is aligned with the tibia and tends to slide posteriorly. A hyperextended position to 9 degrees of flexion is unstable since the femur tilts posteriorly and is minimally restricted (29). At a 9 degree tilt of the tibia, the femur slides anteriorly to a position where it is more stable and supported by the patella and the quadriceps femoris.

Another important support structure surrounding the knee joint is the capsule. One of the largest capsules in the body, it is reinforced by numerous ligaments and muscles, including the medial collateral ligament, the cruciate ligaments, and the arcuate complex (30). In the front, the capsule forms a big pocket that offers a large patellar area and is filled with the infrapatellar fat pad and the infrapatellar bursa. The fat pad offers a stopgap in the anterior compartment of the knee.

The capsule is lined with the largest synovial membrane in the body, and forms embryonically from three separate pouches (38). In 20 to 60% of the population, a permanent fold, called plica, remains in the synovial membrane (39). The common location of plica is normally medial and superior to the patella. It is soft and pliant and passes over the femoral condyle in flexion and extension. If injured, it can become fibrous and create both resistance and pain in motion (39). There are also more than 20 bur-

sae located in and around the knee, serving to reduce friction between muscle, tendon, and bone (30).

The second joint in the region of the knee is the patellofemoral joint, consisting of the articulation of the patella with the trochlear groove on the femur. The patella is a triangular-shaped sesamoid bone encased by the tendons of the quadriceps femoris muscles. The primary role of the patella is to increase the mechanical advantage of the quadriceps femoris (38).

The posterior articulating surface of the patella is covered with the thickest cartilage found in any joint in the body (28). A vertical ridge of bone separates the underside of the patella into medial and lateral facets, each of which can be further divided into superior, middle, and inferior facets. A seventh facet lies on the far medial side of the patella and is called the *odd facet* (30). The structure of the patella and the location of these facets are presented in FIGURE 6–23. During normal flexion and extension activities, typically five of these facets make contact with the femur.

The patella is connected to the tibial tuberosity via the strong patellar ligament and connected to the femur and tibia by small patellofemoral and patellotibial ligaments, which are actually thickenings in the extensor retinaculum surrounding the joint (38).

Positioning of the patella and alignment of the lower extremity in the frontal plane is determined by measuring

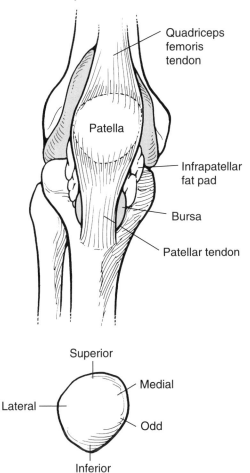

FIGURE 6–23. The patella increases the mechanical advantage of the quadriceps femoris group. The patella has five facets, or articulating surfaces which are the superior, inferior, medial, lateral, and the odd facet.

the Q-angle. Illustrated in FIGURE 6–24, the Q-angle is the angle formed by drawing a line from the anterior superior spine of the ilium to the middle of the patella, and a second line from the middle of the patella down to the tibial tuberosity. The most efficient Q-angle for quadriceps femoris function is one close to 10 degrees (40). Males typically have Q-angles averaging 10 to 14 degrees, while females average 15 to 17 degrees, primarily because of their wider pelvic basins (40).

The Q-angle represents the valgus stress acting on the knee, and if it is excessive, many patellofemoral problems can develop. Any Q-angle over 17 degrees is considered to be excessive and might be termed genu valgum, or knock-knees (40). A very small Q-angle would create bowleggedness, or genu varum.

The third and final articulation for examination is the small, superior tibiofibular joint shown in FIGURE 6–25.

This joint consists of the articulation between the head of the fibula and the posterolateral and inferior aspect of the tibial condyle. It is a gliding joint moving anteroposteriorly, up and down, and rotating in response to rotation of the tibia and the foot (41). The fibula externally rotates and moves out and up with dorsiflexion of the foot and accepts approximately 16% of the static load applied to the leg (41).

The primary functions of the superior tibiofibular joint are to dissipate the torsional stresses applied by the movements of the foot, and to dissipate lateral tibial bending. Both the tibiofibular joint and the fibula absorb and control tensile rather than compressive loads applied to the lower extremity. The middle part of the fibula has more ability to withstand tensile forces than any other part of the skeleton (41).

The function of the knee is complex due to the asymmetrical medial and lateral articulations and the patellar mechanics on the front. The knee flexes through approximately 145 degrees with the thigh flexed and 120 degrees with the thigh hyperextended (10). This difference in range is due to the length-tension relationship in the hamstring muscle group.

When flexion first begins, the femur rolls on the tibia with the medial condyle rolling 10 degrees and the lateral condyle 15 degrees (10). After the initial rolling is completed, the femur both rotates and translates, and finishes off in maximal flexion by just sliding. These movements are illustrated in FIGURE 6–26.

Accompanying the flexion movement is internal rotation of the tibia on the femur that is created, in part, by the greater movement of the lateral condyle on the tibia through almost twice the distance (FIGURE 6–26). Rotation can only occur with the joint in some amount of flexion, so there is no rotation in the extended, locked position. Six to 30 degrees of internal rotation occurs through 90 degrees of flexion at the joint around an axis passing through the medial intercondylar tubercle of the tibial plateau (5,42). Internal rotation also occurs with dorsiflexion and pronation at the foot, and roughly 6 degrees of subtalar motion creates roughly 10 degrees of internal rotation (43).

When the knee flexes, the patella moves down a distance over twice its length, entering the intercondylar notch on the femur (10) (FIGURE 6–27). The movement of the patella is most affected by the joint surface and the length of the patellar tendon, and minimally affected by the quadriceps femoris. In the first 20 degrees of flexion, the tibia internally rotates and the patella is drawn from its lateral positioning down into the groove where first contact is made with the inferior facets (30). In the first

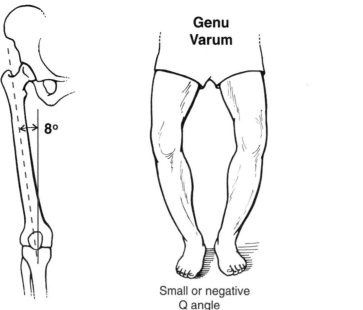
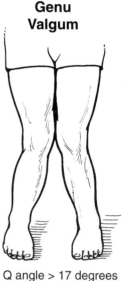

Genu Varum

Small or negative Q angle

Genu Valgum

Q angle > 17 degrees

FIGURE 6–24. The Q-angle is measured from the anterior superior spine to the middle of the patella, down to the tibial tuberosity. Q-angles range from 10 to 14 degrees for males and 15 to 17 degrees for females. Very small Q-angles create a condition known as genu varum, or bowleggedness. Large Q-angles create a genu varum, or knock-kneed position.

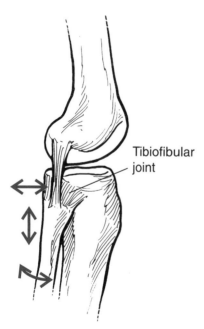

Tibiofibular joint

FIGURE 6–25. The tibiofibular joint is a small joint between the head of the fibula and the tibial condyle. It moves anteroposterior, up and down, and rotates in response to movements of the tibia or the foot.

20 degrees of flexion, stability offered by the lateral condyle is most important since most subluxations or dislocations occur in this early range of motion.

The patella follows the groove up until 90 degrees of flexion, at which point contact is made with the superior facets of the patella (FIGURE 6–27). At that time, the patella again moves laterally over the medial condyle, and if flexion continues until 135 degrees, contact will be made with the odd facet (30). In flexion, the linear/translatory movements of the patella are posterior and downward, but the patella also has some angular movements affecting its position. During knee flexion, the patella also flexes, abducts, and externally rotates. The movements of the patella for one leg cycle in walking are presented in FIGURE 6–28.

Flexion of the patella occurs about a mediolateral axis running through a fixed axis in the distal femur, with flexion representing the upward tilt about this axis. Likewise, patellar abduction involves movement of the patella away from the midline in the frontal plane, and external rotation is rotation of the patella outward about a longitudinal axis (44).

In the movement of extension, the reverse movements occur and the extension range of motion is terminated, usually at approximately 10 degrees of hyperextension.

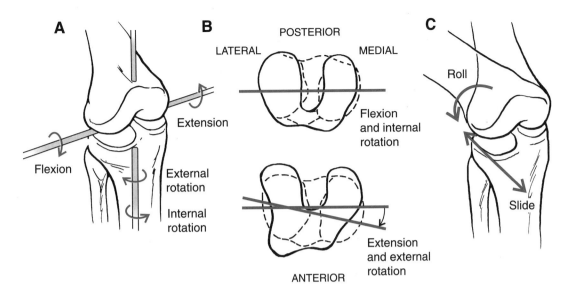

FIGURE 6–26. (A) The movements at the knee joint are flexion and extension, and internal and external rotation. (B) When there is knee joint flexion, there is an accompanying internal rotation of the tibia on the femur (non-weight bearing). In extension, the tibia externally rotates on the femur. (C) There are also translatory movements of the femur on the tibial plateau surface. In flexion, the femur rolls and slides posteriorly.

The contact point in the tibiofemoral joint moves anterior on the tibial plateau as the femur slides, rolls and slides, and then rolls in the last few degrees of extension. Through 120 degrees of extension, the anterior movement is 40% of the length of the tibial plateau (30).

With the extension movement is an accompanying external rotation that terminates in the locking action at the end of extension and is termed the *screw home mechanism*. The screw home mechanism is the point at which the medial and lateral condyles are locked in to form the close-packed position for the knee joint, and it occurs during the last 20 degrees of extension. This moves the tibial tuberosity laterally and produces a medial shift at the knee. Some of the speculative causes of the screw home movement are that the lateral condyle surface is covered first, and a rotation occurs to accommodate the larger surface of the medial condyle, or that the anterior cruciate ligament becomes taut just before rotation, forcing a rotation of the femur on the tibia (45). Finally, it is speculated that the cruciates become taut in early extension and pull the condyles in opposite directions, causing the rotation.

The screw home mechanism is disrupted with injury to the anterior cruciate ligament because the tibia moves more anteriorly on the femur. It is not significantly disrupted with loss of the posterior cruciate ligament, indicating that the anterior cruciate ligament is the main controller (45).

External rotation of the tibia is possible through approximately 45 degrees (10). External rotation of the tibia also accompanies plantar flexion and supination of the foot. With 34 degrees of supination, there will be a corresponding 58 degrees of external rotation (43).

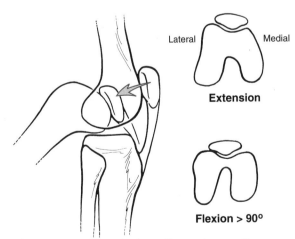

FIGURE 6–27. When the knee flexes, the patella moves down, and posterior, over two times its length. The patella sits in the groove and is held in place by the lateral condyle of the femur. If the knee joint continues into flexion past 90 degrees, the patella will move laterally over the condyle, until, at approximately 135 degrees of flexion, contact is made with the odd facet.

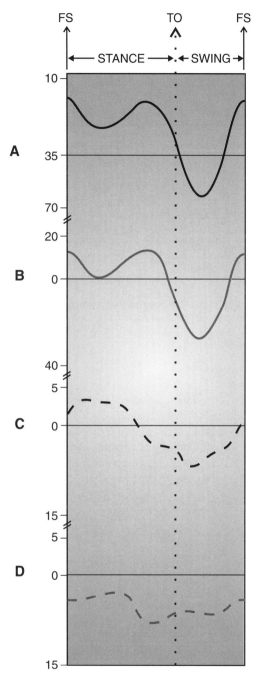

FS   TO   FS

←——— STANCE ———→ • ← SWING →

FIGURE 6–28. The angular movements of the patella are shown for one walking cycle. The top graph (A) represents the tibiofemoral flexion and extension pattern in walking shown with the accompanying patellar movements of (B) flexion/extension, (C) abduction/adduction, and (D) rotation. (Redrawn from Lafortune, M.A., and Cavanagh, P.R.: Three-dimensional kinematics of the patella during walking. *In* Biomechanics X. Edited by B. Jonsson. Champaign, IL, Human Kinetics, 1985.)

In extension, the patella returns to its resting position high and lateral on the femur, where it is above the trochlear groove and resting on the suprapatellar fat pad. The linear movement is reversed as it moves anteriorly and up in the groove. The angular movements of the patella are extension, adduction, and internal rotation (44,46).

There is translatory motion at the knee joint in the medial and lateral direction as the knee flexes. This motion is varus, or abduction and valgus, or adduction. No active varus or valgus motion can take place while the knee joint is extended, even though it is maintained in a valgus position. The valgus motion is maximum with the knee flexed and internally rotated, while the varus motion is maximum as the joint is nearing maximal extension and external rotation.

During walking, the foot strikes the ground with the knee joint almost extended (5 to 8 degrees flexion), externally rotated, and in a maximal varus position (47). During the support phase, the knee undergoes 17 to 20 degrees of flexion, 5 to 7 degrees of internal rotation, 7 to 14 degrees of external rotation, and 3 to 7 degrees of varus movement (42,47,48).

During the support phase, the knee moves from slight flexion at heel strike to extension at toe-off, internal rotation to external rotation, and valgus to varus. During the swing phase, there is 60 to 88 degrees of knee flexion, 12 to 17 degrees of rotation, and 8 to 11 degrees of valgus (42,47).

The knee joint moves from a position of maximum flexion after toe-off to maximum extension, from internal rotation to external rotation, and from valgus to varus. Maximum extension occurs just before heel strike, maximum external rotation just before heel strike, maximum varus at heel strike, and maximum valgus in the swing phase (42,47).

The patella moves through 41 total degrees of flexion and extension, 9.5 degrees of abduction and adduction, and 5 degrees of rotation during the walking action (44). There is lateral movement of the patella from foot strike until maximum flexion in the swing phase. When the foot strikes, the patella also moves posteriorly. As the knee extends prior to foot strike, there is a rapid medial shift of the patella.

During running, there are 80 and 36 degrees of knee flexion in the swing and support phases, respectively. There are 11 and 8 degrees of rotation, and 19 and 8 degrees of varus/valgus motion for the swing and support phases, respectively (49). A sample of the knee joint motions in running is presented in FIGURE 6–29.

Running on a cambered road or track requires more knee flexion, more external rotation, and less internal rotation of the knee joint of the downhill limb since more force is directed medially. The uphill foot is also

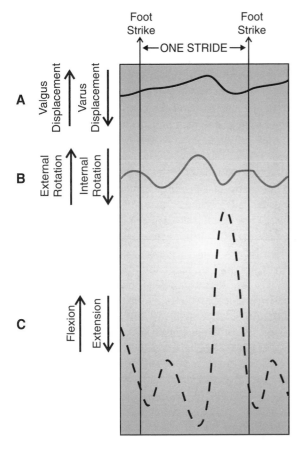

Foot Strike — ONE STRIDE → Foot Strike

A Valgus Displacement / Varus Displacement

B External Rotation / Internal Rotation

C Flexion / Extension

FIGURE 6–29. The pattern of motion at the knee joint for a running stride is presented. The knee joint experiences three movements: (A) linear movement, (B) rotation, and (C) flexion and extension (From Taunton, J.E., et al.: A triplanar electrogoniometer investigation of running mechanics in runners with compensatory overpronation. Canadian Journal of Applied Sports Science. *10*:104–115, 1985).

forced into pronation, and the downhill foot into supination (49).

Daily activities also require a great deal of knee joint motion. Going up stairs requires 83 degrees of flexion, 17 degrees of abduction/adduction, and 16 degrees of rotation (50). Returning down the stairs uses 83 degrees of flexion, 14 degrees of abduction/adduction, and 15 degrees of rotation. Sitting down in a chair requires the use of 93 degrees of flexion, 15 degrees of abduction/adduction, and 14 degrees of rotation. Tying a shoe from a seated position uses the most range of motion at the knee joint with 106 degrees of flexion, 20 degrees of abduction/adduction, and 18 degrees of rotation required to perform the action (50). Any osteoarthritic degeneration of the knee joint can greatly impair function, result-

ing in a 30 to 50% loss of knee flexion function in daily activities (51).

Muscular Actions

The extension of the leg is a very important contributor to the generation of power in the lower extremity for any form of human projection or translation. The musculature producing the extension movement is also used frequently in human movement to eccentrically contract and decelerate a rapidly flexing leg. Fortunately, the quadriceps femoris muscle group, the producer of extension at the knee, is one of the strongest muscle groups in the body. This muscle group may be as much as three times stronger than its antagonistic muscle group, the hamstrings, because of its involvement in negatively accelerating the leg and continuously contracting against gravity (10).

The quadriceps femoris muscle group consists of the rectus femoris and vastus intermedius forming the middle part of the quadriceps femoris, the vastus lateralis on the lateral side, and the vastus medialis on the medial side (39). The specific insertions, actions, and nerve supply are presented in Appendix B. Graphic illustrations are also provided.

The quadriceps femoris connect to the tibial tuberosity via the patellar ligament and contribute somewhat to the stability of the patella. As a muscle group, they also pull the menisci forward in extension via the meniscopatellar ligament. When they contract, they also reduce the strain in the medial collateral ligament and work with the posterior cruciate ligament to prevent posterior displacement of the tibia. They are antagonistic to the anterior cruciate ligament.

The largest and strongest of the quadriceps femoris is the vastus lateralis, a muscle applying a lateral force to the patella. Pulling medially is the vastus medialis. The lower portion of the vastus medialis is referred to as the vastus medialis oblique. This portion of the vastus medialis attaches to the tendon of the adductor magnus tendon. It has more horizontally directed fibers than the superior portion of the vastus medialis, and is a medial stabilizer of the patella (30).

It was previously noted in the literature that the medial quadriceps were selectively activated in the last few degrees of extension; however, this has been proven to be untrue. There is no selective activation of the vastus medialis muscles in the last degrees of extension, and both muscles contract equally throughout the range of motion (52).

The only two-joint quadriceps, the rectus femoris, does not significantly contribute to the knee extension force unless the hip joint is in a favorable position. It is limited as an extensor of the knee if the hip is flexed, and is facilitated as a knee extensor if the hip joint is extended, lengthening out the rectus femoris. In walking and running, the rectus femoris will contribute to the extension force in the toe-off phase when the thigh is extended. Likewise, in kicking, the rectus femoris activity is maximized in the preparatory phase as the thigh is brought back into hyperextension with the leg in flexion.

Flexion of the leg at the knee joint occurs during support when the body lowers down toward the ground; however, this downward movement is controlled by the extensors so that buckling does not occur. The flexor muscles are very active with the limb off the ground, working frequently to negatively accelerate a rapidly extending leg.

The major muscle group contributing to the flexion of the leg is the hamstrings, consisting of the lateral biceps femoris and the medial semimembranosus and the semitendinosus (See Appendix B). The action of the hamstrings can be quite complex because they represent two-joint muscles working to extend the hip. They are also rotators of the knee joint because of their insertions on the sides of the joint. As flexors, the hamstrings can generate the greatest force from a flexion position of 90 degrees (53).

The flexion strength diminishes with extension due to an acute tendon angle that reduces the mechanical advantage. At full extension, the flexion strength is reduced by 50% compared to the position at 90 degrees of flexion (53).

The lateral hamstring, the biceps femoris, has two heads connecting on the lateral side of the knee joint and offering lateral support to the joint. The biceps femoris also produces external rotation of the leg. Additionally, the lateral quadriceps femoris works with the anterior cruciate ligament to stabilize the joint.

The semimembranosus bolsters the posterior and medial capsule and works with the anterior cruciate ligament to prevent anterior displacement of the tibia. In flexion, it pulls the meniscus posteriorly (30). This lateral hamstring also contributes to the production of internal rotation in the joint. The other medial hamstring, the semitendinosus, is part of the pes anserinus muscular attachment on the medial surface of the tibia. It is the most effective flexor of the pes anserinus muscle group, contributing 47% to the flexion force (53). The semitendinosus works with both the anterior cruciate and the medial collateral ligaments in supporting the knee joint. It also contributes to the generation of internal rotation.

The hamstrings operate most effectively as knee flexors from a position of hip flexion that increases the length and tension in the muscle group. If the hamstrings become tight, they offer greater resistance to the exten-

sion of the knee joint by the quadriceps femoris, imposing a greater workload on the quadriceps muscle group.

The two remaining pes anserinus muscles, the sartorius and the gracilis also contribute 19% and 34% to the flexion strength, respectively (53). The popliteus is also a weak flexor supporting the posterior cruciate ligament in deep flexion and drawing the meniscus posterior. Finally, the two-joint gastrocnemius also contributes to knee flexion, especially when the foot is in the neutral or dorsiflexed position.

Internal rotation of the tibia is produced by the medial muscles: sartorius, gracilis, semitendinosus, semimembranosus, and popliteus (Appendix B). Internal rotation force is greatest at 90 degrees of knee flexion and decreases by 59% at full extension (54,55). The internal rotation force can be increased by 50% if it is preceded by 15 degrees of external rotation. Of the three pes anserinus muscles, the sartorius and the gracilis are the most effective rotators, accounting for 34% and 40% of the pes anserinus force in rotation (53). The semitendinosus contributes 26% of the pes anserinus rotation force. There is only one muscle, the biceps femoris, contributing significantly to the generation of external rotation of the tibia. Both internal and external rotation are necessary movements associated with function of the knee joint.

Strength and Forces at the Knee Joint

The extensors at the knee joint are usually stronger than the flexors throughout the entire range of motion. Peak extension strength is achieved at 50 to 70 degrees of knee flexion (33). The position of maximum strength will vary with the speed of movement. For example, if the movement is slow, peak extension strength occurs in the first 20 degrees of knee extension from the 90 degree flexed position. Flexion strength is greatest in the first 20 to 30 degrees of flexion from the extended position (56). This position will also fluctuate with the speed of movement. Greater knee flexion torques can be obtained if the hips are flexed since the hamstring length-tension relationship is improved.

It is quite common in sports medicine to evaluate the isokinetic strength of the quadriceps femoris group and the hamstrings to construct a hamstring-to-quadriceps ratio. An acceptable ratio has been generally 0.5, with the hamstrings at least half as strong as the quadriceps femoris. It has been suggested that anything below this ratio would indicate a strength imbalance between the quadriceps femoris and the hamstrings that would predispose one to injury. Caution must be observed when using this ratio as it only applies to slow isokinetic testing speeds.

At faster testing speeds when the limbs move through 200 to 300 degrees per second, the ratio approaches 1.0, as the efficiency of the quadriceps femoris drops at higher speeds. Even at the isometric testing level, the hamstring-to-quadriceps ratio is 0.7. Thus, a ratio of 0.5 between the hamstrings and the quadriceps would not be acceptable at fast speeds, and would indicate a strength imbalance between the two groups, whereas at a slower speed, it would not (57).

Internal and external rotation torques are both greatest with the knee flexed to 90 degrees, since a greater range of rotation motion can be achieved in that position. Internal rotation strength will increase by 50% from 45 degrees of knee flexion to 90 degrees (55). The position of the hip joint will also influence internal rotation torque with the greatest strength developed at 120 degrees of hip flexion, at which point the gracilis and the hamstrings are most efficient (54). At low hip flexion angles or in the neutral position, the sartorius is the most effective lateral rotator. Peak rotation torques occur in the first 5 to 10 degrees of rotation, and internal rotation torque is greater than in external rotation (55).

The knee experiences, and is subjected to, very high forces during most activities, whether generated in response to gravity, as a result of the absorption of the force coming up from the ground, or as a consequence of muscular contraction. The muscles generate considerable force, with the quadriceps femoris tension force being as high as 1 to 3 times body weight in walking, 4 times body weight in stair climbing, 3.4 times body weight in climbing, and 5 times body weight in the squat (46).

The tibiofemoral compression force can also be quite high in specific activities. For example, muscle forces applied against a low resistance (40 Nm), can create tibiofemoral compression forces of 1100 N during knee extension, acting through knee angles of 30 to 120 degrees. This force increases to 1230 N when extension occurs from the fully extended position (33). Tibio-femoral compression force in the extended position is greater partly because the quadriceps femoris group loses mechanical advantage at the terminal range of motion and thus has to exert more muscular force to compensate for the loss in leverage.

There is also a tibiofemoral shear force that is maximum in the last few degrees of extension. The direction of the shear force changes with the amount of flexion in the joint, and changes direction between the angles of 50 and 90 degrees of flexion. Operating against the same 40 Nm resistance in extension, there is posterior shear of 200 N at 120 degrees of flexion and 600 N of anterior shear in extension (33). This is partially due to the fact that, when nearing extension, the patellar tendon pulls the tibia forward relative to the femur, and in flexion it pulls the tibia back.

The anterior force in the last 30 degrees of extension places a lot of stress on the anterior cruciate ligament, which takes up 86% of the anterior shear force. By moving the contact pad closer to the knee in an extension exercise, the shear force can be directed posteriorly, taking the strain off of the anterior cruciate (33).

Even though the tibiofemoral compression forces are higher in the extended position, the contact area is large, reducing the pressure. There is 50% more contact area at the extended position than in 90 degrees of flexion. Thus, in the extended position, the compressional forces are high but the pressure is less by 25% (33). The forces for women are 20% higher because of a decreased mechanical advantage associated with shorter moment arm. Since females also have less contact area in the joint, more pressure is created, accounting for the higher rate of osteoarthritis in the knees of women, an occurrence not seen in the hip.

The patellofemoral compressive force between the patella and the femur approximates 0.5 to 1.5 times body weight in walking, 3 to 4 times body weight in climbing, and 7 to 8 times body weight in the squat (33). The patellofemoral joint absorbs compressive forces from the femur and transforms them into tensile forces in the quadriceps and patellar tendons. In vigorous activities, in which there are high negative acceleration forces, the patellofemoral force is also high. This force increases with flexion because the angle between the quadriceps femoris group and the patella decreases, requiring greater quadriceps femoris force to resist the flexion or produce an extension.

The patellofemoral compressive force is maximum at 50 degrees of flexion and declines at extension approaching zero as the patella almost comes off the femur. The largest area of contact with the patella is between the angles of 60 to 90 degrees of knee flexion, and 13 to 38% of the patellar surface bears the weight in joint loading (33). Fortunately, there is a large contact area when the patellofemoral compressive forces are large, reducing the pressure. In fact, there is considerable pressure in the extended position even though the patellofemoral force is low because the contact area is very small.

Activities utilizing more pronounced knee angles usually experience large patellofemoral compressive forces. These include descending stairs (4000 N), maximal isometric extension (6100 N), kicking (6800 N), the parallel squat (14900 N), isokinetic knee extension (8300 N), rising from a chair (3800 N), and jogging (5000 N) (33). In activities using lesser amounts of knee flexion, the force is less. Examples include ascending stairs (1400 N), walking (840–850 N), and bicycling (880 N) (33). The activities with high patellofemoral forces should be limited or avoided by individuals who are experiencing patellofemoral pain.

The patellofemoral compressive force and the quadriceps femoris force both increase at the same rate with knee flexion in weight bearing. If the leg extends against a resistance such as a leg extension machine or weight boot, the quadriceps force will increase, but the patellofemoral force will decrease from flexion to extension. Since the function seen in a weight lifting extension exercise is opposite to that seen in daily activities using flexion in the weight-bearing position, the use of a weight bearing, closed kinetic chain activity is preferable. At knee flexion angles greater than 60 degrees, the patellar tendon force is only one half to two thirds that of the quadriceps tendon force (33).

Those experiencing pain in the patellar region should avoid exercising at angles greater than 30 degrees to avoid high flexing moments and high patellofemoral compression forces. However, in extension when the patellofemoral force is low, the anterior shear force is high, making the terminal extension activities contraindicated for any anterior cruciate injury (33). There is a reversal at 50 degrees of flexion, when the shear force is now low, and the patellofemoral compression force is high.

Conditioning

The extensors of the leg are very easy to exercise because they are commonly used to both lower and raise the body weight. Examples of both stretching and strengthening exercises for the extensors are presented in FIGURE 6–30.

A squat exercise is one of the exercises used to strengthen the quadriceps femoris. To lower into a squat, the force coming through the joint, directed vertically in the standing position, is now partially directed across the joint, creating a shear force. This shear force increases as the amount of flexion increases, so that in a deep squat position, most of the original compressive force is directed posteriorly, creating a shear force. With the ligaments and muscles unable to offer much protection in the posterior direction at the full squat position, this is considered a vulnerable position. This position of maximum knee flexion is contraindicated for the beginner or unconditioned lifter.

An experienced and conditioned lifter who has strong musculature and uses good technique at the bottom of the

lift will most likely avoid any injury when in this position. Good technique involves control over the speed of descent and proper segmental positioning. For example, if the trunk is in too much flexion, the low back will be excessively loaded and the hamstrings will perform more of the work and the quadriceps less, as the focus is shifted to control on the posterior side.

The quadriceps femoris group may also be exercised in an open-chain activity such as the leg extension machine. Starting from the position of 90 degrees of flexion, one can exert considerable force since the quadriceps are very efficient throughout the early parts of the extension action. Near full extension, the quadriceps muscles become inefficient and must exert more force to move the same load.

The terminal extension exercise is good for individuals having patellar pain, because the quadriceps are working very hard with minimal patellofemoral compression force. It should be avoided however, in rehabilitation of an anterior cruciate injury since the anterior shear force is so large in this position. No knee extension exercise should be used at any angle less than 64 degrees in order to minimize the stress on the anterior cruciate ligament (58). Any knee extension exercise for individuals with anterior cruciate injuries should be done from a position of considerable knee flexion. A closed-chain squat exercise is preferable for ACL injuries since it minimizes the stress to the ligament and passive structures around the joint (58).

The flexors of the knee are not actively recruited at the knee joint in the performance of a lowering action with gravity in which knee flexion is the movement, because the quadriceps control the flexion action. Fortunately, the hamstrings are also extensors of the hip as well as flexors of the knee joint, so they are active during a squat exercise by virtue of their influence at the hip, since the hip flexion action in the lowering is controlled eccentrically by the extensors. If it were not for the flexors' role as extensors at the hip, the flexor group would be considerably weaker than their respective extensors.

The flexors are best isolated and exercised in a seated position using a leg curl apparatus. The seated position places the hip in flexion, optimizing the performance of the hamstrings. The flexors, especially the hamstrings and the pes anserinus muscles are important for knee stability since they control much of the rotational motion at the knee. As presented earlier in this text, the hamstrings should be half as strong as the quadriceps groups for slow speeds and should be as strong as the quadriceps group at fast speeds. It is also important to maintain flexibility in the hamstrings because if they are tight, the quadriceps

FLEXIBILITY

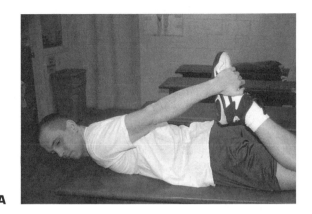

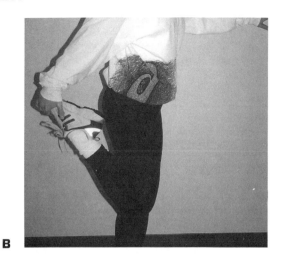

MANUAL RESISTANCE

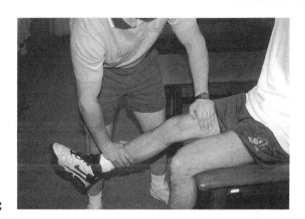

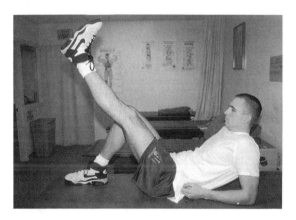

WEIGHT TRAINING

FIGURE 6–30. Exercises for the knee extensors are presented. The extensors can be stretched from the prone or standing position by pulling the leg up to the buttocks (A and B). They can be exercised manually by applying resistance to the leg during extension (C). A low resistance exercise for the extensors is a simple leg extension from the hook lying position (D). Weight-training exercises for the knee extensors are very common and include the squat (E) or leg press (F).

muscles must work harder, and the pelvis develops an irregular posture and function.

The rotators of the knee, since they are all flexors, will be exercised along with the flexion movements. If the rotators are to be selectively stretched or strengthened as they perform the rotation action, it is best to do the exercise from a seated position with the knee flexed to 90 degrees and the rotators in a position of maximum effectiveness. By toeing the foot in from this original starting position, the internal rotators are contracted and the external rotators are stretched. Different levels of resistance can be added to this exercise through the use of elastic bands or cables.

Contribution to Sport Skills or Movements

Walking utilizes the musculature around the knee joint to both propel and stabilize the body, and to absorb significant forces applied to the body. Refer to FIGURE 6–15 for a review and summary of the lower extremity muscles used in the walking gait.

As the heel strikes the ground in the braking phase of walking, the hamstrings reach their peak of muscular activity as they attempt to arrest the movement at the hip joint. The quadriceps femoris muscles then begin to contract to control the load being imposed on the knee joint by the body and force coming up from the ground. The knee is also moving into flexion, eccentrically controlled by the quadriceps. A co-contraction of the hamstrings and the quadriceps muscles continues up until about foot flat, when the activity of the hamstrings drops off. The activity of the quadriceps femoris group diminishes at approximately 30% of stance and is silent through midsupport and into the initial phases of propulsion.

In the propelling phase of walking, the quadriceps femoris muscles become active again around 85 to 90% of stance when they are used to propel the body upward and forward. The hamstrings also become active at approximately the same time to add to the forward propulsion.

In the swing phase of gait, the hamstrings are active after toe-off and again at the end of the swing, prior to foot contact (FIGURE 6–15). Similar activity is seen in the quadriceps femoris muscles, which serve to slow knee flexion after toe-off and initiate a knee extension prior to heel strike.

Running obviously employs more muscular activity at the knee joint to control the vigorous joint actions required in the activity. Refer to FIGURE 6–16 for a review of the lower extremity muscular contributions to running.

Right at the moment of heel strike in running, there is a brief concentric contraction of the hamstrings that flexes the knee to decrease the horizontal, or braking, force being absorbed at impact. The horizontal force travelling through the limb at impact is cushioned by this initial flexion movement. This is followed by activation of the quadriceps femoris muscles, which at first eccentrically slow the negative vertical velocity up to midsupport, and then concentrically produce a vertical velocity of the body. The hamstrings are also active with the quadriceps to generate extension at the hip (20). The period from heel strike to midsupport represents more than half of the energy costs in running.

In the propelling phase, the quadriceps femoris muscles are eccentrically active as the heel lifts off and then become concentrically active up through toe-off. The hamstrings are also concentrically active at toe-off.

During the swing phase, the quadriceps femoris muscles are initially active to slow rapid knee flexion. At the later part of the swing phase, the hamstrings become active to both limit knee extension and hip flexion (20).

For both walking and running, the concentric knee flexor activity increases as the speed increases to lessen the vertical ground reaction force and assist the hip flexor musculature in reducing the hip vertical velocity at contact. Eccentric activity from the quadriceps femoris group also increases with an increase in speed to slow the knee flexion and stop the negative vertical velocity of the body. Finally, the quadriceps muscles increase their activity with speed to contribute to the propelling force through increased extension activity at the knee. Knee extension is more related to sprint speed than any other joint movement or muscle action.

In stair ascent, there is a big peak of musculature activity from the vastus medialis, vastus lateralis, and the rectus femoris at 20% into the stance and a small peak again at 80% of stance (59). A review on the muscular contributions to stair climbing is presented in FIGURE 6–18.

In lifting activities, the contribution of the knee muscles changes depending on the posture. The muscular contribution to a straight-leg lift is presented in FIGURE 6–31. Lifting an object with straight knees and the trunk flexed over to pick up the object, demonstrates high quadriceps femoris group activity at the beginning and at the end of the lift (33). Overall, for the total lift, the quadriceps femoris group activity is lower than one might think. Moderate use of the hamstrings occurs in the middle third of the lift.

In straight knee postures, a working height of 1 m off the ground produces the lowest extensor moment at the knee (33). The horizontal distance from the work station

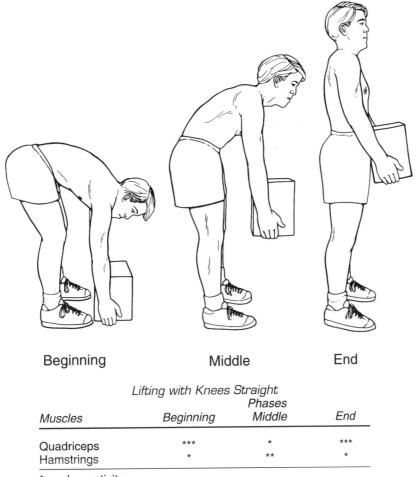

Beginning   Middle   End

Lifting with Knees Straight

Muscles	Beginning	Phases Middle	End
Quadriceps	***	*	***
Hamstrings	*	**	*

*   = low activity
** = moderate activity
*** = high activity

FIGURE 6–31. The contribution of the knee muscles to lifting depends on the posture used. In a straight-leg lift, the quadriceps femoris muscle group contributes significantly at the beginning and the end of the lift. The hamstrings are moderately active during the middle of the lift (Adapted from Nisell, 1985).

is also important, for as one assumes a work posture or stance further away from the station, the extensors must work harder.

If the same lift is performed with the knees flexed, the flexion moment at the beginning of the lift changes to an extension moment (FIGURE 6–32). Unlike the straight-leg lift, the quadriceps activity is moderate to high in this lift. The hamstring activity is low, and a co-contraction takes place in the lift (33). In flexed-knee working postures, there is medium to high activity in the vastus lateralis, less activity in the rectus femoris, and low to medium activities in the biceps femoris.

In cycling, the knee musculature is responsible for a significant portion of the power production. As shown in

FIGURE 6–19, power in the cycling stroke is generated through 25 to 160 degrees (60). In the top of the cycle, from 0 to 90 degrees, the quadriceps femoris muscles are very active, with the rectus femoris active through the arc of 200 degrees through 0 degrees and continuing through 130 degrees, the vastus medialis active from 300 degrees to 135 degrees, and the vastus lateralis active from 315 degrees through 130 degrees (60).

In the lower portion of the cycle, from 90 to 270 degrees, the hamstrings contribute more to power production, with the biceps femoris active from 5 to 265 degrees and the semimembranosus active from 10 to 265 degrees (60). There is co-contraction of the quadriceps and the hamstrings through the early portions of the

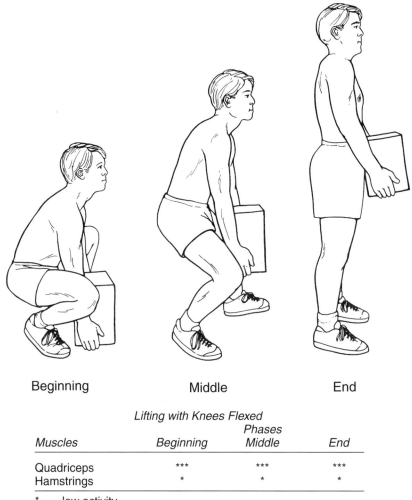

| | Beginning | Middle | End |

Lifting with Knees Flexed

| | | Phases | |
Muscles	Beginning	Middle	End
Quadriceps	***	***	***
Hamstrings	*	*	*

*   = low activity
** = moderate activity
*** = high activity

FIGURE 6–32. A lift starting from a position of flexed knees uses more of the quadriceps femoris group than the straight-leg lift, and the muscle group is active throughout the lift. A co-contraction of the hamstrings accompanies the quadriceps femoris activity (Adapted from Nisell, 1985).

cycle. In the last degrees of the cycle, from 270 to 360/0, the rectus femoris is actively involved as the leg is brought back up into the top position in a movement called "ankling".

Injury Potential in the Knee Joint

The knee joint is a frequently injured area of the body and, depending on the sport, accounts for 25 to 70% of the injuries reported. Approximately 97% of these injuries are associated with athletics or some form of vigorous activity. Only 22% of knee injuries occur as a result of contact (61).

Often the cause of an injury to the knee can be related to poor conditioning or training, or some alignment problem in the lower extremity. Injuries in the knee have been attributable to hindfoot and forefoot varus or valgus, tibial or femoral varus or valgus, limb length differences, deficits in flexibility, strength imbalances between agonists and antagonists, and improper technique or training.

There are a lot of knee injuries associated with running since the knee and the lower extremity are subjected to a force equivalent to approximately 3 times body weight absorbed at every foot contact. If 1500 foot contacts are made per mile of running, the potential for injury is high.

The traumatic injuries to the knee usually involve the ligaments, which are injured as a result of the application of a force causing twisting. High-friction or uneven surfaces are usually associated with increased ligament injury. Any movement fixating the foot while the body continues to move forward, such as occurs in skiing, will likely produce a ligament strain or tear. Simply, any turn on a weight-bearing limb leaves the knee joint vulnerable to ligament injury.

Injury to the anterior cruciate ligament is the most common of the ligament injuries and is usually caused by a twisting action while the knee is flexed and in an internally rotated and varus position while supporting weight. If the trunk and thigh rotate over a lower extremity while supporting weight, the anterior cruciate can be sprained or torn as the lateral femoral condyle moves posteriorly in external rotation (62). Examples from sport in which this ligament is often injured are when skiers catch the edge of the ski, a football player receives a clip, a basketball player lands off-balance from a jump, or a gymnast lands off-balance from a dismount (63).

Loss of the anterior cruciate ligament creates a valgus laxity and single plane or rotatory instability (64). The planar instability is usually in the anterior direction, while the rotatory instabilities can occur in a variety of directions, depending on the other structures injured (65). Instability created by an inefficient or missing anterior cruciate ligament will place added stress on the secondary stabilizers of the knee, the capsule, collateral ligaments, and the iliotibial band. There is also an accompanying deficit in quadriceps musculature. These "side effects" of the anterior cruciate ligament injury are often more debilitating in the long run.

Injury to the posterior cruciate ligament is less common than to the anterior cruciate. The posterior cruciate is injured by receiving an anterior blow to a flexed or hyperextended knee, or by forcing the knee into external rotation when it is supporting weight and is flexed. Damage to the posterior cruciate ligament results in a planar instability in the anterior or posterior direction.

The collateral ligaments on the side are injured upon receipt of a force applied to the side of the joint. The medial collateral ligament torn in an application of force in the direction of the medial side of the joint can also sprain or tear with a violent external rotation or tibial varus (65,66). The collateral is typically injured when the foot is fixed and slightly flexed. A change in direction with the person moving away from the support limb, as when running the bases in baseball, are common scenes for the medial collateral injury.

The lateral collateral ligament is injured upon receipt of a laterally directed force, usually applied with the foot fixed and in slight flexion (65). Injury to the medial or lateral collateral ligaments create planar instabilities in the medial or lateral directions, respectively. A forceful varus or valgus force can also create a distal femoral epiphysitis as the collaterals forcefully pull on their attachment site (67).

Rotatory instabilities created by injury to the ligaments or capsule are usually one of three types. An anteromedial rotatory instability is generated with the external rotation of the tibia with a fixed foot, flexed knee, and abducted thigh. This damages the middle third of the capsule, the semimembranosus attachment, the posterior oblique ligament, and the middle part of the anterior cruciate ligament (65).

An anterolateral rotatory instability is associated with an injury in which a lateral force is applied while the foot is fixed and the knee slightly flexed. This damages the middle third of the lateral capsule, the arcuate complex, the posterior, lateral capsule, and the anterior portion of the anterior cruciate ligament (65).

The third instability is the posterolateral rotatory instability created by a blow to the front of the tibia with the leg externally rotated and in a varus position (65). This damages the posterior third of the lateral capsule, the arcuate complex, and the posterior portion of the anterior cruciate ligament.

Damage to the menisci occurs much the same way that ligament damage is incurred. The menisci can be torn through compression associated with a twisting action in the weight-bearing position, or they can be torn in kicking or other violent extension actions. The tearing of the meniscus by compression is a result of the femur grinding into the tibia and ripping the menisci. A meniscual tear in rapid extension is a result of the meniscus getting caught and torn as the femur moves rapidly forward on the tibia.

Tears in the medial meniscus are usually incurred during moves incorporating valgus, knee flexion, and external rotation in the supported limb, or in the hyperflexed position of the knee (28). The lateral meniscus tear has been associated with a forced axial movement in the flexed position, a forced lateral movement with impact on the knee in extension, a forceful rotational movement, a movement incorporating varus, flexion, and internal rotation of the support limb, and the hyperflexed position (28).

There are many injuries to the knee that are a result of less traumatic, non-contact forces. Muscle strains occur frequently to the quadriceps femoris group or the hamstrings. The strain to the quadriceps muscles usually involves the rectus femoris since it can be placed in a very lengthened position with hip hyperextension and knee flexion. It is injured commonly in a kicking action, especially if the kick is mistimed. The hamstring strain is usually associated with inflexibility in the muscle or stronger quadriceps, which pull the hamstrings into a lengthened position. Sprinting, when the runner is not in condition to handle the stresses of sprinting, could lead to a hamstring strain.

On the lateral side of the knee is the iliotibial band, frequently irritated as the band moves over the lateral epicondyle of the femur in flexion and extension. Iliotibial band syndrome is seen in individuals who run on crowned roads, where the downhill leg is specifically affected. It has also been identified with individuals who run more than 5 miles per session, in stair climbing or downhill running, and in individuals who have a varum alignment in the lower extremity (68). Medial knee pain can be associated with many different structures such as tendinitis of the pes anserinus muscle attachment or irritation of either of the semimembranosus, parapatellar, or pes bursae (68).

Posterior pain in the knee joint is likely associated with popliteus tendinitis, which causes posterior lateral pain and is brought on by hill running or significant amounts of internal work. Posterior pain can also be associated with strain or tendinitis of the gastrocnemius muscle insertion.

Anterior pain accounts for the majority of the overuse injuries to the knee, especially in women. Patellofemoral pain syndrome is pain around the patella seen in individuals exhibiting valgum alignments or femoral anteversion in the extremity (69).

The stress on the patella is associated with the Q-angle, for a higher Q-angle increases the stress and forces on the patella. Patella injury is caused by abnormal tracking which, in addition to an increased Q-angle, can also be created by a functional short leg, tight hamstrings, tight gastrocnemius, a longer patellar tendon (termed *patella alta*), a shortened patellar tendon (termed *patella baja*), tight lateral retinaculum or iliotibial band, or excessive pronation at the foot.

Some patellofemoral pain syndromes are associated with cartilage destruction in which the cartilage underneath the patella becomes soft and fibrillated. This condition is known as chondromalacia patella. Patellar pain similar to that experienced with patellar pain syndrome or chondromalacia patella is also seen with medial retinaculitis in which the medial retinaculum is irritated in running (30).

A subluxated or dislocated patella is common in individuals having predisposing factors. These are patella alta, ligamentous laxity, a small Q-angle with out-facing patella, external tibial torsion, or an enlarged fat pad with patella alta (30). Dislocation of the patella may be congenital. The dislocation occurs in flexion as a result of a faulty knee extension mechanism.

The attachment site of the quadriceps femoris muscles to the tibia at the tibial tuberosity is also a site for injury and the development of anterior pain. The tensile force of the quadriceps femoris can create tendinitis at this insertion site commonly seen in athletes who use vigorous jumping, such as in volleyball, basketball, and track and field (70). In children aged 8 to 15, a tibial tubercle epiphysitis can develop, called Osgood-Schlatter disease. This is an avulsion fracture of the growing tibial tuberosity that can avulse the epiphysis. Bony growths can develop on the site. The cause of both of these conditions is basically overuse of the extensor mechanism (70).

Overuse of the extensor mechanism can also cause an irritation of the plica in individuals who still have plica in the joint. Plica injury can also result from a direct blow, a valgus rotary force applied to the knee, or weakness in the vastus medialis oblique. The plica become thick, inelastic, and fibrous with injury making it difficult to sit for long periods of time and creating pain on the superior knee (39). The medial patella may snap and catch during flexion and extension with injury to the plica.

The Ankle and Foot

The foot and the ankle are a very complex anatomical structure consisting of 26 irregularly shaped bones, 30 synovial joints, over 100 ligaments, and 30 muscles acting on the segment. All of these joints must interact harmoniously and in combination with each other to achieve a smooth motion. Most of the motion in the foot occurs at three of the synovial joints: the talocrural, the subtalar, and the midtarsal joints (71). The foot moves in three planes with most of the motion occurring in the rear-foot.

The foot contributes significantly to the function of the whole lower limb, just as the lower limb contributes to foot function. The foot must be a loose adapter to uneven surfaces at contact. Also, upon contact with the ground, it serves as a shock absorber, attenuating the high forces coming up from the ground. The foot supports the weight of the body in both standing and locomotion, and it must be a rigid lever for effective propulsion at the end of stance. Finally, when the foot is fixed during stance, it must absorb the rotation of the lower extremity. These functions of the foot all occur during a closed kinetic chain as it is receiving frictional and reaction forces from the ground or another surface (71).

Anatomical and Functional Characteristics of the Joints

The proximal joint of the foot is the talocrural, or ankle joint, shown in FIGURE 6–33. It is a uniaxial hinge joint formed by the tibia and fibula (tibiofibular joint) and the tibia and talus (tibiotalar joint). The joint is designed for stability rather than mobility. The ankle is stable when high forces are absorbed through the limb, when stopping and turning, or in many of the different lower limb movements one performs on a daily basis. However, if any of the anatomical support structures around the ankle joint are injured, it can become a very unstable joint (72).

The tibia and fibula form a deep socket for the trochlea of the talus, creating a mortise. The medial side of the mortise is the inner side of the medial malleolus, a projection on the distal end of the tibia. On the lateral side is the inner surface of the lateral malleolus, a distal projection on the fibula. The lateral malleolus projects down more than the medial malleolus and protects the medial ligaments of the ankle, acting as a bulwark against any lateral displacement. Because it projects downward more, it is also more susceptible to fracture with an inversion sprain to the lateral ankle.

The tibia and fibula fit snugly over the trochlea of the talus, a bone wider in front than in back (10). The differ-

ence in width of the talus allows for some abduction and adduction of the foot to occur. The close-packed position for the ankle is the dorsiflexed position, when the talus is wedged in at its widest spot.

The ankle has excellent ligamentous support on the medial and lateral sides. The location and actions of the ligaments are presented in Appendix A. The ligaments surrounding the ankle limit plantar flexion, dorsiflexion, anterior and posterior movement of the foot, tilting of the talus, and inversion or eversion (73).

The stability of the ankle depends on the orientation of the ligaments, the type of loading, and the position of the ankle at the time of stress. The lateral side of the ankle joint is more susceptible to injury, accounting for 85% of all ankle sprains (73).

The axis of rotation for the ankle joint is a line between the two malleoli, running oblique to the tibia and not in line with the body (74). The movement of dorsiflexion occurs at the ankle joint as the foot moves toward the leg, as when lifting the toes and forefoot off the floor, or as the leg moves toward the foot, such as in lowering down with the foot flat on the floor. This is illustrated in FIGURE 6–34.

The range of motion in dorsiflexion is limited by the bony contact between the neck of the talus and the tibia, the capsule and ligaments, and the plantar flexor muscles. The average range of dorsiflexion is 20 degrees, with approximately 10 degrees of dorsiflexion required for efficient gait (75). Healthy elderly individuals typically exhibit less passive dorsiflexion range of motion, but more dorsiflexion in gait than their younger, healthier counterparts.

Any arthritic condition in the ankle also reduces passive and increases active dorsiflexion range of motion. The dorsiflexion increase in the arthritic joint is primarily due to a decrease in flexibility in the gastrocnemius or a weakness in the soleus, with knee flexion maintained through stance causing a collapse into more dorsiflexion (76). With more dorsiflexion and knee flexion, more weight is maintained on the heel.

Plantar flexion is the movement of the foot away from the leg, such as raising up on the toes, or moving the leg away from the foot, such as in leaning back away from the front foot (FIGURE 6–34). The plantar flexion movement is limited by the talus and the tibia, the ligaments and the capsule, and the dorsiflexor muscles. The average range of motion for plantar flexion is 50 degrees, with 20 to 25 degrees of plantar flexion used in gait (75).

In the arthritic or pathologic gait, the plantar flexion range of motion is less for both passive and active measurements. The reduction of plantar flexion in gait is sub-

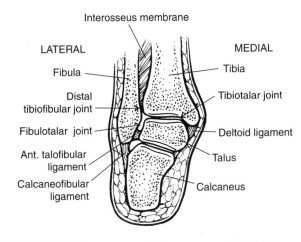

Interosseus membrane

LATERAL — Fibula — Distal tibiofibular joint — Fibulotalar joint — Ant. talofibular ligament — Calcaneofibular ligament

MEDIAL — Tibia — Tibiotalar joint — Deltoid ligament — Talus — Calcaneus

FIGURE 6–33. The talocrural joint, commonly called the ankle joint, refers to the articulations between the tibia and the talus (tibiotalar joint) and the tibia and the fibula (tibiofibular joint). The tibia and fibula create a mortise, making the joint very stable unless the mortise is altered through injury.

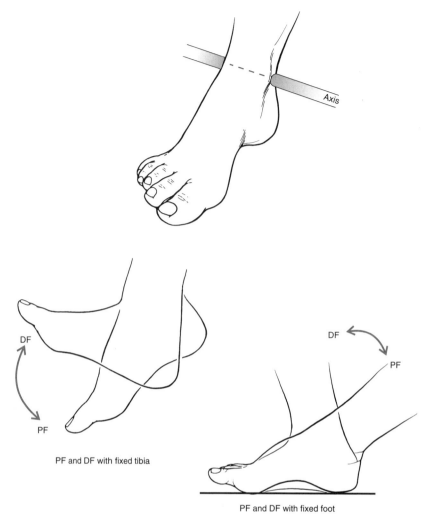

FIGURE 6–34. Plantar flexion and dorsiflexion occur around a mediolateral axis running through the ankle joint. The range of motion for plantar flexion and dorsiflexion is approximately 50 and 20 degrees, respectively. Plantar flexion and dorsiflexion can be produced with the foot moving on a fixed tibia, or with the tibia moving on a fixed foot.

stantial due to weak calf muscles. Healthy elderly people do not demonstrate substantial loss in either passive or active plantar flexion range of motion when compared to young and healthy individuals (76).

During gait, there is 20 to 40 degrees of total plantar flexion and dorsiflexion movement (77,78). The foot strikes the ground with the ankle at 90 degrees, and moves into the loading phase through approximately 10 degrees of plantar flexion to lower the foot to the ground (79). Moving into midstance, the foot moves into approximately 5 degrees of dorsiflexion, increasing to approximately 10 degrees in later stance. At the push-off, the ankle rapidly plantar flexes 20 degrees to unload the limb

and begin the swing phase (79). In the swing phase of gait, the ankle dorsiflexes to move back to the neutral position, and to keep the toes up and prepare for heel strike (80).

In running, there is approximately 10 degrees of dorsiflexion prior to contact, as much as 50 degrees of dorsiflexion through 50% of stance, and a rapid plantar flexion of 25 degrees at toe-off. As the speed of running increases, the amount of plantar flexion decreases (47).

Moving distally from the talocrural joint is the subtalar, or talocalcaneal joint, consisting of the articulation between the talus and the calcaneus, referred to as the *hindfoot*. All of the joints in the foot, including the subtalar joint, are shown in FIGURE 6–35. The talus and the calcaneus are the largest of the weight-bearing bones in the foot. The talus links the tibia and fibula to the foot and is called the "keystone" of the foot. No muscles attach to the talus. It transmits the weight of the whole body.

The prime function of the subtalar joint is to absorb the rotation of the lower extremity in stance. With the foot fixed on the surface and the femur and tibia rotating internally at the beginning of stance and externally at the end of stance, the subtalar joint absorbs the rotation through pronation and supination (81).

A second function is shock absorption, also occurring through pronation at the subtalar joint, which lowers the lower extremity to allow absorption at heel strike. The subtalar movements also allow the tibia to internally rotate at a faster and further pace than the femur, facilitating unlocking at the knee joint.

The talus articulates with the calcaneus at three different sites, anterior, posterior, and medial, where the convex surface of the talus fits into a concave surface on the calcaneus. The joint is supported by five short and powerful ligaments used to resist large forces and severe stresses in gait and lower extremity movements. The location and action of these ligaments are presented in Appendix A. The ligaments supporting the talus limit pronation and supination, and specifically, abduction, adduction, plantar flexion, dorsiflexion, inversion, and eversion.

The axis of rotation for the subtalar joint runs from the posterior, lateral plantar to the anterior, dorsal, medial surface of the talus (FIGURE 6–36). It is tilted vertically, 41 to 45 degrees from the horizontal axis in the sagittal plane, and has a medial slant 16 to 23 degrees in from the longitudinal axis of the tibia in the frontal plane (8). Tri-

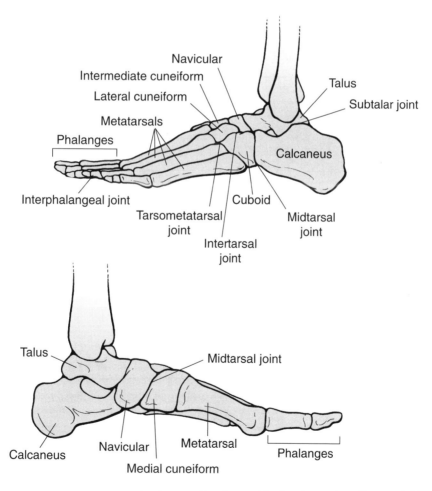

FIGURE 6–35. There are 30 different joints in the foot working in combination to produce the movements of the rearfoot, midfoot, and forefoot. The subtalar and midtarsal joints contribute to the movements of pronation and supination. The intertarsal, tarsometatarsal, metatarsophalangeal, and interphalangeal joints contribute to movements of the forefoot and the toes.

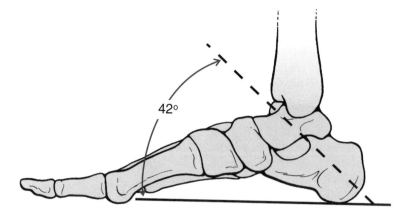

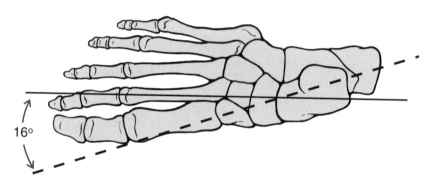

FIGURE 6–36. The axis of rotation for the subtalar joint runs diagonally from the posterior, lateral, plantar surface to the anterior, medial, dorsal surface. The axis is situated approximately 42 degrees in the sagittal plane and 16 degrees in the transverse plane.

plane movement is allowed to occur about the subtalar's single axis due to simultaneous actions in the sagittal, frontal, and transverse planes, since the axis is oblique running through all of the planes.

The tri-plane movements at the subtalar joint are termed pronation and supination. The movement of pronation, occurring in an open-chain system with the foot off the ground, consists of calcaneal eversion, abduction, and dorsiflexion (82). Since the talus is locked in the mortise, the calcaneus moves with respect to the talus. An illustration of differences in subtalar movements between open-chain and closed-chain positioning is shown in FIGURE 6–37.

In the weight-bearing closed kinetic system, the movement of pronation consists of calcaneal eversion, talar adduction, and plantar flexion, and now, the talus

moves on the calcaneus (83). Eversion is the movement in the frontal plane in which the lateral border of the foot moves toward the leg in non-weight bearing, or the leg moves toward the foot in weight bearing as the calcaneus lies on its medial surface (FIGURE 6–37). The transverse plane movement is abduction, with the toes pointing out, and it occurs with external rotation of the foot on the leg and lateral movement of the calcaneus in the non-weight bearing position, or internal rotation of the leg with respect to the calcaneus, and medial movement of the talus in weight bearing. The sagittal plane movement of dorsiflexion occurs as the calcaneus moves up on the talus in non-weight bearing, or as the talus moves down on the calcaneus in weight bearing.

The movement of supination is just the opposite, with calcaneal inversion, adduction, and plantar flexion in the

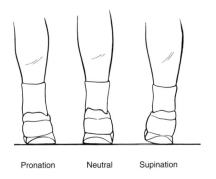

Pronation   Neutral   Supination

Inversion   Eversion

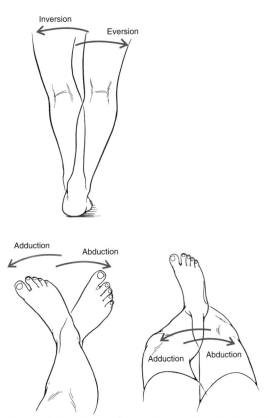

FIGURE 6–37. Movements at the subtalar joint are termed *pronation* and *supination*. With the foot off the ground, the foot moves on a fixed tibia, and the subtalar movement of pronation is produced by eversion, abduction, and dorsiflexion. Supination in the open chain is produced by inversion, adduction, and plantar flexion. In a closed kinetic chain, with the foot fixed on the ground, much of the pronation and supination is produced by the weight of the body acting on the talus. In this weight-bearing position, the tibia moves on the talus to produce the movements of pronation and supination.

non-weight bearing position, and calcaneal inversion and talar abduction and dorsiflexion in the weight-bearing position (83). The frontal plane movement of inversion occurs as the medial border of the foot moves toward the medial leg in non-weight bearing, or as the medial aspect

of the leg moves toward the medial foot in weight bearing as the calcaneus lies on the lateral surface.

In the transverse plane, the adduction, or toeing-in movement, occurs as the foot internally rotates on the leg in non-weight bearing and the calcaneus moves medially, or the leg externally rotates on the foot in weight bearing and the talus moves laterally. The plantar flexion movements in the sagittal plane occur as the calcaneus moves distally while non-weight bearing, or as the talus moves proximally while weight bearing.

Calcaneal movements are the same, regardless of weight-bearing or non-weight-bearing conditions. This makes the calcaneal inversion and eversion measurements very useful in determining subtalar motion (FIGURE 6–37). Passive calcaneal inversion is possible through 20 degrees of motion in young healthy individuals and 18 degrees in healthy elderly individuals (76). Calcaneal varus or inversion is greatly reduced in individuals with osteoarthritis in the ankle joint. Calcaneal eversion, measured passively, averages 5 and 4 degrees for healthy young and elderly individuals, respectively (76). In 84% of arthritic patients, there will be excessive calcaneal eversion, creating what is known as a hindfoot valgus deformity.

In gait, approximately 4 degrees of calcaneal inversion and 6 to 7 degrees of calcaneal eversion are used by healthy individuals. Motion of the foot during a running gait is shown in FIGURE 6–38 for one leg cycle.

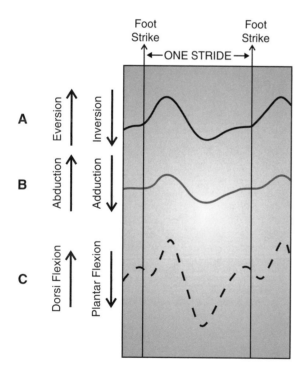

FIGURE 6–38. The pattern of motion of the foot during one running stride is presented. The foot (A) everts and inverts, (B) abducts and adducts, and (C) plantar flexes and dorsiflexes in the running stride (From Taunton, et al., 1985).

Calcaneal inversion, maximum at toe-off, is usually not present in the gait of individuals with arthritis of the ankle. Calcaneal eversion, maximum at mid-stance, is usually excessive, ranging from 11 to 21 degrees, in arthritic gait (76).

The range of motion for pronation and supination is 20 to 62 degrees, with the supination motion about double the range of pronation (83,84).

During the stance phase of gait, the pronation and supination movements should correspond with the rotations of the tibia and femur. At heel strike, the foot typically makes contact with the ground in a slightly supinated position (2 to 3 degrees) and the foot is lowered to the ground in plantar flexion (85). The subtalar joint moves immediately into pronation, accompanying the internal rotation of both the tibia and the femur (86). The talus rotates medially on the calcaneus, initiating the pronation as a result of lateral heel strike, putting the stress on the medial side (87). Pronation continues until it reaches a maximum at approximately 35 to 45% of the stance phase (88). In walking, maximum pronation is in the range of 3 to 10 degrees, and in running 8 to 15 degrees (47,89,90). More than 19 degrees of pronation is considered excessive.

At the stage of foot flat in stance, the tibia begins to externally rotate, and since the forefoot is still fixed on the ground, this external rotation is transmitted to the talus (86). The subtalar joint should begin to supinate in response to the external rotation. Approximately 3 to 10 degrees of supination should occur up until heel-off.

Many injuries of the lower extremity are associated with excessive pronation, not just the maximum amount of pronation, but also the percentage of stance in which pronation is present and the timing of the pronation. Pronation can be present for as much as 55 to 85% of stance, creating problems when the lower limb moves into external rotation (83). This will be discussed in later sections of this chapter.

Of the remaining articulations in the foot, the midtarsal or transverse tarsal joint has the most functional significance (FIGURE 6–35). It actually consists of two joints, the calcaneocuboid on the lateral side, and the talonavicular on the medial side of the foot. In combination, they form an S-shaped joint, with two axes, oblique and longitudinal (8). There are five ligaments supporting this region of the foot (See Appendix A).

Movement at the midtarsal joint is dependent on the subtalar position. When the subtalar joint is in pronation, the two axes of the midtarsal joint are parallel, which unlocks the joint, creating hypermobility in the foot (5). This allows the foot to be very mobile in absorbing the shock of contact with the ground and also in adapting to uneven surfaces. FIGURE 6–39 contains an illustration. When the axes are parallel, the forefoot is also allowed to flex freely and extend with respect to the rearfoot. The motion at the midtarsal joint is unrestricted from heel strike to foot flat in gait, as the foot bends toward the surface.

During supination of the subtalar joint, the two axes running through the midtarsal joint converge and are no longer parallel (FIGURE 6–39). This locks the joint in, creating a rigidity in the foot necessary for efficient force application during the later stages of stance (5). The midtarsal joint becomes rigid and more stable from foot flat to toe-off in gait as the foot supinates. It is usually stabilized, creating a rigid lever, at 70% of the stance phase (83). At this time there is also more load on the midtarsal joint, making the articulation between the talus and the navicular more stable.

The other articulations in the midfoot, the intertarsal articulations, between the cuneiforms and the navicular, and cuboid and intercuneiform, are gliding joints (FIGURE 6–35). At the articulation between the cuneiforms

and the navicular and cuboid, small amounts of gliding and rotation are allowed (10).

At the intercuneiform articulations, there is a small vertical movement taking place, which alters the shape of the transverse arch in the foot (84). These joints are supported by strong interosseus ligaments.

The forefoot comprises the metatarsals and the phalanges, and the respective joints between them. The function of the forefoot is to maintain the transverse metatarsal arch, maintain the medial longitudinal arch, and maintain the flexibility in the first metatarsal. The plane of the forefoot at the metatarsal head, formed by the second, third, and fourth metatarsals, should be oriented perpendicular to the vertical axis of the heel in normal forefoot alignment. This is the neutral position for the forefoot, and is shown in FIGURE 6–40.

If the plane is tilted so that the medial side lifts, it is termed *forefoot supination* or *varus* (91). If the medial side drops below the neutral plane, it is termed *forefoot pronation* or *valgus*, and is not as common as forefoot varus (FIGURE 6–40). Also, if the first metatarsal is below the plane of the adjacent metatarsal heads, it is

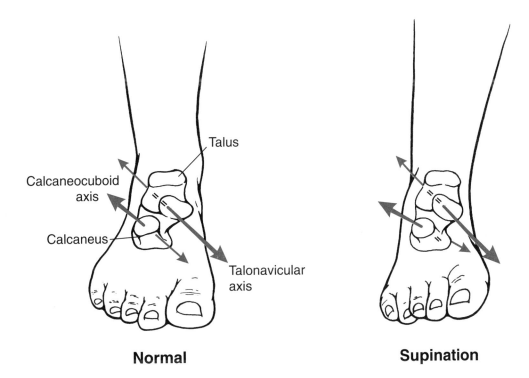

Normal   **Supination**

FIGURE 6–39. The midtarsal joints consist of the articulations between the calcaneus and the cuboid (calcaneocuboid joint) and the talus and the navicular (talonavicular joint). Each joint has an axis of rotation which runs obliquely across the joint. When the two axes are parallel to each other, the foot is flexible, and can freely move. If the axes do not run parallel to each other, the foot is locked in a rigid position. This occurs with the supination movement.

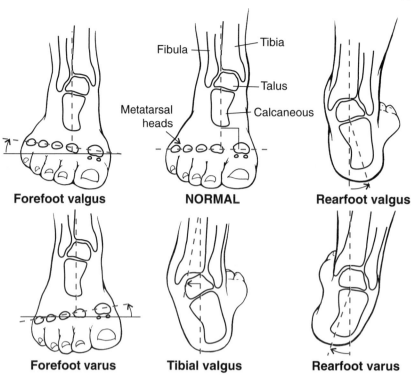

FIGURE 6–40. The metatarsal head should be oriented in a plane that is perpendicular to the heel in a normal alignment in the foot. There are many variations in this alignment, including forefoot valgus in which the medial side of the forefoot drops below the neutral plane; forefoot varus, in which the medial side lifts; rearfoot valgus, in which the calcaneus is everted; and rearfoot varus, in which the calcaneus is inverted. There can also be tibial and subtalar varum or valgus, in which the tibia or talus moves laterally or medially, respectively.

considered to be a plantar flexed first ray, and is commonly associated with high-arched feet (91).

The base of the metatarsals is wedge-shaped, forming a mediolateral or transverse arch across the foot. The tarsometatarsal articulations are gliding or planar joints, allowing limited motion between the cuneiforms and the first, second, and third metatarsals, and the cuboid and the fourth and fifth metatarsals (10).

The tarsometatarsal joint movements change the shape of the arch. When the first metatarsal flexes and abducts as the fifth metatarsal flexes and adducts, the arch deepens or increases in curvature. Likewise, if the first metatarsal extends and adducts and the fifth metatarsal extends and abducts, the arch will flatten.

Flexion and extension movements at the tarsometatarsal articulations also contribute to inversion and eversion of the foot. The most movement is allowed between the first metatarsal and the first cuneiform, and the least movement is allowed between the second metatarsal and the cuneiforms (83). Mobility is an important factor in the first metatarsal since it is significantly involved in weight bearing and propulsion. The limited mobility at the second metatarsal is also significant since it is the peak of the plantar arch and a continuation of the long axis of the foot. The joints are supported by medial and lateral dorsal ligaments.

The metatarsophalangeal joints are biaxial, allowing both flexion and extension, and abduction and adduction (FIGURE 6–35). This joint is loaded during the propulsive phase of gait after heel-off and the initiation of plantar flexion and phalangeal flexion (72). There are two sesamoid bones located under the first metatarsal to reduce the load on one of the hallucis muscles in the propulsive phase. The movements at the metatarsophalangeal joints are similar to those seen in the same joints in the hand, except that more extension occurs in the foot as a result of requirements for the propulsive phase of gait.

The interphalangeal joints are very similar to those found in the hand (FIGURE 6–35). These uniaxial hinge joints allow for the flexion and extension of the toes. The toes are much smaller than the fingers in the hand. They are also less developed, probably due to continual shoe application to the feet (10). The toes are also not as functional as the fingers because they lack a structure like the thumb.

The tarsals and metatarsals of the foot form three arches, two running longitudinally and one running transversely across the foot. This creates an elastic shock-absorbing system. In standing, one half of the weight is borne by the heel and one half by the metatarsals. One third of the weight borne by the metatarsals is on the first metatarsal, while the remaining load is on the other metatarsal heads (72). The arches form a concave surface that is a quarter of a sphere (10). The arches are shown in FIGURE 6–41.

The lateral longitudinal arch is formed by the calcaneus, cuboid, and fourth and fifth metatarsals. It is relatively flat in contour and limited in mobility (72). Since it is lower than the medial arch, it may make contact with the ground and bear some of the weight in locomotion, playing a support role in the foot.

The more dynamic medial longitudinal arch runs across the calcaneus to the talus, navicular, cuneiforms, and first three metatarsals. It is much more flexible and mobile than the lateral arch and plays a significant role in shock absorption upon contact with the ground. Upon heel strike, part of the initial force is absorbed by compression of a fat pad positioned on the inferior surface of the calcaneus. This is followed by a rapid elongation of the medial arch that continues to maximum elongation at toe contact with the ground. The medial arch shortens at midsupport, then slightly elongates, and again rapidly shortens at toe-off (72). Flexion at the transverse tarsal and the tarsometatarsal joints increases the height of the longitudinal arch as the metatarsophalangeal joints extend at push-off (92). The movement of the medial arch is important for it dampens the impact by transmitting the vertical load through deflection of the arch.

Even though the medial arch is very adjustable, it usually does not make contact with the ground unless a person has functional flat feet. The medial arch is supported by the keystone navicular bone, the calcaneonavicular ligament, the long plantar ligament, and the plantar fascia (84,86).

The plantar fascia, illustrated in FIGURE 6–42, is a strong, fibrous plantar aponeurosis running from the cal-

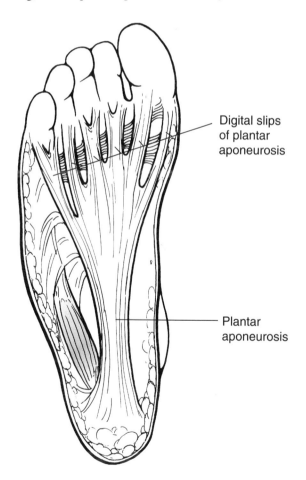

Digital slips of plantar aponeurosis

Plantar aponeurosis

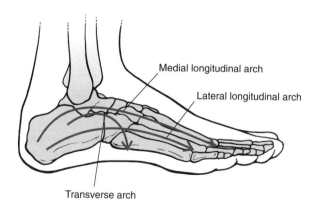

Medial longitudinal arch

Lateral longitudinal arch

Transverse arch

FIGURE 6–41. There are three arches formed by the tarsals and metatarsals: the lateral longitudinal arch, which participates in a support role function during weight bearing; the medial longitudinal arch, which dynamically contributes to shock absorption; and the transverse arch, which supports a significant portion of the body weight during weight bearing.

FIGURE 6–42. The plantar fascia is a strong fibrous aponeurosis that runs from the calcaneus to the base of the phalanges. It supports the arches and protects structures in the foot.

caneus to the metatarsophalangeal articulation. It supports both arches, and protects the underlying neurovascular bundles. The plantar fascia can be irritated as a result of ankle motion through extreme ranges of motion, since the arch is flattened in dorsiflexion and increased in plantar flexion, placing a wide range of tensions on the fascial attachments (84). Also, if the plantar fascia is short, the arch is likely to be higher.

The transverse arch is formed by the wedging of the tarsals and the base of the metatarsals. The bones act as beams for support of this arch, which flattens with weight bearing and can support 3 to 4 body weights (8). The flattening of this arch causes the forefoot to spread a considerable distance in a shoe, indicating the importance of sufficient room in the shoe to allow for this spread.

Individuals can be classified according to the height of the medial arch into foot types that are normal, high-arched or pes cavus, and flat-footed or pes planus. They can be further classified as being rigid or flexible. The high-arched rigid foot does not make any contact with the midfoot in stance and usually has little or no inversion or eversion in stance. It is a foot type poor in shock absorption. The flat foot, on the other hand, is usually hypermobile, with the major percentage of the plantar surface making contact in stance, creating a weakening of the medial side. It is a foot type usually associated with excessive pronation throughout the support phase of gait.

Foot function can be altered significantly with any variation in alignment in the lower extremity or as a result of abnormal motion in the lower extremity linkage. Typically, any varum alignment in the lower extremity will cause an increase in the pronation at the subtalar joint in stance (93). A Q-angle at the knee greater than 20 degrees, tibial varum more than 5 degrees, rearfoot varum more than 2 degrees, and forefoot varum more than 3 degrees are all deemed to be significant enough to produce an increase in subtalar pronation (76).

Rearfoot varus is usually a combination of subtalar varus and tibial varum in which the calcaneus inverts and there is a deviation of the lower third of the tibia in the direction of inversion. Forefoot varus, the most common cause of excessive pronation, is the inversion of the forefoot on the rearfoot with the subtalar joint in the neutral position (75). It is caused by the inability of the talus to derotate, so the foot remains pronated at heel lift, disallowing any supination. This shifts the body weight to the medial side of the foot, creates a hypermobile midtarsal joint, and an unstable first metatarsal.

Both rearfoot and forefoot varus will double the amount of pronation in midstance as compared to normal foot function, and will continue pronation into late stance

(93). In some cases, the pronation will continue until the very end of the support. This is a major injury-producing mechanism because the continued pronation is contrary to the external rotation being produced in the leg. It is the primary cause of discomfort and dysfunction in the foot and leg. The transverse rotation being produced by the hypermobile foot still in pronation late in stance is absorbed at the knee joint, and can create lateral hip pain through an anterior tilt of the pelvis, or strain the invertor muscles (85).

A plantar-flexed first ray can also produce excessive pronation (93). The first ray is usually plantar flexed due to the pull of the peroneus longus muscle and is commonly seen in both rearfoot and forefoot varus alignments. This alignment causes the medial side to load prematurely with greater than normal loads, limiting forefoot inversion and creating a supination in midstance. However, a sudden pronation is generated at heel-off, developing high shear forces across the forefoot, especially at the first and fifth metatarsals (93).

Hypermobility of the first ray is generated because of the peroneus longus's inability to stabilize the first metatarsal. During pronation, the medial side is hypermobile, placing a great amount of load and shear force on the second metatarsal. This is a common etiology for stress fracture of the second metatarsal and subluxation of the first metatarsophalangeal joint (75,94).

Although it is not very common, a person may have a forefoot valgus alignment caused by a bony deformity in which the plantar surface of the metatarsals evert relative to the calcaneus with the subtalar joint in the neutral position (75). Forefoot valgus causes the forefoot to be prematurely loaded in gait, creating supination at the subtalar joint. This alignment is typically seen in the high-arched foot.

Foot type, as mentioned previously, can also affect the amount of pronation or supination. In the normal foot with a subtalar axis of 42 to 45 degrees, the internal rotation of the leg is equal to the internal rotation of the foot (81). In a high-arched foot, the axis of the subtalar joint is more vertical, and is greater than 45 degrees. This has the effect that, for any given internal rotation of the leg, there is less internal rotation of the foot, creating less pronation for any given leg rotation.

In the flat foot, the axis is less than 45 degrees, being closer to the horizontal. This has the opposite effect, and for any given internal rotation of the leg, there will be more internal rotation of the foot, thus creating more pronation (81).

A final alignment consideration is equinus, where the Achilles tendon is short, creating a serious dorsiflexion limitation in gait. The equinus deviation can be reproduced with a tight and inflexible gastrocnemius and soleus, which will also limit dorsiflexion in gait. Since the tibia is unable to move forward on the talus in mid-support, the talus moves anteriorly and pronates excessively to compensate (85). An early heel raise or a toe walker are symptoms of this disorder.

Muscle Actions

There are 23 muscles acting on the ankle and the foot, with 12 of these originating outside the foot, and 11 inside the foot. All of the 12 extrinsic muscles, except for the gastrocnemius, soleus, and the plantaris, act across both the subtalar and midtarsal joints (95). The insertion, actions, and nerve supply of all of the muscles are presented in Appendix B.

The muscles of the foot play an important role in sustaining impacts of very high magnitude. They also provide both kinetic and potential energy as we create movement, and lose energy later as we absorb movement. The ligaments and tendons of the muscles store some of the energy for later return. For example, the tendon of the achilles can store 37 joules (J) of elastic energy and the ligaments of the arch can store 17 J as the foot absorbs the forces and body weight (96).

The plantar flexion movement is used to propel the body forward and upward, contributing significantly to the other propelling forces generated in the heel-off and toe-off phases in stance. Plantar flexor muscles are also used eccentrically to slow down a rapidly dorsiflexing foot, or to assist in the control of the forward movement of the body, and specifically, the forward rotation of the tibia over the foot.

The plantar flexion movement is powerful, created by muscles inserting posterior to the transverse axis running through the ankle joint. The majority of the plantar flexion force is produced by the gastrocnemius and the soleus, also known as the triceps surae muscle group. Since the gastrocnemius also crosses the knee joint, acting as a knee flexor, it is more effective as a plantar flexor with the knee extended and the quadriceps activated.

In the sprint start, the gastrocnemius is maximally activated with the knee extended and the foot placed in full dorsiflexion. The soleus, termed the "workhorse" of plantar flexion, is flatter than the gastrocnemius (84). It is also the predominant plantar flexor during a standing posture. A tight soleus can create a functional short leg, often seen in the left leg of people who drive a lot. As explained in an earlier section, an inflexible or tight soleus limits the dorsiflexion movement and facilitates a compensatory pronation that creates the functionally shorter limb.

The action of these plantar flexors is mediated through a stiff subtalar joint allowing for an efficient transfer of the muscular force. The gastrocnemius and possibly the soleus have also been shown to produce supination when the forefoot is on the floor during the later stages of the stance phase. The movement of plantar flexion is usually accompanied by both supination and adduction.

The other plantar flexors produce only 7% of the remaining plantar flexor force (84). Of these, the peroneus longus and the peroneus brevis are the most significant with minimal plantar flexor contribution from the plantaris, the flexor hallucis longus, the flexor digitorum longus, and the tibialis posterior. The plantaris is an interesting muscle, similar to the palmaris longus in the hand, in that it is absent in some, very small in others, and well developed in yet others. Overall, its contribution is usually very insignificant.

Dorsiflexion at the ankle is actively used in the swing phase of gait to keep the foot up, and in the stance phase of gait to control lowering of the foot to the floor upon heel strike. Dorsiflexion is also present in the middle part of the stance phase as the body lowers and the tibia travels over the foot, but this action is controlled eccentrically by the plantar flexors (97). The dorsiflexor muscles are those muscles inserting anterior to the transverse axis running through the ankle (95) (See Appendix B).

The most medial dorsiflexor is the tibialis anterior whose tendon is the furthest away from the joint, giving it more mechanical advantage and making it the most powerful dorsiflexor (84). The tibialis anterior has a long tendon that begins halfway down the leg. Assisting the tibialis anterior in dorsiflexion are the extensor digitorum longus and the extensor hallucis longus, which pull the toes up in extension. The peroneus tertius also contributes minimally to the dorsiflexion force.

The action of pronation is created primarily by the peroneal muscle group, which lies lateral to the long axis of the tibia. These muscles are known as pronators in the non-weight-bearing position because they create eversion of the calcaneus and abduction of the forefoot. The peroneus longus is an evertor and abductor also responsible for controlling the pressure on the first metatarsal, and some of the finer movements of the first metatarsal and big toe, or hallux.

The lack of stabilization of the first metatarsal by the peroneus longus leads to hypermobility of the medial side of the foot, described in a previous section. The peroneus brevis also contributes through the production of eversion and abduction, and the peroneus tertius contributes through dorsiflexion and eversion actions. Both the peroneus tertius and brevis stabilize the lateral aspect of the foot. It must be pointed out that pronation in the weight-bearing position is primarily generated by weight bearing on the lateral side of the foot in heel strike, which drives the talus medially, producing the pronation. Refer to FIGURE 6–43 for an illustration of how pronation is produced through weight bearing.

The supinators of the foot are those muscles lying medial to the long axis of the tibia, which generate inversion of the calcaneus and adduction of the forefoot (84). Inversion is created primarily by the tibialis anterior and the tibialis posterior with assistance from the toe flexors, the flexor digitorum longus and the flexor hallucis longus. The extensor hallucis longus works with the flexor hallucis longus to adduct the forefoot during supination.

The intrinsic muscles of the foot work as a group and are very active in the support phase of stance. They basically follow the movement of supination and are more active in the later portions of stance to stabilize the foot in propulsion (81). In a foot that excessively pronates, they are also more active as they work to stabilize the midtarsal and subtalar joints. There are 11 intrinsic muscles and 10 of these are on the plantar surface arranged in four layers. Please refer to Appendix B for a full listing of these muscles.

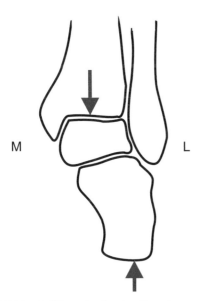

FIGURE 6–43. When the heel strikes the ground on the lateral aspect, a force comes vertically up the outside of the foot. The force of body weight is acting down through the ankle joint. Since these two forces do not line up, the talus is driven medially, initiating and producing the pronation movement.

Strength and Forces of the Ankle and Foot

The strongest movement at the ankle or foot is plantar flexion. This is because of the larger muscle mass contributing to the movement, but it is also related to the fact that the plantar flexors are used more to work against gravity and maintain an upright posture, control lowering to the ground, or add to propulsion. Even standing, the plantar flexors, and specifically, the soleus, contract to control the dorsiflexion present in the standing posture.

Plantar flexion strength is greatest from a position of slight dorsiflexion. A starting dorsiflexion angle of 105 degrees will increase the plantar flexion strength by 16% from the neutral 90 degree position. Plantar flexion strength measured from 75 and 60 degrees of plantar flexion, reduce the plantar flexion strength by 27% and 42%, respectively, when compared to strength measured in the neutral position (8). Additionally, plantar flexion strength can be increased if the knee is maintained in an extended position, placing the gastrocnemius in an advantageous muscle length.

The dorsiflexion movement is incapable of generating a very large force because of its reduced muscle mass, and because it is minimally used in daily activities, other than somewhat in gait. The strength of the dorsiflexors is only about one fourth that of the plantar flexors (8). Dorsiflexion strength can be enhanced by placing the foot in a few degrees of plantar flexion before initiating the dorsiflexion movement.

The ankle and foot are subjected to significant compressive and shear forces in both walking and running. In walking, there is a 0.8 to 1.1 times body weight vertical force coming up through the ground at heel strike. This drops to about 0.8 times body weight in the mid-stance to 1.3 times body weight at toe-off (74,87). This force along with the contraction force of the plantar flexors creates the compression force in the ankle.

In walking, the compression force in the ankle joint can be as high as 3 times body weight at heel strike and 5 times body weight at toe-off. A shear force of 0.45 to 0.8 times body weight is also present, primarily as a result of the shear forces absorbed from the ground and the positioning of the foot relative to the body (74,98). In running, the peak ankle joint forces are predicted to range from 9.0 to 13.3 times body weight, and the peak achilles tendon force in the range of 5.3 to 10.0 times body weight (98). The ankle joint is subjected to forces similar to those seen in the hip and knee joints. Amazingly, the ankle joint has very little incidence of osteoarthritis, which may be due in part to the large weight-bearing surface seen in the ankle, which lowers the pressure on the joint.

The subtalar joint is subjected to forces equivalent to 2.4 times body weight, with the anterior articulation between the talus, calcaneus, and navicular recording forces as high as 2.8 times body weight (74,87). High loads on the talus must be expected since it is the "keystone" of the foot, and loads travel into the foot from the talus to the calcaneus, and then forward to the navicular and cuneiforms.

Forces applied to the foot from the ground are usually applied to the lateral heel, travel lateral to the cuboid, and then transfer over to the second metatarsal and the hallux at toe-off. In FIGURE 6–44, the path of the forces across the plantar surface of the foot is shown. The greatest percentage of support time is spent in contact with the forefoot and the first and second metatarsal. If the second metatarsal is longer than the first, known as *Morton's toe*, the pressure on the head of the second metatarsal is greatly increased (87). This pattern of foot strike and transfer of the forces across the foot is dependent on a variety of factors and can vary with speed, foot type, and the foot contact patterns of individuals.

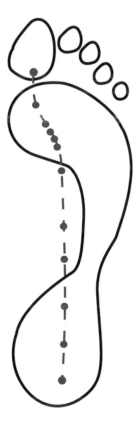

FIGURE 6–44. Forces applied to the plantar surface of the foot during gait normally travel a path from the lateral heel, to the cuboid, and across to the first and second metatarsal.

Forces in running are two times those seen in walking. At foot strike, the forces received from the ground create a vertical force of 2.2 times body weight and a 0.5 times body weight shear force. A 2.8 times body weight vertical force and a 0.5 times body weight shear force is produced at toe off (74,87). With the addition of the muscular forces, the compressive forces can be as high as 8 to 13 times body weight in running, with anterior shear forces in the range of 3.3 to 5.5 times body weight, medial shear at 0.8 times body weight, and lateral shear at 0.5 times body weight (74). Forces are high because the foot must transmit them between the body and the foot as well as the ground and the body. Given the injury record for the ankle and the foot, it is apparent that the foot is resilient and adaptable to the forces it must control with each step in walking or running.

Conditioning

The plantar flexor muscles are exercised in daily living activities to a great extent, being used to walk, get out of chairs, go up stairs, or even drive the car. Strengthening the plantar flexors by using higher resistive exercises is also easy, for any heel-raising exercise offers a significant amount of resistance since the body weight is being lifted with this muscle group. With the weight centered over the foot, the leverage of the plantar flexors is very efficient for handling large loads; thus a heel-raise activity with weight on the shoulders can usually be done with a considerable amount of weight. This exercise is perfect for the gastrocnemius since the strength of this muscle is enhanced with the knee extended and the quadriceps contracting.

To specifically strengthen the soleus, a seated position is best. This position flexes the knee and reduces the contribution of the gastrocnemius significantly. Weight or resistance can be placed on the thigh as plantar flexion is produced.

It is important to maintain flexibility in the plantar flexors, since any inflexibility in the group can create an early heel raise and excessive pronation in gait. Inflexibility in the plantar flexors is more common in women who wear elevated heels a great amount of the time (10). In fact, both men and women are susceptible to strain in the plantar flexors when going from a higher heel to a lower heel in either exercise or activities of daily living. It is better to maintain the flexibility in the muscle group through stretching with the knee extended and the ankle in maximum dorsiflexion.

Flexibility in the gastrocnemius and the soleus can be somewhat isolated. Flexibility of the gastrocnemius can be best tested with the knee extended, while flexibility of the soleus is best tested with the knees flexed 35 degrees. Both stretching and strengthening exercises for the plantar flexors are shown in FIGURE 6–45.

FLEXIBILITY

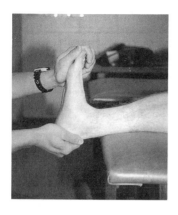
A

B

C

MANUAL RESISTANCE

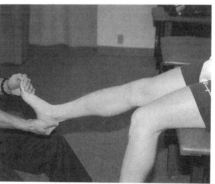
D

E

WEIGHT TRAINING

F

G

FIGURE 6–45. The plantar flexors are easy to both stretch and strengthen. Flexibility exercises include a passive stretch into dorsiflexion (A), a stretch on the plantar flexors from a straight leg position (B), and a wall stretch (C). A manual resistance can be applied across the toes with the knee extended (D) or with the knee flexed (E), isolating the gastrocnemius and soleus, respectively. The plantarflexors can be strengthened in weight training with a heel-raise lift (F). To isolate the soleus more, a seated heel raise exercise with weights on the thigh is used (G).

The strength of the dorsiflexors is very limited, but it should be maintained so that fatigue does not set in during a long walking or running activity. Fatigue in the muscle group would lead to a drop of the foot in swing and slapping of the foot on the surface following heel strike. To strengthen the muscle group, a seated position works best so that a resistance can be applied below the foot with sand bags, weights, or surgical tubing (FIGURE 6–46). There are also ankle machines allowing a full range of dorsiflexion and high-resistance training of this movement. Flexibility of the dorsiflexion action can also be best achieved in the seated position, through maximum plantar flexion activities.

Strength and flexibility of the inverters and evertors of the ankle are important for those athletes participating in activities in which ankle injuries are common. This might include basketball, volleyball, football, soccer, tennis, and a wide variety of other activities. If no ankle machine is available, stretching and strengthening the inversion/eversion actions can be done with the foot flat on the floor on a towel or attached to surgical tubing. Weight can be put on the towel, which can then be pulled toward the foot in either inversion or eversion actions, depending on which side of the weights the foot is placed. Circumduction and figure-eight tracing are good flexibility exercises. FIGURE 6–47 demonstrates some examples of exercises for the muscles creating inversion, eversion, abduction, and adduction of the foot. These exercises would be a part of the sports conditioning program, and would add a preventative measure for ankle sprains. These exercises may be intensified and expanded for individuals with chronic ankle problems.

The intrinsic muscles of the foot are usually very atrophied and weak because we regularly wear shoes. Since the intrinsic muscles support the arch of the foot and stabilize the foot during the propulsive phase of gait, it is worthwhile to give them some attention in terms of conditioning. The best way to exercise the intrinsic muscle group as a whole would be to go barefoot. The movement potential of the foot is best illustrated by individuals who have upper extremity disabilities, and who must use their feet to perform daily functions. These individuals can become very versatile and adept at using the foot to perform a wide range of functions.

During walking or running, impact is the same either with shoes on or barefoot, but it is the manner in which the forces are absorbed that is different between the two. With a shoe on, the foot is more rigid on absorption and dependent on the shoe for support and protection. Absorption with barefeet is more mobile, with more arch deflection upon loading (99). This does not necessarily mean that shoes should not be worn, since the injury rate in barefoot running would be initially high because of the significant change imposed by removing the shoes. There is also a danger associated with barefoot activity and the possibility of injury from sharp objects. However, going barefoot in the summer would be one way of improving the condition of the intrinsic muscles.

A point attesting to the benefits of barefoot activity is the low injury rate in populations that remain largely barefoot. The incidence of injury to barefoot runners is much less than the shod population (99). Finally, the intrinsic musculature in a person with a flat, mobile foot will be much more developed than a person with a high-

FLEXIBILITY   **MANUAL RESISTANCE**   **WEIGHT TRAINING**

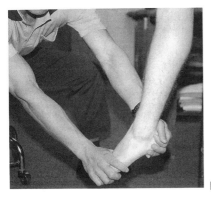
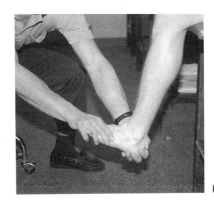
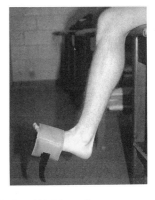

A   B   C

FIGURE 6–46. To stretch the dorsiflexors, the foot is moved slowly into maximum plantar flexion (A). From the same position, a manual resistance is applied to the top of the foot as the foot is pulled into dorsiflexion (B). Weight-training exercises for the dorsiflexors is limited, but weight can be added by sandbag weights or cables as the foot is dorsiflexed (C).

FLEXIBILITY

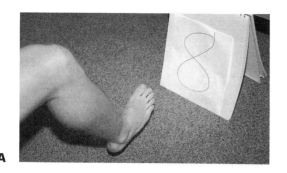

A

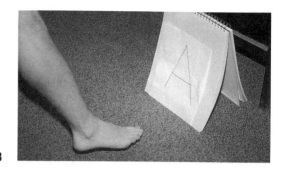

B

MANUAL RESISTANCE

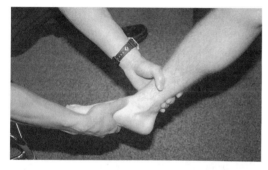

C

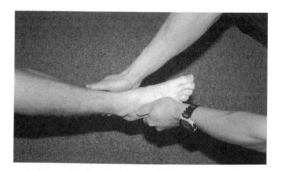

D

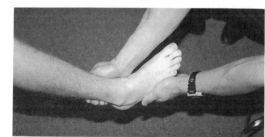

E

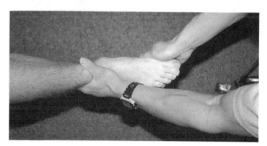

F

WEIGHT TRAINING

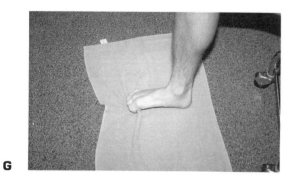

G

H

FIGURE 6–47. To stretch the inverters, everters, abductors, and adductors, simple exercises such as drawing a figure eight (A) or letters of the alphabet (B) are beneficial. Exercises using manual resistance include applying resistance to the inside or outside of the foot (C) during inversion or (D) eversion. Foot adduction and abduction movements can also be resisted by applying a force on the outside (E) or inside (F) respectively. Higher resistance exercises include placing weights on a towel as it is pulled toward or away (G), and using surgical tubing or cables (H).

arched, rigid foot because of the difference in movement characteristics in loading of the foot.

Contribution to Sport Skills or Movements

The ankle and foot muscles are very active in walking, as they control the foot on the ground and generate a significant portion of the propulsive force. Refer to FIGURE 6–15 for a summary of the muscles contributing to the walking gait.

Upon heel strike in walking, there is maximum activity in the dorsiflexors during the attempt to eccentrically control the lowering of the foot to the ground in plantar flexion. The most activity is seen in the tibialis anterior, extensor digitorum longus, and the extensor hallucis longus (96). The activity in this muscle group drops, but maintains activity throughout the total stance phase.

There is no activity in the gastrocnemius and soleus at heel strike. They begin to activate after foot flat and continue into the propulsive phase as they control the movement of the tibia over the foot and generate propelling forces. The intrinsic muscles of the foot are basically inactive at this phase of stance.

In the propelling phase, the dorsiflexors are still active, generating a second peak in the stance phase right before toe-off. The gastrocnemius and soleus reach a peak of muscular activity just prior to toe-off. The intrinsic muscles of the foot are very active in the propelling phase of stance as they work to make the foot rigid and stable, and control depression of the arch. Activity in the gastrocnemius, soleus, and the intrinsic muscles ceases at toe-off.

In the swing phase, the dorsiflexors generate the only significant muscular activity in the ankle and foot. They hold the foot up so that the toes don't drop while the limb is swinging through.

In running, the plantar flexors become much more active than in walking, even though the amount of plantar flexion range of motion is diminished. The plantar flexor activity increases sharply after heel strike and dominates through the total stance period (20) (FIGURE 6–16). In the braking portion of the stance, the plantar flexors work, not to act at the ankle, but to eccentrically halt the vertical descent of the body over the foot. This continues into the propelling phase, when the plantar flexors shift to a concentric contraction, adding to the driving force of the run (20).

In stair climbing, the ankle joint musculature does contribute to the action, even though the quadriceps femoris are the main muscle group. During the pull-up phase of ascent, the leg moves posteriorly, via plantar flexion at the ankle, to increase the vertical position (FIGURE 6–18). The primary ankle muscle producing this motion is the soleus, with some contribution from the gastrocnemius. The most power from the ankle is generated in the forward continuance phase of ascent as the individual continues on to the next step. At this point, there is push-off by the ankle joint, with the plantar flexors active as the body is pushed up to the next step (21).

During descent, the ankle eccentrically absorbs contact initially via eccentric contraction of the plantar flexors (21,100). There is also co-contraction of the soleus and the tibialis anterior early in the absorption phase to stabilize the ankle joint. As a step is taken down to the next step in forward continuance, there is a small eccentric muscular activity at the ankle joint in the soleus as it contributes to the controlled drop and forward movement of the body.

The ankle joint musculature is also a major contributor to cycling, even though the ankle undergoes very little angular change in motion during the 360 degree cycle (22, 60). As shown in FIGURE 6–19, the gastrocnemius contributes through the majority of the power portion of the cycle, being active from 30 to 270 degrees in the revolution. When the activity of the gastrocnemius ceases, the tibialis anterior becomes active, contributing to the lift of the pedal, being active from 280 degrees until slightly past the top. Again, when the tibialis anterior activity ceases, the gastrocnemius becomes active, so there is no co-contraction at the ankle, unlike at the knee and hip.

Injury Potential in the Ankle and Foot

Injuries to the foot and ankle account for a large portion of the injuries to the lower extremity, and in some sports or activities such as basketball, the ankle and foot joint is the most frequently injured in the lower extremity. Injuries to the hindfoot usually occur as a result of vertical compression, while injuries to the midfoot occur with excessive lateral movement or range of motion in the foot (84). Injuries to the forefoot occur similarly to injuries seen in long bones elsewhere in the body, in which both compressive and tensile forces create the injury.

The majority of injuries to ankle and foot occur as a result of overtraining or an excessive training bout, and the joint is injured frequently in activities such as running during which the foot is loaded suddenly and repeatedly (99). Foot and ankle injuries are also associated with anatomical factors, with higher injury seen in people who over-pronate or those with cavus alignment in the lower extremity.

One of the most common injuries to the foot is an ankle sprain, more commonly occurring in the lateral complex of the ankle during inversion. The mechanism of injury is a movement of the tibia sideways, backward, forward, or rotating while the foot is firmly fixed on the surface. Stepping in a pothole, walking off a curb, or losing balance in high heels are other instances in which the ankle can be sprained. However, most ankle sprains in athletics are seen with the cutting maneuver, as a cut is made with the foot opposite to the direction of the run (95), or when landing on another player's foot. For example, the left foot is sprained as it drives in plantar flexion and inversion to the right. The plantar flexion and inversion action is the cause of sprain to the lateral ligamentous structure, with the anterior talofibular ligament most likely to be sprained (101).

If the cut is made with more inversion in the foot, the calcaneofibular ligament is the next ligament damaged (101). The injury is created with a talar tilt as the talus moves forward out of the ankle mortise. Any talar tilt more than 5 degrees will likely cause ligament damage to the lateral ankle (102). With injury to the lateral ankle complex, an anterior subluxation of the talus and talar tilt may be present, creating great instability in the ankle and foot complex.

The medial ligaments of the ankle are not sprained very often because of the support of the strong deltoid ligament and the bulwark created by the longer lateral malleolus. The powerful deltoid ligament can be sprained if the foot is planted and pronated and takes a hit from the lateral side of the leg.

Although not at the ankle, the ligaments holding the tibiofibular joint together can be sprained with a forceful external rotation and dorsiflexion, or a forceful inversion or eversion. The talus pushes the tibia and fibula apart, spraining the ligaments.

There are many other soft tissue injuries to the foot and ankle typically associated with overuse or some other functional malalignment. Posterior or medial tibial syndrome, previously referred to as shin splints, generates a pain above the medial malleolus (91). This condition usually involves the insertion site of the tibialis posterior and can be a tendinitis of the tibialis posterior tendon or a periostitis, in which the insertion of the muscle pulls on the interosseus membrane and periosteum on the bone, causing an inflammation. This muscle is usually irritated through excessive pronation, which places a great deal of tension and stretch on the muscle. Lateral tibial syndrome will cause pain on the anterior lateral aspect of the leg, and is a similar overuse condition of the tibialis anterior muscle.

The achilles tendon is another frequently strained area of the foot. It is injured as a result of overtraining, a vig-

orous contraction of the gastrocnemius, overstretching of the muscle group, in hill running, or in moving to a low-heel shoe from a higher heel (103). The achilles can also be irritated if there is loss of absorption in the heel pad on the calcaneus. This creates a higher amplitude shock at heel strike that is compensated for by an increase in soleus activity and a corresponding increase in the loading on the achilles. Achilles tendinitis can be very painful and difficult to heal since immobilization of the area is difficult. The achilles tendon can also rupture as a result of a vigorous contraction after pushing off with the balls of the feet, or by stepping into a hole or off of the curb.

A condition mimicking the pain associated with achilles tendinitis is retrocalcaneal bursitis, an inflammation of the bursae lying superior to the achilles attachment. It is caused by ill-fitting shoes (103).

Plantar fasciitis, an inflammation of the plantar fascia on the underside of the foot is another common soft tissue injury to the foot (99). The irritation usually develops on the medial plantar fascial attachment to the calcaneus and is caused by stepping into a hole or off the curb, or by training adjustments that include hill running or an increase in mileage. It is more prevalent in high-arched foot types and in individuals with a tight achilles tendon or leg length discrepancy (104). That more tension is placed on the plantar fascia in pronation predisposes this area to this type of injury. The plantar fascia can rupture with a forceful plantar flexion, such as is seen in descending stairs or during rapid acceleration.

At the site of the irritation of the plantar fascia on calcaneus, adolescents can develop a calcaneal apophysitis, an irritation at the site of the epiphysis of the calcaneus (104). Adults may develop a similar irritation at the same site where heel bone spurs develop in response to the pull of the plantar fascia.

Strain to the metatarsals, metatarsalgia, creates a dull burning sensation in the forefoot. Irritation to the ligaments or soft tissue is usually associated with hard-surface running. Injuries to the metatarsal are more prevalent in overpronating feet.

Nerve compression can occur at various sites in the leg and foot. Anterior compartment syndrome is a case in which nerve and vascular compression occur as a result of hypertrophy in the anterior tibial muscles to the point where they are impinged upon in the muscle compartment. Impingement can create tingling sensations or atrophy in the foot.

Injury to the osseus components of the foot will typically occur with overuse or pathological function. Metatarsal fractures are typically found in the middle of the shaft of the second or third metatarsal. This fracture is associated with tight dorsiflexor muscles or forefoot varus. A stress fracture can also be developed on the metatarsals on the lateral side of the foot as a result of a tight gastrocnemius. This prevents dorsiflexion in gait, creating compensatory pronation, an unlocked subtalar joint, more flexibility in the first metatarsal, with the lateral metatarsal absorbing the force. A person is almost five times more likely to acquire a stress fracture if they lack sufficient dorsiflexion in gait.

Fractures to the metatarsals occur with a fall on the foot, avulsion by a muscle such as the site of the attachment of the peroneus brevis on the fifth metatarsal, or as a consequence of compression. Fractures have also been associated with loss of heel pad compressive ability, requiring that a higher amplitude of force be absorbed by the foot. An example of a compression injury is a fracture of the tibia or talus on the medial side that accompanies a lateral ankle sprain. This jamming of the inner ankle can also loosen bony fragments, a condition known as osteochondritis dissecans. An osteochondral fracture of the talus is a shearing type of fracture, occurring with a dorsiflexion—eversion action of the foot where the talus impinges on the fibula in a crouched position.

Summary

The lower extremity absorbs high forces and supports the body's weight. Both lower limbs are connected by the pelvic girdle, making every movement or posture of the lower extremity or trunk interrelated.

The pelvic girdle serves as a base for lower extremity movement, a site for muscular contraction, and is very important in the maintenance of balance and posture. The pelvic girdle consists of two coxal bones (ilium, ischium, pubis), which are joined in the front at the pubic symphysis and connected to the sacrum in the back (sacroiliac joint). The pelvic and sacral movements of flexion, extension, posterior and anterior tilt, and rotation accompany movements of both the thigh and the trunk.

The femur articulates at the acetabulum on the anterolateral surface of the pelvis. This ball-and-socket joint is well reinforced by strong ligaments that restrict all movements of the thigh, except for flexion. The femoral neck is angulated at approximately 125 degrees in the frontal plane and an increase (coxa valga) or decrease (coxa vara) in this angle will influence leg length and lower extremity alignment and function. The angle of anteversion in the transverse plane will also influence the rotation characteristics of the lower extremity.

The hip joint allows considerable movement in flexion (70 to 140 degrees), produced by the hip flexors, the iliopsoas, the rectus femoris, the sartorius, the pectineus, and the tensor fascia latae. Four to fifteen degrees of hyperextension is also possible, and the extension movement is produced by the hamstrings, the semimembranosus, semitendinosus, biceps femoris, and the gluteus maximus. Abduction range of motion is 30 degrees and is produced by the gluteus medius, gluteus minimus, tensor fascia latae, and the piriformis. Adduction (25 degrees) is produced by gracilis, adductor longus, adductor magnus, adductor brevis, and the pectineus. Internal rotation through approximately 70 degrees is produced by the gluteus minimus, gluteus medius, gracilis, adductor longus, adductor magnus, tensor fascia latae, semimembranosus, and semitendinosus. External rotation through 90 degrees is produced by the gluteus maximums, the obturator externus, the quadratus femoris, obturator internus, piriformis, and the inferior and superior gemellus.

Movements of the thigh will usually be accompanied by a pelvic movement and vice versa. For example, flexion of the thigh will produce a posterior tilt of the pelvis, and swinging the leg in a walking or running action will produce rotation of the pelvis to the opposite side.

The hip joint muscles can produce the most strength in the extension movement owing to the large muscle mass from the hamstrings and the gluteus maximus. Extension strength is maximized from a hip flexion position. Strength output in the other movements can also be maximized with accompanying knee flexion for hip flexion strength facilitation, accompanying thigh flexion for the abduction movement, slight abduction for adduction facilitation, and hip flexion for the internal rotators. Loads absorbed by the hip joint can be considerable and may range from 2 to 10 times body weight in activities such as walking, running, or stair climbing.

Conditioning exercises for the lower extremity are easy to implement because they include common movements associated with daily living activities. A closed kinetic chain exercise is beneficial for the lower extremity because of the transfer to daily activities. Because of the many two-joint muscles surrounding the hip joint, the position of adjacent joints is important. The hip flexors are best exercised in the supine or in a hanging position. The extensors are maximally stretched using a hip flexion position with the knees extended. The abductors, adductors, and the rotators require creative approaches to conditioning since they are not easy to isolate.

The muscles of the pelvis and the hip joint are major contributors to a variety of movements and sport activities. In walking, the abductors control the pelvis, the hamstrings control the amount of hip flexion and provide some of the propulsive force, and the hip flexors are very active in the swing phase. In a faster walk, such as race walking, the abductors are more active as the pelvic actions increase. In running, the hip joint motions and the muscular activity increase, but the same muscles used in walking are also used. Hip joint muscles are also important contributors to stair climbing and to cycling.

The hip joint is durable and accounts for only a very small percentage of injuries to the lower extremity. Common soft tissue injuries to the region include tendinitis of the gluteus medius, strain to the rectus femoris, hamstrings, iliopsoas, or piriformis, bursitis, and iliotibial band friction syndrome. Stress fractures are also more prevalent at sites such as the anterior iliac spine, the pubic rami, the ischial tuberosity, the greater and lesser trochanters, and the femoral neck. There are some childhood disorders that are common to the hip joint such as congenital hip dislocation and Legg-Calve-Perthes disease. The hip joint is also a site where osteoarthritis is prevalent in later years.

The knee joint is very complex and is formed by the articulation between the tibia and the femur (tibiofemoral) and the patella and the femur (patellofemoral). In the tibiofemoral joint, the two condyles of the femur rest on the tibial plateau and rely on the collateral ligaments, the cruciate ligaments, the menisci, and the joint capsule for support. The patellofemoral joint is supported by the quadriceps tendon and the patellar ligament. The patella fits into the trochlear groove of the femur, which also offers stabilization to the patella.

An important alignment feature at the knee joint is the Q-angle, the angle representing the positioning of the patella with respect to the femur. An increase in this angle will increase the valgus stress on the knee joint. Higher Q-angles are more common in females owing to their wider pelvic girdles.

Flexion of the leg at the knee joint occurs through approximately 120 to 145 degrees and is produced by the hamstrings, the biceps femoris, semimembranosus, and semitendinosus. Accompanying flexion is internal rotation of the tibia, which is produced by the sartorius, popliteus, gracilis, semimembranosus, and semitendinosus. As the knee joint flexes and internally rotates, the patella also moves down in the groove, and then moves laterally.

Extension at the knee joint is produced by the powerful quadriceps femoris muscle group, which includes the vastus lateralis, vastus medialis, rectus femoris, and vastus intermedius. When the knee extends, the tibia externally rotates via action by the biceps femoris. At the end of extension, the knee joint locks into the terminal position by a screw home movement in which the condyles rotate into their final positions. In extension, the patella moves up in the groove and terminates in a resting position that is high and lateral on the femur.

The strength of the muscles around the knee joint is substantial, with the extensors being one of the strongest muscle groups in the body. The extensors are stronger than the flexors in all joint positions, but not necessarily at all joint speeds. The flexors should not be significantly weaker than the extensors, or the injury potential around the joint will increase.

The knee joint can handle high loads and commonly absorbs 1 to 5 times body weight in activities such as walking, running, or weight lifting. A maximum flexion position should be evaluated for safety, given the high shear forces that are present in the position. Patellofemoral forces can also be high, in the range of 0.5 to 8 times body weight in daily living activities. The patellofemoral force is high in positions of maximum knee flexion.

Conditioning of the knee extensors is an easy task since these muscles control simple lowering and raising movements. Closed-chain exercises are also very beneficial for the extensors because of their relation to daily living activities. The flexors are also exercised during a squat movement because of their action at the hip joint, but can best be isolated and exercised in a seated position.

The knee joint muscles contribute significantly to a wide variety of movements and sport activities. The quadriceps femoris muscle group serves as a force absorption mechanism and a power producer for walking, running, and stair climbing. In cycling, the quadriceps femoris muscles are responsible for a significant amount of the power production.

The knee joint is the most frequently injured joint in the body. There are traumatic injuries resulting in damage to the ligaments or menisci, and there are numerous chronic injuries resulting in tendinitis, iliotibial band syndrome, and general knee pain. Muscle strains to the quadriceps femoris and hamstrings are also common. The patella is a site for injuries such as subluxation or dislocation, and other patellar pain syndromes such as chondromalacia patella.

The foot and ankle consist of 26 bones articulating at 30 synovial joints, supported by over 100 ligaments and 30 muscles. The ankle, or talocrural joint has two main articulations, the tibiotalar and the tibiofibular joints. The tibia and fibula form a mortise over the talus, defined on the medial and lateral sides by the malleoli. Both sides of the joint are strongly reinforced by ligaments, making the ankle very stable.

The foot moves at the tibiotalar joint in two directions, plantar flexion and dorsiflexion. Plantar flexion can occur through approximately 50 degrees and is produced by the gastrocnemius and soleus with some assistance from the peroneal muscles and the toe flexors. Dorsiflexion range of motion is approximately 20 degrees and the movement is created by the tibialis anterior and the toe extensors.

Another important joint in the foot is the subtalar, or talocalcaneal joint, in which the movements of pronation and supination occur. It is at this joint that the rotation of the lower extremity and forces of impact are absorbed. The movement of pronation occurring at the subtalar articulation is a triplane movement consisting of calcaneal eversion, abduction, and dorsiflexion with the foot off the ground, and calcaneal eversion, talar adduction, and plantar flexion with the foot on the ground in a closed chain. Muscles responsible for creating eversion are the peroneals, consisting of the peroneus longus, peroneus brevis, and peroneus tertius. Supination, the reverse movement, is created in the open chain through calcaneal inversion, talar adduction, and plantar flexion, and in the closed chain through calcaneal inversion, talar abduction, and dorsiflexion. Muscles responsible for producing inversion are the tibialis anterior, tibialis posterior, and big toe flexors and extensors. The range of motion for pronation and supination is 20 to 62 degrees.

The midtarsal joint also contributes to pronation and supination of the foot. These two joints, the calcaneocuboid and the talonavicular, allow the foot great mobility if the axes of the two joints lie parallel to each other. This is beneficial in the early portion of support when the body is absorbing forces of contact. When these axes are not parallel, the foot becomes rigid. This is beneficial in the later portion of support when the foot is propelling the body upward and forward. There are numerous other articulations in the foot such as the intertarsal, tarsometatarsal, metatarsophalangeal, and the interphalangeal joints that influence both total foot and toe motion.

The foot has two longitudinal arches that provide both shock absorption and support. The medial arch is higher and more dynamic than the lateral arch. The longitudinal arches are supported by the plantar fascia running along the plantar surface of the foot. There are also transverse arches running across the foot that depress and spread in weight bearing. The shape of the arches and the bony arrangement in the foot determine foot type, which can be normal, flat, or high arched and flexible or rigid. An extremely flat foot is termed *pes planus* and a high-arched foot is called *pes cavus*. There are other foot alignments, such as forefoot and rearfoot varus and valgus, a plantar-flexed first ray, and equinus positions that influence function of the foot.

The plantar flexion movement of the foot is the strongest joint action and is a major contributor to the development of a propulsion force. The dorsiflexion movement is weak and not capable of generating high muscle forces. The foot and ankle can handle high loads and the forces in the ankle joint range from 0.5 to 13 times body weight in walking and running. The subtalar joint also handles forces in the magnitude of 2 to 3 times body weight.

The muscles of the foot and ankle receive a considerable amount of conditioning in daily living activities such as walking. Specific muscles can be isolated through specialized exercises. For example, the gastrocnemius can be strengthened in a standing heel raise and the soleus in a seated heel raise. The intrinsic muscles of the foot can be exercised by drawing the alphabet or drawing figure eights with the foot, or by just going barefoot.

The foot and ankle muscles contribute to a variety of activities, generally serving as a major source of propulsion. Thus, the gastrocnemius and soleus are important contributors to walking, running, stair climbing, and cycling.

The foot and ankle are injured frequently in sports or physical activity. Common injuries are ankle sprains, achilles tendinitis, posterior, lateral, or medial tibial syndrome, plantar fascitis, bursitis, metatarsalgia, and stress fractures.

Review Questions

1. Discuss the lower extremity function in terms of movements and muscular activity for each of the following:
    a. walking
    b. running
    c. cycling
2. Describe the pelvic movements that would accompany:
    a. single-leg lift
    b. curl up
    c. support phase of running
    d. backbend
    e. double-leg raise
3. Is the double-leg raise a contraindicated exercise? Why or why not?
4. How does trunk positioning influence function at the hip and knee?
5. Describe the sacral movements that would accompany trunk flexion and extension, and thigh flexion and extension.

6. Design a set of flexibility exercises for the following muscle(s):
    a. thigh flexors and extensors
    b. thigh abductors and adductors
    c. thigh rotators
    d. leg flexors and extensors
    e. foot plantar flexors and dorsiflexors
    f. foot invertors and evertors
7. Design a set of both low- and high-resistance exercises for the following muscle(s):
    a. thigh flexors and extensors
    b. thigh abductors and adductors
    c. thigh rotators
    d. leg flexors and extensors
    e. foot plantar flexors and dorsiflexors
    f. foot invertors and evertors

Additional Questions

1. Describe how the following alignment variations will influence function at either the hip, knee, ankle, or foot joints:
   a.   coxa vara
   b.   coxa valga
   c.   anteversion
   d.   retroversion
   e.   genu valgum
   f.   genu varum
   g.   excessive Q-angle
   h.   tibial torsion
   i.   rear-foot varum
   j.   excessive pronation
   k.   forefoot varum
   l.   forefoot valgum
   m.   plantar flexed first ray

2. Describe how the abductors and adductors control pelvic action in the walking gait.

3. For each of the following movements or skills, identify the muscles or joint structures that will be more susceptible to injury. State why they are more susceptible, and when in the activity the injury is most likely to occur.
   a.   deep squat
   b.   kicking a ball
   c.   hurdling
   d.   running on uneven surfaces
   e.   hill running
   f.   running on a cambered surface
   g.   jumping

4. Since running accounts for a large percentage of lower extremity injuries, discuss the injury potential at the hip, knee, ankle, and foot, and relate the injury to biomechanics of the running cycle.

5. Compare the movements and muscular actions at the ankle joint between raising up on the toes and lowering the heel below the foot.

6. What are the basic muscular differences between stair ascent and descent?

7. What would be the impact of having inflexible hamstrings? Inflexible plantar flexors?

8. Analyze the following exercises and offer reasons why you think each exercise is a good or bad exercise for the lower extremity.
   a.   double-leg raise
   b.   jumping jacks
   c.   hurdler stretch
   d.   squat thrust or "burpee"
   e.   straddle sit

9. What would be the difference in lower extremity muscular contributions to:
   a.   bicycling with, versus without, foot straps or cleats
   b.   running uphill versus downhill
   c.   jumping up versus landing
   d.   thigh extension with the knee extended versus flexed
   e.   thigh flexion with the knee extended versus flexed
   f.   leg extension with the thigh flexed versus extended
   g.   leg flexion with the thigh flexed versus extended

10. Apply resistance to a fully flexed thigh. What contributes to such a strong resistance?

11. Standing on one foot, palpate above and below the iliac crest on the supported and nonsupported side. Explain the muscle action on each side.

Additional Reading

Alexander, M.J.L.: The relationship between muscle strength and sprint kinematics in elite sprinters. Canadian Journal of Sports Science. *14*:148–157, 1989.

Allard, P., Nagata, S.D., Duhaime, M., and Labelle, H.: Application of steriophotogrammetry and mathematical modelling in the the study of ankle kinematics. *In* Biomechanics X. Edited by B. Jonsson. Champaign, Ill., Human Kinetic Publishers, 1985.

Andrews, J. R.: Posterolateral rotatory instability of the knee. Surgery for acute and chronic problems. Physical Therapy. *60*:1637–1639, 1980.

Andrews, J. R., Sanders, R. A., and Morin, B.: Surgical treatment of anterolateral rotatory instability. The American Journal of Sports Medicine. *13*:112–119, 1985.

Apkarian, J., Naumann, S., and Cairns, B.: A three-dimensional kinematic and dynamic model of the lower limb. Journal of Biomechanics. *22*:143–155, 1989.

Arno, S. A.: The A-angle: A quantitative measurement of patella alignment and realignment. Journal of Orthopaedic and Sports Physical Therapy. *12*:237–242, 1990.

Bach, T. M., Chapman, A. E., and Calvert, T. W.: Mechanical resonance of the human body during voluntary oscillations about the ankle joint. Journal of Biomechanics. *16*:85–90, 1983.

Bates, B., Osterning, L., Mason, B., and James, S.: Functional variability of the lower extremity during the support phase of running. Medicine and Science in Sports and Exercise. *11*:328–331, 1979.

Bobbert, M. F., and Van Ingen Schenau, G. J.: Mechanical output about the ankle joint in isokinetic plantar flexion and jumping. Medicine and Science in Sports and Exercise. *22*:660–668, 1990.

Borzov, V.: The optimal starting position in sprinting. Legkaya Atletika. *4*:173–174, 1978.

Bose, K., Kanagasuntheram, R., and Osman, B. H.: Vastus medialis oblique: An anatomic and physiologic study. Orthopedics. *3*:880–883, 1980.

Brand, R. A., et al.: A model of lower extremity muscular anatomy. Transactions of the ACSM. *104*:304–310, 1982.

Briner, W. W., Carr, D. E., and Lavery, K. M.: Anteroinferior tibiofibular ligament injury: Not just another ankle sprain. The Physician and Sports Medicine. *17*:63–69, 1989.

Brooke, R.: The sacro-iliac joint. Journal of Anatomy. *58*:299–305, 1924.

Cabaud, H. E., and Slocum, D. B.: The diagnosis of chronic anterotational rotary instability of the knee. The American Journal of Sports Medicine. *5*:99–105, 1977.

Cavanaugh, P., and Lafortune, M.: Ground reaction forces in distance running. Journal of Biomechanics. *13*:397–406, 1980.

Caylor, D., Fites, R., and Worrell, T.W.: The relationship between quadriceps angle and anterior knee pain syndrome. Journal of Sports Physical Therapy. *17*:11–15, 1993.

Cibulka, M. T., Rose, S. J., Delitto, A., and Sinacore, D. R.: Hamstring muscle strain treated by mobilizing the sacroiliac joint. Physical Therapy. *66*:1220–1224, 1986.

Clancy, W. G.: Runners' injuries. Part two: Evaluation and treatment of specific injuries. The American Journal of Sports Medicine. *8*:287–289, 1980.

Clark, J. M., and Jaynor, D. R.: Anatomy of the abductor muscles of the hip as studied by computer tomography. The Journal of Bone and Joint Surgery. *69-A*:1021–1031, 1987.

Colville, M. R., Marder, R. A., Boyle, J. J., and Zarins, B.: Strain measurement in lateral ankle ligaments. The American Journal of Sports Medicine. *18*:196–200, 1990.

Conteduca, F., et al.: Chondromalacia and chronic anterior instabilities of the knee. American Journal of Sports Medicine. *9*:119–123, 1991.

Coplan, J. A.: Rotational motion of the knee: A comparison of normal and pronating subjects. The Journal of Orthopaedic and Sports Physical Therapy. *10*:366–369, 1989.

Crowninshield, R. D., Johnston, R. C., Andrews, J. G., and Brand, R. A.: A biomechanical investigation of the human hip. Journal of Biomechanics. *11*:75–85, 1978.

Czerniecki, J. M., Lippert, F., and Olerud, J. E.: A biomechanical evaluation of tibiofemoral rotation in anterior cruciate deficient knees during walking and running. American Journal of Sports Medicine. *16*:327–331, 1988.

Donatelli, R.: Normal biomechanics of the foot and ankle. Journal of Orthopaedic and Sports Physical Therapy. *7*:91–95, 1985.

DonTigny, R. L.: Anterior dysfunction of the sacroiliac joint as a major factor in the etiology of low back pain syndrome. Physical Therapy. *70*:250–262, 1990.

Dostal, W. F., and Andrews, J. G.: A three-dimensional biomechanical model of hip musculature. Journal of Biomechanics. *14*:803–812, 1981.

Doxey, G.: Calcaneal pain: A review of various disorders. Journal of Orthopaedic and Sports Physical Therapy. *9*:25–32, 1987.

Ellison, J. B., Rose, S. J., and Sahrmann, S. A.: Patterns of hip rotation range of motion between healthy subjects and patients with low back pain. Physical Therapy. *70*:537–541, 1990.

Falkel, J.: Plantar flexor strength testing using the cybex isokinetic dynamometer. Physical Therapy. *58*:847–850, 1978.

Ferkel, R. D., Mai, L. L., Ullis, K. C., and Finerman, G. A. M.: An analysis of roller skating injuries. The American Journal of Sports Medicine. *10*:24–30, 1981.

Ficat, P., and Hungerford, D.: Disorders of the patellofemoral joint. Baltimore, Williams & Wilkins, 1979.

Fischer, R. L.: An epidemiological study of Legg-Perthes disease. The Journal of Bone and Joint Surgery. *54-A*, 769–777, 1972.

Fujita, M., et al.: Motion and role of the MP joints in walking. *In* Biomechanics VIII-A. Edited by H. Matsui and K.

Kobayashi. Champagne, IL, Human Kinetics Publishers, 1981, pp. 467–470.

Goodfellow, J., Hungerford, D. S., and Sindel, M.: Patellofemoral joint mechanics and pathology: 1. Functional anatomy of the patellofemoral joint. The Journal of Bone and Joint Surgery. 58-B:287–290, 1976.

Gross, T. S., and Nelson, R. C.: The shock attenuation role of the ankle during landing from a vertical jump. Medicine and Science in Sports and Exercise. 20:506–514, 1988.

Hajak, M. R., and Noble, H. B.: Stress fractures of the femoral neck in joggers. The American Journal of Sports Medicine. 10:112–116, 1982.

Hontas, M. J., Haddad, R. J., and Schlesinger, L. C.: Conditions of the talus in the runner. The American Journal of Sports Medicine. 14:486–490, 1986.

Houtz, S. J., and Fischer, F. J.: An analysis of muscle action and joint excursion during exercise on a stationary bicycle. The Journal of Bone and Joint Surgery. 41-A:123–131, 1959.

Hughes, L. Y.: Biomechanical analysis of the foot and ankle for predisposition to developing stress fractures. The Journal of Orthopaedic and Sports Physical Therapy. 7:96–101, 1985.

Illingworth, C. M.: 128 limping children with no fracture, sprain, or obvious cause. Clinical Pediatrics. 17:139–142, 1978.

Insall, J.: Current concepts review: Patellar pain. The Journal of Bone and Joint Surgery. 64-A:147–152, 1982.

Jones, A. L.: Rehabilitation for anterior instability of the knee: Preliminary report. The Journal of Orthopaedic and Sports Physical Therapy. 3:121–127, 1982.

Jorgensen, U.: Achillodynia and loss of heel pad shock absorbency. American Journal of Sports Medicine. 13:128–132, 1985.

Kannus, P.: Relationships between peak torque, peak angular impulse, and average power in the thigh muscles of subjects with knee damage. Research Quarterly for Exercise and Sport. 61:141–145, 1990.

Kannus, P., and Jarvinen, M.: Maximal peal torque as a predictor of peak angular impulse and average power of thigh muscles. An isometric and isokinetic study. International Journal of Sports Medicine. 11:146–149, 1990.

Klampner, S. L., and Wissinger, A.: Anterior slipping of the capital femoral epiphysis. The Journal of Bone and Joint Surgery. 54-A:1531–1537, 1972.

Leib, F. J., and Perry, J.: Quadriceps function: An anatomical and mechanical study using amputated limbs. The Journal of Bone and Joint Surgery. 50-A:749–758, 1968.

Lentell, G. L., Katzman, L. L., and Walters, M. R.: The relationship between muscle function and ankle stability. The Journal of Orthopaedic and Sports Physical Therapy. 11:605–611, 1990.

Luethi, S. M., Frederick, E. C., Hawes, M. R., and Nigg, B. M.: Influence of shoe construction on lower extremity kinematics and load during lateral movements in tennis. International Journal of Sport Biomechanics. 2:166–174, 1986.

Mann, R., and Inman, V.: Phasic activity of intrinsic muscles of the foot. Journal of Bone and Joint Surgery. 46-A:469–481, 1964.

Mann, R., and Sprague, P.: A kinetic analysis of the ground leg during sprint running. Research Quarterly for Exercise and Sport. 51:334–348, 1980.

Manter, J.: Distribution of compression forces in the joints of the human foot. Anatomical Record. 96:313–325, 1946.

Marquet, P.: Mechanics and osteoarthritis of the patellofemoral joint. Clinical Orthopaedics and Related Research. 144:70–73, 1979.

Martens, M., van Audekercke, R., de Meester, P., and Mulier, J. C.: The mechanical characteristics of the long bones of the lower extremity in torsional loading. Journal of Biomechanics. 13:667–676, 1980.

Massada, J. L.: Ankle overuse injuries in soccer players. The Journal of Sports Medicine and Physical Fitness. 31:447–451, 1991.

Mathews, H. S., Sonstegard, D. A., and Henke, J. A.: Load-bearing characteristics of the patellofemoral joint. Acta Orthopaedica. 48:511–516, 1977.

McBride, I. D., and Reid, J. G.: Biomechanical considerations of the menisci of the knee. Canadian Journal of Sports Science. 13:175–187, 1988.

McLeish, R. D., and Charnley, J.: Abduction forces in the one-legged stance. Journal of Biomechanics. 3:191–209, 1970.

Mero, A., and Komi, P. V.: Force-, EMG-, and elasticity-velocity relationships at submaximal, maximal, and supramaximal running speeds in sprinters. European Journal of Applied Physiology. 55:553–561, 1986.

Messier, S., and Pittala, K.: Etiologic factors associated with selected running injuries. Medicine and Science in Sports and Exercise. 20:501–505, 1988.

Neumann, D. A., Soderberg, G. L., and Cook, T. M.: Comparison of maximum isometric hip abductor muscle torques between hip sides. Physical Therapy. 68:496–502, 1988.

Noyes, F. R., et al.: The diagnosis of knee motion limits, subluxations, and ligament injury. American Journal of Sports Medicine. 19:163–171, 1991.

Noyes, F. R., Grood, E. S., Cummings, J. F., and Wroble, R. R.: An analysis of the pivot shift phenomenon. American Journal of Sports Medicine. 19:148–155, 1991.

Nuber, G. W.: Biomechanics of the foot and ankle during gait. Clinics in Sports Medicine. 7:1–13, 1988.

Offuerski, C. M., and Macnab, I.: Hip-spine syndrome. Spine. 8:316–321, 1983.

Reilly, D. T., and Martens, M.: Experimental analysis of the quadriceps muscle force and patellofemoral joint reaction force for various activities. Acta Orthopedica Scandinavica. 43:126–137, 1972.

Roy, S.: How I manage plantar fasciitis. The Physician and Sportsmedicine. 11:127–131, 1983.

Sanner, W. H., et al.: A study of ankle joint height changes with subtalar joint motion. Journal of the American Podiatry Association. 71:158–161, 1981.

Seto, J. L., et al.: Assessment of quadriceps/hamstring strength, knee ligament stability, functional and sports activity levels 5 years after anterior cruciate ligament reconstruction. The American Journal of Sports Medicine. *16*:170–178, 1988.

Takai, S., et al.: Rotational alignment of the lower limb in osteoarthritis of the knee. International Orthopaedics (SICOT). *9*:209–216, 1985.

Thorstensson, A.: How is the normal locomotor program modified to produce backward walking? Experimental Brain Research. *61*:664–668, 1986.

Vagenas, G., and Hoshizaki, B.: Evaluation of rearfoot asymmetries in running with worn and new running shoes. International Journal of Sport Biomechanics. *4*:220–230, 1988.

Van Eijden, T. M. G., Kouwenhoven, E., and Weijs, W. A.: A mathematical model of the patellofemoral joint. Journal of Biomechanics. *19*:219–229, 1986.

Vaz, M. D., Kramer, J. F., Roraabeck, C. H., and Bourne, R. B.: Isometric hip abductor strength following total hip replacement and its relationship to functional assessments. Journal of Sports Physical Therapy. *18*:526–531, 1993.

Vleeming, A., Volkers, A. C., Snidjers, C. J., and Stoeckart, R.: Relation between form and function in the sacroiliac joint; Part II, Biomechanical aspects. Spine. *15*:133–135, 1990.

Vrahas, M. S., Brand, R. A., Brown, T. D., and Andrews, J. G.: Contribution of passive tissues to the intersegmental moments at the hip. Journal of Biomechanics. *23*:357–362, 1990.

Wang, C., and Walker, P. S.: Rotatory laxity of the human knee joint. The Journal of Bone and Joint Surgery. *56-A*:161–170, 1974.

Warren, B.: Anatomical factors associated with predicting plantar fasciitis in long-distance runners. Medicine and Science in Sports and Exercise. *16*:60–63, 1984.

Winter, D. A., Patla, A. E., Frank, J. S., and Walt, S. E.: Biomechanical walking pattern changes in the fit and healthy elderly. Physical Therapy. *70*:340–347, 1990.

Yoshioka, Y., Siu, D., and Cooke, T. D. V.: The anatomy and functional axes of the femur. The Journal of Bone and Joint Surgery. *69-A*:873–879, 1987.

References

1. Porterfield, J.A.: The sacroiliac joint. *In* Orthopedic and Sports Physical Therapy. Edited by J.A. Gould and G.J. Davies. St. Louis, C.V. Mosby Co., 1985, pp. 550–579.
2. Hole, J.W.: Human Anatomy and Physiology. 5th Ed. Dubuque, Iowa, William C. Brown Pub. Co, 1990.
3. Vleeming, A., Stoeckart, R., Volkers, A.C., and Snidjers, C. J.: Relation between form and function in the sacroiliac joint; Part I, clinical anatomical aspects. Spine. *15*:130–132, 1990.
4. Grieve, G.P.: The sacroiliac joint. Journal of Anatomy. *58*:384–399, 1976.
5. Nordin, M., and Frankel, V.H.: Basic Biomechanics of the Musculoskeletal System. Philadelphia, Lea & Febiger, 1989.
6. Radin, E.L.: Biomechanics of the human hip. Clinical Orthopaedics. *152*:28–34, 1980.
7. Kempson, G.E., Spivey, C.J., Swanson, S.A.V., and Freeman, M.A.R.: Patterns of cartilage stiffness on the normal and degenerative human femoral head. Journal of Biomechanics. *4*:597–609, 1971.
8. Soderberg, G.L.: Kinesiology: Application to Pathological Motion. Baltimore, Williams & Wilkins, 1986.
9. Saudek, C.E.: The hip. *In* Orthopaedic and Sports Physical Therapy. Edited by J. Gould and G.J. Davies. St. Louis, C.V. Mosby Co., 1985, pp. 365–407.
10. Kapandji, I.A.: The Physiology of the Joints. Vol. 2. Edinburgh, Churchill Livingstone, 1970.
11. Godges, J.J., MacRae, H., Longdon, C.T., and MacRae, P.: The effects of two stretching procedures on hip range of motion and gait economy. The Journal of Orthopaedic and Sports Physical Therapy. *10(9)*:350–357, 1989.
12. Apkarian, J., Naumann, S., and Cairns, B.A.: Three-dimensional kinematic and dynamic model of the lower limb. Journal of Biomechanics. *22*:143–155, 1989.
13. Hodge, W.A., et al.: The influence of hip arthroplasty on stair climbing and rising from a chair. *In* Biomechanics of Normal and Prosthetic Gait. Edited by J.L. Stein. New York, The American Society of Mechanical Engineers, 1987, pp. 65–67.
14. Markhede, G., and Stener, G.: Function after removal of various hip and thigh muscles for extirpation of tumors. Acta Orthopedica Scandinavica. *52*:373–395, 1981.
15. Lloyd-Smith, R., Clement, D.B., McKenzie, D.C., and Taunton, J.E.: A survey of overuse and traumatic hip and pelvic injuries in athletes. The Physician and Sports Medicine. *13(10)*:131–141, 1985.
16. Lovejoy, C.O.: Evolution of human walking. Scientific American. *259(5)*:118–125, 1988.
17. MacKinnon, C.D., and Winter, D.A.: Control of whole body balance in the frontal plane during human walking. Journal of Biomechanics. *26*:633–644, 1993.
18. Cairnes, M.A., Burdett, R.G., Pisciotta, J.C., and Simon, S.R.: A biomechanical analysis of racewalking gait.

Medicine and Science in Sports and Exercise. *18*:446–453, 1986.
19. Brody, D.M.: Running Injuries. Clinical Symposium. *32*:2–36, 1980.
20. Mann, R.A., Moran, G.T., and Dougherty, S.E.: Comparative electromyography of the lower extremity in jogging, running, and sprinting. The American Journal of Sports Medicine. *14*:501–510, 1986.
21. McFadyen, B.J., and Winter, D.A.: An integrated biomechanical analysis of normal stair ascent and descent. Journal of Biomechanics. *21*:733–744, 1988.
22. Erickson, M.O., et al.: Power output and work in different muscle groups during ergometer cycling. European Journal of Applied Physiology. *55*:229–235, 1986.
23. Polisson, R.P.: Sports medicine for the internist. Medical Clinics of North America. *70(2)*:469–474, 1986.
24. Metzmaker, J.N., and Pappas, A.M.: Avulsion fractures of the pelvis. The American Journal of Sports Medicine. *13*:349–358, 1985.
25. Tehranzadeh, J., Kurth, L.A., Elyaderani, M.K., and Bowers, K.D.: Combined pelvic stress fracture and avulsion of the adductor longus in a middle distance runner. The American Journal of Sports Medicine. *10*:108–111, 1982.
26. DonTigny, R.L.: Function and pathomechanics of the sacroiliac joint: A review. Physical Therapy. *65*:35–43, 1985.
27. Reid D.C., Burnham, R.S., Saboe, L.A., and Kushner, S.F.: Lower extremity flexibility patterns in classical ballet dancers and their correlation to lateral hip and knee injuries. The American Journal of Sports Medicine. *15(4)*:347–352, 1987.
28. Segal, P., and Jacob, M.: The Knee. Chicago, Year Book Medical Publishers, Inc., 1973.
29. McLeod, W. D., and Hunter, S.: Biomechanical analysis of the knee. Primary functions as elucidated by anatomy. Physical Therapy. *60*:1561–1564, 1980.
30. Wallace, L. A., Mangine, R. E., and Malone, T.: The knee. *In* Orthopaedic and Sports Physical Therapy. Edited by J. Gould and G. J. Davies. St. Louis, C. V. Mosby, 1985, pp. 342–364.
31. Fukubayashi, T., and Kurosawa, H. The contact area and pressure distribution pattern of the knee. A study of normal and osteoarthritic knee joints. Acta Orthopaedica Scandinavica. *51*:871–879, 1980.
32. Yates, J. W., and Jackson, D. W.: Current status of meniscus surgery. The Physician and Sportsmedicine. *12*:51–56, 1984.
33. Nisell, R.: Mechanics of the knee. A study of joint and muscle load with clinical applications. Acta Orthopaedica Scandinavica. *56*:1–42, 1985.
34. Nissan, M.: Review of some basic assumptions in knee biomechanics. Journal of Biomechanics. *13*:375–381, 1979.

35. Wang, C., Walker, P. S., and Wolf, B.: The effects of flexion and rotation on the length patterns of the ligaments of the knee. Journal of Biomechanics. 6:587–596, 1973.

36. Noyes, F. R., Grood, E. S., Butler, D. L., and Raterman, L.: Knee ligament tests: What do they really mean? Physical Therapy. 60:1578–1581, 1980.

37. McClusky, G., and Blackburn, T. A.: Classification of knee ligament instabilities. Physical Therapy. 60:1575–1577, 1980.

38. Blackburn, T. A., and Craig, E.: Knee anatomy: A brief review. Physical Therapy. 60:1556–1560, 1980.

39. Blackburn, T. A., Eiland, W. G., and Bandy, W. D.: An introduction to the plica. The Journal of Orthopaedic and Sports Physical Therapy. 3:171–177, 1982.

40. Lyon, K. K., et al.: Q-Angle: A factor in peak torque occurrence in isokinetic knee extension. The Journal of Orthopaedic and Sports Physical Therapy. 9:250–253, 1988.

41. Radakovich, M., and Malone, T.: The superior tibiofibular joint: The forgotten joint. The Journal of Orthopaedic and Sports Physical Therapy. 3:129–132, 1982.

42. Kettlecamp, D. H., et al.: An electrogoniometric study of knee motion in normal gait. The Journal of Bone and Joint Surgery. 52-A:775–790, 1970.

43. Rubin, G.: Tibial rotation. Bulletin of Prosthetics Research. 10(15):95–101, 1971.

44. Lafortune, M. A., and Cavanagh, P. R.: Three-dimensional kinematics of the patella during walking. In Biomechanics X. Edited by B. Jonsson. Champaign, Ill., Human Kinetics, 1985.

45. Shaw, J. A., Eng. E., and Murray, D. G.: The longitudinal axis of the knee and the role of the cruciate ligaments in controlling transverse rotation. The Journal of Bone and Joint Surgery. 56-A:1603–1609, 1973.

46. Buchbinder, M. R., Napora, N. J., and Biggs, E. W.: The relationship of abnormal pronation to chondromalacia of the patella in distance runners. Podiatric Sports Medicine. 69:159–162, 1979.

47. Taunton, J. E., et al.: A triplanar electrogoniometer investigation of running mechanics in runners with compensatory overpronation. Canadian Journal of Applied Sports Science. 10:104–115, 1985.

48. Lafortune, M. A., Cavanagh, P. R., Sommer, H. J., and Kalenak, A.: Three-dimensional kinematics of the human knee during walking. Journal of Biomechanics. 25:347–357, 1992.

49. Gehlsen, G. M ., Stewart, L. B., van Nelson, C., and Bratz, J. S.: Knee kinematics: The effects of running on cambers. Medicine and Science in Sports and Exercise. 21:463–466, 1989.

50. Laubenthal, K. N., Smidt, G. L., and Kettlekamp, D. B.: A quantitative analysis of knee motion during activities of daily living. Physical Therapy. 52:34–42, 1972.

51. Brinkmann, J. R., and Perry, J.: Rate and range of knee motion during ambulation in healthy and arthritic subjects. Physical Therapy. 65:1055–1060, 1985.

52. Leib, F. J., and Perry, J.: Quadriceps function: An electromyographic study under isometric conditions. The Journal of Bone and Joint Surgery. 53-A:749–758, 1971.

53. Noyes, F. R., and Sonstegard, D. A.: Biomechanical function of the pes anserinus at the knee and the effect of its transplantation. The Journal of Bone and Joint Surgery: 35-A:1225–1240, 1973.

54. Oshimo, T. A., Greene, T. A., Jensen, G. M., and Lopopolo, R. B.: The effect of varied hip angles on the generation of internal tibial rotary torque. Medicine and Science in Sports and Exercise. 15:529–534, 1983.

55. Osternig, L. R., Bates, B. T., James, S. L., and Jones, C. T.: Knee rotary torque patterns in healthy subjects. In Science in Sports. Edited by J. Terauds. Del Mar, Academic Publishers, 1979, pp. 37–43.

56. Osternig, L. R., Bates, B. T., Tseng, Y., and James, S. L.: Relationships between tibial rotary torque and knee flexion/extension after tendon transplant surgery. Archives of Physical and Medical Rehabilitation. 62:381–385, 1981.

57. Murray, S. M., et al.: Torque-velocity relationships of the knee extensor and flexor muscles in individuals sustaining injuries of the anterior cruciate ligament. The American Journal of Sports Medicine. 12:436–439, 1984.

58. Yack, H. J., Collins, C. E., and Whieldon, T. J.: Comparison of closed and open kinetic chain exercise in the anterior cruciate ligament-deficient knee. The American Journal of Sports Medicine. 21:49–53, 1993.

59. Chesworth, B. M., Culham, E. G., Tata, G. E., and Peat, M.: Validation of outcome measures in patients with patellofemoral syndrome. Journal of Sports Physical Therapy. 10(8):302–308, 1989.

60. Jorge, M., and Hull, M. L.: Analysis of EMG measurements during bicycle pedalling. The Journal of Biomechanics. 19:683–694, 1986.

61. Paulos, L., Noyes, F. R., and Malek, M.: A practical guide to the initial evaluation and treatment of knee ligament injuries. The Journal of Trauma. 20:498–506, 1980.

62. Noyes, F. R., Bassett, R. W., Grood, E. S., and Butler, D. L.: Arthroscopy in acute traumatic hemarthrosis of the knee. Incidence of anterior cruciate tears and other injuries. The Journal of Bone and Joint Surgery. 62-A:687–695, 1980.

63. Lubell, A.: Artificial ligaments: Promise or panacea? The Physician and Sportsmedicine. 15: 150–154, 1987.

64. McDaniel, W. J., and Dameron, T. B.: Untreated ruptures of the anterior cruciate ligament. A follow-up study. The Journal of Bone and Joint Surgery. 62-A:696–704, 1980.

65. Davies, G. J., Wallace, L. A., and Malone, T.: Mechanism of selected knee injuries. Physical Therapy. 60:1590–1595, 1980.

66. Slocum, D. B., and Larson, R. L.: Pes anserinus transplantation. A surgical procedure for control of rotatory instability of the knee. The Journal of Bone and Joint Surgery. *50-A*:226–242, 1963.

67. Larson, R. L.: Epiphyseal injuries in the adolescent athlete. Orthopedic Clinics of North America. *4*:839–851, 1973.

68. Grana, W. A., and Coniglione, T. C.: Knee disorders in runners. The Physician and Sportsmedicine. *13*:127–133, 1985.

69. Davies, G. J., Malone, T., and Bassett, F. H. III: Knee examination. Physical Therapy. *60*:1565–1574, 1980.

70. Mital, M. A., Matza, R. A., and Cohen, J.: The so-called unresolved Osgood-Schlatter lesion: A concept based on fifteen surgically treated lesions. The Journal of Bone and Joint Surgery. *62-A*:732–739, 1980.

71. McPoil, T., and Knecht, H.: Biomechanics of the foot in walking: A functional approach. Journal of Orthopaedic and Sports Physical Therapy. *7*:69–72, 1987.

72. Hamilton, J. J., and Ziemer, L. K.: Functional anatomy of the human ankle and foot. *In* Proceedings of the AAOS Symposium on the Foot and Ankle. Edited by R. H. Kiene and K. A. Johnson. St. Louis, C. V. Mosby, 1981, pp. 1–14.

73. Stormont, D. M., Morrey, B. F., An, K., and Cass, J. R.: Stability of the loaded ankle. Relation between articular restraint and primary and secondary static restraints. The American Journal of Sports Medicine. *13*:295–300, 1985.

74. Czerniecki, J. M.: Foot and ankle biomechanics in walking and running. American Journal of Physical Medicine and Rehabilitation. *67*:246–252, 1988.

75. Brown, L. P., and Yavarsky, P.: Locomotor biomechanics and pathomechanics: A review. The Journal of Orthopaedic and Sports Physical Therapy. *9*:3–10, 1987.

76. Locke, M., Perry, J., Campbell, J., and Thomas, L.: Ankle and subtalar motion during gait in arthritic patients. Physical Therapy. *64*:504–509, 1984.

77. Cerny, K., Perry, J., and Walker, J. M.: Effect of an unrestricted knee-ankle-foot orthosis on the stance phase gait in healthy persons. Orthopedics. *13*:1121–1127, 1990.

78. Murray, M. P., Drought, A. B., and Kory, R. C.: Walking patterns of normal men. Journal of Bone and Joint Surgery. *46A*(2):335–360, 1964.

79. Perry, J.: Gait Analysis: Normal and Pathological Function. Thorofare, N.J., SLACK Inc, 1992.

80. Wright, D., DeSai, S., and Henderson, W.: Action of the subtalar and ankle-joint complex during the stance phase of walking. Journal of Bone and JointSurgery. *46-A*:361–383, 1964.

81. Inman, V. T.: The influence of the foot-ankle complex on the proximal skeletal structures. Artificial Limbs. *13*:59–65, 1959.

82. Scott, S. H., and Winter, D. A.: Talocrural and talocalcaneal joint kinematics and kinetics during the stance phase of walking. Journal of Biomechanics. *24*:734–752, 1991.

83. McPoil, T., and Brocato, R.S.: The foot and ankle: Biomechanical evaluation and treatment. *In* Orthopaedic and Sports Physical Therapy. Edited by J. A. Gould and G. J. Davies. St. Louis, C. V. Mosby Co, 1985, pp. 313–341.

84. DiStefano, V.: Anatomy and biomechanics of the ankle and foot. Athletic Training. *16*:43–47, 1981.

85. Donatelli, R.: Abnormal biomechanics of the foot and ankle. Journal of Orthopaedic and Sports Physical Therapy. *9*:11–15, 1987.

86. Halbach, J.: Pronated foot disorders. Athletic Training. *16*:53–55, 1981.

87. Rodgers, M.: Dynamic biomechanics of the normal foot and ankle during walking and running. Physical Therapy. *68*:1822–1830, 1988.

88. Bates, B.: Foot function in running: Researcher to coach. *In* Biomechanics in Sports. Edited by J. Terauds. Del Mar, CA, Academic Publishers, 1983, pp. 293–303.

89. Areblad, M., et al.: Three-dimensional measurement of rearfoot motion during running. Journal of Biomechanics. *23*:933–940, 1990.

90. Clark, T. E., Frederick, E. C., and Hamill, C. L.: The effects of shoe design parameters on rearfoot control in running. Medicine and Science in Sports and Exercise. *5*:376–381, 1983.

91. James, S. L., Bates, B. T., and Osternig, L. R.: Injuries to runners. American Journal of Sports Medicine. *6*:40–50, 1978.

92. Scott, S. H., and Winter, D. A.: Biomechanical model of the human foot: Kinematics and kinetics during the stance phase of walking. Journal of Biomechanics. *26*:1091–1104, 1993.

93. Hunt, G. C.: Examination of lower extremity dysfunction. *In* Orthopaedic and Sports Physical Therapy. Edited by J. Gould and G. J. Davies. St. Louis, C. V. Mosby Co., 1985, pp. 408–436.

94. Adelaar, R.: The practical biomechanics of running. American Journal of Sports Medicine. *14*:497–500, 1986.

95. Fiore, R. D., and Leard, J. S.: A functional approach in the rehabilitation of the ankle and rearfoot. Athletic Training. *15*:231–235, 1980.

96. Salathe, E. P. Jr., Arangio, G. A., and Salathe, E. P.: The foot as a shock absorber. Journal of Biomechanics. *23*:655–659, 1990.

97. Engsberg, J. R., and Andrews, J. G.: Kinematic analysis of the talocalcaneal/talocrural joint during running support. Medicine and Science in Sports and Exercise. *19*:275–284, 1987.

98. Burdett, R. G.: Forces predicted at the ankle during running. Medicine and Science in Sports and Exercise. *14*:308–316, 1982.

99. Robbins, S. E., and Hanna, A. M.: Running-related injury prevention through barefoot adaptations. Medicine and Science in Sports and Exercise. *19*:148–156, 1987.

100. Freedman, W., Wannstedt, B. S., and Herman, R.: EMG patterns and forces developed during step-down. American Journal of Physical Medicine. *55*:275–290, 1976.

101. Hutson, M. A., and Jackson, J. P.: Injuries to the lateral ligament of the ankle: Assessment and treatment. British Journal of Sports Medicine. *4*:245–249, 1982.

102. Drez, D. Jr., et al.: Nonoperative treatment of double lateral ligament tears of the ankle. The American Journal of Sports Medicine. *10*:197–200, 1982.

103. Bazzoli, A., and Pollina, F.: Heel pain in recreational runners. Physician and Sportsmedicine. *17*:55–56, 1989.

104. Kosmahl, E., and Kosmahl, H.: Painful plantar heel, plantar fascitis, and calcaneal spur: Etiology and treatment. Journal of Orthopaedic and Sports Physical Therapy. *9*:17–24, 1987.

Glossary

Abduction:	Sideways movement of the segment away from the midline or sagittal plane.
Abductor Digiti Minimi:	Muscle inserting on the lateral calcaneus and the base of the proximal fifth phalanx; abducts the little toe.
Abductor Hallucis:	Muscle inserting on the medial calcaneus and the proximal phalanx of the first toe; abducts the big toe.
Acetabular Labrum:	Rim of fibrocartilage which encircles the acetabulum, deepening the socket.
Acetabulum:	The concave, cup-shaped cavity on the lateral, inferior, anterior surface of the pelvis.
Adduction:	Sideways movement of a segment toward the midline or sagittal plane.
Adductor Brevis:	Muscle inserting on the inferior rami and the upper, posterior femur; adducts the thigh.
Adductor Hallucis:	Muscle inserting on the second to fourth metatarsals and the proximal phalanx of the big toe; adducts the big toe.
Adductor Longus:	Muscle inserting on the inferior rami and the middle, posterior femur; adducts and internally rotates the thigh.
Adductor Magnus:	Muscle inserting on the anterior pubis, ischial tuberosity, linea aspera, and adductor tubercle; adducts and internally rotates the thigh.
Angle of Anteversion:	Angle of the femoral neck in the transverse plane; anterior inclination of the femoral neck.
Angle of Retroversion:	Reversal of the angle of anteversion in which the femoral neck is angled posteriorly in the transverse plane.
Angle of Inclination:	Angle formed by the neck of the femur in the frontal plane.
Anterior Compartment Syndrome:	Nerve and vascular compression due to hypertrophy of the anterior tibial muscles in a small muscular compartment.
Anterior Cruciate Ligament:	Ligament inserting on the anterior intercondylar area and the medial surface of the lateral condyle; prevents anterior displacement of the tibia and restrains knee extension, flexion, and internal rotation.
Anterior Talofibular Ligament:	Ligament inserting on the lateral malleolus and the neck of the talus; limits anterior movement of the foot and talar tilt, and restrains plantar flexion and inversion of the foot.
Anterior Talotibial Ligament:	Ligament inserting on the anterior margin of the tibia and the front margin of the talus; limits plantar flexion and abduction of the foot.
Anterior Tilt:	Pelvic movement occurring when the superior portion of the ilium moves anteriorly.
Apophysitis:	Inflammation of the apophysis, or bony outgrowth.

Arcuate Ligament:	Ligament inserting on the lateral condyle of the femur and the head of the fibula; reinforces the posterior capsule of the knee.
Biceps Femoris:	Muscle inserting on the ischial tuberosity, the lateral condyle of the tibia, and the head of the fibula; extends the thigh, flexes the leg, and externally rotates the leg.
Bursitis:	Inflammation of the bursae.
Calcaneal Apophysitis:	Inflammation at the epiphysis on the calcaneus.
Calcaneocuboid Joint:	The articulation between the calcaneus and the cuboid bones; part of the mid-tarsal joint.
Calcaneocuboid Ligament:	Ligament inserting on the calcaneus and cuboid; limits inversion of the foot.
Calcaneofibular Ligament:	Ligament inserting on the lateral malleolus and outer calcaneus; limits backward movement of the foot and restrains inversion.
Calcaneonavicular Ligament:	Ligament inserting on the calcaneus and the navicular; supports the arch and limits abduction of the foot.
Chondromalacia Patella:	Cartilage destruction on the underside of the patella; soft and fibrillated cartilage.
Condyle:	A rounded projection on a bone.
Congenital Hip Dislocation:	A condition existing at birth in which the hip joint experiences subluxation or dislocation for no reason.
Coronary Ligament:	Ligament inserting on the meniscus and the tibia; holds the menisci to the tibia.
Coxa Plana:	Degeneration and recalcification (osteochondritis) of the capitular epiphysis (head) of the femur; also called Legg-Calves-Perthes disease.
Coxa Valga:	An increase in the angle of inclination of the femoral neck (greater than 125 degrees).
Coxa Vara:	A decrease in the angle of inclination of the femoral neck (less than 125 degrees).
Deltoid Ligament:	Ligament inserting on the medial malleolus, talus, navicular, and calcaneus; resists valgus forces, and restrains the movements of plantar flexion, dorsiflexion, eversion, and abduction of the foot.
Distal Femoral Epiphysitis:	Inflammation of the epiphysis at the site of the attachment of the collateral ligaments at the knee.
Dorsal Interossei:	Muscles inserting on the sides of the metatarsals and the lateral side of the proximal phalanx; abducts toes 2 to 4, adducts the second toe, and flexes the proximal phalanx.
Dorsiflexion:	Movement of the foot up in the sagittal plane; movement toward the leg.
Epicondyle:	Eminence on a bone above the condyle.
Equinus:	A limitation in dorsiflexion caused by a short achilles tendon or tight gastrocnemius and soleus muscles.

Eversion:	Movement in the foot where the lateral border of the foot is lifted.
Extension:	Movement of a segment away from an adjacent segment so that the relative angle is increased between the two segments.
Extensor Digitorum Brevis:	Muscle inserting on the lateral calcaneus and the proximal phalanx of toes 1–3; extends toes 1–4.
Extensor Digitorum Longus:	Muscle inserting on the lateral condyle of the tibia, fibula, interosseus membrane, and the dorsal expansion of toes 2–5; dorsiflexes and everts the foot, and extends toes 2–5.
Extensor Hallucis Longus:	Muscle inserting on the anterior fibula, interosseus membrane, phalanx of the big toe; extends the big toe and adducts the foot.
External Rotation:	Movement of the anterior surface of a segment away from the midline; also termed lateral rotation.
Facet:	A small plane surface on a bone where it articulates with another structure.
Flexion:	Movement of a segment toward an adjacent segment so that the relative angle between the two is decreased.
Flexor Digiti Minimi Brevis:	Muscle inserting on the fifth metatarsal and the proximal phalanx of the little toe; flexes the little toe.
Flexor Digitorum Brevis:	Muscle inserting on the medial calcaneus and the middle phalanx of toes 2–5; flexes toes 2–5.
Flexor Digitorum Longus:	Muscle inserting on the posterior tibia and the distal phalanx of toes 2–5; flexes toes 2–5 and plantar flexes the foot.
Flexor Hallucis Brevis:	Muscle inserting on the cuboid and the medial side of the proximal phalanx of the big toe; flexes the big toe and adducts the foot.
Forefoot:	Region of the foot that includes the metatarsals and the phalanges.
Forefoot Valgus:	Eversion of the forefoot on the rearfoot, with the subtalar joint in neutral position.
Forefoot Varus:	Inversion of the forefoot on the rearfoot, with subtalar in the neutral position.
Gastrocnemius:	Muscle inserting on the medial and lateral condyles of the femur and the calcaneus; flexes the leg and plantar flexes the foot.
Gemellus Inferior:	Muscle inserting on the ischial tuberosity and the greater trochanter; externally rotates the thigh.
Gemellus Superior:	Muscle inserting on the ischial spine and the greater trochanter; externally rotates the thigh.
Genu Valgum:	A condition at the knee in which the knees are abnormally close together with the space between the ankles increased; knock knees.
Genu Varum:	A condition at the knee in which the knees are abnormally far apart with the space between the ankles decreased; bowlegs.
Gluteus Maximus:	Muscle inserting on the posterior ilium, sacrum, coccyx, gluteal tuberosity, and iliotibial band; extends and externally rotates the thigh.

Gluteus Medius:	Muscle inserting on the anterior, lateral ilium and the greater trochanter; abducts and internally rotates the thigh.
Gluteus Minimus:	Muscle inserting on the outer, lower ilium and the greater trochanter; abducts and internally rotates the thigh.
Gracilis:	Muscle inserting on the inferior rami and the medial tibia; adducts and internally rotates the thigh, and flexes and internally rotates the leg.
Hamstrings:	A group of muscles on the posterior thigh consisting of the semimembranosus, semitendinosus, and the biceps femoris.
Head of Femur:	The proximal end of the femur consisting of a large spherical structure.
Hindfoot:	Region of foot which includes the talus and calcaneus; also called rear foot.
Hyperextension:	Continuation of the movement of extension past the neutral position.
Iliac Apophysitis:	Inflammation of the attachment sites of the gluteus medius and tensor fascia latae on the iliac crest.
Iliacus:	Muscle inserting on the inner surface of the ilium and sacrum and the lesser trochanter; flexes the thigh.
Iliofemoral Ligament:	Ligament inserting on the anterior, superior spine of the ilium and the intertrochanteric line of the femur; supports the anterior hip joint and offers restraint in the movements of extension, and internal and external rotation.
Iliopectineal Bursa:	A fibrous, fluid-filled sac between the tendon of the iliopsoas and the iliopectineal eminence.
Iliopsoas:	The name given to the combination of the iliacus and the psoas muscle.
Iliotibial Band:	A fibrous band of fascia running from the ilium to the lateral condyle of the tibia.
Iliotibial Band Syndrome:	Inflammation of the iliotibial band caused by thigh adduction and internal rotation.
Infrapatellar Bursa:	A bursa between the patellar ligament and the tibia.
Intercondylar Eminence:	Ridge of bone on the tibial plateau that separates the surface into medial and lateral compartments.
Intercondylar Notch:	Convex surface on the distal, posterior surface of the femur.
Internal Rotation:	Movement of the anterior surface of a segment toward the midline; also termed medial rotation.
Interosseus Ligaments:	Ligaments connecting adjacent tarsals; supports the arch and the intertarsal joints.
Interphalangeal Joints:	Articulation between adjacent phalanges of the fingers and the toes.
Intertarsal Joint:	Articulation between adjacent tarsal bones.
Inversion:	Movement in the foot in which the medial border of the foot is lifted.
Ischiofemoral Ligament:	Ligament inserting on the posterior acetabulum and the iliofemoral ligament; restrains the movements of adduction and internal rotation of the thigh.

Ischiogluteal Bursa:	A fibrous, fluid-filled sac between the gluteus maximus and the ischial tuberosity.
Lateral Collateral Ligament:	Ligament inserting on the lateral epicondyle of the femur and the head of the fibula; resists varus forces and is taut in extension.
Lateral Tibial Syndrome:	Pain on the lateral anterior leg caused by tendinitis of the tibialis anterior or irritation to the interosseus membrane.
Legg-Calve-Perthes Disease:	Degeneration and recalcification (osteochondritis) of the capitular epiphysis (head) of the femur; also called coxa plana.
Longitudinal Arch:	Two arches (medial and lateral) formed by the tarsals and metatarsals, which run the length of the foot, and participate in both shock absorption and support while the foot is weight bearing.
Lumbricales:	Muscles inserting on the tendon of the flexor digitorum longus, the base of the proximal phalanx of toes 2–5; flexes toes 2–5.
Medial Collateral Ligament:	Ligament inserting on the medial epicondyle of the femur, the medial condyle of the tibia, and the medial meniscus; resists valgus forces and restrains the knee joint in internal and external rotation; taut in extension.
Medial Tibial Syndrome:	Pain above the medial malleolus caused by tendinitis of the tibialis posterior or irritation of the interosseus membrane or periosteum; previously called shin splints.
Meniscus:	A crescent-shaped fibrocartilage that lies on the articular surface of the knee joint.
Metatarsalgia:	Strain of the ligaments supporting the metatarsals.
Metatarsophalangeal Joints:	Articulations between the metatarsals and the phalanges in the foot.
Midfoot:	Region of the foot which includes all of the tarsals, except for the talus and the calcaneus.
Midtarsal Joint:	The name given the combination of two articulations: the calcaneocuboid and the talonavicular joints; also called the transverse tarsal joint.
Morton's Toe:	A condition where the second metatarsal is longer than the first metatarsal.
Neck of Femur:	Column of bone connecting the head of the femur to the shaft.
Obturator Externus:	Muscle inserting on the pubis, ischium, and upper posterior femur; externally rotates the thigh.
Obturator Internus:	Muscle inserting on the sciatic notch and the greater trochanter; externally rotates the thigh.
Osgood-Schlatter Disease:	Irritation of the epiphysis at the tibial tuberosity, caused by overuse of the quadriceps femoris muscle group.
Osteoarthritis:	Degenerative joint disease characterized by a breakdown in the cartilage and the underlying subchondral bone, narrowing of the joint space, and osteophyte formation.

Osteochondral Fracture:	Fracture at the bone and cartilage junction.
Osteochondritis Dissecans:	Inflammation of bone and cartilage, resulting in splitting of pieces of cartilage into the joint.
Patella:	Triangular-shaped sesamoid bone located on the anterior knee joint; encased by the tendons of the quadriceps femoris muscle group.
Patella Alta:	Longer patella tendon.
Patella Baja:	Shorter patellar tendon.
Patellar Groove:	The convex surface on the distal, anterior surface of the femur, which accommodates the patella; also called the trochlear groove.
Patellar Ligament:	Ligament inserting on the inferior patella and the tibial tuberosity; transfers the quadriceps femoris muscle force to the tibia.
Patellofemoral Joint:	Articulation between the posterior surface of the patella and the patellar groove on the femur.
Patellofemoral Pain Syndrome:	Pain around the patella.
Pectineus:	Muscle inserting on the pectineal line on the pubis and an area below the lesser trochanter; adducts and flexes the thigh.
Pelvic Girdle:	A complete ring of bones composed of two coxal bones, anteriorly and laterally, and the sacrum and coccyx posteriorly.
Periostitis:	Inflammation of the periosteum, marked by tenderness and swelling on the bone.
Peroneus Brevis:	Muscle inserting on the lower, lateral fibula, and the fifth metatarsal; everts and plantar flexes the foot.
Peroneus Longus:	Muscle inserting on the lateral condyle of the tibia, upper fibula, first cuneiform, and lateral first metatarsal; everts, plantar flexes, and abducts the foot.
Peroneus Tertius:	Muscle inserting on the lower, anterior fibula, the interosseus membrane, the base of the fifth metatarsal; everts the foot.
Pes Anserinus:	The combined insertion of the tendinous expansions from the sartorius, gracilis, and semitendinosus muscles.
Pes Cavus:	High-arched foot.
Pes Planus:	Flat foot.
Piriformis:	Muscle inserting on the anterior, lateral sacrum and the greater trochanter; abducts and externally rotates the thigh.
Plantar Fascia:	Fibrous band of fascia running along the plantar surface of the foot from the calcaneus to the metatarsophalangeal articulation.
Plantar Fasciitis:	Inflammation to the plantar fascia.
Plantar-Flexed First Ray:	Positioning of the first metatarsal below the plane of the adjacent metatarsal heads.
Plantar Flexion:	Movement of the foot downward in the sagittal plane; movement away from the leg.

Plantar Interossei: Muscles originating on the medial side of metatarsals 3–5, medial side of proximal phalanx of toes 3–5; abducts toes 3–5.

Plantaris: Muscle inserting on the linea aspera on the femur and the calcaneus; plantar flexes the foot.

Plica: Ridge or fold in the synovial membrane.

Popliteus: Muscle inserting on the lateral condyle of the femur and the proximal tibia; flexes and internally rotates the leg.

Posterior Cruciate Ligament: Ligament inserting on the posterior spine of the tibia and the inner condyle of the femur; resists posterior movement of the tibia on the femur and restrains flexion and rotation of the knee.

Posterior Oblique Ligament: Ligament inserting on the semimembranosus muscle; supports the posterior, medial capsule of the knee joint.

Posterior Tilt: Pelvic movement designated by posterior movement of the superior portion of the ilium.

Pronation: A tri-planar movement at the subtalar and midtarsal joints that includes the movements of calcaneal eversion, abduction, and dorsiflexion.

Psoas: Muscle inserting on the transverse processes and bodies of the L1–L5, T–12 vertebrae, and the lesser trochanter; flexes the trunk and the thigh.

Pubic Ligament: Ligament inserting on the bodies of the right and left pubic bones; maintains the relationship between right and left pubic bones.

Pubic Symphysis: A cartilaginous joint connecting the pubic bones of the right and left coxal bones of the pelvis.

Pubofemoral Ligament: Ligament inserting on the pubic part of the acetabulum, the superior rami, and the intertrochanteric line; restrains the movements of hip abduction and external rotation.

Q-Angle: The angle formed by the longitudinal axis of the femur and the line of pull of the patellar ligament.

Quadratus Femoris: Muscle inserting on the ischial tuberosity and the greater trochanter; externally rotates the thigh.

Quadratus Plantae: Muscle inserting on the medial, lateral calcaneus and the tendon of the flexor digitorum; flexes toes 2–5.

Quadriceps Femoris: A combination of muscles on the anterior thigh, including vastus lateralis, vastus intermedius, vastus medialis, and rectus femoris.

Rearfoot: Region of foot that includes the talus and calcaneus; also called hindfoot.

Rearfoot Varus: Inversion of the calcaneus, with deviation of the tibia in the same direction.

Rectus Femoris: Muscle inserting on the anterior, inferior iliac spine and the tibial tuberosity via the patellar ligament; flexes the thigh and extends the leg.

Retrocalcaneal Bursitis: Inflammation of the bursa lying between the achilles tendon and the calcaneus.

Sacral Extension: Movement of the top of the sacrum posteriorly.

Sacral Flexion:	Movement of the top of the sacrum anteriorly.
Sacral Rotation:	Rotation of the sacrum about an axis running diagonally through the bone; right rotation occurs as the anterior surface of the sacrum faces right.
Sacroiliac Joint:	A strong, synovial joint between the sacrum and the ilium.
Sacroiliitis:	Inflammation at the sacroiliac joint.
Sacrum:	A triangular bone below the lumbar vertebrae that consists of five fused vertebrae.
Sartorius:	Muscle inserting on the anterior, superior iliac spine and the medial tibia; flexes and externally rotates the thigh, and flexes and internally rotates the leg.
Screw Home Mechanism:	The locking action at the end of the knee extension movement, external rotation of the tibia on the femur caused by incongruent joint surfaces.
Semimembranosus:	Muscle inserting on the ischial tuberosity and the medial condyle of the tibia; extends the thigh, flexes the leg, and internally rotates the leg.
Semitendinosus:	Muscle inserting on the ischial tuberosity and the medial tibia; extends the thigh, flexes the leg, and internally rotates the leg.
Slipped Capital Femoral Epiphysitis:	Displacement of the capital femoral epiphysis of the femur caused by external forces that drive the femoral head back and medial to create a tilt in the growth plate.
Snapping Hip Syndrome:	A clicking sound that accompanies thigh movements and is created by the hip joint capsule or the iliopsoas tendon moving on a bony surface.
Soleus:	Muscle inserting on the upper, posterior tibia, fibula, interosseus membrane, and calcaneus; plantar flexes the foot.
Sprain:	An injury to a ligament surrounding a joint; rupture of fibers of a ligament.
Strain:	Injury to the muscle, tendon, or muscle-tendon junction due to overstretching or excessive tension applied to the muscle; tearing and rupture of the muscle or tendon fibers.
Stress Fracture:	Microfracture of the bones developed through repetitive force application exceeding the structural strength of the bone or the rate of remodeling in the body tissue.
Subtalar Joint:	The articulation of the talus with the calcaneus; also called the talocalcaneal joint.
Supination:	Tri-plane movement at the subtalar and midtarsal joints that includes calcaneal inversion, adduction, and plantar flexion.
Talocalcaneal Ligament:	Ligament inserting on the talus and the calcaneus; supports the subtalar joint.
Talocrural Joint:	The articulation of the tibia and fibula with the talus; the ankle joint.
Talofibular Ligament:	Ligament inserting on the lateral malleolus and the posterior talus; limits plantar flexion and inversion; supports the lateral ankle.
Talonavicular Joint:	Articulation between the talus and the navicular bones; part of the midtarsal joint.

Talonavicular Ligament:	Ligament inserting on the neck of the talus and the navicular; limits inversion and stabilizes the talonavicular joint.
Talotibial Ligament:	Ligament inserting on the tibia and the talus; limits plantar flexion and supports the medial ankle.
Tarsometatarsal Joint:	Articulation between the tarsals and the metatarsals.
Tarsometatarsal Ligaments:	Ligaments inserting on the tarsals and metatarsals; supports the arch and maintains stability between the tarsals and the metatarsals.
Tendinitis:	Inflammation of a tendon.
Tensor Fascia Latae:	Muscle inserting on the anterior, superior iliac spine and the iliotibial tract; flexes, abducts, and internally rotates the thigh.
Tibial Plateau:	A level area on the proximal end of the tibia.
Tibialis Anterior:	Muscle inserting on the upper, lateral tibia, interosseus membrane, and the plantar surface of the medial cuneiform; dorsiflexes and inverts the foot.
Tibialis Posterior:	Muscle inserting on the upper, posterior tibia, fibula, interosseus membrane, and the inferior navicular; inverts and plantar flexes the foot.
Tibiofemoral Joint:	Articulation between the tibia and the femur; the knee joint.
Tibiofibular Joint (inferior):	Articulation between the distal end of the fibula and the distal end of the tibia.
Tibiofibular Joint (superior):	Articulation between the head of the fibula and the posterolateral, inferior aspect of the tibial condyle.
Tibiotalar Joint:	Articulation between the tibia and the talus.
Transverse Arch:	An arch formed by the tarsals and metatarsals which runs across the foot, contributing to shock absorption in weight bearing.
Transverse Ligament:	Ligament inserting on the medial and lateral meniscus; connects the menisci to each other.
Trendelenburg Gait:	Alteration in a walking or running gait caused by inefficiency in the abductors of the thigh, causing a drop in the pelvis to the unsupported side.
Triceps Surae:	Name given to the combination of two muscles: gastrocnemius and the soleus.
Trochanteric Bursa:	A fibrous, fluid-filled sac situated between the gluteus maximus and the greater trochanter.
Valgus:	Segment angle bowed medially; medial force.
Varus:	Segment angle bowed laterally; lateral force.
Vastus Intermedius:	Muscle inserting on the anterior, lateral femur and the tibial tuberosity; extends the leg.
Vastus Lateralis:	Muscle inserting on the intertrochanteric line, the linea aspera, and the tibial tuberosity; extends the leg.
Vastus Medialis:	Muscle inserting on the linea aspera, the trochanteric line, and the tibial tuberosity; extends the leg.

CHAPTER 7

Functional Anatomy of the Trunk

1. Identify the four curves of the spine and discuss the factors contributing to the formation of each curve.
2. Describe the structure and motion characteristics of the cervical, thoracic, and lumbar vertebrae.
3. Describe the movement relationship between the pelvis and the lumbar vertebrae for the full range of trunk movements.
4. Compare the differences in strength for the various trunk movements.
5. Describe specific strength and flexibility exercises for all of the movements of the trunk.
6. Explain how loads are absorbed by the vertebrae, and describe some of the typical loads imposed upon the vertebrae for specific movements or activities.
7. Describe some of the common injuries to the cervical, thoracic, and lumbar vertebrae.
8. Identify the muscular contributions of the trunk to the activities of walking and running.
9. Discuss the causes and source of pain for the low back.
10. Evaluate common trunk exercises for their effectiveness and safety.
11. Discuss the influence of aging on trunk structure and function.

The vertebral column acts as a modified elastic rod, providing a rigid support and flexibility (1). There are 33 vertebrae in the vertebral column, 24 of which are movable and contribute to trunk movements. The vertebrae are arranged into four curves that facilitate support of the column by offering a spring-like response to loading (2). These curves provide balance and strengthen the spine.

Seven cervical vertebrae form a convex curve to the anterior side of the body. This curve develops as an infant begins to lift its head, it supports the head, and assumes its curvature in response to head position. The 12 thoracic vertebrae form a curve that is convex to the posterior side of the body. The curvature in the thoracic spine is present at birth. Five lumbar vertebrae form a curve convex to the anterior side, which develops in response to weight bearing and is influenced by pelvic and lower extremity positioning. The last curve is the sacrococcygeal curve, formed by five fused sacral vertebrae and the four to five fused vertebrae of the coccyx. FIGURE 7–1 presents the curvature of the whole spine as seen from both the side and the rear.

The junction where one curve ends and the next one begins is usually a site of greater mobility, which is also more vulnerable to injury. These junctions are the cervicothoracic, the thoracolumbar, and the lumbosacral regions of the spine. Additionally, if the curves of the spine are exaggerated, the column will be more mobile, and if the curves are flat, the spine will be more rigid. The cervical and lumbar regions of the spinal column are the most mobile, and the thoracic and pelvic regions are more rigid (2).

Besides offering support and flexibility to the trunk, the vertebral column has the main responsibility of protecting the spinal cord. As illustrated in FIGURE 7–2, the cord runs down through the vertebrae in a canal formed by the body, pedicles, and pillars of the vertebrae, the disc, and a ligament, the ligamentum flavum. Peripheral nerves exit through the intervertebral foramen on the lateral side of the vertebrae, forming aggregates of nerve fibers.

The trunk, as the largest segment of the body, plays an integral role in both upper and lower extremity function since its positioning can significantly alter the function of the extremities. Trunk movement or position can be examined as a whole, it can be examined by observing the movements or position of the different regions of the vertebral column, or movement at the individual vertebral

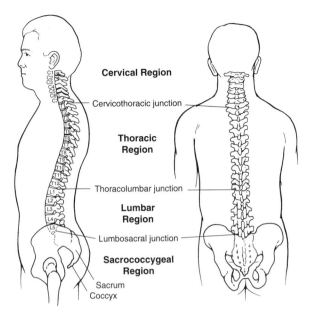

FIGURE 7–1. The vertebral column is both strong and flexible as a result of the four alternating curves. We are born with the thoracic and sacrococcygeal curves. The cervical and lumbar curves form in response to weight bearing or muscular stresses imposed upon them during infancy.

level can be described. In this chapter, both the movement of the trunk as a whole, as well as the movements and function within each region of the spine will be examined. The structural characteristics of the vertebral column will be presented first, followed by an examination of the differences between the three regions of the spine: the cervical, thoracic, and lumbar.

The Vertebral Column

Anatomic and Functional Characteristics of the Joints

The functional unit of the vertebral column, the motion segment, is similar in structure throughout the entire spinal column, except for the first two cervical vertebrae, which have structural differences that are unique. The motion segment consists of two adjacent vertebrae, and a disc that separates them (FIGURE 7–3). The segment can be further broken down into anterior and posterior portions, each playing a different role in vertebral function.

The anterior portion of the motion segment contains the two bodies of the vertebrae, the intervertebral disc, and the anterior and posterior longitudinal ligaments. The two bodies and the disc separating them form a cartilaginous joint that is unique and not found at any other site in the body.

Each vertebral body is cylinder-shaped and thicker on the front side where it absorbs large amounts of compressive forces. It consists of cancellous tissue surrounded by a hard cortical layer and has a raised rim that facilitates attachment of the disc, muscles, and ligaments. Also, the

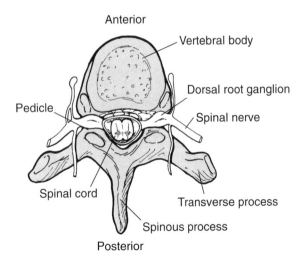

FIGURE 7–2. The vertebral column protects the spinal cord, which runs down the posterior aspect of the column through the vertebral foramen or canal. Spinal nerves exit at each vertebral level.

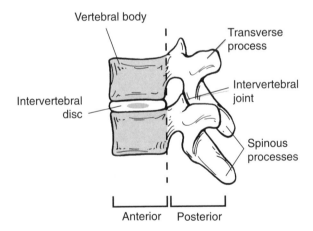

FIGURE 7–3. The vertebral motion segment can be divided into an anterior and posterior portion. The anterior portion of the motion segment contains the vertebral bodies, the intervertebral disc, and ligaments. The posterior portion of the motion segment contains the vertebral foramen, the neural arches, the intervertebral joints, the transverse and spinous processes, and ligaments.

surface of the body is covered with hyaline cartilage, forming articular endplates into which the disc attaches.

Separating the two adjacent bodies is the intervertebral disc, a structure binding the vertebrae together while permitting movement to occur between adjacent vertebrae. The disc is capable of withstanding compressive forces as well as torsional and bending forces applied to the column. The role of the disc is to bear and distribute loads in the vertebral column as well as restrain excessive motion occurring in the vertebral segment. A lateral, superior, and cross-sectional view of the disc is presented in FIGURE 7–4.

Each disc consists of the nucleus pulposus and the annulus fibrosus. The nucleus pulposus is a spherical shaped, gel-like mass located in the central portion of the cervical and thoracic discs, and toward the posterior in the lumbar discs. The nucleus pulposus is 80 to 90% water and 15 to 20% collagen (3), creating a fluid mass that is always under pressure and exerting a preload to the disc. The nucleus pulposus is well suited for withstanding compressive forces applied to the motion segment.

During the day, the water content of the disc is reduced, with compressive forces applied during daily activities, resulting in a shortening of the column. However, at night, the nucleus pulposus imbibes water, restoring height back to the disc. In the elderly, the total water content of the disc is less (approximately 70%) and the ability to imbibe water is reduced, leaving a shorter vertebral column.

The nucleus pulposus is surrounded by rings of fibrous tissue and fibrocartilage, the annulus fibrosus. The fibers of the annulus fibrosus run parallel in concentric layers but, are oriented diagonally at 45 to 65 degrees to the vertebral bodies (4,5). Each alternate layer of fibers runs in a perpendicular direction to the previous layer, creating a criss-cross pattern similar to that seen in a radial tire (6). When a rotation is applied to the disc, half of the fibers will tighten, while the fibers running in the other direction will be loose.

The fibers making up the annulus fibrosus consist of 50 to 60% collagen, providing the tensile strength in the disc (3). Collagen is less abundant in the lateral, posterior portion of the disc, making this area of the disc more vulnerable to injury because of reduced tensile strength. Fibers from the annulus fibrosus attach to the endplates of the adjacent vertebral bodies in the center of the segment, and attach to the actual osseus material at the periphery of the disc (7). Tension is maintained in the annulus fibrosus by the endplates, and by pressure exerted outward from the nucleus pulposus.

The disc is both avascular and aneural, except for some sensory input from the outer layers of the annulus fibrosus. Because of this, healing of a damaged disc is unpredictable and not very promising.

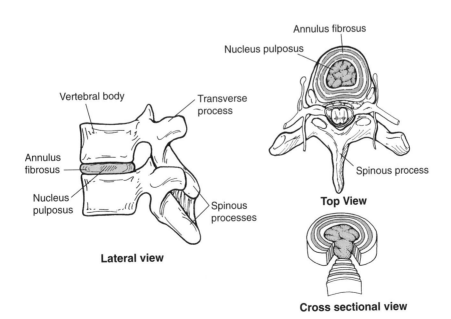

FIGURE 7–4. The intervertebral disc bears and distributes loads imposed upon the vertebral column. The disc consists of a gel-like central portion, the nucleus pulposus, which is surrounded by rings of fibrous tissue, the annulus fibrosus.

The intervertebral disc functions hydrostatically when it is healthy, responding with flexibility under low loads and stiffly when subjected to high loads. When the disc is loaded in compression, the nucleus pulposus uniformly distributes pressure through the disc and acts as a cushion. The disc flattens and widens and the nucleus pulposus bulges laterally as the disc loses fluid. This places tension on the annulus fibers and converts a vertical compression force to a tensile stress in the annulus fibers. The tensile stress absorbed by the annulus fibers is 4 to 5 times the applied axial load (8). Examples of typical loads on the third lumbar vertebrae for a 70 kg individual are presented in Table 7–1. The loads on the disc created by activities such as sitting, coughing, or laughing are larger than those measured in standing, walking or twisting.

The pressure in the disc increases linearly with increased compressive loads, with the pressure 30 to 50% greater than the applied load per unit area (9). The greatest change in disc pressure occurs with compression. During compression, the disc loses fluid and the fiber angle increases (4). The disc is very resilient to the effects of a compressive force and rarely fails under compression. The cancellous bone of the vertebral body will yield and fracture before damage occurs to the disc (4).

In movements such as flexion, extension, and lateral flexion, off-axial compressive loading is developed (10). With this asymmetrical loading, the vertebral body translates toward the loaded side, the fibers are stretched on the other side, and the pressure in the nucleus pulposus returns to the original position.

Table 7–1 Loads on L-3 Disc in 70 Kg Individual

Activity	Load on Disc (N)
Supine	294
Standing	686
Upright Sitting	980
Walking	833
Twisting	882
Bend Sideways	931
Coughing	1078
Jumping	1078
Straining	1176
Laughing	1176
Lifting 20 KG, back straight, knees bent	2058
Lifting 20 KG, back bent, knees straight	3332

(Adapted from Nachemson, A: Lumbar intradiscal pressure. *In* The Lumbar Spine and Back Pain. Edited by M. Jayson. Kent, Pitman Medical Publishing Co Ltd, 1976.)

In flexion, the vertebrae move anteriorly, forcing the nucleus pulposus posteriorly, creating a compression load on the anterior portion of the disc and a tension load on the posterior annulus. In extension the opposite occurs, as the upper vertebrae move posteriorly, driving the nucleus pulposus anteriorly and placing pressure on the anterior fibers of the annulus.

In lateral flexion, there is a tilt of the upper vertebrae to the side of the flexion, generating compression on that side and tension in the opposite side. Refer to FIGURE 7–5 for a graphic illustration of disc behavior in flexion, extension, and lateral flexion.

As the trunk rotates, there is both tension and shear developed in the annulus fibrosus of the disc (FIGURE 7–6). Half of the annulus fibers oriented in the direction of the rotation become taut and the other half of the annulus fibers oriented in the opposite direction slacken. This creates an increase in the intradiscal pressure, narrows the joint space, and creates a shear force in the horizontal plane of rotation, and tension in fibers oriented in the direction of the rotation. The peripheral fibers of the annulus fibrosus are subjected to the greatest stress during rotation (7). However, the disc is most susceptible to injury when the transition from rotation in one direction is made to a rotation in the opposite direction.

The final structures of the anterior portion of the vertebral segment are the longitudinal ligaments running the length of the spine from the base of the occiput to the sacrum. The anterior longitudinal ligament is a very dense, powerful ligament that attaches both to the anterior disc and vertebral bodies of the motion segment. This ligament limits hyperextension of the spine and restrains forward movement of one vertebrae over another. It also maintains a constant load on the vertebral column and supports the anterior portion of the disc in lifting (6).

The posterior longitudinal ligament runs down the posterior surface of the vertebral bodies, inside the spinal canal, and connects to the rim of the vertebral bodies and the center of the disc. The posterolateral aspect of the segment is not covered by this ligament, adding to the vulnerability of this site for disc protrusion. It is broader in the cervical region and narrow in the lumbar region. This ligament offers resistance in flexion of the spine. The location and function of all of the spinal ligaments are presented in Appendix A.

The posterior portion of the vertebral motion segment includes the neural arches, the intervertebral joints, the transverse and spinous processes, and ligaments (FIGURE 7–7). The neural arch is formed by the two pedicles and two laminae, and together, with the posterior side of the vertebral body, they form the vertebral foramen in

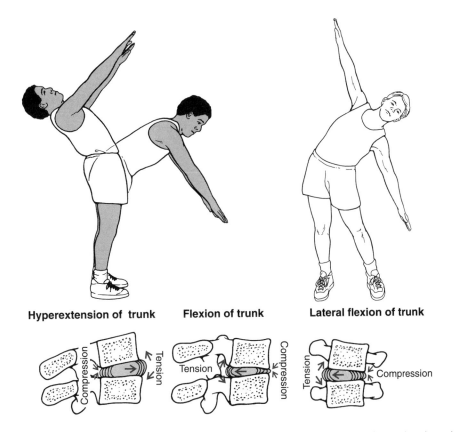

Hyperextension of trunk   **Flexion of trunk**   **Lateral flexion of trunk**

FIGURE 7–5. When the trunk flexes, extends, or laterally flexes, there is a compressive force developed to the side of the bend and a tension force developed on the opposite side.

which the spinal cord travels. The bone in the pedicles and laminae is very hard, providing a good resistance to the large tensile forces that must be accommodated. There are notches above and below each pedicle that form the intervertebral foramen through which the spinal nerves leave the canal.

Projecting sideways at the union of the laminae and the pedicles are the transverse processes, and projecting posteriorly from the junction of the two laminae is the spinous process. Both the spinous and transverse processes serve as attachment sites for the spinal muscles running the length of the column.

There are two synovial joints, termed the apophyseal joints, formed by articulating facets located on the upper and lower border of each laminae. The superior articulating facet is concave and fits into the convex inferior facet of the adjacent vertebrae, forming two joints on each side of the vertebrae. The articulating facets are oriented at different angles in the cervical, thoracic, and lumbar regions of the spine, accounting for most of the functional differences between regions. These differences will be discussed more specifically in a later section of this chapter.

The apophyseal joints are enclosed within a joint capsule and have all of the other characteristics of a typical synovial joint. These joints prevent the forward displacement of one vertebrae over another and also participate in load bearing. In the hyperextended position, these joints bear 30% of the load (1). They also bear a significant portion of the load when the spine is flexed and rotated (11). Highest pressures in the facet joints occur with combined torsion, flexion, and compression of the vertebrae (12).

There are five ligaments supporting the posterior portion of the vertebral segment. These ligaments are presented in Appendix A. The ligamentum flavum connects adjacent vertebral arches longitudinally, attaching laminae to laminae. It is a ligament having elastic qualities, allowing it to deform and return to its original length. It elongates with flexion of the trunk and contracts in extension. In the neutral position, it is under constant tension, imposing a continual tension on the disc.

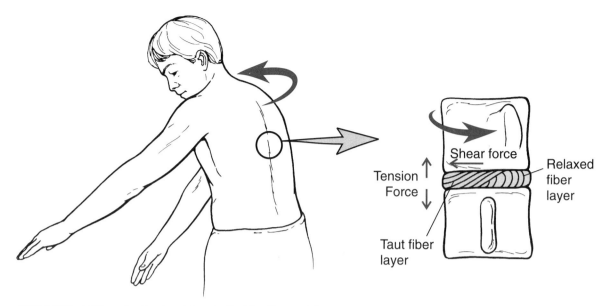

FIGURE 7–6. When the trunk rotates, half of the fibers of the annulus fibrosus becomes taut while the other half relaxes. This creates a tension force in the fibers running in the direction of the rotation and a shear force across the plane of rotation.

The supraspinous and the interspinous ligaments both run from spinous process to spinous process and resist both shear and forward bending of the spine. Finally, the intertransverse ligaments, connecting transverse process to transverse process, resist lateral bending of the trunk.

Motion in the spinal column is very small between each vertebrae, but in combination, the spine is capable of considerable range of motion. Motion is restricted by the discs and the arrangement of the facets, but motion can occur in three planes via active muscular initiation and control (13).

Review the movement characteristics of the total spine presented in FIGURE 7–8. For the total spinal column, flexion and extension occur through approximately 110 to 140 degrees, with free movement in the cervical and lumbar regions and limited flexion and extension in the thoracic region. The axis of rotation for flexion and extension lies in the disc unless there is considerable disc degeneration, which can move the axis of rotation out of the disc. The axis of rotation moves anteriorly with flexion and posteriorly with extension.

Flexion of the whole trunk occurs primarily in the lumbar vertebrae through the first 50 to 60 degrees, and is then moved into more flexion by forward tilt of the pelvis (14). Extension occurs through a reverse movement in which the pelvis first posteriorly tilts, then the lumbar spine extends.

When flexion is first initiated, the top vertebrae slides forward on the bottom vertebrae and the vertebrae tilts, placing a compressive force on the anterior portion of the disc. Both ligaments and the annulus fibers absorb the compressive forces created.

On the back side, the superior portion of the apophyseal joints slide up on the lower facets, creating a compression force between the facets as well as a shear force across the face of the facets. These forces are controlled by the posterior ligaments, the capsules surrounding the apophyseal joints, the posterior muscles, fascia, and the posterior annulus fibers (7). The full flexion position is maintained and supported by the apophyseal capsular ligaments, the intervertebral discs, the supraspinous and interspinous ligaments, the ligamentum flavum, and passive resistance from the back muscles, in that order (15).

Lateral flexion range of motion is around 75 to 85 degrees and occurs mainly in the cervical and lumbar regions (FIGURE 7–8). During lateral flexion, there is a slight movement of the vertebrae sideways, with disc compression to the side of the bend. Lateral flexion is often accompanied by rotation. In a relaxed stance, the accompanying rotation is to the opposite side of lateral flexion, i.e., left rotation accompanying right lateral flexion.

If the vertebrae is in full flexion, the accompanying rotation will occur to the same side, i.e., right rotation accompanying right lateral flexion. However, this can

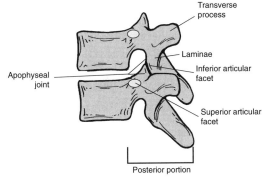

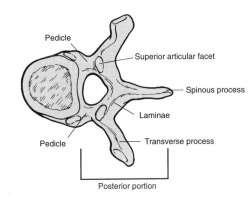

LATERAL VIEW

TOP VIEW

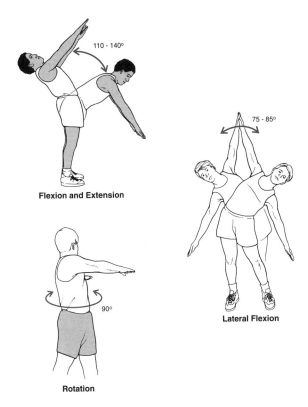

Flexion and Extension

Rotation

Lateral Flexion

FIGURE 7–7. The posterior portion of the spinal motion segment is responsible for a significant amount of spinal support and restriction owing to its ligaments and structure. The posterior portion contains the only synovial joint in the spine, the apophyseal joint, which joins the superior and inferior facets of each vertebrae.

FIGURE 7–8. The range of motion at the individual motion segment level is small, but in combination, the trunk is capable of moving through a significant motion range. Flexion and extension occur through approximately 110 to 140 degrees, primarily in the cervical and lumbar region, with very limited contribution from the thoracic region. The trunk rotates through 90 degrees, with movement occurring freely in the cervical region, and with accompanying lateral flexion in the thoracic and lumbar regions. The trunk laterally flexes through 75 to 85 degrees, with movement limited in the thoracic region.

vary by region of the spine. Also, an inflexible person will usually perform some lateral flexion in order to obtain flexion in the trunk (16).

Rotation occurs through 90 degrees and is free in the cervical region, and occurs in the thoracic and lumbar regions in combination with lateral flexion (FIGURE 7–8). Generally, rotation is limited in the lumbar region. Right rotation in the thoracic or lumbar region will be accompanied by some left lateral flexion in order to perform the action.

The apophyseal joints are in a close-packed position in spinal extension, except for the top two cervical vertebrae, which are in a close-packed position in flexion. The total spine is in a close-packed position and rigid during

the military salute posture with the head up, shoulders back, and the trunk vertically aligned (6).

The flexibility of the regions of the trunk varies, and is determined by the intervertebral discs and the angle of articulation of the facet joints. As pointed out earlier, the mobility is highest at the junction of the regions. Mobility will also increase in a region in response to restriction or rigidity elsewhere in the vertebral column.

The cervical vertebrae have two vertebrae, the atlas (C-1) and the axis (C-2) that have structures unlike any other vertebrae. The atlas has no vertebral body and is shaped like a ring with an anterior and a posterior arch. The atlas has large transverse processes with transverse foramen through which blood supply travels. The atlas has no spinous process. Superiorly, it has a fovea, or dish-like depression for receipt of the occiput of the skull.

The articulation of the atlas with the skull is called the atlanto-occipital joint. It is at this joint that the head nods on the spine, for it allows free sagittal plane movements. This joint allows approximately 10 to 15 degrees of flexion and extension, 8 degrees of lateral flexion, and no rotation (17).

The axis has a modified body with no articulating process on the superior aspect of the body and pedicles. Instead, the articulation with the atlas occurs via a pillar projecting from the superior surface of the axis that fits into the atlas and locks the atlas into a swivel or pivoting joint. The pillar is referred to as the odontoid process or dens.

The articulation between the atlas and the axis is known as the atlantoaxial joint and is the most mobile of the cervical joints, allowing approximately 10 degrees of flexion and extension, 47 degrees of rotation, and no lateral flexion (17). This joint allows us to turn our head and look from one side to the other. In fact, this articulation accounts for 50% of the rotation in the cervical vertebrae (17).

The remainder of the cervical vertebrae have structures in the anterior and posterior compartments similar to those described earlier for the typical vertebrae. The bodies of the cervical vertebrae are small and about half as wide side to side as they are front to back. The cervical vertebrae also have short pedicles, bulky articulating processes, and short spinous processes. The transverse processes of the cervical vertebrae have a foramen where the arteries pass through. This is not found in other regions of the vertebral column. A comparison of the structural differences between the cervical, thoracic, and lumbar vertebrae is presented in FIGURE 7–9.

The articulating facets in the cervical vertebrae face 45 degrees to the transverse plane and lie parallel to the frontal plane (18), with the superior articulating process facing posterior and up, and the inferior articulating processes facing anterior and down. Unlike other regions of the vertebral column, the intervertebral discs are smaller laterally than the bodies of the vertebrae. The cervical discs are thicker ventrally than dorsally, producing a wedge shape and contributing to the lordotic curvature in the cervical region.

Due to the short spinous processes, the shape of the discs, and the orientation of the articulating facets in the backward and downward direction, movement in the cervical region is greater than any other region of the vertebral column. The cervical vertebrae can rotate through approximately 90 degrees, laterally flex 47 degrees to each side, flex through 40 degrees, and extend through 24 degrees (17). Maximum rotation in the cervical vertebrae occurs at C1-C2, maximum lateral flexion between C2-C4, and maximum flexion-extension in the cervical vertebrae occurs between C1-C3 and C7-T1. Also, all cervical vertebrae move simultaneously in flexion.

In addition to the ligaments supporting the whole vertebral column, there are some specialized ligaments found in the cervical region. The location and action of these ligaments are presented in Appendix A.

One of the most restricted regions of the vertebral column is the thoracic vertebrae. Moving down the spinal column, the individual vertebrae increase in size, thus the twelfth thoracic vertebrae is larger than the first thoracic. The bodies become taller and the thoracic vertebrae have longer pedicles than the cervical vertebrae (FIGURE 7–9). The transverse processes on the thoracic vertebrae are long and they angle backwards with the tips of the transverse processes posterior to the articulating facets. On the back of the thoracic vertebrae are long spinous processes overlapping the vertebrae and directed downward rather than posteriorly, as in other regions of the spine.

The connection of the thoracic vertebrae to the rib is illustrated in FIGURE 7–10. The thoracic vertebrae articulate with the ribs via articulating facets on the body of each vertebrae. Full facets are located on the bodies of the first, tenth, eleventh, and twelfth thoracic vertebrae, while demifacets are located on the second to ninth thoracic vertebrae in order to connect with the ribs. The thoracic vertebrae are supported by the same ligaments presented earlier, but have four additional ligaments that serve to support the attachment between the ribs and the vertebral body and transverse processes. These ligaments are presented in Appendix A.

The apophyseal joints between adjacent thoracic vertebrae are angled at 60 degrees to the transverse plane and 20 degrees to the frontal plane, with the superior facets facing posterior and a little up and lateral, and the inferior facets facing anterior, down, and medially (FIGURE 7–9). In comparison to the cervical vertebrae, the thoracic intervertebral joints are oriented more in the vertical plane.

The movements in the thoracic region are limited primarily because of the connection with the ribs and the long spinous processes that overlap in the back. Range of motion in the thoracic region for flexion and extension combined is 3 to 12 degrees, with very limited motion in the upper thoracic (2 to 4 degrees) that increases in the lower thoracic to 20 degrees at the thoracolumbar junction (12,17).

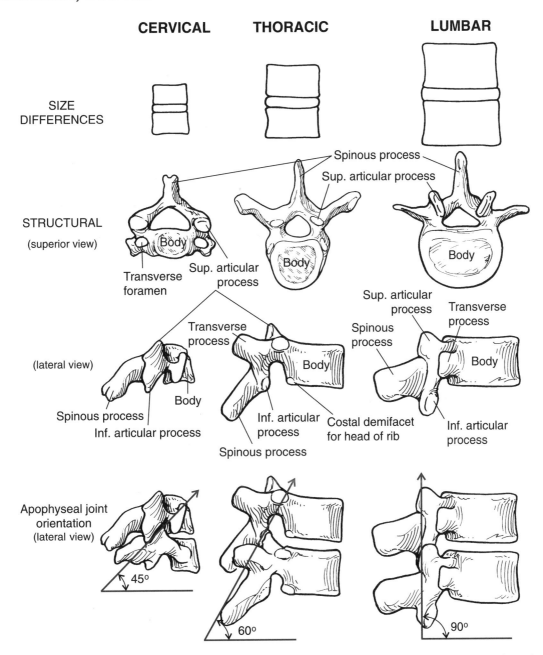

FIGURE 7–9. The cervical, thoracic, and lumbar vertebrae are different from each other. As you move from the cervical down to the lumbar region, the bodies of the vertebrae become larger, and the transverse processes, spinous processes, and the apophyseal joints all change their orientation.

Lateral flexion is also limited in the thoracic verte-brae, ranging from 2 to 9 degrees, and again, increasing as one progresses down through the thoracic vertebrae. In the upper thoracic, lateral flexion is limited to 2 to 4 degrees, while in the lower thoracic it may be as high as 9 degrees (12,17).

Rotation in the thoracic vertebrae ranges from 2 to 9 degrees. Rotation range of motion is opposite to that of flexion and lateral flexion, for it is maximum at the upper levels (9 degrees) and is reduced at the lower levels (2 degrees) (12,17).

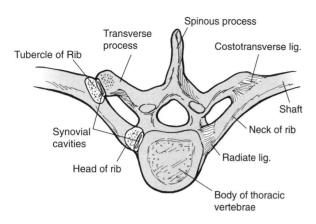

Tubercle of Rib
Transverse process
Spinous process
Costotransverse lig.
Shaft
Neck of rib
Radiate lig.
Synovial cavities
Head of rib
Body of thoracic vertebrae

FIGURE 7–10. The thoracic region is restricted in movement because of its connection to the ribs. The ribs connect to a demifacet on the body of the thoracic vertebrae and a facet on the transverse process.

The intervertebral discs in the thoracic region have a greater ratio of disc diameter to height of the disc than any other region of the spine. This reduces the tensile stress imposed upon the vertebrae in compression by reducing the stress on the outside of the disc (8). Thus, disc injuries in the thoracic region are not as common as other regions of the column.

The large, load-bearing region of the vertebral column is the lumbar vertebrae. Refer to FIGURE 7–9 for an illustration of the structure of the lumbar vertebrae. The lumbar vertebrae are large, with wider bodies side to side than front to back. They also are wider vertically in the front as compared to the back. The pedicles of the lumbar vertebrae are short, the spinous processes broad, and small transverse processes project posteriorly, upward, and laterally. The discs in the lumbar region are thick and, as in the cervical region, are thicker ventrally than dorsally, contributing to the lordotic curve in the region.

The apophyseal joints in the lumbar region lie in the sagittal plane, as the articulating facets are at right angles to the transverse plane and 45 degrees to the frontal plane (17). The superior facets face medially and the inferior facets face laterally. This changes at the lumbosacral junction where the apophyseal joint moves into the frontal plane and the inferior facet on L-5 faces frontally. This change in orientation keeps the vertebral column from sliding forward on the sacrum.

The lumbar region is supported by the same ligaments discussed earlier, which run the full length of the spine and one other ligament, the iliolumbar ligament (See Appendix A). Another important support structure in the region that is not a ligament is the thoracolumbar fascia

running from the sacrum and the iliac crest up to the thoracic cage. This fascia offers resistance and support in full flexion of the trunk. The tension in this fascia also assists with initiating the movement of trunk extension (6).

The range of motion in the lumbar region is large in flexion and extension, ranging from 8 to 20 degrees at the various levels of the vertebrae (12,17). There is limited lateral flexion at the various levels of the lumbar vertebrae, ranging from 3 to 6 degrees, and very little rotation (1 to 2 degrees) at all levels of the lumbar vertebrae (12,17). A review of the range of motion at each level of the vertebral column is presented in FIGURE 7–11.

The lumbosacral joint is the most mobile of the lumbar joints, accounting for a large proportion of the flexion and extension in the region. Seventy-five percent of the flexion and extension in the lumbar vertebrae may occur at this joint, with 20% of the remaining flexion between L4-L5, and 5% at the other lumbar levels (19).

Combined Movements of the Pelvis and Trunk

The relationship of the movements of the pelvis to the movements of the trunk has been discussed in an earlier chapter on the lower extremity. The movement synchronization between the pelvis and the trunk is referred to as the lumbopelvic rhythm. As shown in FIGURE 7–12, the lumbar curve reverses itself, flattens out, and curves in the opposite direction as trunk flexion progresses. This continues to a point at which the low back is rounded in full flexion of the trunk. Accompanying the movements in the lumbar vertebrae are flexion of the sacrum, anterior tilt of the pelvis, and finally, extension of the sacrum. The pelvis also moves backward as weight is shifted over the hips.

As discussed earlier in this chapter, the lumbar activity is maximum through the first 50 to 60 degrees of flexion, after which the pelvic rotation becomes the predominant factor increasing further trunk flexion. On the return extension movement, pelvic posterior tilt dominates the initial stages of the extension, and lumbar activity reverses itself, dominating the later stages of trunk extension. The pelvis also moves forward as weight is shifted.

Movement relationships between the pelvis and the trunk during trunk rotation or lateral flexion are not as clear-cut as seen in flexion and extension, because of restrictions to the movement introduced by the lower extremity. The pelvis will move with the trunk in rotation, and rotate right with trunk right rotation unless the lower extremity is forcing a rotation of the pelvis in the opposite direction. In this case, the pelvis may remain in the neutral position or rotate to the side exerting the greatest force.

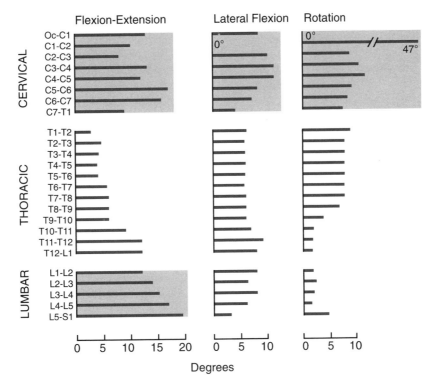

FIGURE 7–11. Range of motion at the individual motion segments of the spine is shown. The cervical vertebrae can produce the most range of motion at the individual motion segments. (Redrawn from White, A.A., and Panjabi, M.M.: Clinical biomechanics of the spine. Philadelphia, J.B. Lippincott, 1978.)

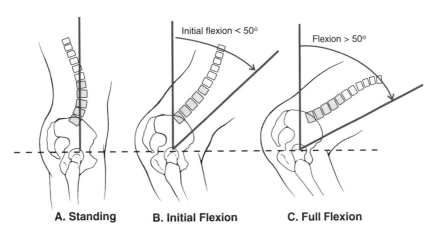

FIGURE 7–12. (A) In the normal standing posture, there is slight curvature in the lumbar region. (B) The first 50 degrees of flexion take place in the lumbar vertebrae as they flatten. (C) The continuation of flexion is a result of an anterior tilt of the pelvis.

Similarly, in lateral flexion of the trunk, the pelvis will lower to the side of the lateral flexion, unless there is resistance offered by the lower extremity. The accompanying pelvic movements will be determined by the trunk movement and the unilateral or bilateral positioning of the lower extremity.

Muscular Actions

Trunk extension is an important movement used to raise the trunk up and to maintain an upright posture. The muscles actively used to extend the trunk also play very dominant roles in trunk flexion; thus it seems logical to review the extensors first. The spinal extensors are graphically presented, and insertion, action, and nerve supply information are provided in Appendix B.

There are numerous small muscles constituting the extensor muscle group; however, they can be classified into two groups, the erector spinae (iliocostalis, longissimus, spinalis) and the deep posterior or the paravertebral muscles (intertransversarii, interspinales, rotatores, multifidus). These muscles run up and down the spinal column in pairs and create extension if activated as a pair, while creating rotation or lateral flexion if activated unilaterally. There is also a superficial layer of muscle that includes the trapezius and the latissimus dorsi. While both the trapezius and the latissimus dorsi can influence trunk motion, they will not be discussed in the present chapter.

The three erector spinae muscles constitute the largest mass of muscles contributing to trunk extension. The extension movement is also produced by contributions from the deep vertebral muscles and other muscles specific to the region. These deep muscles contribute to the generation of trunk extension and other trunk movements, but they also serve to support the vertebral column, maintain rigidity in the column, and produce some of the finer movements in the motion segment (7).

There are some other muscles besides the erector spinae and the deep posterior muscle groups specific to each region. Refer to Appendix B for a full description of these muscles.

The erector spinae muscles are thicker in the cervical and lumbar regions where most of the extension in the spine occurs. The multifidus is also thicker in the cervical and lumbar regions, adding to the muscle mass for generation of a trunk extension force.

The erector spinae and the multifidus muscles are 57 to 62% Type I muscle fibers, but also have proportions of both Type IIa and Type IIb fibers, making them functionally versatile so they can generate rapid, forceful movements while still being fatigue-resistant for the mainte-

nance of postures over long time periods (20). In addition to providing the muscle force for extension of the trunk, these muscles also provide posterior stability to the vertebral column, counteract gravity in the maintenance of an upright or erect posture, and are very important in the control of forward flexion (21).

Flexion of the trunk is free in the cervical region, limited in the thoracic region, and free again in the lumbar region. Unlike the posterior extensor muscles running the length of the vertebral column, the anterior flexors are region-specific and do not run the length of the column. Flexion of the lumbar spine is created by the abdominals. The flexion force of the abdominals also creates what little flexion there is in the thoracic vertebrae as well. The abdominals consist of four muscles: the rectus abdominus, the internal oblique, the external oblique, and the transverse abdominus (See Appendix B).

Besides creating flexion of the trunk, the abdominals also increase the intra-abdominal pressure when they contract. This serves to decrease the compressive force on the spine and reduce the activity of the erector spinae muscles (21). As the trunk flexes, the intra-abdominal pressure can assist in flexion by playing a supportive role. This is done by increasing pressure in the trunk cavities, which produces an extension effect, thereby reducing the extensor muscular forces needed to counteract the resistance produced in trunk flexion (22).

Additionally, the internal and external oblique muscles and the transverse abdominus attach into the thoracolumbar fascia covering the posterior region of the trunk. When they contract, there is tension placed on the fascia that supports the low back and reduces the strain on the posterior erector spinae muscles (21).

The abdominals consist of 55 to 58% Type I fibers, 15 to 23% Type IIa fibers, and 21 to 22% of Type IIb fibers (20). This fiber make-up is similar to that found in the erector spinae muscles, and allows for the same type of versatility in the production of short, rapid movements or prolonged movements of the trunk.

There are two other muscles contributing to flexion in the lumbar region. First is the powerful flexor acting at the hip, the iliopsoas muscle, which attaches to the anterior bodies of the lumbar vertebrae and the inside of the ilium. The iliopsoas can initiate trunk flexion and pull the pelvis forward, creating a lordotic posture in the lumbar vertebrae. Additionally, if this muscle is tight, an exaggerated anterior tilt of the pelvis could be developed. If the tilt is not counteracted by the abdominals, lordosis increases and a compressive stress is developed on the facet joints and the intervertebral disc is pushed posteriorly.

The second muscle found in the lumbar region is the quadratus lumborum, which forms the lateral wall of the abdomen and runs from the iliac crest to the last rib (See Appendix B). Although positioned to be more of a lateral flexor, the quadratus lumborum will also contribute to the flexion movement. It is also responsible for maintaining pelvic position on the swing side in gait (6).

When a person is standing or sitting upright, there is intermittent activity in both the erector spinae muscles and the internal and external obliques. The iliopsoas, on the other hand, is continuously active in the upright posture, while the rectus abdominus is inactive (10).

Movement into the fully flexed position from a standing posture is first initiated by the abdominals and the iliopsoas muscles. Once the movement begins, it is continued by the force of gravity acting on the trunk and controlled by the eccentric action of the erector spinae muscles. There is a gradual increase in the level of activity in the erector spinae muscles up to 50 to 60 degrees of flexion as the trunk flexes at the lumbar vertebrae (23).

As the lumbar vertebrae discontinue their contribution to trunk flexion, the movement continues as a result of the contribution of anterior pelvic tilt. The posterior hip muscles, the hamstrings and the gluteus maximus, eccentrically work to control this forward tilt of the pelvis. As the trunk moves deeper into flexion, the activity in the erector spinae diminishes to total inactivity in the fully flexed position. In this position, the posterior ligaments and the passive resistance of the elongated rector spinae muscles are controlling and resisting the trunk flexion (1). The load on the ligaments in this fully flexed position is close to their failure strength (14), placing additional importance on loads sustained by the thoracolumbar fascia and the lumbar apophyseal joints.

As the trunk raises back to the standing position through extension, the movement is initiated by a contraction of the posterior hip muscles, the gluteus maximus and the hamstrings, which flex and rotate the pelvis posteriorly. The erector spinae are active initially, but are most active through the last 45 to 50 degrees of the extension movement (21).

The erector spinae are more active in the raising than in the lowering phase, being very active in the initial parts of the movements and again at the end of the extension movement, with some diminished activity in the middle of the movement. The abdominals can also be active in the return movement as they serve to control the extension movement (1).

Flexion in the thoracic region is limited and takes place as a result of flexion developed by the muscles of the lumbar and cervical regions. In the cervical region, there are five pairs of muscles producing flexion, if both muscles in the pair are contracting. If only one of the muscles in the pair contracts, the result will be motion in all three directions, including flexion, rotation, and lateral flexion (7). The insertion, action, and nerve supply of these muscles are presented in Appendix B.

Lateral flexion of the spine is created by contraction of muscles on both sides of the vertebral column, with most activity on the side to which the lateral flexion occurs. The most activity in lateral flexion of the trunk occurs in the lumbar erector spinae muscles and the deep intertransversarii and interspinales muscles on the contralateral side. The multifidus muscle is inactive during lateral flexion. If load is held in the arm during lateral flexion, there is also an increase in the thoracic erector spinae muscles on the opposite side.

The quadratus lumborum and the abdominals also contribute to lateral flexion. The quadratus lumborum on the side of the bend is in a position to make a significant contribution to lateral flexion. The abdominals on the opposite side of the bend will contract as the lateral flexion is initiated, and the abdominals on the same side as the bend will contract to continue modifying the lateral flexion.

In the cervical region of the spine, the movement of lateral flexion is further facilitated by unilateral contractions of the sternocleidomastoid, the scalenes, and the deep anterior muscles. The movement of lateral flexion is quite free in the cervical region.

The rotation of the trunk is more complicated in terms of muscle actions because it is produced by muscle actions on both sides of the vertebral column. In the lumbar region, the multifidus and rotatores muscles on the side to which the rotation occurs are active, while the longissimus and iliocostalis on the other side are also active (24). The abdominals exhibit a similar pattern as the internal oblique on the side of the rotation is active, while the external oblique on the opposite side of the rotation is also active.

Strength and Forces at the Vertebral Joints

The greatest strength output in the trunk can be developed in the extension movement, averaging values of 210 Nm for males (25). Trunk flexion strength was 150 Nm, or approximately 70% of the strength of the extensors. Lateral flexion values were 145 Nm, 69% of the extensor strength, and rotation strength values were 90 Nm, or 43% of the extensor values (25). Female strength values were approximately 60% of the values recorded for males. In fact, other studies have shown women to be capable of generating only 50% of the lifting force of men for lifts low to the ground, and 33% of the male lifting force for lifts high off the ground (26).

Taking all things such as forces generated by intra-abdominal pressure, ligaments, and other structures into consideration, the total extensor moment is slightly greater than the flexor moments (27). The abdominals contribute to one third of the flexor moments and the erector spinae contribute one half of the extensor moments. In rotation, the abdominals dominate, with some contribution by the small posterior muscles (27).

Trunk positioning plays a significant role in the development of strength output in the various movements. Trunk flexion strength has been shown to improve by approximately 9% when measured from a position of 20 degrees of hyperextension (7). Likewise, isometric trunk extension values increased by almost 22% when the starting position of trunk flexion was increased from 20 to 40 degrees (7). Higher trunk flexion and extension strength values can also be achieved if the measurement is made in a seated position rather than a supine or prone position.

Strength output while lifting an object using the trunk extensors will diminish when there is a greater horizontal distance between the feet and the hands placed on the object (28). In fact, the forces applied vertically to an object held away from the body are about one half that of a lift completed with the object close to the body. Additionally, the increase in the width of a box will decrease lifting capacity, while the increase in the length of a box has been shown to have no influence (28).

When lifting an object by pulling up at an angle, the loads are reduced at the elbow, shoulder, lumbar, and hip regions, but increased at the knees and ankles. This type of lift decreases the compressive force on the lumbar vertebrae 9 to 15% (29). Also, 16% more weight can be lifted by a more freestyle lift than the traditional straight-back, bent-knee lift (29).

Loads applied to the vertebral column are produced by body weight, muscular force acting on each motion segment, prestress forces present due to the disc and ligament forces, and external loads being handled or applied (1). The lumbar vertebrae handle the largest load, primarily because of their positioning and greater body weight acting at the lumbar region than other regions of the spine. Eighteen percent of the compressional load carried by the lumbar vertebrae are a result of the weight of the head and trunk (30).

The axial load on the lumbar vertebrae in standing is 700 N. This can be quickly increased to values greater than 3000 N when lifting a heavy load from the ground and can be reduced by almost half in the supine posi-tion (300 N) (9). Fortunately, the lumbar spine can resist approximately 9800 N of vertical load before fracturing (31).

The intradiscal pressure in the nucleus at the L3-L4 level for a variety of postures or movements is presented in Table 7–2. Basically, the load on the lumbar vertebrae during lifting or activity will be increased the further the load is from the body rather than by any one technique or another (31). For example, the magnitude of the compressive force acting on the lumbar vertebrae in a half squat is 6 to 10 times body weight (32). Any adjustment increasing the amount of flexion will increase these loads significantly. Even if adjustments are made, such as flattening the lumbar curve in sitting, the loads on the lumbar vertebrae will still increase (33).

Loads acting at the lumbar vertebrae can be as high as 2 to 2.5 times body weight in an activity such as walking (34). These loads are maximum at toe-off and also increase with an increase in walking speed. Loads on the vertebrae in an activity such as walking are a result of muscle activity in the extensors and the amount of trunk lean in the walker (1).

The direction of the force or load acting on the vertebrae is influenced by positioning. In a standing posture with the sacrum inclined 30 degrees to the vertical, there is a shear force acting across the lumbosacral joint that is approximately 50% of the body weight above the joint (FIGURE 7–13). If the sacral angle increases to 40 degrees, the shear force increases to 65% of the body weight, and with a 50 degree sacral inclination, the force acting across the joint is 75% of the body weight above the joint (19).

Table 7–2. L3-L4 Intradiscal Pressure and Compressive Forces.

Activity	Intradiscal Pressure (kPA)	Spinal Compressive Force (N)
Standing	270	380
Flexion (414 Nm)	710	990
Extension (28 Nm)	720	1010
Lateral flex (43 Nm)	620	870
Twisting (28 Nm)	480	670
Flexion to 30 degrees with 4 KG in each hand	1620	2270

(Adapted from Nachemson, A., and Morris, J.M.: In vivo measurements of intradiscal pressure. Discometry, a method for the determination of pressure in the lower lumbar discs. Journal of Bone and Joint Surgery. *46-A*: 1077–1092, 1964.)

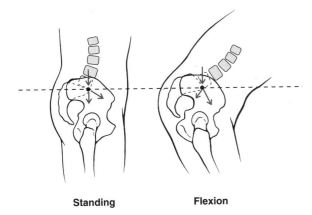

Standing   **Flexion**

FIGURE 7–13. (A) The shear force acting across the lumbosacral joint in standing is approximately 50% of the body weight. (B) If one flexes to where the sacral angle increases to 50 degrees, the shear force can increase to as much as 75% of the body weight above the joint.

Lumbosacral loads are also very high in exercises such as the squat, in which maximum forces are generated at the so-called sticking point of the ascent. These loads are higher than loads recorded at either the knee or the hip for the same activity (35).

There are even loads applied to the lumbar vertebrae in a relaxed supine position of repose. The loads are significantly reduced because of the loss of the body weight forces, but still present as a result of muscular and ligamentous forces. In fact, the straight-leg lying position imposes load on the lumbar vertebrae because of the pull of the psoas muscle in this position. If the thigh can be flexed by placing a pillow under the knees, this load can be reduced.

Loads imposed upon the vertebrae are carried by the various structural elements of the segment. The intervertebral discs absorb and distribute a great proportion of the load imposed upon the vertebrae. The intradiscal pressure is 1.3 to 1.5 times the compressional load applied per unit area of disc (31,36) and the pressure increases linearly with loads up to 2000 N (8). The load on the third lumbar vertebrae in standing is approximately 60% of total body weight (37).

Pressures in sitting are 40% more than those is standing, though the standing interdiscal pressures can be reduced by placing one foot in front of the other and elevating it (7). Pressures within the disc are large with flexion and lateral flexion movements of the trunk, and small with extension and rotation (38). The lateral bend produces larger pressures than the flexion movement and even more pressure if rotation is added to the side bend (24).

Intervertebral discs have been shown to withstand compressive loads in the range of 2500–7650 N (39). In older individuals, the range is much smaller, and in individuals younger than 40, the range is much larger (39).

The posterior elements of the spinal segment assist with load bearing. When the spine is under compression, the load is supported partially by the pedicles, the pars interarticularis, and somewhat by the apophyseal joints. When compression and bending loads are applied to the spine, 25% of the load is carried by the apophyseal joints. Only 16% of the loads imposed by compressive and shear forces are carried by the apophyseal joints (30). With any extension of the spine, there is a respective increase in the compressive strain on the pedicles, increase in both compressive and tensile strain in the pars articularis, and an increase in the compressive force acting at the apophyseal joints (40).

In full trunk flexion, the loads are maintained and absorbed by the apophyseal capsular ligaments, the intervertebral disc, the supraspinous and interspinous ligaments, and the ligamentum flavum, in that order (15). The erector spinae muscles also offer some resistance passively.

In compression, most of the load is carried by the disc and the vertebral body. The vertebral body is susceptible to injury before the disc and will fail at compressive loads of only 3700 N in the elderly and 13000 in a young, healthy adult (41). In rotation, where torsional forces are applied, the apophyseal joints are more susceptible to injury, and during a forward bend movement, the disc and the apophyseal joints are at risk for injury because of compressive forces created on the anterior motion segment, and tensile forces created on the posterior elements.

Loads in the cervical region of the spine are lower than the thoracic or lumbar regions and vary with position of the head. The loads are minimal in the neutral position but become significant in extreme positions of flexion and extension (18). Various postures and exercises are presented in FIGURE 7–14.

Posture

Efficiency of motion and stresses imposed upon the spine are very much determined by the postural positions maintained in the trunk. Positioning of the vertebral segments is so important that a special section on posture is warranted.

To maintain an upright posture in standing, the S-shaped spine acts as an elastic rod in supporting the weight. There is a continuous forward bending action imposed upon the trunk in standing, since the center of

A   **B**   **C**

LOWEST

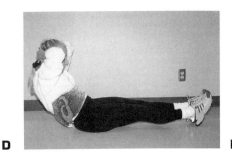

D   **E**   **F**

FIGURE 7–14. The representative postures or movements shown above are presented in order of load imposed upon the lumbar vertebrae with the (A) standing posture imposing the least amount of load, followed by the (B) double straight-leg raise, (C) back hyperextension, (D) sit-ups with knees straight, (E) sit-ups with knees bent, and (F) bending forward with weight in the hands.

gravity lies in front of the spine. As a result of the forward bending action on the trunk, the posterior muscles and ligaments must control and maintain the standing posture.

In an erect standing posture, there is more erector spinae activity than in a slouched standing posture. In the slouched posture, most of the responsibility for maintaining the posture is passed on to the ligaments and capsules. Any disruption in the standing posture or any postural swaying is controlled and brought back into alignment by the erector spinae, the abdominals, and the psoas muscles (42). All of these muscles are slightly active in standing, with more activity in the thoracic region than the other two regions (24).

Posture in the sitting position requires less energy expenditure and imposes less load on the lower extremity compared to standing. However, prolonged sitting positions can have deleterious effects on the lumbar spine (7). Unsupported sitting is similar to standing, such that there is higher muscle activity in the thoracic regions of the trunk with accompanying low levels of activity in the abdominals and the psoas muscles (43).

The unsupported sitting position places more load on the lumbar spine than standing because it creates a backward tilt, a flattening of the low back, and a corresponding forward shift in the center of gravity (1). This places load on the discs and the posterior structures of the vertebral segment. Sitting long periods of time in the flexed position can also overstretch and weaken the erector spinae muscles (21).

Continuous flexion positions are a cause of both lumbar and cervical flexion injuries in the workplace. These postures can be eliminated by raising the height of the work station so that no more than 20 degrees of flexion is present (7).

In supported sitting, the load on the lumbar vertebrae is lessened. A chair back inclined slightly backward, and including a lumbar support, creates a seated posture that produces the least load on the lumbar region of the spine. Both the intradiscal pressure and the muscle activity in the trunk are decreased in this position (10).

Postural deviations in the trunk are common in the general population (FIGURE 7–15). In the cervical region, the curve is concave to the anterior side. This curve should be small and lie over the shoulder girdle. The head should be above the shoulder girdle. If the head is held too far forward, the cervical curve will bend forward into an accentuated curve. When the curve is accentuated to the anterior side, a lordosis is said to be present. Thus, cervical lordosis would appear to be an increase in the curve in the cervical region, often associated with the forward head. It is less common for the cervical curve to flatten and curve to the posterior side. This condition would be termed a kyphosis.

In the thoracic region, the curvature is concave to the posterior side. As a result of maintaining a rounded-shoulder posture, one might develop thoracic kyphosis, a common postural disorder in this region. The kyphotic thoracic region is also associated with osteoporosis and several other disorders.

The lumbar region, curving anteriorly, is subjected to forces that may be created by an exaggerated lumbar curve, termed lumbar lordosis or hyperlordosis. This accentuated sway-back position is often created by anterior positioning of the pelvis or by weak abdominals. In the lumbar region, it is also not uncommon to have a flat back, or lumbar kyphosis, which can be developed through pelvic positioning or rigidity in the spine.

The most serious of the postural disorders affecting the spine is scoliosis, a lateral deviation of the spine. The lateral flexion of the spine will usually occur in the thoracic region and the lumbar region, forming an S-shaped vertebral column. Rotation usually accompanies the lateral flexion, creating a very complex postural malalignment. The cause of scoliosis is unknown, and it is more prevalent in females.

Conditioning

The muscles around the trunk are active during most activities as they stabilize the trunk, move the trunk into an advantageous position for supplementing force pro-

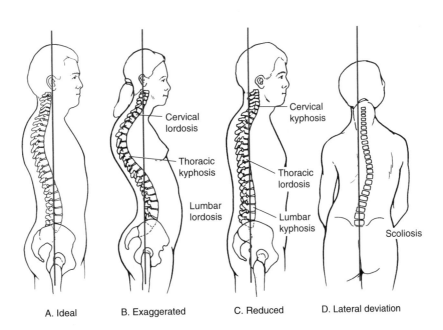

A. Ideal   B. Exaggerated   C. Reduced   D. Lateral deviation

FIGURE 7–15. (A) The ideal posture should be one in which the curves are balanced, yet not exaggerated. Curves can become (B) exaggerated, or (C) reduced. (D) Lateral deviation of the spine, scoliosis, can create very serious postural malalignment throughout the whole body.

TABLE 7–3. Loads of the Lumbar Region

Activity	Loads on the Lumbar Vertebrae (N)
Standing	686
Bend forward 20 degrees with 10 KG in each hand	1813
Supine position	294
Supine with 30 KG traction	98
Double straight leg raise	1176
Sit-up with knees bent	1764
Sit-up with knees straight	1715
Isometric abduction	1078
Back hyperextension	1470

(Adapted from Nachemson, A: Lumbar intradiscal pressure. In The Lumbar Spine and Back Pain. Edited by M. Jayson. Kent, Pitman Medical Publishing Co Ltd, 1976.)

duction, or assist a limb movement. Since the low back is a common site of injury in sport and in the workplace, special attention should be given to those exercises that will strengthen and stretch this part of the trunk. Trunk exercises should also be evaluated in terms of their negative impact on trunk function and structure.

The trunk flexors are usually exercised with some form of trunk or thigh flexion exercise from the supine position so that these muscles can work against gravity. Trunk flexion exercises should be evaluated for their effectiveness and safety. The sit-up has been done for many years, and involves trunk flexion from the supine position with the legs fully extended in the long-lying position. Sometimes the feet are secured by a partner or an external support. This exercise is performed by a contraction of both the rectus abdominus and the obliques during the first 37 to 40 degrees of trunk flexion, followed by hip flexor activity that continues the sit-up through the remainder of the motion (7,21).

This long-lying position creates an anterior movement of the fifth lumbar vertebrae on the sacrum. Also, the muscular force of the flexors in the midrange of motion creates additional stress on the lumbar vertebrae by increasing the lordosis. If the feet are held by a partner, the pull of the iliopsoas creates even more lordosis.

The curl-up or hook-lying position, with the knees flexed and the feet on the floor close to the buttocks, reduces the amount of lordosis created in the lumbar spine, by forcing the pelvis posteriorly. The trade-off in the reduction of lordosis is an increase in the intervertebral disc pressure over those pressures measured in the long-lying sit-up (21). If a trunk raise is performed from

this position, the abdominals are still the most active muscles through the initial 30 to 45 degrees, and the hip flexors are active through the remainder of the flexion movement (7,21). The rectus abdominus appears to be more active in the hook-lying curl-up than the long-lying sit-up (7). As in the sit-up, the contribution from the hip flexors is greater if the feet are held.

This exercise is often performed quickly in a predetermined time limit to measure abdominal endurance and strength. This type of test should be used with caution, especially with individuals having low back pain. By repeating this exercise quickly, the hip flexors will become more active as the abdominals fatigue (21).

The double leg raise has also been used to strengthen the abdominals, and does activate both the rectus abdominus and the external obliques (7). This exercise engages the hip flexors to a greater degree than the other abdominal exercises, creating a great deal of lordosis in the lumbar spine. The strain on the low back is significant enough to label this exercise as contraindicated for those individuals with weak abdominals. The exercise can be modified by producing a single leg raise with the contralateral limb flexed at the knee and the foot flat on the floor, as in the hook-lying position.

Exercises creating excessive lordosis or hyperextension of the lumbar vertebrae should be avoided since they put excessive pressure on the posterior element of the spinal segment and can cause disruption of the facets or the posterior arch. Examples of such exercises are the double leg raise, double leg raise with scissoring, thigh extension from the prone position, donkey kicks, back bends, or ballet arches. When selecting an exercise for the trunk, pay attention to the risks involved. An example of some of the imposed loads incurred with specific exercises are presented in Table 7–3. The supine position produces the least amount of load on the lumbar vertebrae. However, the load is increased substantially in the supine position if the abdominals and the illiopsoas are activated such as in the leg-raise or sit-up.

The best strengthening exercise for the abdominals is a "crunch", a hook-lying curl-up in which the trunk is only raised 30 to 40 degrees off the floor. This exercise creates little strain on the lumbar region, since the hip flexors are minimally involved through this range of motion. The abdominals are very active and receive a good workout. Finally, an isometric abdominal exercise is also effective and creates the least amount of load on the back. Trunk flexion stretching and strengthening exercises are demonstrated in FIGURE 7–16.

Extension of the trunk is usually developed through some type of lift using the legs and back. Refer to FIG-

FLEXIBILITY

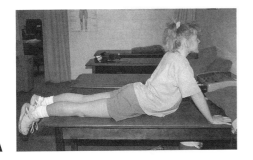

A   B   C

MANUAL OR LOW RESISTANCE

E   F

WEIGHT TRAINING

G   H

FIGURE 7–16. Flexibility, manual or low resistance, and high resistance exercises are presented for the abdominals: whole group (left), the obliques (middle), and the cervical vertebrae (right). To stretch the flexors, a back hyperextension, (A) rotation, (B) or neck hyperextension exercise (C) is used. A low or manual resistance exercise for the flexors includes a curl-up (abdominals), (D) trunk twisting exercise with the shoulders barely off the floor (obliques), (E) and a manual resistance applied to the forehead during a neck flexion movement (cervical). (F) Higher resistance weight-training exercises include flexion (G) or rotation (H) on a trunk machine. Weight training for cervical flexion is also available on a neck machine (not shown).

URE 7–17 for examples of some extensor exercises for stretching and strengthening these muscles.

There are two basic types of lifts that will activate the erector spinae: the leg lift and the back lift. The leg lift is the squat or deadlift exercise in which the back is maintained in an erect or slightly flexed posture and the knees are flexed. This lift has the least amount of erector spinae activity and imposes the lowest shear or compressive

FLEXIBILITY

A   B   C

MANUAL OR LOW RESISTANCE

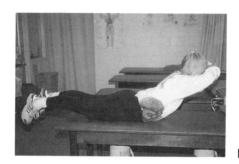

D   E   F

WEIGHT TRAINING

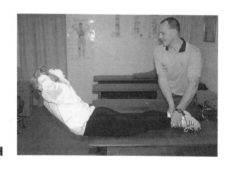

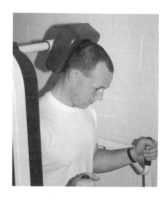

G   H   I

FIGURE 7–17. Flexibility, manual or low resistance, and high resistance or weight-training exercises are presented for the low back and for the cervical regions of the trunk. The low back can be stretched with back rounding (A) or a knee-tuck exercise (B), and the neck can be stretched with a simple flexion movement (C). Manual or low resistance exercises include a diagonal leg and arm lift (D), back hyperextensions (E), and a manual resistance during neck hyperextension (F). High resistance or weight-training exercises for the extensors include exercises on a trunk machine (G), back hyperextensions with the feet supported (H), and neck extension on a machine (I). High resistance exercises for the trunk extensors should be performed with caution so that the low back is not injured.

forces on the spine (44). The leg lift is begun with posterior tilt of the pelvis initiated by the gluteus maximus and the hamstrings. The erector spinae can be delayed and not involved until later in the leg lift when the extension is increased. The delay is related to the magnitude of the weight being lifted, and the muscles usually do not become active until the initial acceleration is completed (23). Since there is considerable stress on the ligaments at the beginning of the lift, it is suggested that the erector spinae activity begin at the initial part of the lift to stabilize the back (1).

The back lift is an exercise in which the person bends over at the waist with the knees straight, as in the good morning exercise. This exercise creates the highest shear and compressive stress on the lumbar vertebrae, but the erector spinae are much more active in this type of exercise (44). In the back lift, the movement is initiated by the hamstrings and the gluteus maximus and then followed up by activity from the erector spinae. Extension of the spine begins approximately 1/3 of the way into the lift (23). In performing the back lift with no load, the erector spinae becomes active after the beginning of the lift, but if the lift is performed with weight, the erector spinae is active before the start of the lift (45).

In comparing the difference between a leg lift-versus a back lift-type of exercise, one must consider both the risks and the gains. The back lift imposes a greater risk of injury to the vertebrae because of the higher forces imposed upon the system. Any stooping posture of the trunk will impose greater compression forces on the spine; consequently, a trunk flexion posture in a lift should be discouraged (46). The disc pressures are much higher in the back lift than the leg lift, mainly because of the trunk position and distance (37). The erector spinae activity in the back lift is greater than that in the leg lift.

Trunk extensor activity increases with greater trunk lean, while knee extension activity decreases with increased trunk lean (47). Maximal erector spinae activity also occurs later in the back lift than the leg lift. Finally, the abdominal activity is lower in the back lift compared to the leg lift (45). Consequently, the back is not as supported in the back lift as it is in the leg lift, creating additional potential for injury.

It is believed that some of the load on the vertebrae in a lifting activity can be partially reduced by intra-abdominal pressure. The intra-abdominal pressure in lifting increases with an increase in the weight being lifted, it is greater in dynamic lifting compared to static lifting, and it increases with forward bending of the trunk (23). It is also greater in leg lifting than in back lifting (46). This may be due to the increased activity in the abdominals

seen in the leg lift, or because there is greater pressure developed in the push movement of the leg lift as compared to the pull movement of the back lift (46).

Maximal intra-abdominal pressure occurs at the bottom of the lift and is increased if one uses a weight belt. The hypothesized effect of greater intra-abdominal pressure is that it creates an extensor moment that reduces some of the load from the erector spinae muscles. The net effect is smaller forces on the lumbosacral joint (41).

To work the extensors from the standing position by hyperextending the trunk, there is an initial contribution from the erector spinae. This activity drops off and then picks up again later in the hyperextension movement (48). If a resistance is offered to the movement, the activity in the lumbar erector spinae movement increases dramatically (49).

The rotation and lateral flexion movements of the trunk are not usually emphasized in an exercise program. Some examples of rotation and lateral bending exercises are provided in FIGURE 7–18. There is some benefit to including some of these exercises in a training routine, since rotation is an important component in many movement patterns. Likewise, lateral flexion is an important movement component for activities such as throwing, diving, and gymnastics.

Some athletes and individuals try to isolate the obliques by performing trunk rotation exercises against an external resistance. The obliques are not isolated in this type of exercise since the erector spinae are also actively involved. If a rotation exercise is added to an exercise set, caution should be used in its implementation. No combined exercises should be performed in which the trunk is flexed or extended and then rotated. This loads the vertebrae excessively and is unnecessary. If rotation is to be included as an exercise, it should be done in isolation and not in combination. The same holds true for lateral flexion exercises that can be performed against a resistance from a stand or from the side-lying position.

Stretching of the trunk muscles is easy to do and can be done from a standing or lying position. The lying positions offer stabilization of the lower extremity and the pelvis that will otherwise contribute to the movements if performed from a stand. All stretching of the trunk muscles should be done through one plane only, since, as pointed out earlier, movements through more than one plane concomitantly will excessively load the vertebral segments.

Use caution is prescribing maximum trunk flexion exercises such as touching the toes for the stretch of the extensors. Remember, the trunk is supported by the liga-

FLEXIBILITY

FIGURE 7–18. Flexibility exercises are provided for the trunk rotators, cervical rotators, trunk lateral flexors, and cervical lateral flexors. The trunk rotators can be stretched by applying a passive force to rotate the trunk (A) or by moving through maximum trunk or head rotation (B). Likewise, the trunk or neck can be stretched in lateral flexion either passively (C) or through an active movement (D) sideways with the head or trunk.

ments and the posterior elements of the segment in this position, and the loads on the discs are large, so choose an alternative exercise.

Similarly, the sit-reach test is often used as a measure of both low back and hamstring flexibility. It has been recently suggested that the sit-reach is primarily an assessment of hamstring, and not low back flexibility (21). The sit-reach position has also been shown to increase the strain on the low back as a result of exaggerated posterior tilt of the pelvis. It is recommended that the sit-reach stretching exercise be done with one hip flexed in order to stabilize the pelvis (21).

Inflexibility in the trunk or posterior thigh will influence the load and strains incurred during exercise. If the low back

is inflexible, the reversal of the lumbar curve is restrained in forward flexion movements. This places an additional strain on the hamstrings. If the hamstrings are inflexible, the rotation of the pelvis is restricted, placing additional strain on the low back muscles and ligaments. Additionally, inhibition of forward rotation of the pelvis will increase the overall compressive stress on the spine (21).

Contribution of Trunk Muscles to Sports Skills or Movements

The contribution of the back muscles to lifting has been presented in an earlier discussion on exercises. Likewise, the contribution of the abdominals to a sit-up or curl-up exercise was evaluated. The trunk muscles

MANUAL OR LOW RESISTANCE

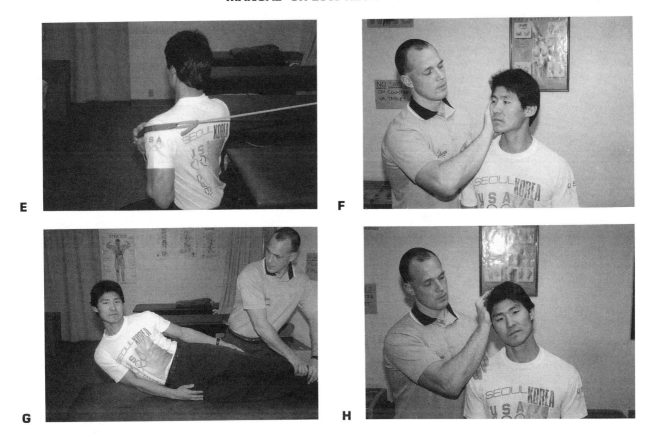

E

F

G

H

WEIGHT TRAINING

I

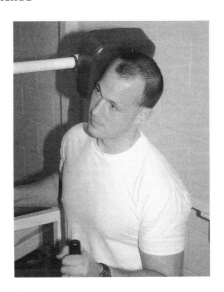

J

FIGURE 7–18. (continued) Manual or low resistance exercises for the rotators and lateral flexors include the use of surgical tubing (E), manual resistance to the side of the head during a rotation or lateral flexion movement (F,H), and a lateral raise of the trunk with the feet stabilized (G). High-resistance or weight-training exercises for the rotators or lateral flexors are usually done on specialized machines, such as the trunk (I) or neck machine (J) shown above.

also contribute to activities such as walking and running. A summary of the contributions of the trunk muscles to walking and running are presented in FIGURE 7–19.

In walking, the trunk moves as one segment in a side-to-side direction. At touchdown, the trunk laterally flexes toward the side of the limb making contact with the ground. It also moves back, and both of these movements are maximum at the end of the double support phase. After moving into single support, the trunk moves forward while still maintaining lateral flexion toward the support limb (50).

For running, the movements in the support phase are much the same, with trunk flexion and a lateral flexion to the support side. One difference is that, in walking, there is trunk extension at touchdown, whereas in running, the trunk will be flexed at touchdown in faster running speeds (13). At slower speeds, the trunk will be extended at touchdown. For a full cycle in both running and walking, the trunk will move forward and backward two times per cycle.

Another difference between walking and running is the amount and duration of lateral flexion in the support phase. In running, the amount of lateral flexion is greater, but in walking, the lateral flexion is held longer in the maximal position than running (50). There is one full oscillation of lateral flexion from one side to the other for every walking and running cycle.

As contact is made with the ground in both running and walking, there is a burst of activity in the longissimus and the multifidus muscles. This muscle activity can begin just before the instant of contact, usually as an ipsilateral contraction to control the lateral bending of the trunk. It is followed up with a contraction of the contralateral erector spinae muscles, so that a co-contraction of both sides occurs (13).

There is a second burst of activity in these muscles in the middle of the cycle occurring with the contact of the other limb. Here, both the longissimus and the multifidus are again active. In the first burst of muscular activity, the ipsilateral muscles are more active, while in this second burst, the contralateral muscles are more active (13). The activity of the erector spinae muscles coincides with extensor activity at the hip, knee, and ankle joints.

Basically, the lumbar muscles serve to restrict locomotion by controlling the lateral flexion and the forward flexion of the trunk (13). Cervical muscles serve to maintain the head in an erect position on the trunk and are not as active as other portions of the spine.

Injury Potential in the Trunk

The incidence of injury to the trunk is very high, for in the general population, 60 to 80% will suffer from back pain at some time in their life (12). Low back pain is a chronic problem for 1 to 5% of the population, and is a recurring injury in 30 to 70% of those experiencing an initial low back problem (21). The sexes are affected equally, and low back pain is more common in the age range of 25 to 60, with 40 years being at the age at which the

Trunk extensor muscle activity during walking and running.

Muscles	Swing Phase (Ipsilateral)			Support Phase (Ipsilateral)		
WALKING (1.5 m/s)						
	R	FS	FD	HS	MS	TO
Multifidus						
Ipsilateral	***	*	**	**	*	*
Contralateral	***	*	*	***	**	*
Longissimus						
Ipsilateral	***	*	*	*	*	*
Contralateral	**	*	*	***	*	*

Muscles	Swing Phase (Ipsilateral)			Support Phase (Ipsilateral)		
RUNNING (5 m/s)						
	R	FS	FD	HS	MS	TO
Multifidus						
Ipsilateral	***	*	**	***	*	**
Contralateral	***	*	*	***	*	**
Longissimus						
Ipsilateral	*	**	**	***	*	**
Contralateral	***	*	*	***	*	**

* = low activity
** = moderate activity
*** = high activity

FIGURE 7–19. The trunk extensors are active in both walking and running as they work to control the flexion and lateral flexion of the trunk (Adapted from Thorstensson, et al., 1980; 1982).

incidence of low back pain is the highest (21). Back pain is uncommon in children and athletes. The incidence of back sprain in athletes accounts for only 2 to 3% of the total sprains in the athletic population (51), but the injury is very debilitating.

Back pain can be caused by compression on the spinal cord or nerve roots from an intervertebral disc protrusion or disc prolapse. Disc protrusions occur more frequently at the intervertebral junctions of C-5 and C-6, C-6 and C-7, L-4 and L-5, and L-5 and S-1 (52). Lumbar disc protrusions occur at a significantly higher rate than in any other region of the trunk. As shown in FIGURE 7–20, the disc protrusion may impinge upon the nerve exiting from the cord, causing problems all throughout the back region and the lower extremity.

A disc injury commonly occurs to a motion segment that is compressed while being flexed slightly more than the normal limits of motion (15). Also, a significant amount of torsion, or rotation, of the trunk has been shown to tear fibers in the annulus fibrosus of the disc. Pure compression to the spine does not usually injure the disc, injuring the vertebral bodies and endplates instead. Likewise, maximal flexion of the trunk, without compression, will injure the posterior ligaments of the arch rather than the disc (15).

There are two weak points where injury is more likely when subjected to high loads. First, the cartilage endplates, into which the disc is attached, is only supported by a very thin layer of bone and thus is subject to fracture. Second, the posterior annulus is thinner and not attached as firmly as other portions of the disc, making it more vulnerable to injury (5).

In the case of a disc prolapse, the nucleus pulposus extrudes into the annulus fibrosus either laterally or vertically. A vertical prolapse is more common than a posterior prolapse, and the result is the creation of an anterolateral bulge of the annulus. This causes the bodies to tilt forward and pivot on the apophyseal joints, placing stress on the facets (5). A posterior or posterolateral prolapse of the disc into the spinal canal creates back pain and neurological symptoms via nerve impingement. Schmorl's nodes is a condition in which a vertical prolapse of part of the nucleus pulposus protrudes into an endplate lesion of the adjacent vertebrae (53). Damage to the disc is created through excessive load, failure of the inner posterior annulus fibers, or through disc degeneration (12).

Disc degeneration occurs in the early elderly years and consists of a gradual process during which splits and tears develop in the disc tissue. The progression of disc degeneration is illustrated in FIGURE 7–21. Although the symptoms of disc degeneration may not appear until the early elderly years, the process may begin much ear-

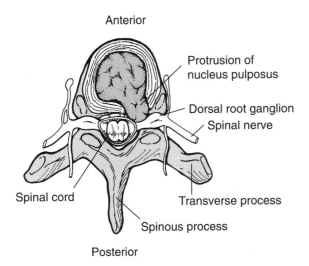

FIGURE 7–20. Injury to a disc can be caused by extreme trunk flexion while the trunk is compressed or loaded. Rotation movements can also tear the disc. When a disc is ruptured, pressure can be put on the spinal nerves as shown above.

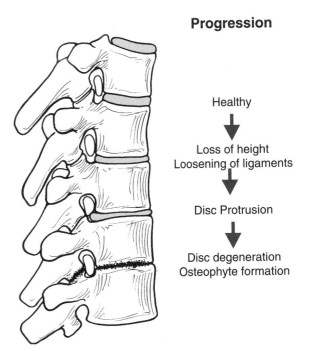

Progression

FIGURE 7–21. Disc degeneration narrows the joint space, causing a shortening of the ligaments, increased pressure on the disc, and stress on the apophyseal joints.

lier in life. It is common for disc degeneration to begin as the posterior muscles and ligaments relax, forcing a compression of the anterior portion, and a tension on the posterior portion of the disc. The tears in the disc are usually parallel to the endplates and located halfway between the endplate and the middle of the disc (5). As these tears get larger, there is potential for separation of the central portion of the disc. The splits and tears usually occur in the posterior and posterolateral portions of the disc along the posterior border of the marginal edges of the vertebral bodies (5). Eventually, the tears may be filled with connective tissue and later with bone. Osteophytes develop on the periphery of the vertebral bodies and cancellous bone is gradually laid down in the anterior portion of the disc where the pressure is great.

This condition can progress to a point where there is an osseus connection between two vertebral bodies that leads to further necrosis of the disc. Osteoarthritis of the apophyseal joints is also a by-product of disc degeneration as added stress is placed on these joints. Additionally, a disc that has undergone slight degeneration is also more susceptible to prolapse (15).

There can be fractures of the various osseus components of the vertebrae. The fractures can be in the spinous processes, transverse processes, laminae, or they can be compressive fractures of the vertebral body itself. There is an injury called spondylolysis, shown in FIGURE 7–22, that involves a fatigue fracture of the posterior neural arch at a site called the pars interarticularis. This injury is more

common in sports requiring repeated flexion, extension, and rotation actions, such as in gymnastics, weight lifting, football, dance, and wrestling (54). There is a 20.7% incidence of spondylolysis in athletes (55).

A typical example of an athlete who may fracture the neural arch is in the football lineman position and activity. The lineman assumes a starting position in a 3 or 4 point stance with the trunk flexed. This flattens the low back, compresses and narrows the anterior portion of the disc, and placing stress on the transverse arch. When the lineman drives up with trunk extension and makes contact with an opponent, there is a large shear force created across the apophyseal joint (54).

Another example of a spondylolysis-causing activity is in the pole vaulter who extends the trunk at the plant, followed by a rapid flexion of the trunk (56). The large range of motion occurring with rapid acceleration and deceleration is responsible for the development of the stress fracture. This condition is usually associated with repetitive activities, and is seldom associated with a single traumatic event (12).

When spondylolysis develops on both sides, a condition called spondylolisthesis can develop (FIGURE 7–22). With a bilateral defect of the neural arch, the motion segment is unstable, and there is a separation of the anterior and posterior elements. There is anterior slippage of the top vertebrae over the bottom vertebrae. This condition is more common in the lumbar vertebrae, especially at the site of L5-S1 where shear forces are often

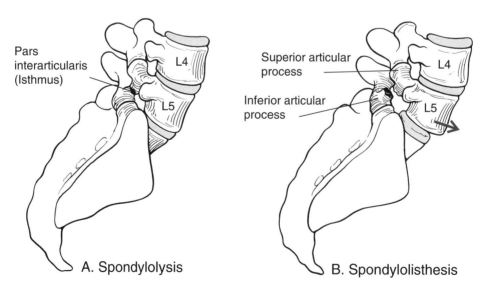

FIGURE 7–22. A fatigue fracture to the pars interarticularis is called (A) spondylolysis. When the fracture occurs bilaterally, (B) spondylolisthesis develops.

high. The condition is worsened with flexion of the spine, which adds to the amount of anterior shear being imposed upon the motion segment (57).

The cervical, thoracic, and lumbar regions of the trunk incur specific injuries more common within each region. In the cervical region of the spine, flexion and extension injuries, or whiplash injuries, are more common. In the whiplash injury, the head is rapidly flexed, straining the posterior ligaments, or even dislocating the posterior apophyseal joints if the force is large (18). The whiplash action can also fracture the vertebral bodies through the wedging action in the flexion movement, which compresses the bodies together. The seventh cervical vertebrae is a likely site for fracture in a flexion-type injury.

Injuries caused to the cervical vertebrae as a result of forceful extension include rupture of the anterior longitudinal ligament or actual separation of the annulus fibers of the disc from the vertebrae. A forceful hyperextension of the spine can be a part of the whiplash injury described earlier, and usually affects the sixth cervical vertebrae (7).

The cervical vertebrae is susceptible to injury in certain activities that subject the region to repeated forces. In diving, high jumping, and other activities having unusual landing techniques, individuals will be subjected to repeated extension and flexion forces acting on the cervical vertebrae which may cause an injury (12).

The thoracic region of the spine is not injured as frequently as the cervical and lumbar regions, probably due to its stabilization and limited motion as a result of interface with the ribs. There is a condition called Scheuermann's disease that is commonly found in the thoracic region. This disease creates an increase in the kyphosis of the thoracic region from wedging of the vertebrae. The cause of Scheuermann's disease is unknown, but it appears to be more prevalent in individuals who handle heavy objects. It is also more common among competitive butterfly-stroke swimmers (12).

The lumbar region of the spine is the most injured primarily because of the magnitude of the loads it carries. The source of low back pain can be located in a number of different sites in the lumbar area. It is believed that in a sudden onset of pain, muscles are usually the problem, irritated through some rapid twisting or reaching movement. If the pain is of the low-grade chronic type, overuse is seen as the culprit (57).

Myofascial pain is common in the low back and involves the muscle sheaths and tendons that have been strained as a result of some mechanical trauma or reflex spasm in the muscle (58). Muscle strain in the lumbar region is also related to the high tensions created while lifting from a stooped position.

Muscle spasms over a prolonged period of time will produce a dull aching pain in the lumbar region. Likewise, a dull pain can be caused by distorted postures maintained for long periods of time. The muscles fatigue, the ligaments are stressed, and connective tissue can become inflamed as a result of poor postural positioning.

There can be irritation of the joints in the lumbar region, occurring more often in activities in which there is frequent stooping, such as gardening or construction. Abnormal stress on the apophyseal joints is also common in activities such as gymnastics, ballet, and figure skating (58). Both spondylolysis and spondylolisthesis occur more frequently in the lumbar region than any other region of the trunk.

The intervertebral discs in the lumbar region experience a greater incidence of disc prolapse than any other segment of the spinal column. A disc protrusion, as in any other area of the trunk, may impinge upon nerve roots exiting the spinal cord, creating numbness, tingling, or pain in the adjacent body segments. Sciatica is such a condition. In it, the sciatic nerve is compressed, sending pain down the lateral aspect of the lower extremity.

The etiology of low back pain is not clearly defined because of the multiple risk factors associated with the disorder. Some of these factors are repetitive work, bending and twisting, pushing and pulling, tripping, slipping, and falling, and sitting or static work postures (21). A low back injury can be created through some uncoordinated or abnormal lift, or through repetitive loading over extended periods of time.

Low back pain associated with standing postures is related to positions maintaining hyperextension of the knees, hyperlordosis of the lumbar vertebrae, rounded shoulders, or hyperlordosis of the cervical vertebrae. In the seated posture, it is best to avoid crossing the legs at the knees, for this position places stress on the low back. Likewise, positions that maintain the legs in an extended position with the hips flexed should also be avoided since they accentuate the lordosis in the low back.

Low back injuries as a result of lifting are primarily a consequence of the magnitude of the load lifted and the distance of the weight from the body. A correct lifting posture, as mentioned earlier in this chapter, is one with the back erect, knees bent, weight close to the body, and movement through one plane only (FIGURE 7–23). This lifting technique will minimize the load imposed on the low back.

Muscle strength and flexibility are also seen as predisposing factors for low back pain. Tight hamstrings or an inflexible iliotibial band have both been associated with low back pain (21). Weak abdominals are also related to

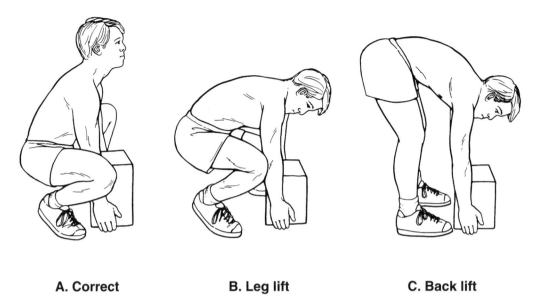

A. Correct   **B. Leg lift**   **C. Back lift**

FIGURE 7–23. Low back injury can be reduced if proper lifting techniques are used. The most important consideration is not whether you use your legs, but where the weight is with respect to your body. (A) Proper lifting technique has the weight close to the body with the head up and the back arched. The (B) leg lift technique is no better than the (C) back lift if the weight is held far away from the body.

low back pain. If the abdominals are weak, control over the pelvis is lacking and a hyperlordosis position will prevail. The hyperlordotic position puts undue stress on the posterior apophyseal joints and the intervertebral disc. This is an important consideration in an activity such as a sit-up or curl-up.

Erector spinae muscle activity has also been shown to relate to incidence of low back pain. In individuals having low back pain, there is increased electrical activity and fatigue in the erector spinae muscle group (21). Even though there are relationships between strength and flexibility and low back pain, the strength and flexibility of an individual may not predict whether low back pain will be experienced. However, strength, flexibility, and fitness are predictive of the recurrence of low back pain (21).

Effects of Aging on the Trunk

The effects of aging on the spine may predispose someone to an injury or painful condition. During the process of aging, the flexibility of the spine decreases by as much as ten times that of younger individuals (31). There is also a corresponding loss of strength in the trunk muscles of approximately 1% per year (12). Between the ages of 30 and 80, the strength in cartilage, bone, and the ligaments reduce by approximately 30%, 20%, and 18%, respectively (12).

The shape and length of the spine also change with aging. The discs lose height and create a shorter spine. Thus, individuals become shorter with increasing age. There is also an increase in lateral bending of the trunk, an increase in thoracic kyphosis, and a decrease in lumbar lordosis (59). In the lumbar region specifically, there is a loss of mobility in the L5-S1 segment with an accompanying increase in the mobility of the other segments (53).

It is not clear whether these age-related changes are a normal process of aging or whether the changes are associated with abuse of the trunk, disuse of the trunk, or are disease-related. It is clear that there is benefit in maintaining strength and flexibility in the trunk well into the elderly years.

Summary

The vertebral column provides both flexibility and stability to the body. There are four curves, cervical, thoracic, lumbar, and sacral, which form a modified elastic rod. The cervical and lumbar curves are mobile, and the thoracic and sacral curves are rigid.

The spinal column movement as a whole is created by small movements occurring at each motion segment. Each individual motion segment consists of two adjacent vertebrae and the disc separating them. The anterior portion of the motion segment includes the vertebral body, the intervertebral disc, and ligaments. Movement is allowed as the disc compresses. Within the disc itself, the gel-like mass in the center, the nucleus pulposus, absorbs the compression and creates a tension force in the annulus fibrosus, the concentric layers of fibrous tissue surrounding the pulposus.

The posterior portion of the motion segment includes the neural arches, the intervertebral joints, the transverse and spinous processes, and ligaments. This portion of the motion segment must accommodate large tensile forces.

The range of motion in each motion segment is only a few degrees, but in combination, the trunk is capable of moving through considerable range of motion. Flexion occurs freely in the lumbar region through 50 to 60 degrees. Lateral flexion range of motion is approximately 75 to 85 degrees, and occurs mainly in the cervical and lumbar regions. Rotation occurs through 90 degrees and is free in the cervical region. Rotation takes place in combination with lateral flexion in the thoracic and lumbar region.

Most lumbar spine movements are accompanied by pelvic movements, termed the *lumbopelvic rhythm*. In trunk flexion, the pelvis tilts anteriorly and moves backward. In trunk extension, the pelvis moves posteriorly and shifts forward. The pelvis will move with the trunk in rotation and lateral flexion.

The extension movement of the trunk is produced by the erector spinae and the deep posterior muscles running up and down the spinal column in pairs. The extensors also are very active and control the flexion of the trunk through the first 50 to 60 degrees of a lowering action with gravity. The abdominals produce flexion of the trunk against gravity or a resistance. They also produce rotation and lateral flexion of the trunk with assistance from the extensors.

The trunk muscles can generate the greatest amount of strength in the extension movement, but the total extensor moment supplemented by intra-abdominal pressure and ligaments is only slightly greater than the flexor moment. The strength output is also influenced by trunk positioning. In lifting, the extensor contribution will diminish the further the object is horizontally placed from the body. The contribution of the various segments and muscles is also influenced by the angle of pull and the width of the object being lifted.

The loads on the vertebrae are substantial in lifting and in different postures. The loads on the lumbar vertebrae can range from 2 to 10 times body weight in activities such as walking or weight lifting. Loads on the actual intervertebral discs are influenced by a change in posture.

For example, the pressures on the disc are 40% more in sitting than standing.

Posture is an important consideration in the maintenance of a healthy back. Postures that should be avoided include a slouched standing posture, prolonged sitting positions, the unsupported sitting position, and continuous flexion positions.

Postural deviations are common in the general population. Some of the common postural deviations occurring in the trunk are lordosis, kyphosis, and scoliosis. The most serious of these is scoliosis.

Conditioning of the trunk muscles should always include exercises for the low back. Additionally, trunk exercises should be evaluated in terms of safety and effectiveness. For example, the trunk flexors can be strengthened using the long-lying sit up, the curl-up or hook-lying sit-up, or the double straight-leg raise. All of these exercises work the abdominals, but only through the first 30 to 45 degrees, and then the hip flexors take over. A good alternative exercise that does not place undue stress on the low back is the crunch.

Conditioning of the extensors can be done through the use of various lifts. Both the leg lift and the back lift are commonly used to strengthen the extensors. The back lift imposes more stress on the vertebrae and produces more disc pressure than the leg lift.

Stretching of the trunk muscles can be done from either a standing or lying position. The lying positions offer more support for the trunk. A toe-touch exercise for flexibility should be avoided because of the strain to the posterior elements of the vertebral column.

The contribution of the muscles of the trunk to sport skills or movements is important for balance and stability. Trunk muscles are active in both walking and running as the trunk laterally flexes, flexes and extends, and rotates. There is also considerable activity in the cervical region of the trunk as the head and upper body are maintained in an upright position.

The incidence of injury to the trunk is very high, and it is predicted that 60 to 80% of the population will suffer from back pain at some time in their life. Back pain can be caused by disc protrusion or prolapse on a nerve, but is more likely to be associated with soft tissue sprain or strain. Disc degeneration occurs with aging and eventually may lead to reduction of the joint space and nerve compression. The spinal column can also experience fractures in the vertebral body as a result of compressive loading, or in the posterior neural arch that are associated with hyperlordosis (spondylolysis). When the defect occurs on both sides of the neural arch, a condition known as spondylolisthesis develops whereby the vertebrae slip anteriorly over each other. There are some injuries that are specific to regions of the trunk, such as whiplash in the cervical region and Scheuermann's disease in the thoracic vertebrae.

There are changes in the spine associated with the aging process, including decreased flexibility, loss of strength, loss of height in the spine, and an increase in lateral bending and thoracic kyphosis. It is not clear whether these changes are a normal consequence of aging or if they are related to disuse, misuse, or a specific disease process.

Review Questions

1. What are the motion characteristics of the cervical, thoracic, and lumbar regions of the spine? Why are there differences between the regions?
2. Indicate a possible cause for the postural deviations listed below. Describe a prescriptive exercise for each condition.
   a. cervical lordosis
   b. thoracic kyphosis
   c. lumbar lordosis
   d. lumbar kyphosis
   e. scoliosis
3. Design a set of both low- and high-resistance exercises for the following muscles:
   a. trunk extensors
   b. trunk flexors
   c. trunk rotators
4. Design a set of flexibility exercises for the following:
   a. abdominals

   b. erector spinae
   c. cervical region of the spine
   d. lumbar region of the spine
5. Analyze the following exercises and offer reasons why you think each exercise is good or bad for the trunk.
   a. toe-touch
   b. donkey kick
   c. double leg raise
   d. sit-up with feet held
   e. curl-up (no feet held)
   f. crunch
   g. single leg raise
   h. V-ups
   i. sit-reach
   j. neck circles
   k. back arch

Additional Questions

1. Compare the three lifting techniques of back lifting with bent knees, back lifting with straight legs, and a leg lift. What are the differences in the muscle actions? What are the differences in terms of loads on the vertebrae?
2. How would each of the following influence trunk function?
   a. weak abdominals
   b. weak erector spinae
   c. inflexible hamstrings
   d. inflexible erector spinae
   e. inflexible psoas
3. Discuss the potential for trunk injury in the activities or movements listed below.
   a. running
   b. weight lifting

   c. jumping
   d. kicking
   e. football
4. Describe the basic trunk muscle contributions to the movements or activities listed below.
   a. cycling
   b. weight lifting—squat
   c. weight lifting—dead lift
   d. forward pike dive
   e. lowering into a chair
   f. bending over and tying your shoes from a standing position
   g. power phase of the discus throw
   h. backbend
   i. cartwheel

Additional Reading

Adams, M. A., and Hutton, W. C.: The effects of posture on the fluid content of lumbar intervertebral discs. Spine. 8:665–671, 1983a.

Adams, M. A., and Hutton, W. C.: The mechanical function of the lumbar apophyseal joints. Spine. 8:327–329, 1983b.

Andersson, G. B. J.: Epidemiologic aspects on low-back pain in industry. Spine. 6:53–60, 1981.

Andersson, G. B. J., Ortengren, R., and Nachemson, A.: Quantitative studies of back loads in lifting. Spine. 1:178–185, 1977.

Andersson, G. B. J., Ortengren, R., and Schultz, A. B.: Analysis and measurement of the loads on the lumbar spine during work at a table. Journal of Biomechanics. 13:513–520, 1980.

Bailes, J. B., and Maroon, J. C.: Management of cervical spine injuries in athletes. Clinics in Sports Medicine. 8:43–58, 1989.

Bartelink, D. L.: The role of abdominal pressure in relieving the pressure on the lumbar intervertebral discs. Journal of Bone and Joint Surgery. 39-B:718–725, 1957.

Battie, M. C., et al.: A prospective study of the role of cardio-vascular risk factors and fitness in industrial back pain complaints. Spine. 14:141–147, 1989.

Battie, M. C., et al.: The role of spinal flexibility in back pain complaints within industry: A prospective study. Spine. 15:768, 1990.

Battie, M., Bigos, S., Sheehy, A., and Wortley, M.: Spinal flex-ibility and individual factors that influence it. Physical Therapy. 67:653–658, 1987.

Bogduk, N.: A reappraisal of the anatomy of the human lum-bar erector spinae. Journal of Anatomy. 131:525–540, 1980.

Bogduk, N.: The innervation of the lumbar spine. Spine. 8:286–293, 1983.

Bogduk, N., and Macintosh, J. E.: The applied anatomy of the thoracolumbar fascia. Spine. 9:164–170, 1984.

Brady, I. A., Cahill, B. R., and Bodmar, D. M.: Weight train-ing-related injuries in the high school athlete. American Journal of Sports Medicine. 10:1–9, 1982.

Burton, A. K., Tillotson, K. M., and Troup, J. D. G.: Prediction of low-back trouble frequency in a working population. Spine. 14:939–946, 1989a.

Burton, A, K., Tillotson, K. M., and Troup, J. D. G.: Variation in lumbar sagittal mobility with low-back trouble. Spine. 14:584–590, 1989b.

Carlsoo, S.: The static muscle load in different work positions: An electromyographic study. Ergonomics. 4:193, 1961.

Carter, D. R., and Frankel, V. H.: Biomechanics of hyperex-tension injuries to the cervical spine in football. American Journal of Sports Medicine. 8:302, 1982.

Crisco, J. J., Panjabi, M. M., and Dvorak, J.: A model of the alar ligaments of the upper cervical spine in axial rotation. Journal of Biomechanics. 27:607–614, 1991.

Cyron, B. M., and Hutton, W. C.: The fatigue strength of the lumbar neural arch in spondylolysis. Journal of Bone and Joint Surgery. 60B:234–238, 1978.

Dempster, D.W., et al.: Relationships between bone structure in the iliac crest and bone structure and strength in the lumbar spine. Osteoporosis International. 3:90–96, 1993.

Dolan, P., and Adams, M.A.: The relationship between EMG activity and extensor moment generation in the erector spinae muscles during bending and lifting activities. Journal of Biomechanics. 26:513–522, 1993.

Eie, N.: Load capacity of the low back. Journal of the Oslo City Hospitals. 16:73–98, 1966.

Ekholm, J., et al.: Activation of abdominal muscles during some physiotherapeutic exercises. Scandinavian Journal of Rehabilitative Medicine. 11:75–84, 1979.

Ekholm, J., Arborelius, U. P., and Nemeth, G.: The load of the lumbo-sacral joint and trunk muscle activity during lifting. Ergonomics. 24:145–161, 1982.

Erickson, M. F.: Aging on the lumbar spine II, L-1 and L-2. American Journal of Physiological Anthropology. 48:241–246, 1978.

Farfan, H. F., Huberdeau, R. M., and Dubow, H. I.: Lumbar intervertebral disc degeneration: The influence of geomet-rical features on the pattern of disc degeneration. Journal of Bone and Joint Surgery. 54A:492–510, 1972.

Good, C. J., and Mikkelsen, G. B.: Intersegmental sagittal motion in the lower cervical spine and discogenic spondy-losis: A preliminary study. Journal of Manipulative and Physiological Therapeutics. 15:556–563, 1992.

Gracovetsky, S., Farfan, H., and Helleur, C.: The abdominal mechanism. Spine. 10:317–324, 1985.

Gracovetsky, S., Farfan, H. F., and Lamy, C.: Mathematical model of the lumbar spine using an optimized system to control muscles and ligaments. Orthopedic Clinics of North America. 8:135–153, 1977.

Gracovetsky, S., et al.: Analysis of spinal and muscular activ-ity during flexion/extension and free lifts. Spine. 15:1333–1339, 1990.

Gravetsky, S., et al.: The importance of pelvic tilt in reducing compressive stress in the spine during flexion/extension exercises. Spine. 14:412–416, 1989.

Grieve, D. W.: Dynamic characteristics of man during crouch and stoop lifting. In Biomechanics IV. Edited by R.C. Nelson and C.A. Morehouse. Baltimore, University Park Press, 1974, pp. 19–29.

Grillner, S., Nilsson, J., and Thorstensson, A.: Intra-abdominal pressure changes during natural movements in man. Acta Physiologica Scandinavia. 103:275–283, 1978.

Gunzburg, R., et al.: A cadaveric study comparing discogra-phy, magnetic resonance imaging, histology, and mechani-cal behavior of the human lumbar disc. Spine. 17:417–423, 1992.

Haggmark, T., and Thorstensson, A.: Fibre types in human abdominal muscles. Acta Physiologica Scandinavia. *107*:319–325, 1979.

Hansson, T., Roos, B., and Nachemson, A.: The bone mineral content and ultimate compressive strength of lumbar vertebrae. Spine. *5*:46–55, 1980.

Jackson, A. W., and Baker, A. A.: The relationship of the sit and reach test to criterion measures of hamstring and back flexibility in young females. Research Quarterly for Exercise and Sport Science. *57*:183–186, 1986.

Jordan, B. D., et al.: Acute cervical radiculopathy in weight lifters. Physician and Sportsmedicine. *8*:73–76, 1990.

Kalimo, H., Rantanen, J., Viljanen, T., and Einola, S.: Lumbar muscles: Structure and function. Annals of Medicine. *21*:353–359, 1989.

Kapandji, I.: The Physiology of the Joints. New York, Churchill Livingstone, 1974.

Karvonen, M. J., et al.: Occupational health studies on air transport workers II. Muscle strength of air transport workers. International Archives of Occupational and Environmental Health. *47*:233–244, 1980.

Kelsey, J. L., et al.: An epidemiological study of acute prolapsed intervertebral disc. Journal of Bone and Joint Surgery. *66A*:907–914, 1984.

Kelsey, J. L., et al.: An epidemiologic study of lifting and twisting on the job and risk for acute prolapsed lumbar intervertebral disc. Journal of Orthopaedic Research. *2*:61–66, 1984.

King, A. I., Prasad, P., and Ewing, C. L.: Mechanism of spinal injury due to caudocephalad acceleration. Orthopaedic Clinics of North America. *6*:19–31, 1975.

Kishenbaum, K. J., Nadimpallii, S. R., Fantus, R., and Cavallino, R. P.: Unsuspected upper cervical spine fractures associated with significant head trauma: Role of CT. Journal of Emergency Medicine. *8*:183–198, 1990.

Kraus, D. R., and Shapiro, D.: The symptomatic lumbar spine in the athlete. Clinics in Sports Medicine. *8*:59–69, 1989.

Lander, J. E., Bates, B. T., and Devita, P.: Biomechanics of the squat exercise using a modified center-of-mass bar. Medicine and Science in Sports and Exercise. *18*:469–478, 1986.

Langrana, N. A., and Lee, C. K.: Isokinetic evaluation of trunk muscles. Spine. *9*:171–175, 1984.

Lin, H. S., Liu, Y. K., and Adams, K. H.: Mechanical response of the lumbar intervertebral joint under physiological (complex) loading. Journal of Bone and Joint Surgery. *60A*:41, 1978.

Marks, M. R., Bell, G. R., and Boumphrey, P. R. S.: Cervical spine fractures in athletes. Clinics in Sports Medicine. *9*:13–29, 1990.

Maroon, J., Onik, G., and Day, A.: Percutaneous automated discectomy in athletes. The Physician and Sportsmedicine. *16*:61–74, 1988.

McCarrol, J., Miller, J., and Ritter, M.: Lumbar spondylolysis and spondylolisthesis in college football players. The American Journal of Sportsmedicine. *14*:404–406, 1986.

McGlashen, K. M., Miller, J. A. A., Schultz, A. B., and Andersson, G. B. J.: Load-displacement behavior of the human lumbosacral joint. Journal of Orthopaedic Research. *5*:488–496, 1986.

Moll, J., and Wright, V.: Measurement of spinal movement. *In* The Lumbar Spine and Low Back Pain. Edited by M. Jayson. Kent, Pitman Medical Publishing Company Ltd, 1976, pp. 157–183.

Monkhouse, W. S., and Khalique, A.: Variations of the composition of the human rectus sheath: A study of the anterior abdominal wall. Journal of Anatomy. *145*:61–66, 1986.

Moore, K., Dumas, G. A., Reid, J. G., and Stevenson, J. M.: A longitudinal study of the mechanical changes in posture associated with pregnancy: A preliminary report. *In* C. E. Cotton, M. Lamontagne, D. G. E. Robertson, & J. P. Stothart (Eds.), Proceedings of the Fifth Biennial Conference and Human Locomotion Symposium of the Canadian Society for Biomechanics. Edited by C. E. Cotton, M. Lamontagne, D. G. E. Robertson, and J. P. Stothart. London, Ontario Spodym Publishers, 1988, pp. 114–115.

Moroney, S. P., and Schultz, A. B. Analysis and measurement of loads on the neck. Transcripts of the Orthopaedic Research Society. *10*:329, 1985.

Morris, J. M., Lucas, D. B., and Bresler, B.: Role of the trunk in stability of the spine. Journal of Bone and Joint Surgery. *43*:327–351, 1961.

Mutoh, Y., Mori, T., Nakamura, Y., and Miyashita, M.: The relationship between sit-up exercises and the occurrence of low-back pain. *In* Biomechanics VIII-A. Edited by H. Matsui and K. Kobayshi. Champaigne, IL. Human Kinetics Publishers, 1981, pp. 180–185.

Nachemson, A.: The effect of forward leaning on lumbar intradiscal pressure. Acta Orthopaedica Scandinavia. *35*:314–328, 1965.

Nachemson, A.: Electromyographic studies on the vertebral portion of the psoas muscle. Acta Orthopaedica Scandinavia. *37*:177, 1966.

Nachemson, A. L.: Towards a better understanding of low-back pain: A review of the mechanics of the lumbar disc. Rheumatological Rehabilitation. *14*:129–143, 1975.

Nemeth, G.: On hip and lumbar biomechanics: A study of joint load and muscular activity. Scandinavian Journal of Rehabilitative Medicine. *10*:1984.

Noble, L.: Effects of various types of situps on EMG of the abdominal musculature. Journal of Human Movement. *7*:124–130, 1981.

Oddsson, L., and Thorstensson, A.: Fast voluntary trunk flexion movements in standing: Primary movements and associated postural adjustments. Acta Physiologica Scandinavia. *128*:341–349, 1986.

Oddsson, L. I. E., and Thorstensson, A. T.: Reaction time and pattern of muscle activation in trunk flexion and extension movements. *In* Biomechanics X. Edited by B. Johnson, et al. Champaigne, IL., Human Kinetics Publishers, 1986, pp. 1–10.

Ortengren, R., and Andersson, G. B. J.: Electromyographic studies of trunk muscles, with special reference to the lumbar spine. Spine. *2*:44–52, 1977.

Ortengren, R., Andersson, G. B. J., and Nachemson, A.: Lumbar back loads in fired spinal postures during flexion and rotation. *In* Biomechanics VI. Edited by E. Asmussen and K. Jorgenson. Baltimore, University Park Press, 1978, pp. 159–166.

Panjabi, M. M., Brand, R. A., and White, A. A.: Mechanical properties of the human thoracic spine. Journal of Bone and Joint Surgery. *58A*:642–652, 1976.

Pintar, F. A., et al.: Biomechanical properties of human lumbar spine ligaments. Journal of Biomechanics. *25*:1351–1356, 1992.

Pope, M. H.: Risk indicators in low back pain. Annals of Medicine. *21*:387–392, 1989.

Reuber, M., Schultz, A., Denis, F., and Spencer, D.: Bulging of lumbar intervertebral discs. Journal of Biomechanical Engineering. *104*:187–192, 1982.

Ricci, B., Marchetti, M., and Figura, F.: Biomechanics of sit-up exercises. Medicine and Science in Sports and Exercise. *13*:54–59, 1981.

Roaf, R.: Vertebral growth and its mechanical control. Journal of Bone and Joint Surgery. *42B*:40–59, 1960.

Roy, S. H., DeLuca, C. J., and Casavant, D. A.: Lumbar muscle fatigue and chronic lower back pain. Spine. *14*:992–1001, 1989.

Roy, S. H., et al.: Fatigue, recovery, and low back pain in varsity rowers. Medicine and Science in Sports and Exercise. *22*:463–469, 1990.

Ryden, L., Molgarrd, C., and Bobbitt, S.: Benefits of a back care and light duty health promotion program in a hospital setting. Journal of Community Health. *13*:220–230, 1988.

Shirazi-Adl, A., and Dronin, G.: Load-sharing function of lumbar intervertebral disc and facet joints in compression, extension, and flexion. Advances in Bioengineering. *2*:18–19, 1986.

Schultz, A. B., Warwick, D. N., Berkson, M. H., and Nachemson, A. L.: Mechanical properties of human lumbar spine motion segments, Part 1. Responses in flexion, extension, lateral bending and torsion. Journal of Biomechanical Engineering. *101*:46–52, 1979.

Snijders, C. J., Hoek van Dijke, G. A., and Roosch, E. R.: A biomechanical model for the analysis of the cervical spine in static postures. Journal of Biomechanics. *24*:783–792, 1991.

Stokes, I. A., and Abery, J. M.: Influences of the hamstring muscles on lumbar spine curvature in sitting. Spine. *5*:525–528, 1980.

Sullivan, M.: Back support mechanisms during manual lifting. Physical Therapy. *69*:38–45, 1989.

Thorstensson, A., and Arvidson, A.: Trunk muscle strength and low back pain. Scandinavian Journal of Rehabilitative Medicine. *14*:69–74, 1982.

Warwick, D., Novak, G., and Schultz, A.: Maximum voluntary strengths of male adults in some lifting, pushing and pulling activities. Ergonomics. *23*:49–54, 1980.

White, A. A., and Panjabi, M. M.: Clinical biomechanics of the spine. Philadelphia, J. B. Lippincott, 1978.

Wilberger, J. B., and Maroon, J. C.: Cervical spine injuries in athletes. The Physician and Sportsmedicine. *18*:56–70, 1990.

Williams, J. G.: Biomechanical factors in spinal injuries. British Journal of Sports Medicine. *14*:14–17, 1980.

References

1. Lindh, M.: Biomechanics of the lumbar spine. *In* Basic Biomechanics of the Musculoskeletal System. Edited by M. Nordin and V. H. Frankel. Philadelphia, Lea & Febiger, 1989, pp. 183–208.
2. Haher, T. R., O'Brien. M., Kauffman, C., and Liao, K. C.: Biomechanics of the spine in sports. Clinics in Sports Medicine. *12*:449–463, 1993.
3. Beard, H. K., and Stevens, R. L.: Biochemical changes in the intervertebral disc. *In* The Lumbar Spine and Back Pain. Edited by M. Jayson. Kent, Pitman Medical Publishing Company Ltd, 1976, pp. 407–433.
4. Hickey, D. S., and Hukins, D. W. L.: Relation between the structure of the annulus fibrosus and the function and failure of the intervertebral disc. Spine. *5*:106–116, 1980.
5. Vernon-Roberts, B.: The pathology and interrelation of intervertebral disc lesions, osteoarthrosis of the apophyseal joints, lumbar spondylosis and low back pain. *In* The Lumbar Spine and Back Pain. Edited by M. Jayson. Kent, Pitman Medical Publishing Company Ltd, 1976, pp. 83–113.
6. Gould, J. A.: The spine. *In* J. A. Gould and G. J. Davies (Eds.), Orthopaedic and Sports Physical Therapy. Edited by J. A. Gould and G. J. Davies. St. Louis, C. V. Mosby Co, 1985, pp. 518–549.
7. Soderberg, G. L.: Kinesiology: Application to Pathological Motion. Baltimore, Williams & Wilkins, 1986.
8. Nachemson, A.: Lumbar intradiscal pressure. Experimental studies of post-mortem material. Acta Orthopedica Scandinavia. *43*:10–104, 1960.
9. Broberg, K. B.: On the mechanical behavior of intervertebral discs. Spine. *8*:151–165, 1983.
10. Shah, J. S.: Structure, morphology, and mechanics of the lumbar spine. *In* The Lumbar Spine and Back Pain. Edited by M. Jayson. Kent, Pitman Medical Publishing Company, Ltd, 1976, pp. 339–405.
11. El-Bohy, A. A., and King, A. I. Intervertebral disc and facet contact pressure in axial torsion. *In* Advances in Bioengineering. Edited by S. A. Lantz and A. I. King. New York, American Society of Mechanical Engineers, 1986, pp. 26–27.
12. Ashton-Miller, J. A., and Schultz, A. B.: Biomechanics of the human spine and trunk. *In* Exercise and Sport Sciences Reviews. Edited by K. B. Pandolf. New York, Macmillan Publishing Company, 1988, pp. 169–204.
13. Thorstensson, A., Carlson, H., Zomlefer, M. R., and Nilsson, J.: Lumbar back muscle activity in relation to trunk movements during locomotion in man. Acta Physiologica Scandinavia. *116*:13–20, 1982.
14. Farfan, H. F.: Muscular mechanism of the lumbar spine and the position of power and efficiency. Orthopedic Clinics of North America. *6*:135, 1975.
15. Adams, M. A., Hutton, W. C., and Stott, J. R. R.: The resistance to flexion of the lumbar intervertebral joint. Spine. *5*:245–253, 1980.
16. Adams, M. A., and Hutton, W. C.: Prolapsed intervertebral disc: A hyperflexion study. Spine. *7*:184–191, 1982.
17. White, A. A., and Panjabi, M. M.: The basic kinematics of the spine. Spine. *3*:12–20, 1978.
18. Shapiro, I., and Frankel, V. H.: Biomechanics of the cervical spine. *In* Basic Biomechanics of the Musculoskeletal System. Edited by M. Nordin and V. H. Frankel. Philadelphia, Lea & Febiger, 1989.
19. Saunders, H. D.: Evaluation, Treatment and Prevention of Musculoskeletal Disorders. Minneapolis, MN, Viking Press Inc. 1985.
20. Thorstensson, A., and Carlson, H.: Fiber types in human lumbar back muscles. Acta Physiologica Scandinavia. *131*:195–202, 1987.
21. Plowman, S. A.: Physical activity, physical fitness, and low back pain. *In* Exercise and Sport Sciences Reviews. Edited by J. O. Holloszy. New York, Williams and Wilkins, 1992, pp. 221–242.
22. Andersson, G. B. J., Ortengren, R., and Nachemson, A.: Intradiscal pressure, intra-abdominal pressure and myoelectric back muscle activity related to posture and loading. Clinical Orthopaedics. *129*:156–164, 1977.
23. Andersson, G. B. J., Herberts, P., and Ortengren, R.: Myoelectric back muscle activity in standardized lifting postures. *In* Biomechanics 5-A. Edited by P. V. Komi. Baltimore, University Park Press, 1976, pp. 520–529.
24. Andersson, G. B. J., Ortengren, R., and Herberts, P.: Quantitative electromyographic studies of back muscle activity related to posture and loading. Orthopaedic Clinics of North America. *8*:85–96, 1977.
25. McNeill, T., Warwick, D., Andersson, G., and Schultz, A.: Trunk strengths in attempted flexion, extension, and lateral bending in healthy subjects and patients with low-back disorders. Spine. *5*:529–538, 1980.
26. Yates, J. W., Kamon, E., Rodgers, S. H., and Champney, P. C.: Static lifting strength and maximal isometric voluntary contractions of back, arm, and shoulder muscles. Ergonomics. *23*:37–47, 1980.
27. Rab, G. T., Chao, E. Y. S., and Stauffer, R. N.: Muscle force analysis of the lumbar spine. Orthopaedic Clinics of North America. *8*:193, 1977.
28. Garg, A.: A comparison of isometric strength and dynamic lifting capability. Ergonomics. *23*:13–27, 1980.
29. Garg, A., Sharma, D., Chaffin, D. B., and Schmidler, J. M.: Biomechanical stresses as related to motion trajectory of lifting. Human Factors. *25*:527–539, 1983.
30. Miller, J. A. A., Haderspeck, K. A., and Schultz, A. B.: Posterior element loads in lumbar motion segments. Spine. *8*:331–337, 1983.
31. Nachemson, A.: Lumbar intradiscal pressure. *In* The Lumbar Spine and Back Pain. Edited by M. Jayson. Kent, Pitman Medical Publishing Co Ltd, 1976, pp. 257–269.

32. Capazzo, A., Felici, F., Figura, F., and Gazzani, F.: Lumbar spine loading during half squat exercises. Medicine and Science in Sports and Exercise. *17*:613–620, 1985.

33. Eklund, J. A. E., Corlett, E. N., and Johnson, F.: A method for measuring the load imposed on the back of a sitting person. Ergonomics. *26*:1063–1076, 1983.

34. Capazzo, A.: Compressive loads in the lumbar vertebral column during normal level walking. Journal of Orthopaedic Research. *1*:292, 1984.

35. Nisell, R., and Ekholm, J.: Joint load during the parallel squat in powerlifting and force analysis of in vivo bilateral quadriceps tendon rupture. Scandinavian Journal of Sports Science. *8*:63–70, 1986.

36. Schultz, A. B., et al.: Loads on the lumbar spine: Validation of a biomechanical analysis by measurements of intradiscal pressures and myoelectric signals. Journal of Bone and Joint Surgery. *64-A*:713–720, 1982.

37. Nachemson, A., and Morris, J. M.: In vivo measurements of intradiscal pressure. Discometry, a method for the determination of pressure in the lower lumbar discs. Journal of Bone and Joint Surgery. *46-A*:1077–1092, 1964.

38. Nachemson, A., Schultz, A. B., and Berkson, M. H.: Mechanical properties of human lumbar spine motion segments. Influence of age, sex, disc level and degeneration. Spine. *4*:1–8, 1979.

39. Perey, O.: Fracture of the vertebral endplates in the lumbar spine. An experimental biomechanical investigation. Acta Orthopaedica Scandinavia. *25*:1–101, 1957.

40. Jayson, M.I.V.: Compression stresses in the posterior elements and pathologic consequences. Spine. *8*:338–339, 1983.

41. Lander, J. E., Simonton, R. L., and Giacobbe, J. K. F.: The effectiveness of weight-belts during the squat exercise. Medicine and Science in Sports and Exercise. *22*:117–126, 1990.

42. Oddsson, L., and Thorstensson, A.: Fast voluntary trunk flexion movements in standing: Motor patterns. Acta Physiologica Scandinavia, *129*:93–106, 1987.

43. Andersson, G. B. J., Jonsson, B., and Ortengren, R. Myoelectric activity in individual lumbar erector spinae muscles in sitting. Scandinavian Journal of Rehabilitative Medicine. *3*:91, 1974.

44. Roozbazar, A.: Biomechanics of lifting. *In* Biomechanics IV. Edited by R.C. Nelson and C.A. Morehouse. Baltimore, University Park Press, 1974, pp. 37–43.

45. Takala, E., Leskinen, T. P. J., and Stalhammar, H. R.: Electromyographic activity of hip extensor and trunk mus-

cles during stooping and lifting. *In* Biomechanics X-A. Edited by B. Jonsson. Champaign, Ill., Human Kinetics, 1987.

46. Davies, P. R.: The use of intra-abdominal pressure in evaluating stresses on the lumbar spine. Spine. *6*:90–92, 1981.

47. McLaughlin, T. M., Lardner, T. J., and Dillman, C. J.: Kinetics of the parallel squat. Research Quarterly. *49*:175–189, 1978.

48. Pauly, J.E.: An electromyographic analysis of certain movements and exercises. I. Some deep muscles of the back. Anatomical Record. *155*:223, 1966.

49. Jonsson, B.: The functions of individual muscles in the lumbar part of the erector spinae muscle. Electromyography. *10*:5, 1970.

50. Thorstensson, A., Nilsson, J., Carlson, H., and Zomlefer, M. R.: Trunk movements in human locomotion. Acta Physiologica Scandinavia. *121*:9–22, 1984.

51. Dehaven, K. E., and Lintner, D. M.: Athletic injuries: Comparison by age, sport, and gender. American Journal of Sports Medicine. *14*:218–224, 1986.

52. Kelsey, J. L., et al.: Acute prolapsed lumbar intervertebral disc: An epidemiological study with special reference to driving automobiles and cigarette smoking. Spine. *9*:608–613, 1984.

53. Hilton, R. C.: Systematic studies of spinal mobility and Schmorl's nodes. *In* The lumbar spine and back pain. Edited by M. Jayson. Kent, Pitman Medical Publishing Company Ltd, 1976, pp. 115–131.

54. Saal, J.: Rehabilitation of football players with lumbar spine injury (Part 1). The Physician and Sportsmedicine. *16*:61–67, 1988.

55. Hoshina, H.: Spondylolysis in athletes. The Physician and Sportsmedicine: *3*:75–78.

56. Rossi, F.: Spondylolysis, spondylolisthesis and sports. Journal of Sports Medicine and Physical Fitness. *18*:317–340, 1978.

57. Weiker, G. G.: Evaluation and treatment of common spine and trunk problems. Clinics in Sports Medicine. *8*:399–417, 1989.

58. Wyke, B. The neurology of lower back pain. *In* The lumbar spine and back pain. Edited by M. Jayson. Kent, Pitman Medical Publishing Company Ltd, 1976, pp. 266–315.

59. Milne, J. S., and Lauder, I. J.: Age effects in kyphosis and lordosis in adults. Annals of Human Biology. *1*:327–337, 1974.

Glossary

Abdominals:	A name given to a combination of muscles including: rectus abdominus, internal oblique, external oblique, and transverse abdominus; flexors and rotators of the trunk.
Alar Ligament:	Ligament inserting on the apex of the dens and the medial occipital; limits lateral flexion and rotation of the head, holds dens into the atlas.
Annulus Fibrosus:	Rings of fibrocartilage that run in concentric layers around the nucleus pulposus in the intervertebral disc; absorbs tensile stress as the disc is compressed.
Anterior Longitudinal Ligament:	Ligament inserting from the sacrum, the anterior vertebral body and disc, up to the atlas; limits hyperextension; limits forward sliding of vertebrae.
Apical Ligament:	Ligament inserting on the apex of dens and foramen magnum; holds dens into atlas.
Apophyseal Joints:	Synovial joints between adjacent vertebrae, connected at the superior and inferior facets located on the laminae.
Atlantoaxial Joint:	Articulation between the atlas and the axis.
Atlanto-occipital Joint:	The articulation between the atlas with the occipital bone of the skull.
Atlas:	The first cervical vertebrae, which articulates with the occipital bone.
Axis:	The second cervical vertebrae.
Cervical:	The neck region of the trunk consisting of seven vertebrae.
Cervicothoracic Junction:	The vertebral region where the cervical curve ends and the thoracic curve begins; C-7 and T-1.
Costotransverse Ligament:	Ligament inserting on the tubercles of the ribs, transverse process of the vertebrae; supports rib attachment to thoracic vertebrae.
Cruciform Ligament:	Ligament inserting on the odontoid bone and the arch of the atlas; stabilizes the odontoid and atlas; prevents posterior movement of dens in atlas.
Dens:	Tooth-like process projecting from the superior surface of the axis; articulating surface with the atlas; also called odontoid process.
Disc Degeneration:	Gradual breakdown of the intervertebral disc in which splits and tears develop in the disc.
Disc Prolapse:	Injury to the intervertebral disc in which the nucleus pulposus extrudes into the annulus fibrosus.
Erector Spinae:	A name given to a combination of muscles including: iliocostalis, longissimus, and spinalis muscles; extensors of the trunk.
External Oblique:	Muscle inserting on ribs 9-12, anterior, superior spine, pubic tubercle, anterior iliac crest; flexes, laterally flexes and rotates the trunk to the opposite side.

Iliocostalis Cervices:	Muscle inserting on ribs 3-6, transverse process of C4-C6; extends, laterally flexes, and rotates the cervical region of the trunk to the same side.
Iliocostalis Lumborum:	Muscle inserting on the sacrum, spinous processes of L1-L5, T-11, T-12, iliac crest, lower six ribs; extends, laterally flexes and rotates the thoracic region of the trunk to the same side.
Iliocostalis Thoracis:	Muscle inserting on the lower six ribs, upper six ribs, transverse process of C-7; extends, laterally flexes, and rotates the thoracic region of the trunk of the same side.
Iliolumbar Ligament:	Ligament inserting on the transverse process of L-5 to the iliac crest; limits lumbar flexion and rotation.
Iliopsoas:	Two muscles, the iliacus and the psoas, which insert on the bodies of T-12, L1-L5, the transverse processes of L1-L5, and the inner surface of ilium, sacrum, and lesser trochanter; flexes the trunk and thigh.
Internal Oblique:	Muscle inserting on the iliac crest, lumbar fascia, ribs 8-10, and linea alba; flexes, laterally flexes, and rotates the trunk to the same side.
Interspinales:	Muscle inserting on the spinous processes; extends and hyperextends the trunk.
Interspinous Ligament:	Ligament inserting on the spinous processes; limits flexion of trunk; limits shear forces on the vertebrae.
Intertransversarii:	Muscle inserting on the transverse processes; extends and laterally flexes the trunk.
Intertransverse Ligament:	Ligament inserting on the transverse processes; limits lateral flexion of the trunk.
Intervertebral Disc:	Layers of fibrocartilage between the adjacent bodies of the vertebrae; a fibrous ring with a pulposus center.
Intervertebral Foramen:	A passage through the vertebrae, formed by the inferior and superior notches on the pedicles; pathway for spinal nerves.
Kyphosis:	Increase in the convexity of the vertebral curve to the posterior.
Laminae:	One of the paired dorsal parts of the vertebral arch, connecting to the pedicles.
Ligamentum Flavum:	Ligament inserting on the laminae; limits flexion of the trunk, creates extension of the trunk, creates tension in the disc.
Ligamentum Nuchae:	Ligament inserting on the laminae, connects with the supraspinous ligament; limits cervical flexion, assists in cervical extension, creates tension in the disc.
Longissimus Capitis:	Muscle inserting on the transverse processes of T1-T5, C4-C7, and mastoid process; extends, laterally flexes, and rotates the trunk.
Longissimus Cervices:	Muscle inserting on the transverse processes of T1-T5 and C4-C6; extends, laterally flexes, and rotates the trunk to the same side.
Longissimus Thoracic:	Muscle inserting on the transverse processes of L1-L5, thoracolumbar fascia, transverse process of T1-T12; extends, laterally flexes, and rotates the trunk to the same side.

Longus Capitis:	Muscle inserting on the transverse processes of C3-C6, and the occipital bone; flexes the head and the cervical region of the trunk, laterally flexes the trunk.
Longus Cervices and Colli:	Muscle inserting on the transverse processes of C3-C5, bodies of T1-T2, bodies of C5-C7, T1-T3, atlas, transverse processes of C5-C6, and bodies of C2-C4; flexes and laterally flexes the cervical region of the trunk.
Lordosis:	Increase in the anterior concavity of the vertebral curve.
Lumbar:	The lower region of the trunk between the thorax and the pelvis consisting of five vertebrae.
Lumbar Kyphosis:	Decrease in the lumbar curve; flat back.
Lumbar Lordosis:	Increase in the lumbar curve; sway back.
Lumbopelvic Rhythm:	The movement relationship and synchronization between the pelvis and the lumbar vertebrae.
Lumbosacral Junction:	The site on the vertebrae where the lumbar curve ends and the sacral curve begins; L-5 and S-1.
Multifidus:	Muscle inserting on the sacrum, iliac spine, transverse processes of L5-C4, and spinous processes; extends, laterally flexes, and rotates the trunk to the opposite side.
Neural Arch:	Protective arch for the spinal cord, formed by the laminae and the pedicles; also called the vertebral arch.
Nucleus Pulposus:	Spherical-shaped, gel-like mass in the middle of the intervertebral disc; resists compressive forces applied to the spine.
Odontoid Process:	Tooth-like process projecting from the superior surface of the axis; articulating surface with the atlas; also called dens.
Pars Interarticularis:	A site on the posterior neural arch.
Pedicle:	A paired stem that connects the lamina to the vertebral body; part of the vertebral, or neural arch.
Posterior Longitudinal Ligament:	Ligament inserting on the posterior vertebral bodies and discs of the vertebrae; limits flexion of the trunk.
Quadratus Lumborum:	Muscle inserting on the iliac crest, transverse process of L1-L5, and the last rib; laterally flexes the trunk.
Radiate Ligament:	Ligament inserting on the head of the ribs and the body of the vertebrae; maintains the ribs to the vertebrae.
Rotatores:	Muscle inserting on the transverse processes and laminae; extends and rotates the trunk to the opposite side.
Scaleni:	Muscle inserting on the transverse process of cervical vertebrae, and ribs 1 and 2; flexes and laterally flexes the cervical region of the trunk.
Scheuermann's Disease:	Necrosis and recalcification of the vertebrae; increase in kyphosis of the thoracic region because of vertebral wedging.

Schmorl's Nodes:	Vertical prolapse of part of the nucleus pulposus into an endplate lesion of an adjacent vertebrae.
Scoliosis:	A lateral curve of the spine.
Semispinalis Capitis:	Muscle inserting on the facets of C4-C6, transverse processes of C-7 to base of occipital; extends and laterally flexes the trunk.
Semispinalis Cervicis:	Muscle inserting on the transverse processes of T1-T6, and spinous processes of C1-C5; extends, laterally flexes, and rotates the trunk.
Semispinales Thoracis:	Muscle inserting on the transverse processes of T6-T10, spinous processes of T1-T4, C-6, and C-7; extends, laterally flexes, and rotates the trunk.
Spinales Thoracis:	Muscle inserting on the spinous processes of L-1, L-2, T-11, T-12, and spinous processes of T1-T8; extends and laterally flexes the trunk.
Spinales Cervices:	Muscle inserting on the spinous process of C-7 to the spinous process of the axis; extends and laterally flexes the trunk.
Spinous Process:	A projection posteriorly from each vertebrae, exiting at the arch.
Splenius Capitis:	Muscle inserting on the ligamentum nuchae, spinous processes of C7, T1-T3, mastoid process, and occipital bone; extends, laterally flexes, and rotates the cervical region of the trunk to the same side.
Splenius Cervicis:	Muscle inserting on the spinous processes of T3-T6, and the transverse processes of C1-C3; extends, laterally flexes, and rotates the cervical region of the trunk to the same side.
Spondylolisthesis:	Forward displacement of one vertebrae over another; bilateral defect at the pars interarticularis site.
Spondylolysis:	Fatigue fracture of the posterior neural arch of the vertebrae at the pars interarticularis site.
Sternocleidomastoid:	Muscle inserting on the sternum, clavicle, and mastoid process; flexes the head and cervical vertebrae, laterally flexes and rotates the cervical region of the trunk to the same side.
Supraspinous Ligament:	Ligament inserting on the spinous processes; limits trunk flexion, resists forward shear forces on the spine.
Thoracic:	The region of the trunk which refers to the chest or rib area, consisting of 12 vertebrae.
Thoracic Kyphosis:	Increase in the thoracic curve; hunch back.
Thoracolumbar Junction:	The region of the vertebrae where the thoracic curve ends and the lumbar curve begins; T-12 and L-1.
Transverse Process:	Projection on both sides of each vertebrae; projects from the junction of the laminae and the pedicles.
Transverse Abdominus:	Muscle inserting on the last six ribs, iliac crest, inguinal ligament, lumbodorsal fascia, linea alba, and pubic crest; increases the intra-abdominal pressure.

PART 3

Mechanical Analysis of Human Motion

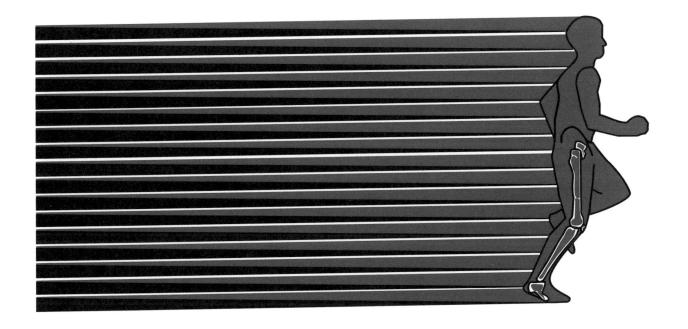

CHAPTER 8

Linear Kinematics

After reading this chapter, this student will be able to:
1. Describe how kinematic data are collected.
2. Distinguish between vectors and scalars.
3. Discuss the relationship among the kinematic parameters of position, displacement, velocity, and acceleration.
4. Distinguish between average and instantaneous quantities.
5. Conduct a numerical calculation of velocity and acceleration using the first central difference method.
6. Conduct a numerical calculation of the area under a parameter-time curve.
7. Sketch the general shape of the derivative of a curve.
8. Discuss various research studies that have utilized a linear kinematic approach.
9. Demonstrate knowledge of the three equations of constant acceleration.
10. Calculate the range of a projectile using the equations of constant acceleration.

The branch of mechanics that deals with the description of the spatial and temporal components of motion is called *kinematics*. The description involves the position, velocity, and acceleration of a body with no concern for the forces causing the motion. A kinematic analysis of motion may be either *qualitative* or *quantitative*. A qualitative kinematic analysis is a non-numerical description of a movement based on a direct observation. The description can range from a simple dichotomy of performance—good or bad—to a rather sophisticated identification of the joint actions. The key is, however, that it is non-numerical and subjective. Qualitative analyses are conducted primarily by teachers and coaches, among others.

In the field of biomechanics, there is greater interest in a quantitative analysis. The word "quantitative" implies a numerical result. In a quantitative analysis, the movement is analyzed numerically based on measurements from data collected during the performance of the movement. Movements may then be described precisely. Quantitative analyses are conducted by researchers but rarely by coaches and teachers. The researcher then uses this type of analysis to fully describe the movement techniques to interested parties. For example, a quantitative analysis may be conducted by a clinician on an individual who has cerebral palsy to determine this individual's gait pattern. The clinician quantifies the gait parameters for the surgeon who then decides what type of surgery is required to enable the patient to walk more effectively.

A sub-set of kinematics that is particular to motion in a straight line is called *linear kinematics*. *Translation* or *translational motion* is referred to as "straight line" motion and occurs when all points on a body or an object move the same distance over the same time. In Figure 8-1a, an object undergoes translation. The points A1 and B1 move to A2 and B2 respectively, in the same time. The distance from A1 to A2 and B1 to B2 is the same, thus translation. For example, a skater gliding across the ice maintaining a pose is an example of translation. While it appears that translation can occur only in a straight line, there are also instances when linear motion can occur along a curved path. This is known as *curvilinear motion* and is presented in Figure 8-1b. While the object undergoes a curved path, the distance from A1 to A2 and B1 to B2 is the same and is accomplished in the same amount of time. For example a sky-diver falling from an airplane prior to opening the parachute undergoes curvilinear motion.

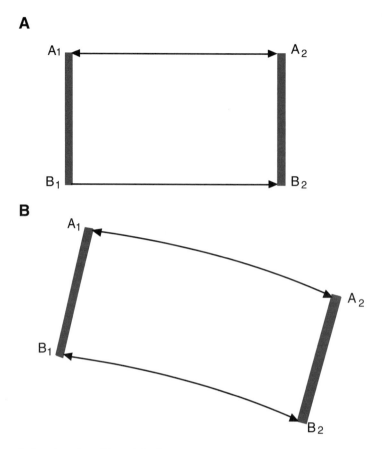

FIGURE 8-1. Types of translational motion: A) straight-line or rectilinear motion; and A) curvilinear motion. In both A) and B), the motion from A^1 to A^2 and B^1 to B^2 is the same and occurs in the same amount of time.

Collection of Kinematic Data

There are several methods by which kinematic data are collected for use in a quantitative analysis. Biomechanics laboratories, for example, may use accelerometers which measure the accelerations of body segments directly. The most common method of obtaining kinematic data, however, is high speed cinematography or high speed video. High speed video is presently a more popular technique than cinematography and will be referred to exclusively throughout, although the same tasks can be accomplished with high speed cinematography. The data obtained from high speed video result in the location of positions of body segments with respect to time. These data are acquired from the videotape by means of *digitization*, a computer-assisted technique that allows the movement to be analyzed frame-by-frame.

Reference Systems

Before analyzing the film, a spatial reference system is determined. There are many options for the biomechanist

in regards to a reference system; however, most laboratories use a *Cartesian coordinate system*. A Cartesian coordinate system is also referred to as a *rectangular reference system*. This system may either be two-dimensional or three-dimensional.

A two-dimensional reference system has two imaginary axes arranged perpendicular to each other (Figure 8-2a). The two axes (x, y) are usually positioned so that one is vertical (y) and the other is horizontal (x), although they may be oriented in any manner. For example, in certain circumstances, the axes may be re-oriented such that one axis (y) runs down the long axis of a segment. As the segment moves, the y-axis corresponding to the long axis of the bone also moves with the result that the y-axis may not necessarily be vertically aligned (Figure 8-2b).

An ordered pair of numbers is used to designate any point with reference to the axes with the intersection or origin of the axes designated as (0, 0). This pair of numbers is always designated in the order of the x-value followed by the y-value; these are referred to as the hori-

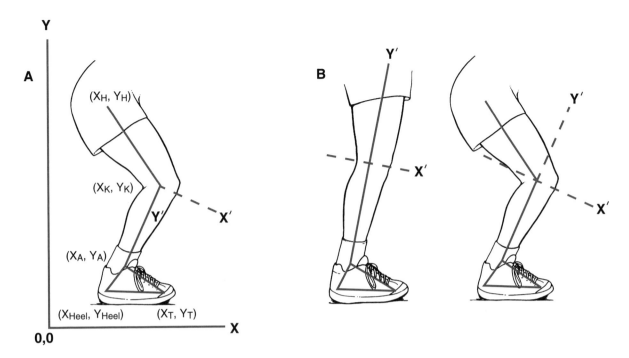

FIGURE 8-2. A two-dimensional reference system that defines the motion of all digitized points in a frame A) and a two-dimensional reference system placed at the knee joint center with the y-axis defining the long axis of the tibia.

zontal and vertical components, respectively. The x-value refers to the distance from the vertical axis and the y-value refers to the distance from the horizontal axis. The coordinates are usually written as (x, y) and can be used to designate any point on the x-y plane. A two-dimensional reference system is used when the motion being described is planar in nature. For example, if the object or body can be seen to move up or down (vertically) and to the right or to the left (horizontally) as viewed from one direction, the movement is planar. A two-dimensional reference system results in four quadrants, although the first quadrant is generally employed when digitizing because both the x- and y-values are positive (Figure 8-3).

If an individual was to flex and abduct the thigh as they swung it forward and out to the side, the movement would not be planar but would be three-dimensional in nature. In any physical space, three pieces of information are required to accurately locate parts of the body or any point of interest because the concept of depth (medial and lateral) must be added to the two-dimensional components of height (up and down) and width (forward and backward). Consequently, a three-dimensional coordinate system must be used to describe the movement in this instance. This reference system has three axes, each

perpendicular to the other to describe a position relative to the horizontal or x-axis, to the vertical or y-axis, and to

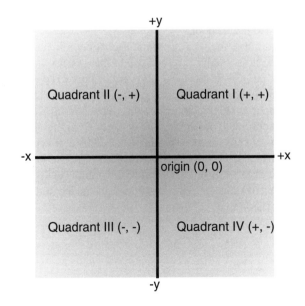

FIGURE 8-3. The quadrants and the signs of the coordinates in a two-dimensional coordinate system.

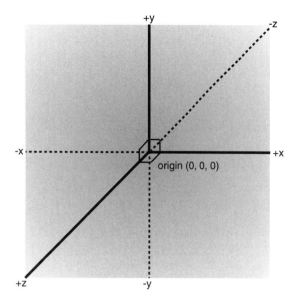

FIGURE 8-4. A three-dimensional coordinate system.

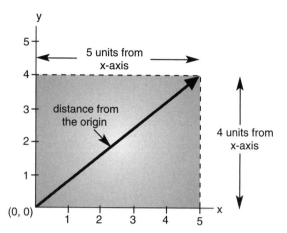

FIGURE 8-5. A 2-dimensional coordinate system illustrating the ordered pair of numbers defining a point relative to the origin.

the medial/lateral or z-axis. In a three-dimensional system (Figure 8-4), the coordinates would be written as (x, y, z). The intersection of the axes or the origin is defined (0, 0, 0) in three-dimensional space. All coordinate values are positive in the first quadrant of the reference system. In this system, the coordinates can designate any point on a surface—not just a plane—as in the two-dimensional system. A three-dimensional kinematic analysis of human motion is much more complicated than a two-dimensional analysis and thus will not be addressed in this book.

Figure 8-5 shows a two-dimensional coordinate system and how a point is referenced in this system. In this figure, point A is located 5 units from the y-axis and 4 units from the x-axis. The designation of point A is (5, 4). It is important to remember that the number designated as the x-coordinate determines the distance from the y-axis and the y-coordinate determines the distance from the x-axis.

The distance from the origin to the point is called the resultant (r) and can be determined using the Pythagorean Theorem as follows:

$$r = \sqrt{x^2 + y^2}$$

In this case:

$$r = \sqrt{5^2 + 4^2}$$
$$= 6.40$$

Prior to filming or videotaping the movement, the biomechanist may place markers on the subject. These markers are usually placed on the end points of the body segments to be analyzed. For example, if the biomechanist were interested in a sagittal view of the individual's leg, markers would be placed on the lateral condyle of the knee and the lateral malleolus of the ankle. Segments in question are generally identified in this manner. Figure 8-6 is a photograph of a single frame of a high speed video illustrating a sagittal view of a runner's leg. In certain circumstances, such as during a performance or competition, markers cannot be placed on the individual. In this case the biomechanist must estimate the end points of the body segments during the digitization process.

To analyze the video, the biomechanist projects the videotape one picture or frame at a time. The coordinate system is imposed on each frame with the origin placed at the lower left corner of the frame. Once placed, the coordinate system remains constant until all frames of the movement are digitized. In this way, each segment end point that is digitized can be referenced according to the x-y axes and identified in each frame for the duration of the movement.

Movements Occur Over Time

The analysis of the temporal or timing factors in human movement is an initial approach to a biomechanical analysis. In human locomotion, factors such as cadence, stride duration, duration of the stance or support (when the body is supported by a limb), duration of swing (when the limb is swinging through to prepare for the next ground contact) phases, and the period of non-

FIGURE 8-6. Photograph of a runner marked for a sagittal kinematic analysis of the right leg.

support, may be investigated. The knowledge of the temporal patterns of a movement is critical in a kinematic analysis since changes in position occur over time.

In a video analysis, the time interval between each frame is determined by the frame rate of the camera. This forms the basis for timing the movement. Video cameras generally operate at 30 fields or frames per second (fps). High speed video cameras however, typically used in biomechanics, can operate at 60 fps, 120 fps, 180 fps or 200 fps. For example at 60 fps, the time between each picture or frame is $\frac{1}{60}$ s (0.01667 s) or $\frac{1}{200}$ s (0.005 s) when the video has been obtained at 200 fps. Usually a key event at the start of the movement is designated as the beginning frame for digitization. For example, in a gait analysis, the first event may be considered to be the ground contact of the heel of the camera side foot. With camera side foot contact occurring at time 0.00 s, all subsequent events in the movement would be timed from this event.

Units of Measurement

If a quantitative analysis is conducted, it is necessary to report the findings in the correct units of measurement. In biomechanics, the *metric system* is used extensively in scientific research literature. The metric system is employed for everyday use in most countries in the world. In the United States, however, the English system is still employed. The metric system is based on the Systeme International d'Unites or the SI system. Every quantity of a measurement system has a dimension associated with it. The term dimension represents the nature of a quantity. In the SI system the base dimensions are mass, length, time, and temperature. Each dimension has a unit associated with it. The base units of the SI system are the kilogram (mass), the meter (length), the second (time), and degrees Kelvin (temperature). All other units used in biomechanics are derived from these base units. The SI units and their abbreviations are presented in Appendix D. Since SI units are used most often in biomechanics, they will be used exclusively in this text.

Vectors and Scalars

Certain quantities such as mass, distance and volume may be described fully by their amount or their magnitude. These quantities are called *scalar* quantities. For example, when one runs a race that is 5 kilometers (km) long, the distance or the magnitude of the race is 5 km. Other quantities, however, cannot be completely described by their magnitude alone. In these cases, the quantities are called *vectors* and are described by both their magnitude and direction. For example, when an object undergoes a displacement, the distance and the direction are important. Many of the quantities calculated in a kinematic analysis are vectors, so a thorough understanding of vectors is necessary.

In linear kinematics, vectors are represented by an arrow with the magnitude represented by the length of the line with the arrow pointing in the appropriate direction (Figure 8-7). Vectors are equal if their magnitudes are equal and they are pointed in the same direction.

Vectors can be added together. Graphically, vectors may be added by placing the tail of one vector at the head of the other vector (Figure 8-8a). In Figure 8-8b, the vectors are not in the same direction but the tail of b can still be placed at the head of a. By joining the tail of a to the head of b, the vector c, which is the sum of a + b or the resultant of the two vectors, can be determined. Subtracting vectors is accomplished by adding the negative of one of the vectors. That is :

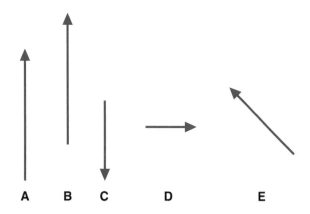

A   B   C   D   E

FIGURE 8-7. Examples of vectors. Only vectors A) and B) are equal because they are equivalent in magnitude and direction.

or
$$c = a + (-b)$$

This is illustrated in Figure 8-8c.

   Vectors may also undergo forms of multiplication that are used mainly in a three-dimensional analysis and will not be described in this book. Multiplication by a scalar, however, will be discussed. Multiplying a vector by a scalar changes the magnitude of a vector, but not its

direction. Therefore, the number "3" (a scalar) times the vector a is the same as adding a + a + a (Figure 8-8d).

   A vector may also be resolved or broken down into its horizontal and vertical components. In Figure 8-9a, the vector a is illustrated with its horizontal and vertical components. The vector may be resolved into these components using the trigonometric functions of the sine and cosine. A right angle triangle can be constructed consisting of the two components and the vector itself. Consider a right angle triangle with sides x, y, a in which a is the hypotenuse of the triangle (Figure 8-9b). The sine of the angle theta (θ) is defined as:

$$\sin \theta = \frac{\text{length of side opposite } \theta}{\text{hypotenuse}}$$

or

$$\sin \theta = \frac{y}{r}$$

The cosine of the angle θ is defined as:

$$\cos \theta = \frac{\text{length of side adjacent to } \theta}{\text{hypotenuse}}$$

or

$$\cos \theta = \frac{x}{r}$$

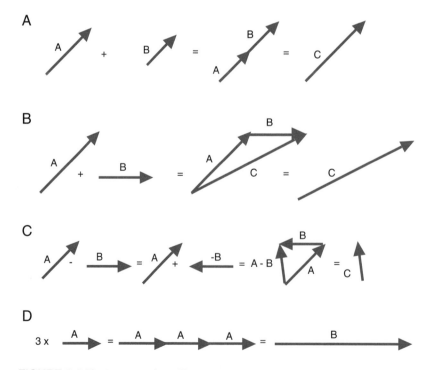

FIGURE 8-8. Vector operations illustrated graphically: A) and B) addition, C) subtraction, D) multiplication by a scalar.

a)

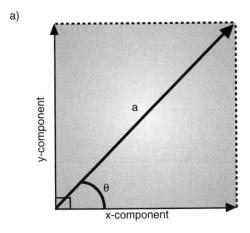

b)

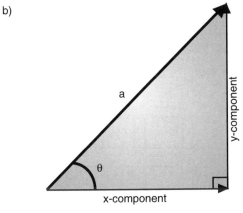

$$\sin \theta = \frac{\text{y-component}}{a} \qquad \cos \theta = \frac{\text{x-component}}{a}$$

FIGURE 8-9. Vector a resolved into its horizontal (x) and vertical (y) components using the trigonometric functions, sine and cosine. Components are illustrated in a). The components and the vector form a right angle triangle in b).

If the vector components, x and y, and the resultant r form a right-angle triangle, the sine and cosine functions can be used to solve for the components if the length of the resultant vector and the angle, [θ], that this vector makes with the horizontal are known.

If the resultant vector has a length of 7 units and the vector is at an angle of 43°, the horizontal component is found, using the definition of the cosine of the angle. That is:

$$\cos 43° = \frac{x}{a}$$

Re-arranging this equation to solve for the horizontal component:

$$\begin{aligned} x &= a \cos 43° \\ &= 7 * 0.7314 \\ &= 5.12 \end{aligned}$$

where cos 43° is 0.7314 (Appendix E).

The vertical component is found using the definition of the sine of the angle. That is:

$$\sin 43° = \frac{y}{a}$$

and re-arranging this equation to solve for the vertical component y:

$$\begin{aligned} y &= a \sin 43° \\ &= 7 * 0.6820 \\ &= 4.77 \end{aligned}$$

where sin 43° is 0.6820 (Appendix E).

The respective lengths of the horizontal and vertical components are, therefore, 5.12 and 4.77. These two values identify the point relative to the origin of the coordinate system.

Position and Displacement

The *position* of an object refers to its location in space relative to some reference. Units of length are used to measure the position of an object from a reference axis. Since the metric system is used almost exclusively in biomechanics, the most commonly used unit of length is the meter (m). For example, a platform diver standing on a 10 m tower is located 10 m from the surface of the water. The reference is the water surface and the diver's position is 10 m above the reference. The position of the diver may be determined throughout the dive with a height measured from the water surface. As previously mentioned, the process of digitization determines the position of a body or segment end point relative to two references in a two-dimensional reference system, the x-axis and the y-axis.

When the diver leaves the platform, motion occurs as they travel through the air towards the water. *Motion* occurs when an object or body changes position. Objects cannot instantaneously change position, thus time is of concern when considering the motion of an object. Motion, therefore, may be thought of as a progressive change of position over a period of time. In this example, the diver experienced a 10 m displacement from the diving board to the water. *Displacement* is measured in a straight line from one position to the successive position. Displacement is not to be confused with distance.

The *distance* an object travels may or may not be a straight line. In Figure 8-10, a runner starts the race, runs to point A, turns right to point B, left to point C, right to

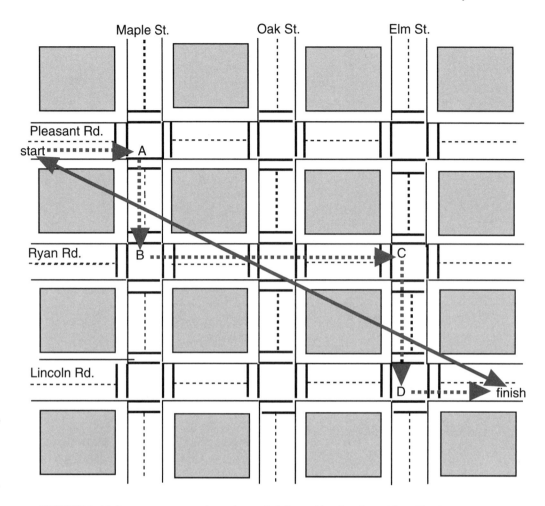

FIGURE 8-10. A runner moves along the path followed by the dotted line. The length of this path is the distance traveled. The length of the solid line represents the displacement of the object.

point D, and left to the finish. The distance run is the actual length of the path traveled. Displacement on the other hand is a straight line between the start and the finish of the race.

Displacement is defined by both how far the object has moved from its starting position and by the direction in which it has moved. Since displacement inherently describes the magnitude and the direction of the change in position, it is a vector quantity. Distance, because it refers only to how much an object moved, is a scalar quantity.

The Greek letter delta (Δ) refers to a change in a parameter; thus Δs would mean a change in s. If s represents the position of a point, then Δs is the displacement of that point. The subscripts f and i refer to the final position and the initial position respectively with the implication that the final position occurred after the initial position in

time. Mathematically, displacement (Δs) is for the general case:

$$\Delta s = s_f - s_i$$

where s_f is the final position and s_i is the initial position. Displacement for each component of position may also be calculated as follows:

$$\Delta x = x_f - x_i \text{ for the horizontal displacement}$$
$$\Delta y = y_f - y_i \text{ for the vertical displacement}$$

The resultant displacement may also be calculated by:

$$r = \sqrt{\Delta x^2 + \Delta y^2}$$

For example, if an object were located at position A (1.0, 2.0) at time 0.02 s and position B (7.0, 7.0) at time

0.04 s (Figure 8-11a), the horizontal and vertical displacements are:

$$\Delta x = 7.0 \text{ m} - 1.0 \text{ m}$$
$$= 6.0 \text{ m}$$
$$\Delta y = 7.0 \text{ m} - 2.0 \text{ m}$$
$$= 5.0 \text{ m}$$

The object would have been displaced 6 m horizontally and 5 m vertically. The movement may also be described as to the right and upwards relative to the origin of the reference system. The resultant displacement or the length of the vector from A to B may be calculated as:

$$r = \sqrt{6.0^2 \text{ m} + 5.0^2 \text{ m}}$$
$$= 7.81 \text{ m}$$

The point therefore is displaced a distance of 7.81 m in a direction up and to the right from the origin.

Consider Figure 8-11b. In a successive position to B, the object moved to position C (11.0, 3.0). The displacement would then be:

$$\Delta x = 11.0 \text{ m} - 7.0 \text{ m}$$
$$= 4.0 \text{ m}$$
$$\Delta y = 3.0 \text{ m} - 7.0 \text{ m}$$
$$= -4.0 \text{ m}$$

The object would have been displaced 4.0 m horizontally and – 4.0 m vertically or 4 m to the right away from the y-axis and 4 m downwards towards the x-axis. The resultant displacement between points B and C may be calculated as:

$$r = \sqrt{4.0^2 \text{ m} + 4.0^2 \text{ m}}$$
$$= 5.66 \text{ m}$$

The displacement from point B to C is a distance of 5.66 m in a direction to the right and downward towards the x-axis from point B.

Velocity and Speed

When the concepts of displacement and time are combined, the concept of velocity is of concern. *Velocity* is a vector quantity defined as the time rate of change of position. *Speed*, on the other hand, is a term that is used every day but it is a scalar quantity. In automobiles, for example, speed is measured constantly as one travels from place to place. Speed is defined as the distance an object has traveled divided by the time it took the object to travel the distance. Thus:

$$\text{speed} = \frac{\text{distance}}{\text{time}}$$

In everyday use the terms velocity and speed are interchangeable, but you can see from the definitions that velocity, a vector quantity, describes the magnitude and the direction while speed, a scalar quantity, only describes the magnitude. In road races, the start is usually close to the finish and the velocity over the whole race may be quite small. In this case, speed may be a more critical parameter to the participant.

In biomechanics, however, velocity is generally of more interest than speed. Velocity is usually designated by the lower case letter, v, and time by the lower case letter, t. Velocity can be determined by:

$$v = \frac{\text{displacement}}{\text{time}}$$

A

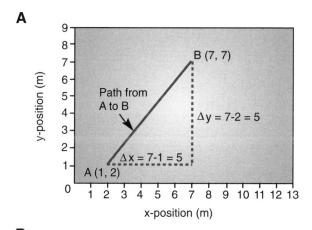

B

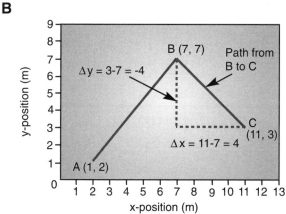

FIGURE 8-11. The horizontal and vertical displacements in a coordinate system of the path from a) A to B and b) B to C.

Specifically, velocity is:

$$v = \frac{\text{position}_{\text{final}} - \text{position}_{\text{initial}}}{\text{time at final position} - \text{time at initial position}}$$
$$= \frac{\text{change in position}}{\text{change in time}}$$
$$= \frac{\Delta s}{\Delta t}$$

Since the metric system of measurement is the recommended system in biomechanics, the most commonly used unit of velocity in biomechanics is m/s (meters per second) although any unit of length divided by a unit of time is correct as long as it is appropriate to the situation. The units for velocity can be determined by using the formula for velocity and dividing the units of length by units of time.

$$\text{velocity} = \frac{\text{displacement (m)}}{\text{time (s)}}$$
$$= \text{m/s}$$

Consider the position of an object that is at point A (2, 4) at time 1.5 s and moved to point B (4.5, 9) at time 5.0 s. The horizontal velocity (v_x) is:

$$v_x = \frac{4.5 \text{ m} - 2 \text{ m}}{5.0 \text{ s} - 1.5 \text{ s}}$$
$$= \frac{2.5 \text{ m}}{3.5 \text{ s}}$$
$$= 0.71 \text{ m/s}$$

The vertical velocity (v_y) could be similarly determined by:

$$v_y = \frac{9 \text{ m} - 4 \text{ m}}{5.0 \text{ s} - 1.5 \text{ s}}$$
$$= \frac{5 \text{ m}}{3.5 \text{ s}}$$
$$= 1.43 \text{ m/s}$$

The resultant or overall velocity can be calculated using the Pythagorean relationship as follows:

$$v = \sqrt{0.71^2 + 1.43^2}$$
$$= \sqrt{2.55}$$
$$= 1.60 \text{ m/s}$$

Figure 8-12 is an illustration of the change in horizontal position or position along the x-axis as a function of time. In this graph, the geometric expression describing the change in horizontal position (Δx) is called the *rise*. The expression which describes the change in time (Δt) is called the *run*. The slope of a line is:

$$\text{slope} = \frac{\text{rise}}{\text{run}} = \frac{\Delta x}{\Delta t}$$

The *slope* of a line indicates the relationship between two parameters, in this case, horizontal displacement and time. Therefore, the slope of the line plotted on a displacement-time graph is the relationship between displacement and time and represents the average velocity over the time interval.

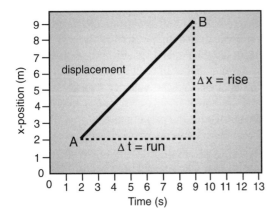

FIGURE 8-12. Horizontal position plotted as a function of time. The slope of the line from A to B is $\frac{\Delta x}{\Delta t}$

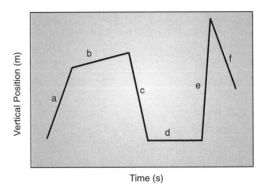

FIGURE 8-13. An illustration of different slopes on a vertical position versus time graph. Slopes a, b, and e are positive. Slopes c and f are negative while d has a zero slope.

The steepness of the slope gives a very clear picture regarding the velocity. If the slope is very steep, that is a large number, then the position is changing very rapidly and the velocity is great. If the slope is equal to zero, then the object has not changed position and the velocity is zero. However, since velocity is a vector, it can have both positive and negative slopes. In Figure 8-13 an illustration of positive, negative, and zero slopes is presented. Lines a and b have positive slopes, implying that the object was displaced away from the origin of the reference system. However, line a has a steeper slope than b, indicating that the object was displaced a greater distance per unit time. Line c illustrates a negative slope indicating the object was moving towards the origin. Line d shows a zero slope meaning that the object was not displaced either away from or towards the origin over that time interval. Lines e and f have identical slopes but e has a positive slope while f is a negative slope.

First Central Difference Method

The kinematic data that are collected in certain biomechanical studies are based on positions of the segment end points generated from each frame of video with a time interval based on the frame rate of the camera. This presents the biomechanist with all of the information that is needed to calculate velocity. However, when velocity over a time interval is calculated, the velocity at either end of the time interval is not generated; that is, the calculated velocity cannot be assumed to occur at the time of the final position nor at the time of the initial position.

The position of an object can change over a period of time less than the time interval between video frames. Thus the velocity calculated between two video frames represents an average of the velocities over the whole time interval between frames. An *average velocity*, therefore, is used to estimate the change in position over the time interval. This is not the velocity at the beginning of the time interval nor is it the velocity at the end of the time interval. If this is the case there must be some point in the time interval between frames when the calculated velocity occurs. The best estimate for the occurrence of this velocity is at the mid-point of the time interval. For example, if the velocity is calculated using the data at frames 4 and 5, the calculated velocity would occur at the mid point of the time interval between frames 4 and 5 (Figure 8-14a).

If data are collected at 60 fps, the positions at video frames 1, 2, 3, 4, and 5 would occur at the times 0 s, 0.0167 s, 0.0334 s, 0.0501 s, and 0.0668 s. The velocities calculated using this method would occur at the times 0.0084 s, 0.0251 s, 0.0418 s, and 0.0585 s. This means that after using the general formula for calculating velocity, the positions obtained from the video and velocities calculated are not exactly matched in time. While this is a problem that can be overcome, it may be inconvenient in certain calculations. To overcome this problem, the most often used method for calculating velocity is referred to as the *first central difference method*. This method employs the difference in positions over two frames as the numerator. The denominator in the velocity

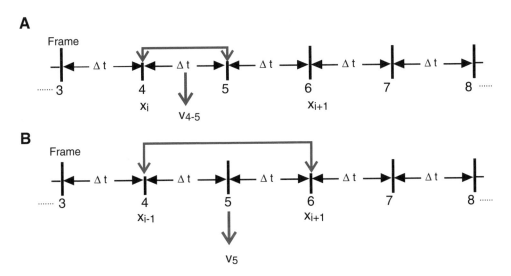

FIGURE 8-14. Illustration of the location in time of velocity calculated using: A) traditional method over a single time interval; B) first central difference method.

calculation is the change in time over two time intervals. The formula for this method is:

$$v_{xi} = \frac{x_{i+1} - x_{i-1}}{2\Delta t} \text{ for the horizontal component}$$

$$v_{yi} = \frac{y_{i+1} - y_{i-1}}{2\Delta t} \text{ for the vertical component}$$

This means that the velocity at frame i is calculated using the following position at frame i+1 and the previous position at frame i–1. By using 2Δt, the velocity calculated is at the same time as frame i since that is the mid-point of the time interval. For example, if the velocity at frame 5 is calculated, the data at frames 4 and 6 are used. If the time of frame 4 is 0.0501 s and frame 6 is 0.0835 s, the velocity calculated using this method would occur at time 0.0668 s or at frame 5 (Figure 8-14b).

Similarly, if the velocity at frame 3 is calculated, the positions at frame 2 and frame 4 are used. Since the time interval between the two frames is the same, the change in time would be two times the Δt. If the horizontal velocity at the time of frame 13 is calculated, the following equation would be used:

$$v_{\mathbf{x13}} = \frac{x_{14} - x_{12}}{t_{14} - t_{12}}$$

The location of the calculated velocity would be at t_{13} or the same point in time as the frame 13. Using this method of computation the position and velocity data are exactly aligned in time. It is assumed that the time intervals between frames of data are constant. As pointed out previously, this usually is the case in biomechanical studies.

The first central difference method uses the data point prior to and after the point where velocity is calculated. One problem that should be apparent is that data will be missing at the beginning of the video trial and the end of the video trial. This means that either the velocity at the beginning and end of the trial are estimated or some other means are used to evaluate the velocity at these points. A simple method is to digitize several frames prior to the event that begins the movement of interest and several frames after the event that ends the movement. If a walking stride, for example, was digitized, the first contact of the right foot on the ground might be picked as the beginning event for our video trial. In that case at least one frame prior to that event would be digitized in order to calculate the velocity at the instant of right foot contact. Similarly if the ending event in the trial was the subsequent right foot contact, at least one frame beyond that event would be digitized in order to calculate the velocity at the end event. In practice biomechanists generally digitize several frames prior to the initial event and several frames after the final event of the trial.

Numerical Example

The following data represent the vertical movement of an object over a 0.167 second time interval. In this set of data, the frame rate of the camera was 60 frames per second so that the Δt was ⅟₆₀ s or 0.0167 s. The object starts at rest, first moves vertically upward for 0.1002 s and then downward beyond the starting position before returning to the starting position. Table 8-1 shows the time at each frame, the vertical position for each frame, and the calculated velocity for each frame.

Using the formula for the first central difference method, if the velocity at the time for frame 3 is to be calculated, the computation would be as follows:

$$v_{y3} = \frac{y_4 - y_2}{t_4 - t_2}$$

$$= \frac{0.27 \text{ m} - 0.15 \text{ m}}{0.0501 \text{ s} - 0.0167 \text{ s}}$$

$$= 3.59 \text{ m/s}$$

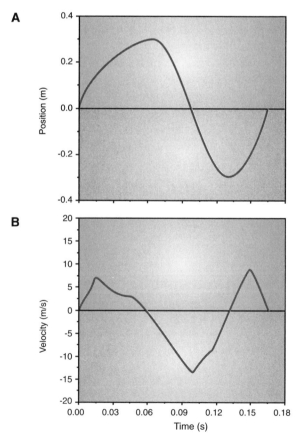

FIGURE 8-15. Position-time profile A) and velocity-time profile B) of the data in Table 8-1.

As another example, if the velocity at the time for frame 9 is to be calculated, the computation would be as follows:

$$v_{y9} = \frac{y_{10} - y_8}{t_{10} - t_8}$$

$$= \frac{-0.22 \text{ m} - -0.26 \text{ m}}{0.1503 \text{ s} - 0.1169 \text{ s}}$$

$$= 1.20 \text{ m/s}$$

Figure 8-15 is a graph of the position and velocity profiles of this movement. Each of these calculated velocities represents the slope of the straight line representing the rate of position change within that time interval or the average velocity over that time interval. Note that as the position changes rapidly, the slope of the velocity curve becomes steeper, as the position changes less rapidly the slope is less steep.

Instantaneous Velocity

Even when using the first central difference method, an average velocity over a time interval is computed. In some instances it may be necessary to calculate the velocity at a particular instant in time. When this velocity is calculated, it is called the *instantaneous velocity*. When the change in time, Δt, becomes smaller and smaller, the calculated velocity is the average over a much briefer

Table 8-1. Calculation of velocity from a set of position-time data.

Frame	Time (s)	Vertical Position (y) (m)	Velocity (v_y) (m/s)
1	0.0000	0.00	0.00
2	0.0167	0.15	6.59
3	0.0334	0.22	3.59
4	0.0501	0.27	2.40
5	0.0668	0.30	-2.10
6	0.0835	0.20	-8.98
7	0.1002	0.00	-13.77
8	0.1169	-0.26	-8.98
9	0.1336	-0.30	1.20
10	0.1503	-0.22	8.98
11	0.1670	0.00	0.00

time interval. The calculated value then approaches the velocity at a particular instant in time. In the process of making the time interval progressively smaller, the Δt will eventually approach zero. In a branch of mathematics called calculus, this calculation is called a *limit*. A limit occurs when the change in time approaches zero. The concept of the limit is graphically illustrated in Figure 8-16. If the velocity is calculated over the time interval from t_1 to t_2, as is done using the first central difference method, the slope of a line called a *secant* is cal-

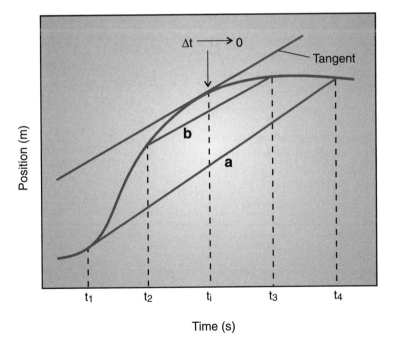

FIGURE 8-16. Slope of the secant a is the average velocity over the time interval t_1 to t_4. The slope of secant b is the average velocity over the time interval t_2 to t_3. The slope of the tangent is the instantaneous velocity at the time interval t_1 when the time interval is so small it is, in effect, zero.

culated. A secant line intersects a curved line at two points on the curve. The slope of this secant is the average velocity over the time interval t_1 to t_2. When change in time becomes very small, and approaches zero, however, the slope line actually touches the curve at only one point. This slope line is actually a *line tangent* to the curve or a line that touches the curve at only one point. The slope of the tangent represents the instantaneous velocity since the time interval is so small, that in effect, it is zero.

Instantaneous velocity, therefore, is the slope of a line tangent to the position-time curve. In calculus, instantaneous velocity is expressed as a limit. The numerator in a limit is represented by dx or dy meaning a small change in position in the horizontal or vertical positions, respectively. The denominator is referred to as dt, meaning a very small change in time. For the horizontal and vertical cases, the formulae for instantaneous velocity expressed as limits are:

$$\text{limit } \mathbf{v_x} = \frac{dx}{dt}$$
$$dt -> 0$$

$$\text{limit } \mathbf{v_y} = \frac{dy}{dt}$$
$$dt -> 0$$

For the instantaneous horizontal velocity, this is read as $^{dx}/_{dt}$ or the limit of v_x as dt approaches zero. It is also known as the *derivative* of x with respect to t. Similarly, the instantaneous vertical velocity, $^{dy}/_{dt}$ is the limit of V_y as dt approaches zero or the derivative of y with respect to t.

Graphical Example

It is possible to graph an estimation of the shape of a velocity curve based on the shape of the position-time profile. The ability to do this is critical to demonstrate our understanding of the concepts previously discussed. Two such concepts will be used to construct the graph: 1) the concept of the slope and 2) the concept of the local extremum. The point at which a curve changes direction (when it reaches a maximum or a minimum), is called a *local extremum*. The slope at this point is zero, thus the derivative of the curve at that point in time will be zero (Figure 8-17). That is, when the position changes direction, the velocity at the point of the change in direction will be instantaneously zero.

In Figure 8-18a the horizontal position of an object is plotted as a function of time. The local extrema, the points at which the curve changes direction, are indicated

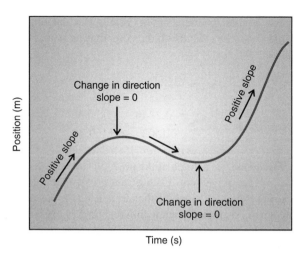

FIGURE 8-17. Illustration of local extrema (slope = 0) on a position-time graph.

as P_1, P_2, and P_3. At these points the velocity, by definition, will be zero. If the velocity curve is to be constructed on the same time line, these points can be projected to the velocity time line knowing that the velocity at these points will be zero. The slopes of each section of the position-time curve are 1) positive; 2) negative; 3) positive; and 4) negative. From the beginning of the motion to the local extremum P_1, the object was moving in a positive direction but at the local extremum P_1, the velocity was zero. The corresponding velocity curve in this section must increase positively and then become less positive, thereby returning to zero. In section 2) of the position-time curve, the slope is negative, indicating that the velocity must be negative. The local extrema, P_1 and P_2, however, indicate that the velocity at these points will be zero. Thus, in section 2) the corresponding velocity curve starts at zero, increases negatively, and then becomes less negative, returning to zero at P_2. Similarly, the shape of the velocity curve can be generated for sections 3) and 4) on the position curve (Figure 8-18b).

Acceleration

In human motion the velocity of a body or a body segment is rarely constant. The velocity often continually changes throughout a movement. Even when the velocity is constant, it may be so only when averaged over a large time interval. For example, in a distance race, the runner may run consecutive 300 m distances in 65 s, indicating a constant velocity over each distance. A detailed analysis, however, would reveal that the runner actually increased velocity and decreased velocity with the average over the 300 m being constant. In fact, it has been

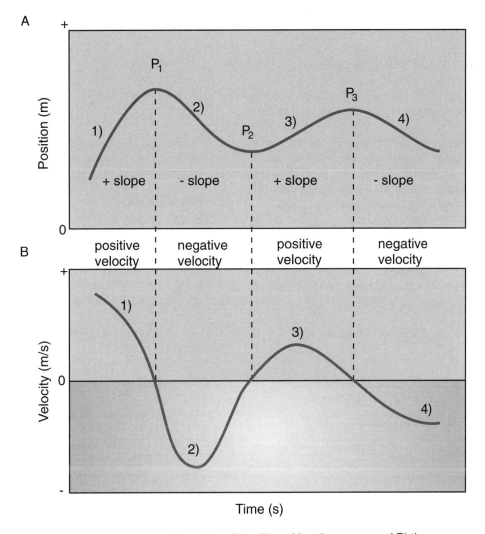

FIGURE 8-18. Graphical illustration of the A) position-time curve and B) the
respective velocity-time curve drawn using the concepts of local extrema and slopes.

shown that runners decrease their velocity and then increase velocity during each ground contact with each foot (1). If velocity continually changes, it would appear that these variations in velocity must be described. In addition, the rate at which velocity changes can be related to the forces that cause movement.

The rate of change of velocity with respect to time is called *acceleration*. In everyday usage, accelerating means speeding up. In a car when the accelerator is depressed, the speed of the car is increased. When the accelerator is released, the speed of the car decreases. In both instances the direction of the car is not of concern since speed is a scalar. Acceleration, however, refers to both increasing and decreasing velocities. Since velocity is a vector, acceleration must also be a vector.

Acceleration, usually designated by the lower case letter *a*, can be determined by:

$$a = \frac{\text{change in velocity}}{\text{change in time}}$$

More generally, acceleration is:

$$a = \frac{\text{velocity}_{final} - \text{velocity}_{initial}}{\text{time at final position} - \text{time at initial position}}$$
$$= \frac{\text{change in velocity}}{\text{change in time}}$$
$$= \frac{\Delta v}{\Delta t}$$

The units of acceleration are the unit of velocity (m/s) divided by the unit of time (s) resulting in m/s/s or m/s^2.

$$acceleration = \frac{velocity \ (m/s)}{time \ (s)}$$

This is the most common unit of acceleration used in biomechanics.

Since acceleration represents the rate of change of a velocity with respect to time, the concepts regarding velocity also apply to acceleration. Thus, acceleration may be represented as a slope indicating the relationship between velocity and time. On a velocity-time graph, the steepness and direction of the slope will indicate whether the acceleration is positive, negative or zero. In addition, instantaneous acceleration may be defined in an analogous fashion to instantaneous velocity. Instantaneous acceleration may be defined as the slope of a line tangent to a velocity time graph or as a limit:

$$limit \ \boldsymbol{a_x} = \frac{dv_x}{dt} \ \text{for horizontal acceleration}$$
$$dt - > 0$$

$$limit \ \boldsymbol{a_y} = \frac{dv_y}{dt} \ \text{for vertical acceleration}$$
$$dt - > 0$$

The term dv refers to a change in velocity. Horizontal acceleration is the limit of v_x as dt approaches zero and vertical acceleration is the limit of v_y as dt approaches zero.

The first central difference method is also used to calculate acceleration in many biomechanical studies. The use of this method means that the calculated acceleration is associated with a time in the movement in which a calculated velocity and a digitized point are also associated.

The first central difference formula for calculating acceleration is analogous to that for calculating velocity:

$$\boldsymbol{a_{xi}} = \frac{Vx_{i+1} - Vx_{i-1}}{2\Delta t} \ \text{for the horizontal component}$$

$$\boldsymbol{a_{yi}} = \frac{Vy_{i+1} - Vy_{i-1}}{2\Delta t} \ \text{for the vertical component}$$

For example, to calculate the acceleration at frame 7, the velocity values at frames 8 and 6, and 2 times the time interval between individual frames would be used.

Numerical Example

The velocity data previously calculated from the data in Table 8-1 representing the vertical (y) position data of an object will be used to illustrate the first central difference method in acceleration calculations. Table 8-2 presents the time at each frame, the vertical position, the vertical velocity, and the calculated vertical acceleration for each frame.

To calculate the acceleration at the time of frame 4, the computation is as follows:

$$\begin{aligned} \boldsymbol{a_{y4}} &= \frac{v_5 - v_3}{t_5 - t_3} \\ &= \frac{-2.10 \ m/s - 3.59 \ m/s}{0.0668 \ s - 0.0334 \ s} \\ &= -170.36 \ m/s^2 \end{aligned}$$

As another example, if the acceleration at the time of frame 8 is calculated, the computation is as follows:

$$\begin{aligned} \boldsymbol{a_{y8}} &= \frac{v_9 - v_7}{t_9 - t_7} \\ &= \frac{1.20 \ m/s - 13.77 \ m/s}{0.1336 \ s - 0.1002 \ s} \\ &= 448.20 \ m/s^2 \end{aligned}$$

Table 8-2. Calculation of acceleration from a set of velocity-time data.

Frame	Time (s)	Vertical Position (y) (m)	Velocity (v_y) (m/s)	Acceleration (a_y) (m/s^2)
1	0.0000	0.00	0.00	0.0000
2	0.0167	0.15	6.59	107.49
3	0.0334	0.22	3.59	−125.45
4	0.0501	0.27	2.40	−170.36
5	0.0688	0.30	−2.10	−340.72
6	0.0835	0.20	−8.98	−349.40
7	0.1002	0.00	−13.77	0.00
8	0.1169	−0.26	−8.98	448.20
9	0.1336	−0.30	1.20	537.72
10	0.1503	−0.22	8.98	−35.93
11	0.1670	0.00	0.00	0.00

Figure 8-19 is a graph of the velocity and acceleration profiles of the complete movement. Note, that as the velocity increases rapidly, the slope of the acceleration curve becomes steeper, and as the velocity changes less rapidly, the slope is less steep.

Graphical Example

Previously, an estimation of the shape of the relationship between position and velocity was graphed using the concepts of slope and local extrema. It is also possible to graph an estimation of the shape of an acceleration curve

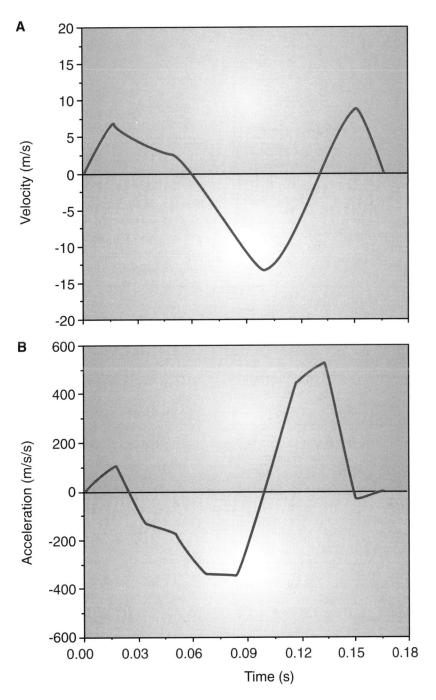

FIGURE 8-19. Velocity-time profile (A) and acceleration-time profile (B) for Table 8-2.

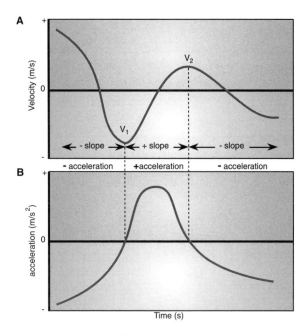

FIGURE 8-20. Graphical illustration of the relationship between a velocity-time and the acceleration-time curve drawn using the concepts of local extrema and slopes.

based on the shape of the velocity-time profile. Once again the two concepts of the slope and the local extrema are used, this time on a velocity-time graph. Figure 8-20a represents the horizontal velocity of the data previously presented in Figure 8-18. The local extrema of the velocity curve, where the curve changes direction, are indicated as V_1, and V_2. At these points the acceleration is zero. Constructing the acceleration curve on the same time line as the velocity curve, the occurrence of these local extrema from the velocity curve time line can be projected to the acceleration time line.

The slopes of each section of the velocity-time curve are 1) to V_1, negative; 2) V_1 to V_2, positive; and 3) beyond V_2, negative. The velocity curve to V_1 has a positive slope, but the curve reaches the local extremum at V_1. The corresponding acceleration curve of this section (Figure 8-20b) is negative, but it becomes zero at the local extremum V_1. Between V_1 and V_2 the velocity curve has a positive slope. The acceleration curve between these points in time will begin with a zero value at the time corresponding to V_1, become more positive and eventually return to zero at a time corresponding to V_2. Similar logic can be used to describe the construction of the remainder of the acceleration curve.

Acceleration and the Direction of Motion

One complicating factor in understanding the meaning of acceleration relates to the direction of motion of an object. The term "accelerate" is often used to indicate an increase in velocity and the term "decelerate" to describe a decrease in velocity. These terms are satisfactory when the object under consideration is moving in the same direction continually. Even if velocity, and therefore acceleration change, the direction in which the object is traveling may not necessarily change. For example, a runner in a 100 m sprint race starts from rest or from a zero velocity. When the race begins, the runner increases velocity up to the 70 m point in the race, thereby changing their velocity to some greater value than at rest and their acceleration is positive. After the 70 m mark, their velocity may not change for a period of the race, resulting in zero acceleration. After the runner crosses the finish line, the runner reduces their velocity, resulting in a negative acceleration. Eventually the runner comes to rest, at which point their velocity equals zero. Throughout the race, the runner moved in the same direction but had positive, zero, and negative accelerations. Thus it can be seen that acceleration may be considered to be independent of the direction of motion.

Consider an athlete completing a shuttle run that consists of one 10 m run away from a starting position, followed by a 10 m run back to the starting position. The two sections of this run are illustrated in Figure 8-21. The first 10 m section of the run may be considered a run in a positive direction. The runner, running in a positive direction, increases their velocity and then, as they approach the turn-around point, they must decrease their positive velocity. Thus the runner initially has a positive acceleration followed by a negative acceleration. At the turn-around point, the runner, now running in a negative direction, increases their negative velocity. As they approach the finish line, they must decrease their negative velocity in order to have a positive acceleration. Thus, since positive and negative accelerations occur in positive and negative directions, it may be seen that acceleration is independent of the direction of motion.

In Figure 8-22, an idealized horizontal velocity profile and the corresponding horizontal acceleration profile of a shuttle run are presented. From t_0 to t_2, the velocity is positive because the change in position was constantly away from the y-axis. In addition, the slope of the velocity curve from t_0 to t_1 is positive, indicating a positive acceleration, while the slope of the velocity curve from t_1 to t_2 is negative, resulting in a negative

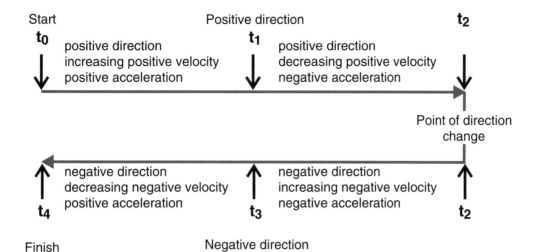

FIGURE 8-21. Motion to the right is regarded as positive and to the left is negative. Positive and negative velocity is based on the direction of motion. Acceleration may be positive, negative, or zero based on the change in velocity.

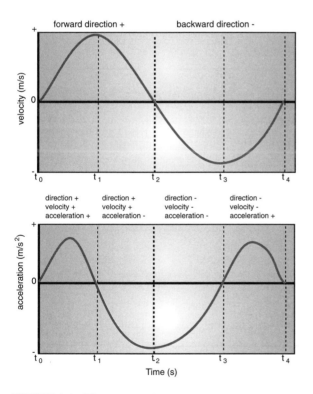

FIGURE 8-22. The graphical relationship between acceleration and direction of motion during a shuttle run (t_2 denotes when the runner changed direction).

acceleration. Thus, both positive and negative accelerations can result without the object changing direction.

From t_2 to t_4, the velocity is negative because the object moved back towards the y-axis or the reference point. The slope of the velocity curve from t_2 to t_3 is negative, indicating a negative acceleration. However, the slope of the velocity curve from t_3 to t_4 is positive, resulting in a positive acceleration. Once again, even though the direction did not change, positive and negative accelerations resulted.

It should be noted that if the final velocity is greater than the initial velocity, the acceleration is positive. For example:

$$a = \frac{v_f - v_i}{t_f - t_i}$$
$$= \frac{10 \text{ m/s} - 3 \text{ m/s}}{3 \text{ s} - 1 \text{ s}}$$
$$= \frac{7 \text{ m/s}}{2 \text{ s}}$$
$$= 3.5 \text{ m/s}^2$$

If, however, the final velocity is less than the initial velocity, the acceleration is negative. For example:

$$a = \frac{v_f - v_i}{t_f - t_i}$$
$$= \frac{4 \text{ m/s} - 10 \text{ m/s}}{5 \text{ s} - 3 \text{ s}}$$
$$= \frac{-6 \text{ m/s}}{2 \text{ s}}$$
$$= -3 \text{ m/s}^2$$

In the first case it is said that the object is accelerating, and in the latter, decelerating. These terms become confusing, however, when the object actually changes direction. For the sake of easing confusion, it is best that the terms "acceleration" and "deceleration" be avoided, and the use of positive and negative acceleration is encouraged.

Differentiation and Integration

Thus far a kinematic analysis has been described that is based on a process whereby position data are accumulated first. When velocity is calculated from the combination of displacement and time, or when acceleration is calculated from the combination of velocity and time, the mathematical process involved is called *differentiation*. The solution of the process of differentiation is called a *derivative*. A derivative is simply the slope of a line, either a secant or tangent, as a function of time. Thus, when velocity is calculated from position and time, differentiation is the method used to calculate the derivative of position. Velocity is called the derivative of displacement and time. Similarly, acceleration is the derivative of velocity and time.

In certain situations, however, acceleration data may be collected. From these data, velocities and positions may be calculated based on a process that is opposite to that of differentiation. This mathematical process is known as *integration*. Integration is often referred to as anti-differentiation. The result of the process of integration is called the *integral*. Velocity, then, is the time integral of acceleration. The following equation describes the above statement:

$$v = \int_{t1}^{t2} a\, dt$$

This expression reads that velocity is the integral of acceleration from time 1 to time 2. The terms t1 and t2 define the beginning and ending points at which the velocity is evaluated. Likewise, position is the integral of velocity and is expressed as:

$$s = \int_{t1}^{t2} v\, dt$$

The meaning of the integral is not quite as obvious as that of the derivative, however. The process of integration requires calculating the area under a velocity-time curve to calculate the average displacement or the area under an acceleration-time curve to calculate the average velocity. The integration sign,

$$\int_{t1}^{t2}$$

is an elongated "s" and indicates a summation of areas between time t1 and time t2. Figure 8-23 illustrates the concept of the area under the curve. For purposes of illustration, two rectangles have been drawn representing a constant acceleration of 3 m/s^2 for a period of 6 s in the first portion of the curve and a constant acceleration of 7 m/s^2 for a period of 2 s. To calculate the area of a rectangle, the product of length times width is calculated. Thus, the area under the first rectangle is 3 m/s^2 times 6s or 18 m/s. In the latter rectangle, the area is 7 m/s^2 times 2 s or 14 m/s. The total area is 32 m/s.

The area under an acceleration-time curve is the change in velocity over the time interval. This can be demonstrated by an analysis of the units in calculating the area under the curve. For example, taking the area under an acceleration-time curve involves multiplying an acceleration value by a time value:

$$\text{Area under the curve} = \text{acceleration} * \text{time}$$
$$= \frac{m}{s^2} * s$$
$$= \frac{m}{s * s} * s$$
$$= m/s$$

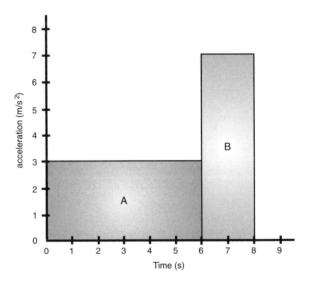

FIGURE 8-23. An idealized acceleration-time curve. Area A equals 3 m/s^2 * 6 s or 18 m/s. This represents the change in velocity over the time interval from 0 s to 6 s. The change in velocity for area B is 14 m/s.

The area under the curve would have units of velocity. Thus, a measure of velocity is the area under an acceleration-time curve. This area represents the change in velocity over the time interval in question. Similarly, the change in displacement is the area under a velocity-time curve.

Velocity-time or acceleration-time curves do not generally form rectangles as in the previous examples and thus the computation of the integral is not quite so simple. The technique generally used is called a *riemann sum* and is dependent on the size of the time interval, dt. If dt is small enough, and it generally is in a kinematic study, the integral or area under the curve can be calculated by progressively summing the product of each data point along the curve and dt. For example, if the curve to be integrated is a horizontal velocity-time curve, then the integral equals the change in position. If the horizontal velocity-time curve is made up of 30 data points each 0.005 s apart, the integral would be:

$$\int_{t_1}^{t_{30}} v_{xi} dt = ds$$

and to find the area under the curve:

$$ds = \sum_{i=1}^{30} \left(v_{xi} * dt \right)$$

The Riemann Sum calculation generally gives an excellent estimation of the area under the curve.

Kinematics of Running

A kinematic analysis describes the positions, velocities, and accelerations of bodies in motion. It is one of the most basic types of analyses that may be conducted, because it is used only to describe the motion with no reference to the causes of motion. Kinematic data are usually collected using high speed video cameras and the video frames are digitized to generate the positions of the body segments. In order to illustrate kinematic analyses in biomechanics, the study of human gait will be used as an example. The most studied forms of human gait are walking and running. Since there are a number of research papers on the kinematics of gait, this discussion will be limited only to the kinematics of running.

In both locomotor forms of movement, the body actions are cyclic, involving sequences in which the body is supported by first one leg and then the other. These sequences are defined by certain parameters. Typical parameters such as the *stride* and *step* are presented in Figure 8-24. A locomotor cycle or stride is defined by events in these sequences. A stride is defined from one event on one leg until the same event on the same leg in the following contact. Usually an event such as the first instant at foot contact will define the beginning of a stride. For example, a stride could be defined from heel contact of the right limb to subsequent heel contact on the right limb. The stride is further sub-divided into a step. A step is a portion of the stride from an event occurring on one leg to the same event occurring on the opposite leg. For example, a step could be defined as foot contact on

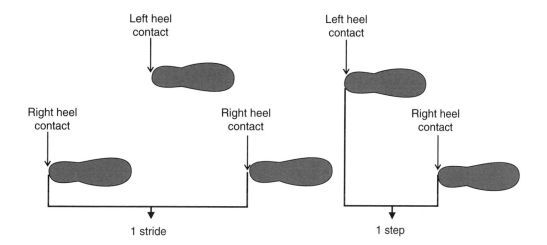

FIGURE 8-24. Illustration of stride parameters during gait.

the right limb to foot contact of the left limb. Thus, two steps equal one stride.

Stride length and *stride rate* are among the most commonly studied kinematic parameters. The distance covered by one stride is the stride length and the number of strides per minute is called the stride rate. Running velocity is the result of the relationship between stride rate and stride length. That is:

Running speed = stride length * stride rate

Runners can increase their running velocity by increasing their stride length or their stride rate or increasing both stride rate and stride length.

A number of studies have investigated this relationship as running velocity increased from a slow jog to a sprint (2, 3, 4, 5). These studies have shown that there is an increase in both stride rate and stride length with increasing velocity. This is illustrated in Figure 8-25. For velocities up to 7 m/s the increases that have been reported are linear while at higher speeds there is a smaller increment in stride length and a greater increment in stride rate. This indicates that when sprinting, runners increase their velocity by increasing their stride rate more than their stride length. A runner initially increases their velocity by increasing their stride length. However, there is a physical limit to how much an individual can increase their stride length. To run faster, therefore, the runner must then increase their stride rate.

The running stride can be further sub-divided into phases known as *support* or *stance* and *non-support* or *swing*. The support or stance phase occurs when the foot is in contact with the ground, that is, from the point of foot contact until the foot leaves the ground. During support, the point at which the center of mass of the runner is directly over the base of support is referred to as mid-stance. The non-support or swing phase occurs from the point that the foot leaves the ground until the same foot contacts the ground again. It has also been reported that, as running speed increases, the time for a running cycle decreases (6). In addition, the absolute time and the relative time (a percent of the total stride time) spent in support decrease as running speed increases (1, 7). Typical changes in relative time range from 68% at a jogging pace to 54% at a moderate sprint pace to 47% at a full sprint.

The velocity of the runner during the race has also been studied by biomechanists. In several cases, runners were considered as single points and no consideration was given to the movement of the arms and legs as individual units. Over the years, a number of researchers have tried to measure the velocity curve of a runner during a sprint race (8). A. V. Hill, who later won the Nobel Prize in physiology, proposed a simple mathematical

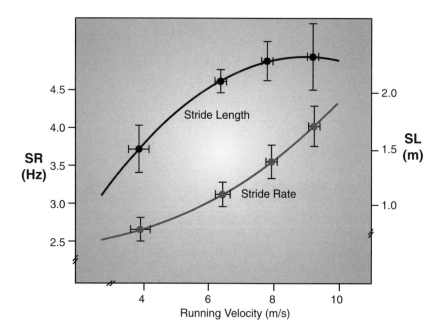

FIGURE 8-25. Changes in stride length and stride rate as a function of running velocity (after Luhtanen, P. and Komi, P. V. Mechanical factors influencing running speed. In E. Asmussen and K. Jorgensen (eds.). Biomechanics VI-B. Baltimore: University Park Press, 1973.).

model to represent the velocity curve, and subsequent investigations have confirmed this model (Figure 8-26). Most sprinters conform relatively closely to this model. At the start of the race the runner's velocity is zero. The velocity increases rapidly at first but then levels off to a constant value. This means that the runner accelerates rapidly at first, but that the acceleration decreases toward the end of the run. The sprinter cannot increase velocity indefinitely throughout the race. In fact, the winner of a sprint race is usually the runner whose velocity decreases the least at the end of the race. In a study of female sprinters (9), it was reported that the sprinters reached their maximum velocity between 23 and 37 m in a 100 m race. It was also reported that these sprinters lost an average of 7.3% from their maximum velocity in the final 10 m of the race.

The fastest instantaneous velocity of a runner during a race has not yet been measured during competition. Average speed can be readily calculated, however. Carl Lewis, during his gold medal performance at the World Championships in Rome in 1991, covered 100 m in 9.85 s for an average speed of 10.15 m/s or a speed equivalent to 22.7 miles/hour.

When calculating the average velocity over a race, it is important to remember that this was not the velocity of the runner at every instant during the race. During a race, a runner contacts the ground numerous times and it is important to note what occurs to the horizontal velocity during these ground contacts. The horizontal velocity of a runner during the support phase of the running stride from a study by Bates et al. (1) is presented in Figure

8-27a. An analysis of runners in this study indicated that the horizontal velocity decreased immediately at foot contact and continued to decrease during the first portion of the support period. As the runner's leg is extending in the latter portion of the support period, the velocity increases. The corresponding acceleration-time graph of a runner during the support phase (Figure 8-27b) shows distinct negative and positive accelerations. It can be seen that the runner instantaneously has a zero acceleration during the support phase, representing the transition from a negative acceleration to a positive acceleration. This results from the runner slowing down during the first portion of support and speeding up in the latter portion. To maintain a constant average velocity, the runner must gain as much speed in the latter portion of the support phase as was lost in the first portion.

Projectile Motion

Previously, situations that arise when objects undergo a change in velocity have been discussed. There are, however, special circumstances when the rate of change in velocity is constant. If the change in velocity remains the same, the slope, and therefore the acceleration, is zero. This situation occurs in projectile motion. Projectile motion refers to the motion of bodies that have been projected into the air. This type of motion occurs in many activities such as baseball, diving, figure skating, basketball, golf, and volleyball. Teachers, coaches, and athletes who perform in such activities should have some knowledge of the factors that influence projectile motion.

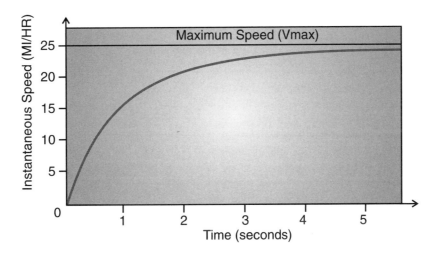

FIGURE 8-26. Graphical representation of Hill's proposed mathematical model of a sprint race velocity curve (after Brancazio, P. J. Sport Science. New York: Simon and Schuster, 1984.).

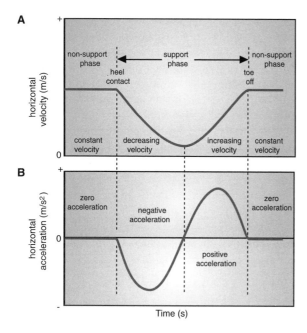

FIGURE 8-27. Changes in A) velocity and B) acceleration during the support phase of a running stride (after Bates, B. T., Osternig, L. R., Mason, B. R. Variations of velocity within the support phase of running. In J. Terauds and G. Dales (eds.). Science in Athletics. Del Mar: Academic Publishers, 1979.).

When no other forces are acting on a body, the force of gravity on a projectile results in constant acceleration. The acceleration due to gravity is approximately 9.81 m/s² at sea level and results from the attraction of two masses—the earth and the object. Only gravity and air resistance act on an object when the object is moving through the air unassisted. Objects in this situation are called *projectiles*. Gravity uniformly accelerates a projectile toward the earth's surface and air resistance retards its progress. It should be noted, however, that not all objects that fly through the air are projectiles. Objects such as airplanes are not projectiles because they also are influenced by forces from their engines.

For the following discussion, however, air resistance will be considered negligible since it is relatively small compared to gravity. Depending on the projectile, different kinematic questions may be asked. For example, in the long jump or the shot put, the horizontal displacement is critical. In the high jump or pole vaulting, however, the vertical displacement must be maximized. In biomechanics, it is important to understand the nature of projectile motion.

Trajectory of a Projectile

The flight path that a projectile takes through the air is called its *trajectory* (Figure 8-28a). The instant at which

an object becomes a projectile—such as when a pitcher releases the baseball—is known as the *instant of release*. Gravity continually acts to change the motion of the object once it has been released. If gravity did not act on the projectile, it would continue to travel indefinitely with the same velocity it had when it was released (Figure 8-28b). In space, when a space vehicle is beyond the earth's gravitational pull, a short firing burst of the vehicle's engine will result in a change in velocity. When the engine ceases to fire, the velocity at that instant remains constant, resulting in a zero acceleration. Since there is no gravity, the vehicle will continue on this path until the engine is fired again.

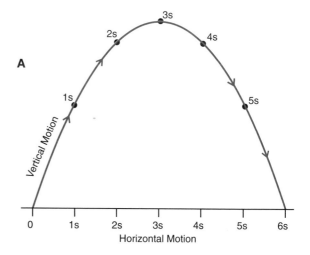

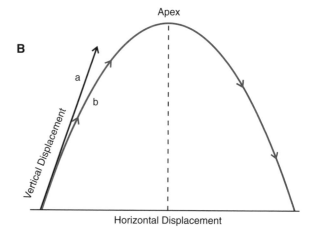

FIGURE 8-28. A) The parabolic trajectory of a projectile; B) path a represents the trajectory of a projectile without the influence of gravity while path b is a trajectory with gravity acting. Path b forms a parabolic trajectory.

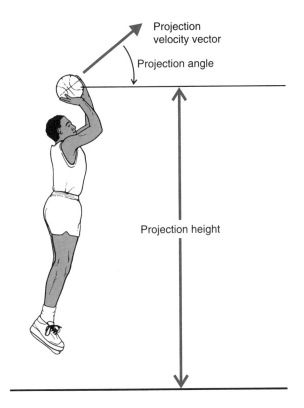

FIGURE 8-29. The factors influencing the trajectory of a projectile are: 1) projection velocity; 2) projection angle; and 3) projection height.

The flight path followed by a projectile in the absence of air resistance is in the shape of a *parabola* (Figure 8-28a). A parabola is a curved shape that is symmetrical about an axis through its highest point. The highest point of a parabola is called its *apex*.

Factors Influencing Projectiles

There are three primary factors that influence the trajectory of a projectile: *the projection angle*, *projection velocity*, and the *projection height* (Figure 8-29).

Projection Angle. The angle at which the object is released determines the shape of the projectile's trajectory. Projection angles generally vary from 0° (parallel to the ground) to 90° (perpendicular to the ground), although in some sporting activities, such as ski-jumping, the projection angle is negative. If the projection angle is 0° (parallel to the horizontal), the trajectory becomes essentially the latter half of a parabola because it has zero vertical velocity, and is immediately acted upon by gravity to pull it to the earth's surface. On the other hand, if the projection angle is 90°, the object would be projected straight up into the air with zero horizontal velocity. In this case, the parabola would be so narrow as to form a straight line.

If the projection angle is between 0° and 90°, the trajectory would be truly parabolic in shape. Figure 8-30 displays theoretical trajectories for an object projected at different projection angles with the same speed and height of projection.

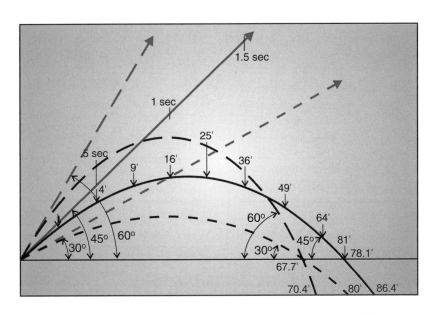

FIGURE 8-30. Theoretical trajectories of a projectile projected at different angles keeping projection velocity (15.2 m/s) and projection height (2.4 m) constant (after Broer, M. R., and Zernike, R. F. Efficiency of Human Movement (4th edition). Philadelphia: Saunders College, 1979.).

Activity	Angle	Reference
racing dive	5°–22°	Heusner 14
ski jumping	–4°	Komi et al. 15
tennis serve	–3°–15°	Owens and Lee 16
discus	–35°–15°	Terauds 17
high jump (flop)	40°–48°	Dapena 12

The optimal angle of projection for a given activity is based on the purpose of the activity. Intuitively, it would appear that if someone tried to jump over a relatively high object like a high jump bar, their projection angle would be quite steep. This has proved to be the case as high jumpers have a projection angle of 40 to 48° using the Flop high jump technique (12). On the other hand, if one tried to jump for maximal horizontal distance such as in a long jump, the projection angle would be a quite smaller angle. This has also proved to be the case as long jumpers have projection angles of 18 to 27° (13). Table 8-3 illustrates the projection angles reported in the research literature for several activities. Positive angles of projection indicate angles greater than zero degrees, where the object is projected above the horizontal. Negative angles of projection refer to those less than zero degrees or those below the horizontal. For example, in a tennis serve, the serve is actually projected downward from the point of impact.

Projection Velocity. The velocity of the projectile at the instant of release will determine the height and the length of the trajectory as long as all other factors are held constant. The resultant velocity of projection is usually calculated and given when discussing the factors influencing the projectile. The resultant velocity of projection is the vector sum of the horizontal and vertical velocities. It is necessary, however, to focus on the components of the velocity vector since they will dictate the height of the trajectory and the distance the projectile will travel. Like other vectors, the velocity of projection has a vertical component (v_y) and a horizontal component (v_x).

The magnitude of the vertical velocity is reduced by the effect of gravity (9.81 m/s for every second of upward flight). Gravity reduces the vertical velocity of the projectile until the velocity equals zero. At that point, the projectile reaches its highest point in flight. The vertical velocity component, therefore, determines the height of the apex of the trajectory. The vertical velocity is likewise increased by the effect of gravity on the downward flight. Thus, the vertical velocity affects the height that the projectile achieves, the time the projectile takes to reach that height, and consequently the time to return to the earth's surface (Figure 8-31).

The horizontal component of the projection velocity is constant throughout the flight of the projectile. It is known that:

$$v_x = \frac{dx}{dt}$$

and re-arranging this equation:

$$dx = v_x * dt$$

Recalling that dx represents the change in position of the projectile during its flight, dx is then determined by the product of the horizontal velocity and the flight time to that position. The magnitude of dx is the distance that the projectile travels and is called the range of the projectile.

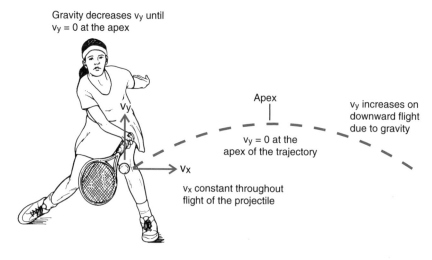

Gravity decreases v_y until $v_y = 0$ at the apex

v_y

v_x

Apex

$v_y = 0$ at the apex of the trajectory

v_y increases on downward flight due to gravity

v_x constant throughout flight of the projectile

FIGURE 8-31. Projection velocity components during the flight of a projectile.

For example, if a projectile is released at a horizontal velocity of 13.7 m/s, the projectile will have traveled 13.7 m in the first second, 27.4 m after 2 seconds, 40.1 m after 3 seconds, and so on. In another example, if a shot putter releases the shot with a horizontal velocity of 10.5 m/s and it travels through the air for 2.2 s, the range of the throw would be:

$$\text{Range} = 10.5 \text{ m/s} * 2.2 \text{ s}$$
$$= 23.1 \text{ m}$$

The angle of projection will affect the relative magnitude of the horizontal and vertical velocity components. If the angle of projection is 40° and the projection velocity is 13.7 m/s, the horizontal component of the projection velocity is:

v_x = projection velocity * cosine of projection angle
$$= 13.7 \text{ m/s} * \cos 40°$$
$$= 10.49 \text{ m/s}$$

and the vertical component is:

v_y = projection velocity * sine of projection angle
$$= 13.7 \text{ m/s} * \sin 40°$$
$$= 8.81 \text{ m/s}$$

If the angle is altered to 35°, the horizontal component becomes:

$$v_x = 13.7 \text{ m/s} * \cos 35°$$
$$= 11.22 \text{ m/s}$$

and the vertical component becomes:

$$v_y = 13.7 \text{ m/s} * \sin 35°$$
$$= 7.86 \text{ m/s}$$

To understand in general how the angle of projection affects the velocity components, consider that the cosine of 0° is 1 and decreases to zero as the angle increases. If the cosine of the angle is used to represent the horizontal velocity, then the horizontal velocity decreases as the angle of projection increases from 0 to 90° (Figure 8-32) Also, the sine of 0° is zero and increases to 1 as the angle increases. Consequently, if the sine of the angle is used to represent the vertical velocity, then the vertical velocity increases as the angle increases from 0 to 90° (Figure 8-32) It can readily be seen that as the angle gets closer to 90°, the horizontal velocity becomes smaller and the vertical velocity becomes greater. As the angle gets closer to 0°, the horizontal velocity becomes greater and the vertical velocity gets smaller.

At 45°, however, the sine of the angle equals the cosine of the angle. For any given velocity, therefore, the horizontal velocity equals the vertical velocity. It would appear

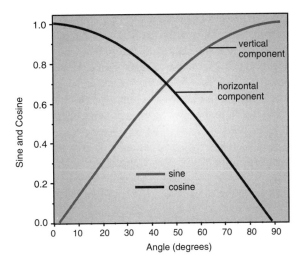

FIGURE 8-32. Graph of sine and cosine values at angles from 0° to 90°. Note that sin 45° = cosine 45°.

that 45° would be the optimum angle of projection since, for any velocity, the horizontal and vertical velocities would be equal. This is true under certain circumstances that shall be discussed in relationship to the projection height. Generally, if the maximum range of the projectile is critical, then an angle to optimize the horizontal velocity or an angle less than 45° would be appropriate. Thus, in activities such as the long jump or shot putting, the optimal angle of projection is less than 45°. If the height of the projectile is important, an angle greater than 45° would be chosen. This is the case in activities such as high jumping.

Projection Height. The height of projection of a projectile is the difference in height between the vertical take-off position and the vertical landing position. There are three situations that may occur that greatly affect the shape of the trajectory. In each case, the trajectory is parabolic but the shape of the parabola may not be completely symmetrical; that is, the initial half of the parabola may not be equal in shape to the latter half of the parabola. In the first case, the projectile is released and lands at the same height. The shape of the trajectory is symmetrical and thus the time for the projectile to reach the apex from the point of release equals the time for the projectile to reach the ground from the apex. If a ball is kicked from the surface of a field and lands on the field's surface, the relative projection height is zero and thus time up to the apex is equal to time down from the apex. In the second situation, the projectile is released from a point higher than the surface on which it lands. The parabola is asymmetrical with the initial portion to the apex less than the latter portion. In this case, time for the projectile to reach the apex is less than the time to reach the ground from the apex. For example, if a shot putter releases the shot from a height of 2.2 m above the ground and the shot lands on the ground, the height of projection is 2.2 m. In the third situation, the projectile is released from a point below the surface on which it lands. Once again the trajectory is asymmetrical, but now the initial portion to the apex of the trajectory is greater than the latter portion. Thus, time for the projectile to reach the apex is greater than time for the projectile to reach the ground from the apex. For example, if a ball is thrown from a height of 2.2 m and lands in a tree at a height of 4 m, the height of projection is 1.8 m (Figure 8-33).

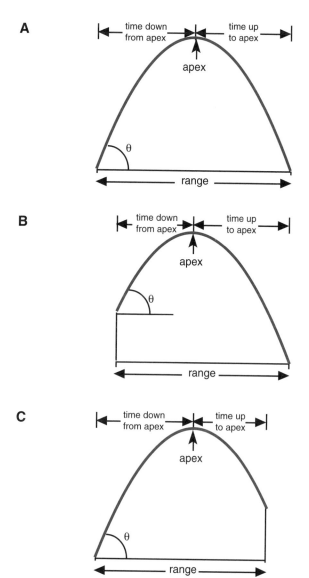

FIGURE 8-33. Influence of projection height on the shape of the trajectory of a projectile.

Generally, when the projection velocity and the angle of projection are held constant, the higher the projection height of release, the longer the flight time. If the flight time is longer, the range is greater. Also, for maximum range, when the relative height of projection is zero, the optimum angle is 45°; when the projection height is above the landing height, the optimum angle is less than 45°; and when the projection height is below the landing height, the optimum angle is greater than 45°.

Optimizing Projection Conditions

To optimize the conditions for the release of a projectile, the purpose of the projectile must be considered. As discussed previously, the three primary factors that affect the flight of a projectile are inter-related and will affect both the height of the trajectory and the distance traveled by the projectile. It may be intuitive that, since the height of the apex and the length of the trajectory of the projectile are both affected by the projection velocity, then increasing the projection velocity would increase both of these parameters. However, this common perception is incorrect. The choice of an appropriate projection angle is what will dictate whether the vertical or the horizontal velocity will be increased with increasing projection velocity. In addition, the angle of projection can be affected by the height of projection.

The relative importance of these factors is illustrated in the following example. If an athlete puts the shot with a velocity of 14 m/s at an angle of 40° from a height of 2.2 m, the resulting distance of the throw is 22.0 m. If each factor is increased by a given percentage (10% in this case), as the other two factors are held constant, the relative importance of each factor may be calculated. Increasing the velocity to 15.4 m/s results in a throw of 26.2 m; increasing the angle to 44° results in a throw of 22.0 m; and increasing the height of projection to 2.4 m results in a throw of 22.2 m. It is readily evident that increasing the velocity of projection increases the range of the throw more substantially than increasing either the angle or the height of projection. It should be emphasized, however, that the three factors are inter-related and any change in one results in a change in the others.

Equations of Constant Acceleration

When a projectile is traveling through the air, only gravity and air resistance act upon it. If, as stated earlier, the effects of air resistance are ignored, then only gravity acts on the projectile. The acceleration due to gravity is constant and thus the projectile undergoes a constant acceleration. Using the concepts from the previous section, equations of constant acceleration motion, or projectile motion, can be determined based on the definitions of velocity and acceleration. There are three such expressions involving the inter-relationships of the kinematic parameters of time, position, velocity, and acceleration. These expressions are often referred to as the equations of constant acceleration.

The first equation expresses final velocity as a function of the initial velocity, acceleration, and time.

$$v_f = v_i + at$$

The variables v_f and v_i refer to the final velocity and the initial velocity, respectively. As the velocity components are considered individually, it can be seen that many of the points previously discussed are described by this equation.

As described previously, the horizontal component of velocity, v_x, is constant throughout the flight of the projectile and thus the horizontal acceleration is zero. Indeed if a = 0 is substituted into this equation it can be seen that:

$$vx_f = vx_i$$

The final velocitFy equals the initial velocity or the horizontal velocity of projection is constant throughout the flight period.

In the earlier discussion of the vertical velocity component, v_y, it was suggested that gravity immediately acted to decrease the magnitude of v_y after the release of the projectile. If the coordinate system being used is oriented such that positive is up and negative is down, then the acceleration downward must be negative. Since gravity accelerates a projectile towards the ground, the acceleration due to gravity must be negative. Therefore, the acceleration due to gravity is –9.81 m/s². For the vertical component, the first equation of constant acceleration equation is:

$$vy_f = vy_i - 9.81t$$

This simply reaffirms that the vertical velocity at any instant is equal to the initial vertical velocity of release decreased by a value of 9.81 m/s for every second of flight until the final velocity is zero. At the point in the trajectory where the velocity is zero, the projectile is at the apex of its trajectory.

In the second equation, position is expressed as a function of initial velocity, acceleration and time.

$$s = v_i t + \frac{1}{2}at^2$$

The variable s in this expression may refer to the horizontal or vertical case and is the change in position or the distance that the object travels from one position to

another. This equation is derived by integrating the first equation. If the horizontal acceleration is zero, this equation becomes:

$$x = vx_i t$$

This expression reaffirms that the horizontal distance the object travels—the range—is the product of the horizontal velocity and the time of the flight.

In considering the vertical case of this equation, note that the variable "a" representing the acceleration due to gravity equals – 9.81 m/s². The expression, therefore, is not as simple as the horizontal component case. If, for example, an object begins at rest and is dropped from some height, the initial vertical velocity is zero. This expression then becomes:

$$y = \frac{1}{2}at^2$$
$$y = \frac{1}{2}*9.81*t^2$$

The variable "t" represents the time it takes for the projectile to reach the ground from the point at which it was dropped. This expression indicates that the height may be calculated by simply knowing the length of time it took for the object to contact the ground once it was released, given that the acceleration due to gravity is constant.

The last equation expresses final velocity as a function of initial velocity, acceleration, and position.

$$v_f^2 = v_i^2 + 2as$$

Each of the kinematic variables in this expression appeared in one or both of the previous equations. Simplifying this expression for the horizontal case (that is $a_x = 0$), it is again reaffirmed that the horizontal velocity is constant throughout the projectile's flight. That is:

$$vx_f^2 = vx_i^2$$

For the vertical case, the example of an object dropping from some height, by beginning with a vertical velocity of zero, may be used. The expression becomes:

$$vy_f^2 = 2ay \text{ (where a = 9.81 m/s}^2)$$
$$vy_f^2 = 2*9.81*y$$

The final vertical velocity is, therefore, a function of the acceleration due to gravity and the height, y, from which the object was dropped.

Numerical Example

The equations of constant acceleration all employ parameters that are basic to linear kinematics. The three equations of constant acceleration thus provide a useful method of analyzing projectile motion. If calculating the range of a projectile, for example, the following expression can be used:

$$\text{Range} = \frac{v^2 * \sin\theta * \cos\theta + v_x * \sqrt{(v_y)^2 + 2gh}}{g}$$

where v = velocity of projection, θ = angle of projection, h = height of projection, and g = acceleration due to gravity. This is a rather complicated expression, however. Instead, the equations of constant acceleration can be used to logically work through a problem of calculating the range of a projectile.

In the following example, a shot putter releases the shot at an angle of 40° from a height of 2.2 m with a velocity of 13.3 m/s. Figure 8-34 illustrates what is known about the conditions of the projectile at the instant of projection and the shape of the trajectory based on our previous discussion.

Remember, in order to find the range of a projectile, the horizontal velocity and the length of time that the shot was in the air must be known. The problem can be solved in seven steps, utilizing the equations of constant acceleration.

Step 1. Calculate the initial vertical and horizontal velocities. (Use Appendix E for sine and cosine values).

$$v_x = v * \cos\theta$$
$$= 13.3 \text{ m/s} * \cos 40°$$
$$= 13.3 \text{ m/s} * 0.766$$
$$= 10.19 \text{ m/s}$$
$$v_y = v * \sin\theta$$
$$= 13.3 \text{ m/s} * \sin 40°$$
$$= 13.3 \text{ m/s} * 0.643$$
$$= 8.55 \text{ m/s}$$

Step 2. Use the first equation of constant acceleration to calculate the time required for the projectile to reach the apex of its trajectory. Since it is necessary to calculate the time to the apex, the equation using the vertical veloc-

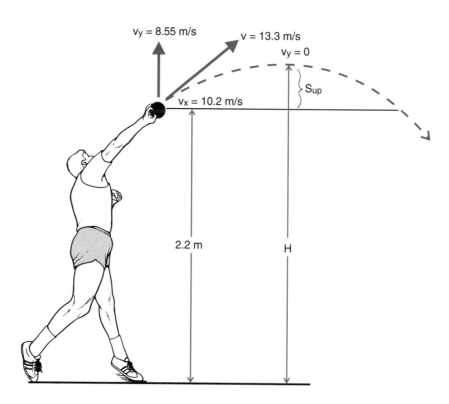

FIGURE 8-34. Conditions during the flight of the shot. Initial conditions are: v = 13.3 m/s; projection angle = 40° ; and projection height = 2.2 m.

ity must be used. Thus, the acceleration due to gravity, a, will act, bringing the vertical velocity to zero at the apex.

$$v_y = v_{yi} + at_{up}$$
$$0 = 8.55 \text{ m/s} - 9.81 \text{ m/s}^2 * t_{up}$$
$$t_{up} = \frac{8.55 \text{ m/s}}{9.81 \text{ m/s}^2}$$
$$t_{up} = 0.87 \text{ s}$$

Step 3. Use the third equation of constant acceleration to calculate the height of the apex of the trajectory above the release height. Remember that the height of the trajectory is dependent on the vertical velocity, the acceleration due to gravity is -9.81 m/s^2, and the vertical velocity at the apex is zero.

$$v_{yf^2} = v_{fi}^2 + 2ay$$
$$0 = (8.55 \text{ m/s})^2 - 2 * 9.81 \text{ m/s}^2 * y_{up}$$
$$y_{up} = \frac{(8.55 \text{ m/s})^2}{2 * 9.81 \text{ m/s}^2}$$
$$y_{up} = 3.72 \text{ m}$$

Step 4. Calculate the total height to the apex of the trajectory.

$$H = \text{projection height} + y_{up}$$
$$= 2.2 \text{ m} + 3.72 \text{ m}$$
$$= 5.92 \text{ m}$$

Step 5. Calculate the time for the projectile to reach the ground from the apex of the trajectory. Since the projectile will land at a level lower than the release point, then time for the projectile to reach the apex of the trajectory must be less than the time for the projectile to reach the ground from the apex. The time down, t_{down}, can be calculated using the second equation of constant acceleration. In this equation, note that the displacement is negative, since it measures the distance from the apex down to the ground. In addition, the initial vertical velocity of the downward flight is zero.

$$y = v_i t + \frac{1}{2} at^2$$
$$-5.92 \text{ m} = 0 + \frac{1}{2} * -9.81 \text{ m/s}^2 * t_{down^2}$$
$$-t_{down} = \sqrt{\frac{2 * 5.92 \text{ m}}{-9.81 \text{ m/s}^2}}$$

Since time cannot be negative, each side of this equation must be multiplied by –1 resulting in the following:

$$t_{down} = 1.10 \text{ s}$$

Step 6. Calculate the total time the projectile is in the air by adding the time for the projectile to reach the apex of the trajectory to the time for the projectile to reach the ground from the apex.

$$T_{total} = t_{up} + t_{down}$$
$$= 0.87 \text{ s} + 1.10 \text{ s}$$
$$= 1.97 \text{ s}$$

Step 7. Calculate the range of the projectile.

$$\text{Range} = v_x * T_{total}$$
$$= 10.19 \text{ m/s} * 1.97 \text{ s}$$
$$= 20.07 \text{ m}$$

The distance that the shot traveled under these conditions is 20.1 m. In our initial discussion of projectiles and the equations of constant acceleration, air resistance was considered to be negligible. The mathematics of considering air resistance are well beyond the scope of this book, since it would require solving differential equations. However, it may be of interest to know that if these initial conditions were used to solve this problem taking into account air resistance, a reduction of 1.46% in the range of the throw would be found. Considering air resistance, the throw would be 0.3 m less or a total of 19.8 m.

Summary

Biomechanics is a quantitative discipline. One type of quantitative analysis involves linear kinematics. Linear kinematics is the study of linear motion with respect to time, and involves the vector quantities position, velocity, and acceleration, as well as the scalar quantities displacement and speed. Velocity is defined as the time rate of change of position and is calculated in biomechanics, using the first central difference method as follows:

$$v = \frac{s_{i+1} - s_{i-1}}{2\Delta t}$$

Acceleration is defined as the time rate of change of velocity and is calculated as follows:

$$a = \frac{v_{i+1} - v_{i-1}}{2\Delta t}$$

The process of calculating velocity from position and time or acceleration from velocity and time is called differentiation. Calculating the derivative via differentiation entails finding the slope of a line tangent to the parameter-time curve. The opposite process of differentiation is called integration. Velocity may be calculated as the integral of acceleration and position as the integral of velocity. Integration implies calculating the area under the parameter-time curve. The method of calculating the area under a parameter-time curve is called the Riemann Sum.

Projectile motion involves an object that undergoes a constant acceleration because it is uniformly accelerated by gravity. The flight of a projectile, its height and distance, are affected by conditions at the point of release: the angle of projection, velocity of projection, and the relative height of projection. There are three equations governing constant acceleration. The first expresses final velocity, v_f, as a function of initial velocity, v_i, acceleration, a, and time, t. That is:

$$v_f = v_i + at$$

The second equation expresses position, s, as a function of initial velocity, v_i, acceleration, a, and time, t. That is:

$$s = v_i t + \frac{1}{2} a t^2$$

The third equation expresses final velocity, v_f, as a function of initial velocity, v_i, acceleration, a, and position, s.

$$v_f^2 = v_i^2 + 2as$$

These equations may be used to calculate the range of a projectile.

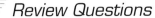
Review Questions

1. List the dimensions and the units associated with those dimensions of the SI system that are commonly used in biomechanics.
2. Use a dimensional analysis to determine the units of the following kinematic parameters.
   a) speed b) position c) velocity d) acceleration
3. A swimmer completes four laps in a 50 m swimming pool, finishing where they started. a) What was the linear distance traveled? b) What was the linear displacement? (Answer: a) 200 m; b) 0 m).
4. Suppose an individual moves from point s_1 (3, 5) to point s_2 (6, 8). What are the a) horizontal, b) vertical, and c) resultant displacements? (Answer: a) 3 units; b) 3 units; c) 4.24 units).
5. An individual drives from point A to point B, a distance of 33 km, in a period of 55 minutes. What was the average speed driven in m/s? (Answer: 10 m/s).

6. When velocity is at a maximum, how can you describe the corresponding acceleration?
7. A runner starts from rest and reaches a maximum velocity of 4.7 m/s in 3.2 s. What was their average acceleration from rest to maximum velocity? (Answer: 1.47 m/s^2).
8. What is the sign of the acceleration vector if the object is moving in a positive direction?
9. Sketch a position-time graph, illustrating points on the graph where the velocity is positive, negative, and zero.
10. A golf ball is driven straight out from a tee box that is 2 m above the fairway. At the instant the club contacts the ball, another golfer drops a golf ball from a height of 2 m. Which ball will contact the ground first? Why?

Additional Questions

1. An object begins at rest and linearly changes its horizontal position over a period of 5 s to a position 4 m from the position of rest. Sketch the position, velocity and acceleration-time graphs corresponding to the motion.

2. Calculate the velocity in frames 5, 6, and 7, given the following data.

Frame	Time (s)	Position (m)
4	0.020	1.034
5	0.025	1.041
6	0.030	1.050
7	0.035	1.041
8	0.040	1.044

(Answer: frame 5: 1.60 m/s; frame 6: 0.0 m/s; frame 7: –0.60 m/s).

3. Calculate the acceleration in frame 6, using your calculations from question #2. (Answer: –220.0 m/s^2).

4. Calculate the integral of the velocity-time curve over the time period from frame 5 to frame 7, using the data calculated in question #2. (Answer: 0.01 m).

5. If the resultant velocity is 29.7 m/s and is directed at an angle of 22°, what are the a) vertical and b) horizontal components of the velocity vector? (Answer: a) 11.13 m/s; b) 27.54 m/s).

6. If the horizontal acceleration is -0.51 m/s^2 and the vertical acceleration is –7.683 m/s^2, what is the resultant acceleration? (Answer: 7.70 m/s^2).

7. Sketch the acceleration-time profile corresponding to the following velocity-time profile.

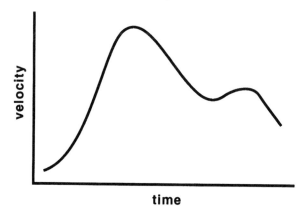

8. What is the change in position based on the following velocity-time graph?

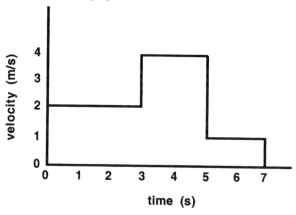

9. A softball is thrown with a velocity of 22.5 m/s at an angle of 56° from a height of 1.7 m. Calculate :
   a) the vertical and horizontal velocity components
   b) the time to peak trajectory
   c) the height of the trajectory from the point of release
   d) the total height of the parabola
   e) the time from the apex to the ground
   f) the total flight time
   g) the range of the throw
   Answer: a) 18.65 m/s, 12.58 m/s; b) 1.90 s; c) 17.73 m; d) 19.43 m; e) 1.99 s; f) 3.89 s; g) 48.94 m.

References

1. Bates, B.T., Osternig, L.R., Mason, B.R. Variations of velocity within the support phase of running. In J. Terauds and G. Dales (eds.). Science in Athletics. pp. 51–59. Del Mar: Academic Publishers, 1979.

2. Elliott, B.C. and Blanksby, B.A. A biomechanical analysis of the male jogging action. Journal of Human Movement Studies 5:42–51, 1979.

3. Hoshikawa, T., Matsui, H., Miyashita, M. Analysis of running patterns in relation to speed. In Medicine and Sport Vol. 8: Biomechanics lll. pp. 342–348. Basel: Karger, 1973.

4. Luhtanen, P. and Komi, P.V. Mechanical factors influencing running speed. In E. Asmussen and K. Jorgensen (eds.). Biomechanics Vl-B. pp. 23–29. Baltimore: University Park Press, 1973.

5. Sinning, W.E. and Forsyth, H.L. Lower limb actions while running at different velocities. Medicine and Science in Sports 2:28–34, 1970.

6. Saito, M., Kobayashi, K., Miyashita, M., Hoshikawa, T. Temporal patterns in running. In R.C. Nelson and C.A. Morehouse (eds.). Biomechanics IV. pp. 106–111. Baltimore: University Park Press, 1974.

7. Bates, B.T. and Haven, B.H. Effects of fatigue on the mechanical characteristics of highly skilled female runners. In R.C. Nelson and C.A. Morehouse (eds.). Biomechanics lV. pp. 119–125. Baltimore: University Park Press, 1974.

8. Henry, F.M. and Trafton, I. The velocity curve of sprint running. Research Quarterly 23: 409–422, 1951.

9. Chow, J. W. Maximum speed of female high school runners. International Journal of Sports Biomechanics 3:110–127, 1987.

10. Brancazio, P.J. Sport Science. New York: Simon and Schuster, 1984.

11. Broer, M.R. and Zernike, R.F. Efficiency of Human Movement (4th edition). Philadelphia: Saunders College, 1979.

12. Dapena, J. Mechanics of translation in the Fosbury flop. Medicine and Science in Sports and Exercise 12:37–44, 1980.

13. Hay, J.G. The biomechanics of the long jump. Exercise and Sport Science Review. pp. 401–446, 1986.

14. Heusner, W.W. Theoretical specifications for the racing dive : optimum angle for take-off. Research Quarterly 30:25–37, 1959.

15. Komi, P.V., Nelson, R.C., Pulli, M. Biomechanics of ski-jumping. Jyvaskyla: University of Jyvaskyla. pp. 25–29, 1974.

16. Owens, M.S. and Lee, H.Y. A determination of velocities and angles of projection for the tennis serve. Research Quarterly 40:750–754, 1969.

17. Terauds, J. Some release characteristics of international discus throwing. Track and Field Review 75:54–57, 1975.

Additional Reading

1. Cavanagh, P.R., Kram, R. Stride length in distance running: Velocity, body dimensions, and added mass effects. In P.R. Cavanagh (ed.). Biomechanics of Distance Running. pp. 35–63. Champaign, IL : Human Kinetics Publishers, 1990.

2. Daish, C.B. The Physics of Ball Games. London: The English Universities Press Ltd, 1972.

3. Dillman, C.J. Kinematic Analysis of Running. In Exercise and Sports Science Review. pp. 193–218, New York : MacMillan Publishing Company, 1974.

4. Williams, K.R. Biomechanics of Running. In Exercise and Sports Science Review. pp. 389–441. New York: MacMillan Publishing Company, 1985.

Glossary

apex:	the highest point of a parabola. It is the highest point that a projectile reaches in its trajectory.
calculus:	a method of calculating the derivative or the integral of a function.
Cartesian coordinate system:	an x, y, z reference system, with either two or three axes, in which a point may be located as a distance from each of the axes.
cosine of an angle:	in a right angle triangle, the ratio of the side adjacent to the angle and the hypotenuse.
curvilinear motion:	linear motion along a curved path.
derivative:	the result of the process of differentiation—the slope of a line—either a secant or a tangent, on a parameter-time curve.
dimension:	a term denoting the nature of a measurable quantity.
differentiation:	the mathematical process of calculating a derivative.
digitization:	the process of applying x-y coordinates to points on a video frame.
first central difference method:	a method of calculating the average slope over two time intervals, as in generating velocity from position-time data or acceleration from velocity-time data.
instantaneous linear acceleration:	the slope of a line tangent to a velocity-time curve.
instantaneous linear velocity:	the slope of a line tangent to a position-time curve.
integral:	the result of the process of integration—the area under a parameter-time curve.
integration:	the mathematical process of calculating an integral.
kinematics:	the area of study that examines the spatial and temporal components of motion.
limit:	the derivative of a function when the change in time approaches zero.
linear acceleration:	the time rate of change of linear velocity.
linear distance:	the length of the actual path traveled.
linear displacement:	a vector representing the straight line distance and direction from one position to another.
linear motion:	see translation.
linear kinematics:	the description of linear motion involving position, velocity, and acceleration.
linear velocity:	the time rate of change of linear position.
midstance:	the point during locomotor support when the center of mass of the individual is directly over the foot.
motion:	the progressive change in position of an object.

non-support:	a phase of the gait cycle in which the leg is not supported on the ground.
parabola:	a curve that describes the shape of the trajectory of a projectile.
projectile:	an object that has been projected into the air.
projectile motion:	the motion of projectiles.
projection angle:	the angle at which a projectile is released.
projection height:	the difference between the height at which a projectile is released and which it lands.
projection velocity:	the velocity at which a projectile is released.
Pythagorean Theorem:	a mathematical relationship describing the relationship among the sides of a right angle triangle. That is: $$a^2 = b^2 + c^2$$ where a is the hypotenuse and b and c are the other sides of the triangle.
qualitative analysis:	a non-numeric description or evaluation of movement that is based on direct observation.
quantitative analysis:	a numeric description or evaluation of movement based on data collected during the performance of the movement.
range:	the distance that a projectile travels.
resultant:	the sum of two vectors.
Riemann Sum:	a mathematical process by which the area under a parameter-time curve can be calculated, given that the time interval, dt, is small.
rise:	the change in a parameter between two successive time intervals.
run:	the change in time between two successive locations of parameter.
scalar:	a quantity that is defined by its magnitude alone.
secant:	a line that intersects a curve at two places on the curve.
sine of an angle:	in a right angle triangle, the ratio of the side opposite the angle and the hypotenuse.
slope:	the ratio of the rise to the run.
speed:	the magnitude of the velocity vector.
stance:	see support.
step:	a portion of a stride from an event occurring on one leg to the same event on the opposite leg.
stride:	a cycle lasting from an event during the motion of one limb to the next occurrence of that same event on the same limb.
stride length:	the distance traveled during one stride.

stride rate:	the number of strides per minute.
support:	the phase of the gait cycle when the foot is in contact with the ground.
swing:	see non-support.
tangent	a line that touches a curve at only one place.
translation:	motion in a straight or curved path where different regions of the object move the same distance in the same time interval.
trajectory:	the flight path of a projectile.
vector:	a quantity that is defined by both its magnitude and direction.

Angular Kinematics

Angular Kinematics

After reading this chapter, the student should be able to:
1. Distinguish between linear, angular, and general motion.
2. Determine relative and absolute angles.
3. Determine the direction of angular motion vectors.
4. Discuss the relationship among the angular kinematic quantities of angular distance and displacement, angular velocity, and angular acceleration.
5. Discuss the conventions for the calculation of lower extremity angles.
6. Discuss the relationship between angular and linear motion, particularly the relationships between angular and linear displacement, angular and linear velocity, and angular and linear acceleration.
7. Interpret angle-angle diagrams.
8. Discuss selected research studies that have used an angular kinematics approach.
9. Solve quantitative problems that employ angular kinematic principles.

Angular motion occurs when all parts of a body move through the same angle but do not undergo the same linear displacement. The sub-set of kinematics that deals with angular motion is called *angular kinematics*. Angular kinematics is the description of angular motion without regard to the causes of the motion. Consider a bicycle wheel as an example (Figure 9-1). Pick any point close to the center of the wheel and any point close to the edge of the wheel. The point close to the edge undergoes a greater linear displacement than the point close to the center as the wheel spins. Thus, the wheel will undergo a *rotation*. The motion of the wheel is called *angular motion*.

Angular motion occurs about an axis of rotation that is a line perpendicular to the plane in which the rotation occurs. For example, the bicycle wheel spins about its axle which is its axis of rotation. The axle of the wheel is perpendicular to the rim of the wheel that describes the plane of rotation (Figure 9-1).

An understanding of angular motion is critical to comprehend how one moves. Nearly all human movement involves the rotations of body segments. The segments rotate about the joint centers that form the axes of rotation for these segments. When an individual moves, the segments generally undergo both rotation and translation. When the combination of rotation and translation occurs, it is described as *general motion*. Figure 9-2 illustrates the combination of linear and rotational motions. The gymnast undergoes translation as they move across the ground. At the same time, the gymnast is rotating.

Measurement of Angles

An *angle* is composed of two lines that intersect at a point called the *vertex*. In a biomechanical analysis, the intersecting lines are generally body segments. If you consider the longitudinal axis of the leg segment as one side of an angle and the longitudinal axis of the thigh segment as the other side, the vertex would be the joint center of the knee. Angles can be determined from the same coordinate points as were described in Chapter 8. Coordinate points describing the joint centers determine the sides and the vertex of the angle. For example, an angle at the knee can be constructed using the thigh and leg segments. The coordinate points describing the ankle and knee joint centers define the leg segment, while the coordinate points describing the hip and knee joint centers define the thigh segment. The vertex of the angle is the knee joint center.

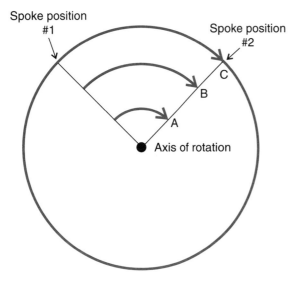

FIGURE 9-1. A bicycle wheel as an example of rotational motion. Points A, B, and C undergo the same amount of rotation but undergo different linear displacements with C undergoing the greatest.

The definition of a segment by placing markers on the subject at the joint centers makes a technically incorrect assumption. This incorrect assumption is that the joint

center at the vertex of the angle does not change throughout the movement. Because of the asymmetries in the shape of the articulating surfaces in most joints, one or both bones constituting the joint may displace relative to *each* other. For example, while the knee is often considered a hinge joint, it is not. At the knee joint, the medial and lateral femoral condyles are asymmetric, causing the tibia to rotate along its long axis and about an axis through the knee from front to back as the knee flexes and extends. The location of the joint center, therefore, changes throughout any motion of the knee. The center of rotation of a joint at an instant in time is called the *instantaneous joint center* (Figure 9-3). It is difficult to locate this moving axis of rotation unless special techniques such as x-ray measurements are used. These measurements are not practical in most situations; thus, the assumption of a static instantaneous joint center must be made.

Units of Measurement

In angular motion, there are three units used to measure angles. It is important to use the correct units in order for biomechanists to communicate the results of their work clearly, and to compare values from study to study. It is also essential to use the correct units because angle measurements may be used in further calculations. The first and most commonly used is the degree (°). A circle, describing one complete rotation, transcribes an arc of 360° (Figure 9-4a). An angle of 90°, for example,

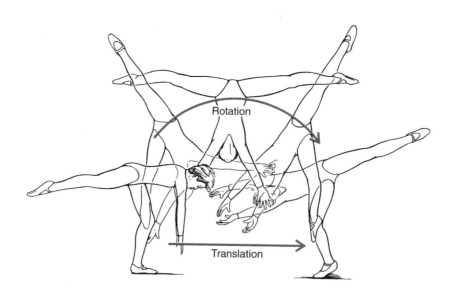

FIGURE 9-2. A gymnast completing a cartwheel as an example of general motion. The gymnast simultaneously undergoes both translation and rotation.

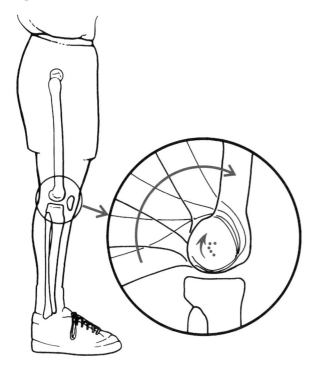

FIGURE 9-3. Instantaneous center of rotation of the knee (after Nordin, M. and Frankel, V.H. (eds.). Biomechanics of the Musculoskeletal System (2nd ed.). Philadelphia: Lea & Febiger, 1979.).

would result in sides that are perpendicular to each other. A straight line would have a 180° angle (Figure 9-4b).

The second unit of measurement describes the number of rotations or revolutions about a circle (Figure 9-4a). One revolution is a single 360° turn. For example, a triple jump in skating requires the skater to complete three and one-half revolutions in the air. The skater completes a rotation of 1260°. This unit of measurement is useful in qualitative descriptions of movements such as those in figure skating, gymnastics, and diving but is not useful in quantitative analyses.

While the degree is most commonly understood and the revolution is often used, the most appropriate unit for angular measurement in biomechanics is the *radian*. A radian is defined as the measure of an angle at the center of a circle described by an arc equal to the length of the radius of the circle (Figure 9-4c). That is:

$$\theta = \frac{s}{r} = 1 \text{ radian}$$

where θ = the angle equaling 1 radian, s = arc of length r, and r = radius of the circle. Since both s and r have units of length (m), the units in the numerator and denominator cancel each other out with the result that the radian is dimensionless.

In further calculations, the radian is not considered in determining the units of the result of the calculation. Degrees have a dimension and must be included in the unit of the product of any calculation. It is necessary, therefore, to use the radian as a unit of angular measurement instead of the degree in any calculation involving angular motion because the radian is dimensionless. One radian is the equivalent of 57.3°. To convert an angle in degrees to radians, divide the angle in degrees by 57.3° For example, 72° in radians is:

$$72° = \frac{72°}{57.3°} = 1.26 \text{ rad}$$

To convert radians to degrees, simply multiply the angle in radians by 57.3°. For example, 0.67 radians in degrees is:

$$0.67 \text{ rad} = 0.67 \text{ rad} * 57.3 = 38.4°$$

Angular measurement in radians is often determined in multiples of pi (π = 3.1416). Since there are 2π radians in a complete circle, 180° may be represented as p radians, 90° as $^\pi/_2$) radians and so on.

Although the unit of angular measurement in the SI system is the radian and this unit must be used in further calculations, the angular motion concepts presented in the remainder of this chapter will use the degree for ease of understanding.

Types of Angles

Relative Angle

Two types of angles are generally calculated in biomechanics. The first angle is called the *relative angle* (Figure 9-5a). This angle defines the included angle between longitudinal axes of two segments. For example, the relative angle at the elbow describes the amount of flexion or extension at the joint. Relative angles, however, do not describe the position of the segments or the sides of the angle in space. If an individual has a relative angle of 90° at the elbow and that angle is maintained, the arm could be in any number of positions (Figure 9-5b).

Relative angles can be calculated using the *Law of Cosines*. This law is simply a more general case of the Pythagorean Theorem and describes the relationship between the sides of a triangle. For our purposes, the triangle is made up of the two segments (b and c) and a line (a) joining the distal end of one segment to the proximal end of the other (Figure 9-6).

In Figure 9-6, the coordinate points for two segments describing the thigh and the leg are given. To calculate the relative angle at the knee (θ), the lengths a, b, and c would be calculated using the Pythagorean relationship.

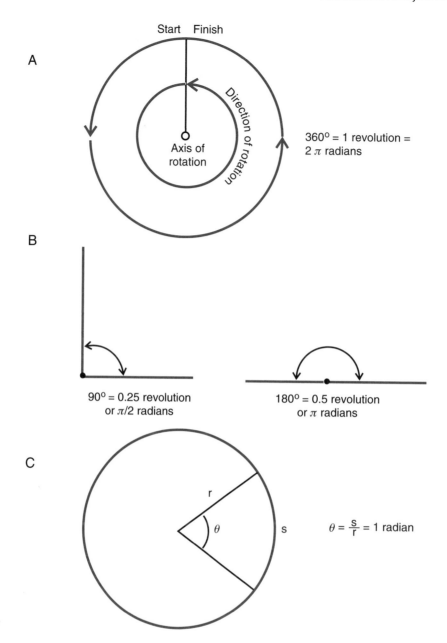

FIGURE 9-4. Units of angular measurement: A) the revolution; B) perpendicular and straight lines; C) radian.

$$a = \sqrt{(x_h - x_a)^2 + (y_h - y_a)^2}$$
$$= \sqrt{(1.14 - 1.09)^2 + (0.80 - 0.09)^2}$$
$$= \sqrt{0.0025 + 0.5041}$$
$$= 0.71$$
$$b = \sqrt{(x_h - x_k)^2 + (y_h - y_k)^2}$$
$$= \sqrt{(1.14 - 1.22)^2 + (0.80 - 0.51)^2}$$

$$= \sqrt{0.0064 + 0.0841}$$
$$= 0.30$$
$$c = \sqrt{(x_k - x_a)^2 + (y_k - y_a)^2}$$
$$= \sqrt{(1.22 - 1.09)^2 + (0.51 - 0.09)^2}$$
$$= \sqrt{0.0169 + 0.1764}$$
$$= 0.44$$

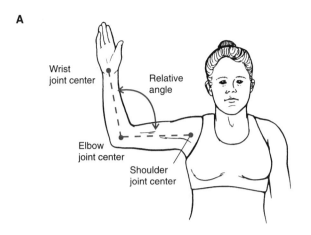

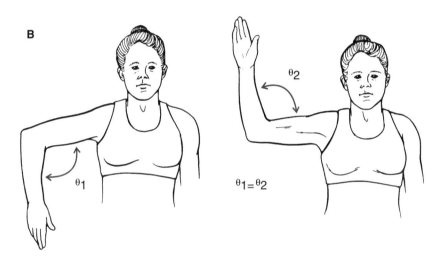

FIGURE 9-5. A) Relative elbow angle; B) the same relative elbow angle with the arm and forearm in different positions.

The next step would be to substitute these values into the Law of Cosines equation and solve for the cosine of the angle θ.

$$a^2 = b^2 + c^2 - 2*b*c*\cos\ \theta$$
$$0.71^2 = 0.30^2 + 0.44^2 - 2*0.30*0.44*\cos\ \theta$$
$$\cos\ \theta = \frac{0.30^2 + 0.44^2 - 0.71^2}{2*0.30*0.44}$$
$$\cos\ \theta = \frac{0.09 + 0.19 - 0.50}{0.26}$$
$$\cos\ \theta = -0.833$$

To find the angle θ, the angle whose cosine is –0.833 can be determined using either trigonometric tables (Appendix E) or a calculator with trigonometric func-

tions. This is the process known as finding the inverse cosine and is written as follows.

$$\theta = \cos^{-1} -0.833$$
$$\theta = 146.4°$$

Therefore the relative angle at the knee is 146.4°. In this case, the knee is slightly flexed (180° representing full extension).

Absolute Angle

The other type of angle calculated in biomechanics is the *absolute angle*. An absolute angle is the angle of inclination of a body segment. This type of angle describes the orientation of a segment in space. There are two primary conventions for calculating absolute angles.

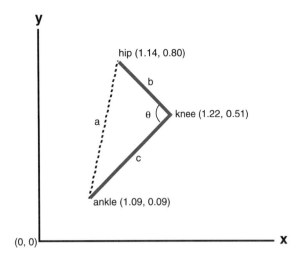

FIGURE 9-6. Coordinate points describing the hip, knee, and ankle joint center and the relative angle of the knee (θ).

One involves placing a coordinate system at the proximal endpoint of the segment. The angle is then measured in a counter-clockwise direction from the right horizontal. The most frequently used convention for calculating absolute angles, however, places a coordinate system at the distal endpoint of the segment (Figure 9-7). The angle

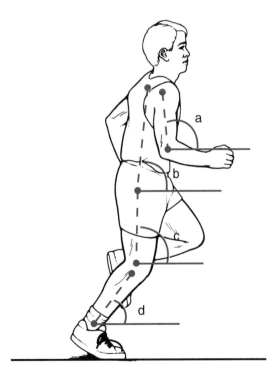

FIGURE 9-7. Absolute angles of: a) the arm, b) the trunk, c) the thigh and d) the leg of a runner.

using this convention is also measured in a counter-clockwise direction from the right horizontal. The absolute angles calculated using these two conventions are related and give comparable information. When calculating absolute angles, however, the convention used must be stated clearly.

Absolute angles are calculated using the trigonometric relationship of the tangent. The tangent is defined based on the sides of a right angle triangle. It is the ratio of the side opposite the angle in question and the side adjacent to the angle. The angle in question is not the right angle in the triangle. If the same leg and thigh segment coordinate positions as in Figure 9-6 are considered, the absolute angles of both the thigh and leg segments can be calculated (Figure 9-8).

To calculate the absolute leg angle, the coordinate values of the segment endpoints of the leg are substituted into the formula to define the tangent of the angle:

$$\tan \theta_{leg} = \frac{y_{proximal} - y_{distal}}{x_{proximal} - x_{distal}}$$
$$= \frac{y_{knee} - y_{ankle}}{x_{knee} - x_{ankle}}$$
$$= \frac{0.51 - 0.09}{1.22 - 1.09}$$
$$= \frac{0.42}{0.13}$$
$$= 3.23$$

Next, the angle whose tangent is 3.23 is again determined using either the trigonometric tables (Appendix E) or a

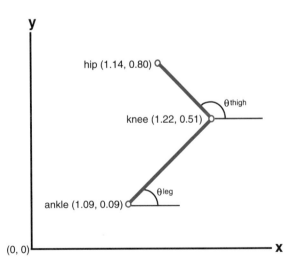

FIGURE 9-8. Absolute angles of the thigh and leg as defined in a coordinate system.

calculator. This is called finding the inverse tangent and is written as follows:

$$\theta_{leg} = \tan^{-1} 3.23$$
$$= 72.8°$$

The absolute angle of the leg, therefore, is 72.8° from the right horizontal. This orientation indicates that the leg is positioned such that the knee is further from the vertical (y) axis of the coordinate system than the ankle. That is, the knee joint is to the right of the ankle joint.

Similarly, to calculate the thigh angle, the coordinate values are substituted into the following:

$$\tan \theta_{thigh} = \frac{y_{hip} - y_{knee}}{x_{hip} - x_{knee}}$$
$$= \frac{0.80 - 0.51}{1.14 - 1.22}$$
$$= \frac{0.29}{-0.08}$$
$$= -3.625$$

Once again, the angle whose tangent is −3.625 is determined as follows:

$$\theta_{thigh} = \tan^{-1} -3.625$$
$$= 105.4°$$

If we consider a coordinate system with the origin at the knee joint, an angle greater than 90° and less than 180° is not in the first quadrant but is in the second quadrant. If the angle is greater than 180° and less than 270°, it is in the third quadrant. Angles greater than 270° and less than 360° are in the fourth quadrant. Since the absolute thigh angle is 105.4°, the angle must be in the second quadrant. This means that the thigh is oriented such that the hip joint is closer to the vertical (y) axis of the coordinate system than is the knee joint. In this case, the thigh would be oriented with the knee to the right of the hip in this reference system.

In clinical situations, the relative angle is most often calculated. In quantitative biomechanical analyses, however, absolute angles are calculated more often than relative angles because they are used more often in a number of subsequent calculations conducted in biomechanics. Regardless of the type of angle calculated, however, a consistent frame of reference must be used. Unfortunately, many different systems of defining angles have been used in biomechanics, resulting in difficulty comparing values from study to study. Several organizations, such as the Canadian Society of Biomechanics and the International Society of Biomechanics, are now attempting to standardize the calculation and representation of angles to provide consistency in biomechanics research.

Representation of Angular Motion Vectors

It is difficult to represent angular motion vectors graphically as lines with arrows as was done in linear kinematics. It is essential, however, to determine how the direction of rotation is determined. The direction of rotation of an angular motion vector is referred to as the *polarity* of the vector. The polarity of an angular motion vector is determined by a convention known as the *Right Hand Rule*. The direction of an angular motion vector is determined using this rule by placing the curled fingers of the right hand in the direction of the rotation. The angular motion vector is defined by an appropriate length arrow that coincides with the direction of the extended thumb of the right hand (Figure 9-9). The convention generally used is that, in the sagittal plane, all segments moving in a counter-clockwise (CCW) direction from the right horizontal have a positive polarity and all segments rotating in a clockwise (CW) direction have a negative polarity.

Angular Motion

The relationships discussed in Chapter 8 on linear kinematics are comparable to those that will be discussed in the angular case. The angular case is simply an analog of the linear case.

Angular Distance and Displacement

The concepts of distance and displacement in the angular case must be discerned. Consider a simple pendulum swinging in the x-y plane through an arc of 70° (Figure 9-10). If the pendulum swings though a single arc, the angular distance is 70°, but if it swings through 1.5 arcs, the angular distance is 105°. *Angular distance* is the total of all angular changes measured following its exact path. As in the linear case, however, angular distance is not the same as angular displacement.

Angular displacement is the difference between the initial and final positions of the rotating object (Figure 9-11). In the example of the pendulum, if the pendulum swings through two complete arcs, the angular displacement would be zero, since its final position is the same as the starting position. In discussing angular displacement, it is necessary to designate the direction of the rotation. A counter-clockwise rotation is considered to be positive (+) and a clockwise rotation is negative (−).

If the absolute angle of a segment, theta (θ), was calculated for successive positions in time, the angular displacement (Δθ) would be:

$$\Delta\theta = \theta_{final} - \theta_{initial}$$

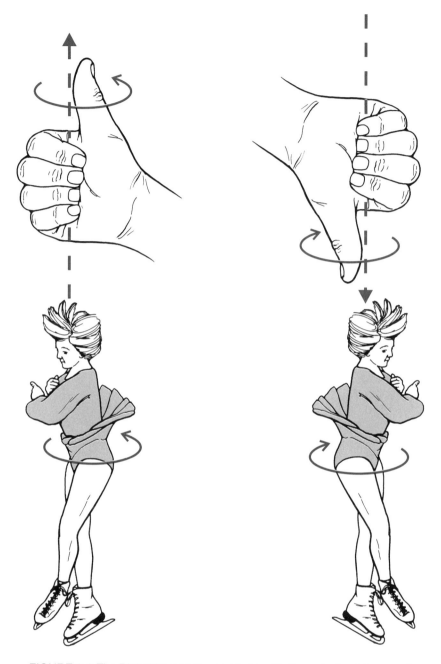

FIGURE 9-9. The Right Hand Rule used to identify the polarity of the angular
velocity of a figure skater during a spin. The fingers of the right hand point in
the direction of the rotation and the right thumb points in the direction of the
angular velocity vector. Note that the angular velocity vector is perpendicular
to the plane of rotation.

The polarity or the sign of the angular displacement is determined by the sign of $\Delta\tau$ as calculated and may be confirmed by the Right Hand Rule.

Angular Velocity

The definitions for *angular speed* and *angular velocity* are analogous with linear speed and linear velocity in

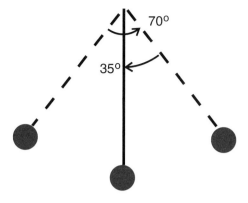

FIGURE 9-10. A swinging pendulum illustrating the angular distance over 1.5 arcs of swing.

both definition and meaning. Angular speed is the angular distance traveled per unit of time.

$$\text{angular speed} = \frac{\text{angular distance}}{\text{time}}$$

Angular speed is a scalar quantity and is generally not of critical importance in a biomechanical analysis because it is not used in any further calculations.

Angular velocity, characterized by the Greek letter omega (ω), is a vector quantity that describes the time rate of change of angular position. If the measured angle is θ, then the angular velocity would be:

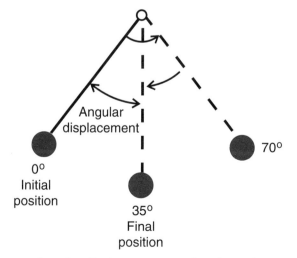

Angular displacement = 35° - 0° = 35°

FIGURE 9-11. Angular displacement is the difference between the final position and the initial position.

$$\omega = \frac{\text{change in angular position}}{\text{change in time}}$$
$$= \frac{\theta_{final} - \theta_{initial}}{\text{time}_{final} - \text{time}_{initial}}$$
$$= \frac{\Delta\theta}{\Delta t}$$

If the initial angle of a segment was 34° at time 1.25 s and the segment moved to an angle of 62° at time 1.30 s, the angular velocity would be:

$$\omega = \frac{\Delta\theta}{\Delta t}$$
$$= \frac{62° - 34°}{1.30 \text{ s} - 1.25 \text{ s}}$$
$$= \frac{28°}{0.05 \text{ s}}$$
$$= 560 \text{ }^{°}\!/_{s}$$

The units of angular speed and angular velocity are generally presented in degrees per second (°/s). If, however, as noted previously, any further computation is to be done using the angular velocity, the units must be radians per second (rad/s).

In the previous example, the average angular velocity was calculated over the time interval from 1.25 s to 1.30 s. According to the discussion in the previous chapter, this angular velocity would represent the slope of a secant on an angular position-time graph over this time interval. The instantaneous angular velocity would represent the slope of a tangent to an angular position-time graph and would be calculated as a limit.

$$\text{limit } \omega = \frac{d\theta}{dt}$$
$$dt \rightarrow 0$$

Angular velocity is thus the first derivative of angular position.

As in the linear case, the direction of the slope on an angle-time profile will determine if the angular velocity is positive or negative, and the steepness of the slope will indicate the rate of change of angular position. If θ_{final} is greater than $\theta_{initial}$, then ω is positive (i.e. the slope is positive), but if θ_{final} is less than $\theta_{initial}$, ω is negative (i.e. slope is negative). Both situations can be confirmed using the Right Hand Rule. If, however, there is no change in the angle, the slope is zero and ω is zero.

The method used to calculate angular velocity over a series of video frames of a kinematic analysis is the first central difference method. This method calculates the angular velocity at the same instant in time at which data for angular position is available. For angular velocity, this formula is:

$$\omega_i = \frac{\theta_{i+1} - \theta_{i-1}}{t_{i+1} - t_{i-1}}$$

where θ_i is the angle at time t_i.

Angular Acceleration

Angular acceleration is the rate of change of angular velocity with respect to time and is symbolized by the Greek letter alpha (α).

$$\text{angular acceleration} = \frac{\text{change in angular velocity}}{\text{change in time}}$$

$$\alpha = \frac{\omega_{final} - \omega_{initial}}{\text{time}_{final} - \text{time}_{initial}}$$

$$\alpha = \frac{\Delta\omega}{\Delta t}$$

For ease of understanding, biomechanists generally present their results in degrees/s² but the most commonly used unit for angular acceleration is rad/s².

As with the linear case and with angular velocity, angular acceleration is the derivative of angular velocity and represents the slope of a line, either a secant or a tangent. If α is the slope of a secant to an angular velocity-time profile, it represents an average acceleration over a time interval. If α is the slope of a tangent, the instantaneous angular acceleration has been calculated. This also implies that the slope may be positive (ω_{final} is greater than $\omega_{initial}$), negative (ω_{final} is less than $\omega_{initial}$), or zero (ω_{final} equals $\omega_{initial}$). The direction of the angular acceleration vector may be confirmed using the Right Hand Rule. The instantaneous angular acceleration is calculated by:

$$\text{limit } \alpha = \frac{d\omega}{dt}$$
$$dt \to 0$$

Once again, in a kinematic analysis, the usual method of calculating angular acceleration is the first central difference method. The formula for angular acceleration for this method would be:

$$\alpha_i = \frac{\omega_{i+1} - \omega_{i-1}}{t_{i+1} - t_{i-1}}$$

where ω_i is the angular acceleration at time t_i.

It should be noted that, as in the case for linear acceleration, the sign or polarity of angular acceleration does not indicate the direction of rotation. For example, a pos-

itive angular acceleration may mean an increasing angular velocity in the positive direction or a decreasing angular velocity in the negative direction. Also, a nega-

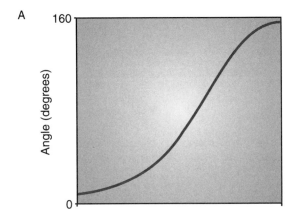

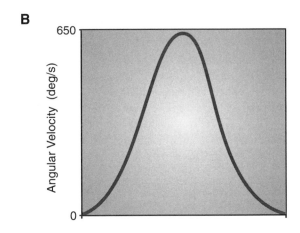

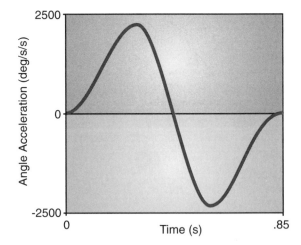

FIGURE 9-12. Graphic representations of: A) elbow flexion angle; B) angular velocity; and C) angular acceleration as a function of time.

tive angular acceleration may indicate a decreasing angular velocity in the positive direction or an increasing angular velocity in the negative direction. Figure 9-12 presents the angular position, angular velocity, and angular acceleration of elbow flexion. It can be seen that even though the motion is in only one direction, the angular acceleration is both positive and negative.

Lower Extremity Joint Angles

Sagittal Plane Angles

In discussing the angle of a joint such as the knee or ankle joint, it is imperative that a meaningful representation of the action of the joint be made. Although by definition all joint angles are relative, the lower extremity joint angles can be calculated, using the absolute angles and one can derive reasonable estimates of the respective segments in space. A system of lower extremity angle conventions was presented by Winter (2). These lower extremity angle definitions are for use in a two-dimensional analysis only. This lower extremity angle convention was later accepted as a standard by the Canadian Society for Biomechanics for use in gait studies. To this date, no other standard for a two-dimensional angular analysis has been presented.

In Winter's system, digitized points describing the trunk, thigh, leg, and foot are used to calculate the absolute angles of each (Figure 9-13). In such a biomechanical analysis, it is assumed that a right side sagittal view is being analyzed. That is, the right side of the subject's body is closest to the camera and is considered to be in the x-y plane. If not, the data must be converted to represent a right side view. All segments of the lower extremity, therefore, rotate according to the Right Hand Rule.

Based on the absolute angles of the trunk and the thigh calculated, the hip angle is:

$$\theta_{hip} = \theta_{thigh} - \theta_{trunk}$$

In this scheme, if the hip angle is positive, the action at the hip is flexion, while if the hip angle is negative, the action is extension. If the angle is zero, the thigh and the trunk are aligned vertically in a neutral position. In human walking at a moderate pace, the hip angle oscillates $\pm20°$ about $0°$, while in running, the hip angle oscillates $\pm35°$.

Using the absolute angle of the thigh and the leg, the knee angle is defined as:

$$\theta_{knee} = \theta_{thigh} - \theta_{leg}$$

In human locomotion, the knee angle is always positive, that is, in some degree of flexion, and usually varies

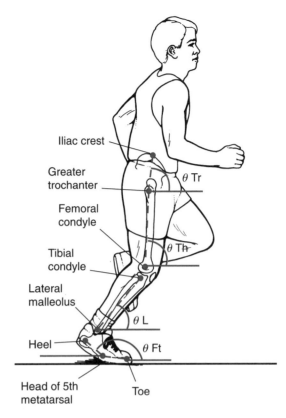

FIGURE 9-13. Definition of the sagittal view absolute angles of the trunk, thigh, leg, and foot (after Winter, D.A. The Biomechanics and Motor Control of Gait. Waterloo, Ont., Canada: University of Waterloo Press, 1987.).

from 0 to $50°$ throughout a walking stride and from 0 to $80°$ during a running stride. Since the knee angle is positive, the knee is always in some degree of flexion. If the knee angle gets progressively greater, the knee is flexing, while if it gets progressively smaller, the knee is extending. A zero knee angle is a neutral position while a negative angle would indicate a hyperextension of the knee.

The ankle angle is calculated using the absolute angles of the foot and the leg:

$$\theta_{ankle} = \theta_{foot} - \theta_{leg} - 90°$$

This may seem more complicated than the other lower extremity joint angle calculations. Without subtracting the additional $90°$, however, the ankle angle would oscillate about $90°$, making interpretation of the ankle angle difficult. By subtracting the additional $90°$, the ankle angle then oscillates about $0°$. Thus, a positive angle represents dorsiflexion and a negative angle represents plantar flexion. The ankle angle generally oscillates $\pm20°$ during a natural pace walking stride and $+35°$ during a running stride. Lower

A

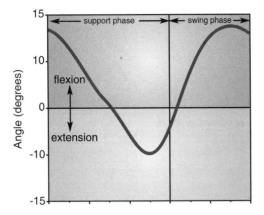

B

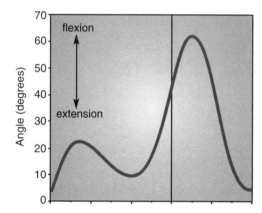

extremity angles calculated for a walking stride using Winter's convention are presented in Figure 9-14.

Rearfoot Angle

Another lower extremity angle that is often calculated in biomechanical analyses is the rearfoot angle. The motion of the sub-talar joint in a two-dimensional analysis is considered to be in the frontal plane. The rearfoot angle represents the motion of the sub-talar joint. The rearfoot angle thus approximates calcaneal eversion and calcaneal inversion in the frontal plane. Calcaneal eversion/inversion is one of the motions in the pronation/ supination action of the sub-talar joint. In the research literature, calcaneal eversion is often referred to as pronation, while calcaneal inversion is referred to as supination.

Rearfoot angle is calculated using the absolute angles of the leg and the calcaneus in the frontal plane. Two segment markers are placed on the rear of the leg to define the longitudinal axis of the leg. Two markers are also placed on the calcaneus (or the rear portion of the shoe) to define the longitudinal axis of the calcaneus (Figure 9-15). These markers are used to calculate the absolute angles of the leg and heel and thus the rearfoot angle is:

$$\theta_{RF} = \theta_{leg} - \theta_{calcaneus}$$

By this calculation, a positive angle represents calcaneal inversion, a negative angle represents calcaneal eversion, and a zero angle is the neutral position.

During the support phase of the gait cycle, the position of the rearfoot, as defined by the rearfoot angle, is in an

C

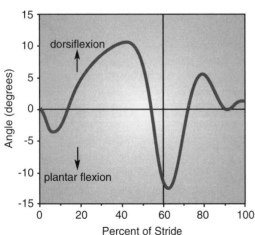

FIGURE 9-14. Graphs of: A) the hip; B) the knee; and C) ankle joint angles during a walking stride.

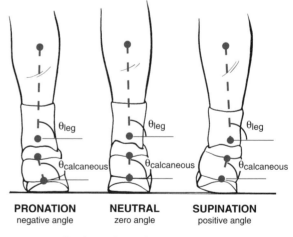

FIGURE 9-15. Definition of the absolute angles of the leg and calcaneus in the frontal plane. These angles are used to constitute the rearfoot angle of the right foot.

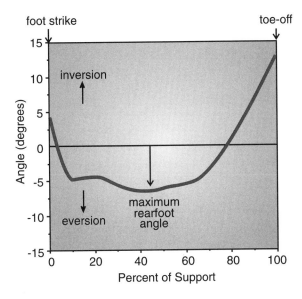

FIGURE 9-16. A typical rearfoot angle-time graph during running (FS - initial foot touchdown; TO - toe-off). Maximum rearfoot angle is indicated.

inverted position at the initial foot contact with the ground. At this instance, the rearfoot angle is positive. From that point onward, during support until midstance, the rearfoot moves to an everted position. Thus, the rearfoot angle is negative. At midstance position, the foot becomes less everted and moves to an inverted position at toe-off. The rearfoot angle becomes less negative and eventually positive at toe-off. Figure 9-16 is a representa-

tion of a typical rearfoot angle curve during the support phase of a running stride.

Relationship Between Angular and Linear Motions

In many human movements, the result of the movement is linear, while the motions of the segments constituting the movement are angular in nature. For example, a baseball pitcher throws a baseball that travels linearly. However, the motions of the pitcher's segments resulting in the throw are rotational. In many instances it is necessary to know the linear motion of the hand that depends on the angular motion of the segments of the upper extremity. This example suggests a mechanical relationship between linear and angular motion.

Linear and Angular Displacement

When the angular measure of an angle, the radian, was defined it was noted that:

$$\theta = \frac{s}{r}$$

where θ was the angle subtended by an arc of length s that is equal to the radius of the circle. The length of the arc can be presented as:

$$s = r\theta$$

Suppose the forearm, with length r_1, rotates about the elbow joint (Figure 9-17). The arc described by the rota-

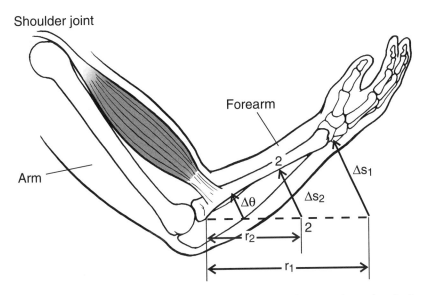

FIGURE 9.17. Illustration of the relationship between linear and angular displacement.

tion—the distance that the wrist moves—is Δs_1 and the angle is $\Delta\theta$. The linear distance that the wrist travels is then described as:

$$\Delta s_1 = r_1 \, \Delta\theta$$

Therefore, the linear distance that any point on the segment moves can be described if the distance of that point to the axis of rotation and the angle through which the segment rotates is known. Suppose another point on the arm is marked as s_2 with a distance of r_2 to the axis of rotation. The distance this point travels during the same angular motion is:

$$\Delta s_2 = r_2 \, \Delta\theta$$

Since r_1 is longer than r_2, the distance traveled by s_1 must be greater than s_2. Thus, the most distal points on a segment travel a greater distance than points closer to the axis of rotation. The value for the expression r is called the radius of rotation and refers to the distance of a point from the axis of rotation.

Consider that the change in the angle, $\Delta\theta$, is very small; then the length of the arc, Δ's, can be approximated as a straight line. Therefore, a relationship between angular and linear displacement can be formulated. That is, when r is the radius of rotation, then:

linear displacement = radius of rotation * angular displacement

or

$$\Delta s = r \, \Delta\theta$$

or using calculus (that is, when $d\theta$ is very small)

$$ds = r \, d\theta$$

For example, if the arm segment of length 0.13 m rotated about the elbow an angular distance of 0.23 radians, the linear distance that the wrist traveled is:

$$\Delta s = r \, \Delta\theta$$
$$\Delta s = 0.23 \text{ rad} * 0.13 \text{ m}$$
$$\Delta s = 0.03 \text{ m}$$

Note that Δs has a unit of length (m) which is the correct unit since it is a linear distance. Remember that radians are dimensionless, so that radians times meters results in units of meters.

Linear and Angular Velocity

The relationship between linear and angular velocity is similar to the relationship between linear and angular displacement. In the example in the last section, the arm, with length r, rotates about the elbow that was used. The

linear displacement of the wrist is the product of the distance r, the radius of rotation, and the angular displacement of the segment. Differentiating this equation with respect to time:

$$ds = r \, d\theta$$
$$\frac{ds}{dt} = r \, \frac{d\theta}{dt}$$

Thus, the linear velocity of a point on a rotating body is the product of the distance of that point from the axis of rotation and the angular velocity of the body. The linear velocity vector in this expression is instantaneously tangent to the path of the object and is referred to as the tangential velocity or v_T (Figure 9-18). That is, the linear velocity vector would behave as a tangent, only touching the curved path at one point. The vector, therefore, would be perpendicular to the rotating segment.

For example, if the arm segment of length r = 0.13 m rotated with an angular velocity of 2.4 rad/s, the velocity of the wrist would be:

$$v_T = r\omega$$
$$v_T = 0.13 \text{ m} * 2.4 \text{ rad/s}$$
$$v_T = 0.31 \text{ m/s}$$

It should be noted that linear velocity has units of m/s which results in this instance from meters times rad/s since radians are dimensionless.

The relationship between linear velocity and angular velocity is critical in a number of human movements, particularly those in which the performer throws or strikes an object. To increase the linear velocity of the ball, for example, the soccer player can either increase the angular velocity of the lower extremity segments, or

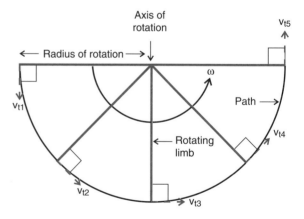

FIGURE 9-18. Tangential velocity of a rotating segment at different instants in time. Note the tangential velocity is perpendicular to the radius of rotation.

increase the length of the extremity by extending at the joints, or both, to gain the maximum range of a kick. For an individual, the major alternative is to increase the angular velocities of these segments. For example, Plagenhoef (3) reported foot velocities prior to impact of 16.33 m/s to 24.14 m/s for several types of soccer kicks for the same individual. Since the segment lengths did not change substantially, it can be stated that if the foot velocity changed, then the angular velocity certainly must have varied for each type of kick.

In some activities, however, the length r can change. In golf, the clubs have varying lengths and club head lofts according to the desired distance that the ball should travel (Figure 9-19). For example, the 2-iron is longer than the 9-iron and has a different club head loft, with the 9-iron having a steeper club head loft than the 2-iron. If both clubs had the same loft, the 2-iron shot would go further than a 9-iron shot, given the same angular velocity of the golf swing as it does for most expert golfers. Golfers often use the same club but vary the length, r, by "choking up" on the handle or gripping the club closer to the middle of the shaft. Using this technique, the golfer may swing with the same angular velocity but vary the length, and they can then vary the linear velocity of the club head.

Linear and Angular Acceleration

Remember that the linear velocity vector calculated from the product of the radius and the angular velocity is tangent to the curved path and can be referred to as the tangential velocity. As previously stated:

$$v_T = \omega \, r$$

If the time derivative of this expression is determined, a relationship exists which expresses the *tangential acceleration* in terms of the radius of rotation and the angular acceleration. The expression of the derivative is:

$$a_T = \alpha \, r$$

where a_T is the tangential acceleration, r is the radius of rotation, and a is the angular acceleration. The tangential acceleration, like the tangential velocity vector, is a vector tangent to the curve and perpendicular to the rotating segment (Figure 9-20). In any activity, such as the discus, in which the performer spins in order to throw the implement, the purpose is to throw an implement as far as possible. An understanding of tangential velocity and tangential acceleration is therefore necessary. The time rate of change in the tangential velocity of the discus along its curved path is the tangential acceleration. The peak tangential velocity is ideally reached just prior to the release

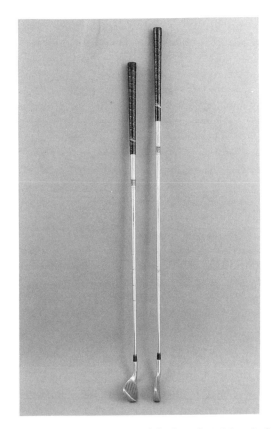

FIGURE 9-19. A comparison of the lengths of the shafts of a 2-iron and a 9-iron golf club.

of the discus at which time the tangential acceleration must be zero.

Consider a softball pitcher using an underhand pitch; further insight into another component of linear acceleration acting during rotational movement can be gained. As

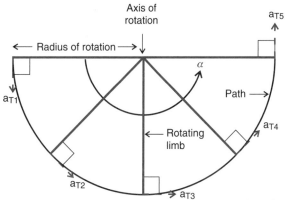

FIGURE 9-20. Illustration of the tangential acceleration of a swinging segment. Notice that it is perpendicular to the swinging limb.

the pitcher moves his arm to the point of release of the pitch, the ball follows a curved path. Because the pitcher's arm is attached to the shoulder, the ball must follow the curved path produced by the rotation of the arm. Therefore, to continue on this path, the ball moves slightly inward and slightly vertically downward at each instant in time until the ball is released (Figure 9-21). That is, the ball is incrementally accelerated downward and it is also accelerated inward toward the shoulder or the axis of rotation.

Two components of acceleration produced by the rotation of a segment have been discussed: one tangential to the path of the segment and one along the segment towards the axis of rotation These two accelerations are necessary for the ball in the pitcher's hand to continue on its curved path. The forward movement is the result of the tangential acceleration that has been previously discussed. Acceleration towards the axis or center of rotation, however, is called the *centripetal acceleration* (Figure 9-22). The adjective centripetal means "center seeking". Centripetal acceleration is also known as radial acceleration. Either name is correct, although for the remainder of this discussion, the term centripetal acceleration will be used.

To derive the formula for centripetal acceleration, it must be noted that the resultant linear acceleration of a segment endpoint, such as the wrist, of a rotating segment is:

$$a = \frac{dv}{dt}$$

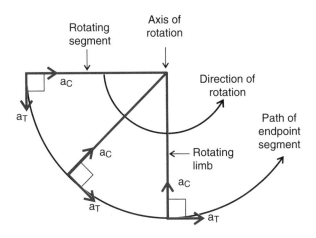

FIGURE 9-22. Illustration of the tangential acceleration (a_T) and the centripetal acceleration (a_C). Note that they are perpendicular to each other. The tangential acceleration accelerates the segment endpoint downwards and the centripetal acceleration accelerates the endpoint towards the center of rotation. The result is motion along a curved path.

Since the segment is rotating, the linear velocity is:

$$v_T = \omega\, r$$

Substituting this into the acceleration equation, the acceleration becomes:

$$a = \frac{d(\omega\, r)}{dt}$$

If certain computational rules from calculus are applied, this equation becomes:

$$a = \omega * \frac{dr}{dt} + \frac{d\omega}{dt} * r$$

Since $\frac{dr}{dt}$ is the linear velocity and $\frac{d\omega}{dt}$ is the angular acceleration of the segment, then this expression becomes:

$$a = \omega\, v + \alpha\, r$$

Note that the linear velocity, v, is equal to $\omega\, r$ and thus the expression for the linear acceleration of the segment endpoint is, therefore:

$$a = \omega\, \omega\, r + \alpha\, r$$

or

$$a = \omega^2\, r + \alpha\, r$$

Remember that the resultant acceleration has two components that are perpendicular to each other. This expression illustrates these two components. This explanation,

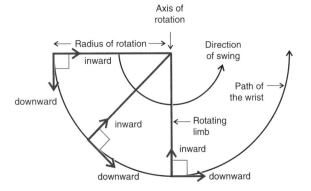

FIGURE 9-21. The directions of the acceleration components of the wrist of a softball pitcher during the downswing of the arm to the release of the ball. The wrist is accelerated inward towards the shoulder and downward tangential to the path of the wrist. These two vectors are perpendicular to each other.

however, requires the use of vector calculus and is much more complicated in derivation than presented. It should be noted that the addition (+) sign in this expression means vector addition. It was previously determined that αr was the tangential acceleration; thus $\omega^2 r$ is the centripetal acceleration. The expression for centripetal acceleration is:

$$a_C = \omega^2 r$$

Centripetal acceleration may also be expressed in the following form as a function of the tangential velocity and the radius of rotation. That is, if $v = \omega r$ is substituted into the centripetal acceleration equation, the equation becomes:

$$a_C = \frac{v^2}{r}$$

From this expression, it can be seen that the centripetal acceleration will increase if the tangential velocity increases or if the radius of rotation decreases. For example, the usual difference between an indoor running track and an outdoor track is that the indoor track is smaller and thus has a smaller radius. If a runner attempted to maintain the same velocity around the indoor turn as they did on an outdoor track, the centripetal acceleration would need to be necessarily greater for the runner to accomplish the turn. Generally, the runner cannot accomplish the turn at the same velocity as outdoors, thus the race times on indoor tracks are somewhat slower than on outdoor tracks.

Since the centripetal acceleration and the tangential acceleration are components of the linear acceleration, they must be perpendicular to each other. The resultant acceleration vector of these components may then be constructed. The resultant acceleration (Figure 9-23) is computed using the Pythagorean relationship:

$$a = \sqrt{a_T{}^2 + a_C{}^2}$$

In computing either the tangential or the centripetal acceleration, the units of angular velocity and angular acceleration are rad/s and rad/s^2, respectively. The units of linear acceleration (m/s^2) can only result when a radian-based unit is used in the computation.

Angle-Angle Diagrams

In most graphical presentations of human movement, usually some parameter (e.g. position, angle, velocity, etc.) is graphed as a function of time. In certain activities, such as locomotion, the motions of the segments are

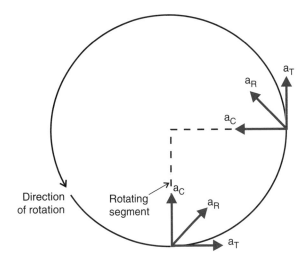

FIGURE 9-23. The resultant linear acceleration vector (aR) composed of the centripetal and tangential acceleration components.

cyclic, in that they are repetitive with the end of one cycle at the beginning of the next. In these instances, an *angle-angle diagram* may be useful to represent the relationship between two angles during the movement. An angle-angle diagram is the plot of one angle as a function of another angle. That is, one angle is used for the x-axis and one for the y-axis. In an angle-angle diagram, one angle is usually a relative angle and the other is an absolute angle. For the angle-angle graph to be meaningful, a functional relationship between the angles should exist (Figure 9-24). For example, in studying an individual running, the relationship between the sagittal view ankle and knee angles may be meaningful, whereas the relationship between the elbow angle and the ankle angle may not.

One problem with this type of diagram is that time cannot be easily represented on the graph. It can be presented, however, by placing marks on the angle-angle curve to represent each instant in time at which the data were calculated. These marks are placed at equal time intervals and give an indication of the angular distance through which each joint has moved in equal time intervals. Thus angular velocity of the movement is represented, since the further apart the marks are on the curve, the greater the velocity of the movement. Conversely, the closer together the marks, the less the velocity (Figure 9-25).

Angle-angle diagrams have proven very useful in the examination of the relationship between the rearfoot angle and the knee angle (5, 6). This relationship is based on the related anatomical motions of the sub-talar and knee joints. During the support phase of gait, the knee

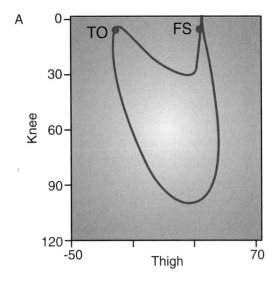

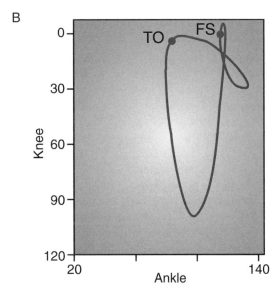

FIGURE 9-24. Angle-angle diagrams of the knee angle as a function of the thigh angle (A) and the knee angle plotted as a function of the ankle angle (B) for one complete running stride of an individual running at 3.6 m/s. TO represents toe-off and FS represents initial foot contact (after Williams, K.R. Biomechanics of Running. In Exercise and Sports Sciences Review 14, 1985.).

flexes at touchdown and continues to flex until midstance. At the same time, the foot lands in an inverted position and immediately begins to evert until midstance. Both of these actions—knee flexion and sub-talar eversion—are associated with internal tibial rotation. After midstance, the knee extends and the sub-talar joint inverts. Both of these joint actions result in external tibial rotation. These actions are presented in Figure 9-25 with

the knee angle expressed as a relative angle and sub-talar joint inversion/eversion expressed as an absolute angle. Figure 9-26 is an angle-angle diagram presented in a paper by van Woensel and Cavanagh (6) and illustrates the relationship between the knee angle and the rearfoot angle in different shoe conditions. One of the three shoes used in this study was specifically designed to force the runner to pronate during support, another to force the runner to supinate during support, and the third pair was a neutral shoe.

Angular Kinematics of Running

Many researchers have reported on how the lower extremity joint angles vary throughout the running stride, particularly during the support portion of the stride. Generally, these angles are reported at discrete instants in time, such as prior to contact, midstance, and toe-off. There have been numerous biomechanical analyses of the lower extremity angular kinematics. Thus, for this brief review, only selected literature describing particular lower extremity angles—the rearfoot angle and the knee angle—during running, will be presented.

The knee angle is flexed at touchdown and has been reported in the literature to be in the range of 21 to 30° (5, 7, 8, 9). After touchdown, the knee flexes to values ranging from 38 to 50° with the greater flexion occurring at faster running speeds (5, 10). Maximal knee flexion occurs at midstance, after which the knee extends until toe-off. Full extension is not achieved at toe-off with values ranging from 27° (7) to 18° (11), depending on running speed. Greater extension values at toe-off are generally associated with faster running speeds. While the magnitude of the knee angles at these specific instants in time during the support phase of running vary, the basic profile of the curve does not. The profile of the knee joint angle appears to be relatively stable and immune to distortion from influences such as running shoe construction (6, 9) or delayed onset muscle soreness (12).

A number of investigations have described the rearfoot angle during the support phase of running. Excessive rearfoot motion has been hypothesized to cause a variety of lower extremity injuries although there is little evidence to directly relate excessive rearfoot motion and injury (8, 13). In fact, a valid definition of excessive rearfoot motion has not yet been determined. From a functional standpoint, eversion of the calcaneus is a necessary movement because it allows the foot to assume a flat position on the ground. Typically, maximum rearfoot angle values from –6 to –17° at midstance have been reported in the literature (8, 14). This wide range in maximum values can be due to differences in the foot struc-

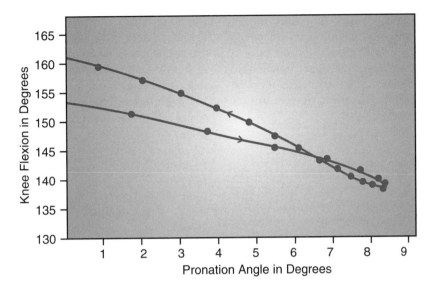

FIGURE 9-25. Angle-angle diagram of knee flexion as a function of sub-talar pronation angle for an individual running at a 6:00 minute/mile pace on a treadmill. The plus signs on the curve indicate equal time intervals (after Bates, B.T., James, S.L., Osternig, L.R. Foot function during the support phase of running. Running 24:29, Fall/1978.).

ture of individuals as well as the influence of footwear. It has been reported that more extreme rearfoot angles occurred at midstance when the subjects ran in racing shoes than in training shoes (14). More extreme rearfoot eversion angles have also been reported for runners in a very soft midsole shoe than in a firmer midsole shoe (9). Although the rearfoot angle is related in motion to the knee angle by the action of tibial rotation, it is, unlike the knee angle, highly variable and certainly can be influenced by a number of factors.

The simultaneous actions of these two lower extremity angles has been a topic of several investigations. Since internal tibial rotation accompanies knee flexion and sub-talar joint eversion, and both reach a maximum at midstance, the mis-timing of these joint actions has been suggested as a possible mechanism for lower extremity injury (5). Hamill et al. (9) illustrated that the rearfoot angle could be changed by a very soft midsole running shoe, while the knee angle could not. They reported that, in a soft midsole running shoe, the maximum rearfoot angle occurred sooner in the support period than did maximum knee flexion. The sub-talar joint also stayed at this maximum while the knee began to extend. Thus, it appeared that a twisting action may be applied to the tibia by the differential speeds at which the tibia rotated early in support and late in support. Since the tibia is a rigid structure and may be difficult to twist, the tibia may continue to rotate internally at the knee, even though it should externally rotate. This undesirable action at the

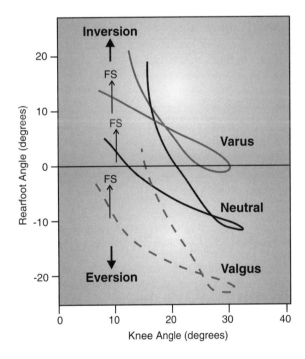

FIGURE 9-26. Knee-rearfoot angle-angle diagram of an individual in three types of running shoes (FS = foot strike). The VARUS shoe has a medial wedge, thus mediating rearfoot pronation; the VALGUS shoe has a lateral wedge, thus enhancing rearfoot pronation; and the NEUTRAL shoe is a normal running shoe (after van Woensel, W. and Cavanagh, P.R. A perturbation study of lower extremity motion during running. International Journal of Sports Biomechanics 8:30–47, 1992.).

knee may possibly cause knee pain in the runner. If these actions are repeated with each foot-ground contact and the runner experiences many foot-ground contacts, a knee injury could result that would keep the runner from training. This type of injury is often referred to as an *overuse* injury. It results from an accumulation of stresses and not a single high-level stress.

Summary

Nearly all purposeful human movement involves the rotation of segments about axes passing through the joint centers, and thus a knowledge of angular kinematics is necessary to understand human movement. Angles may be measured in units of degrees, revolutions, or radians. If the angular measurement is to be used in further calculations, then the radian must be used. A radian is equal to $57.3°$.

Angles may be defined as relative and absolute, and both may be used in biomechanical investigations. A relative angle measures the angle between two segments but cannot determine the orientation of the segments in space. An absolute angle measures the orientation of a segment in space relative to the right horizontal axis placed at the distal end of the segment. How segment angles are defined must be clearly stated when presenting the results of any biomechanical analysis.

The kinematic quantities of angular position, displacement, velocity, and acceleration have the same relationship to each other as their linear analogs. Thus, angular velocity is calculated using the first order central difference method as follows:

$$\omega_i = \frac{\theta_{i+1} - \theta_{i-1}}{2\Delta t}$$

Likewise angular acceleration is defined as:

$$\alpha_i = \frac{\omega_{i+1} - \omega_{i-1}}{2\Delta t}$$

The techniques of differentiation and integration apply to angular quantities as well as linear quantities. Thus, angular velocity is the first derivative of angular position with respect to time, and angular acceleration is the second derivative. The concept of the slope of a secant and a tangent also applies in the angular case to distinguish between average and instantaneous quantities.

It is difficult to represent angular motion vectors in the manner in which linear motion vectors were represented. The Right Hand Rule is used to determine the direction of the angular motion vector. Rotations that are counterclockwise are positive (+) rotations while clockwise rotations are negative (–).

Sagittal view lower extremity angles were defined using a system suggested by Winter (2). In this convention, ankle, knee, and hip angles were defined using the absolute angles of the foot, leg, thigh, and pelvis segments. Rearfoot angle measures the relative motion of the leg and the calcaneus in the frontal plane and is calculated from the absolute angles of the calcaneus and the leg.

There is a relationship between linear and angular motion. Comparable quantities of the two forms of motion may be related when the radius of rotation is considered. The linear velocity of the distal end of a rotating segment is called the tangential velocity and is calculated as:

$$v_T = \omega r$$

where Ω is the angular velocity of the segment and r is the length of the segment. The derivative of the tangential velocity, the tangential acceleration is:

$$a_T = \alpha r$$

where a is the angular acceleration of the rotating segment. The other component of the linear acceleration of the end point of the rotating segment is the centripetal or radial acceleration. This is expressed as:

$$a_C = \omega^2 r$$

The tangential and centripetal acceleration components are perpendicular to each other.

A useful tool in biomechanics is the presentation angular motion in angle-angle diagrams. These diagrams generally present angles of joints that are anatomically functionally related. Time can only be indirectly presented on this type of graph, however.

Review Questions

1. Which of the following activities illustrate translational motion, curvilinear motion, angular motion, or general motion?
   a) cycling
   b) gliding on skates
   c) leg action during running
   d) sky-diving
   e) runner during a race
   f) motion of a pool cue during a shot
   e) glide during a swim breast-stroke

2. A skater completes a double twisting jump followed by a triple twisting jump. a) How many revolutions were completed on each jump? b) What was the angular distance in degrees and radians completed for each jump? (Answer: a) 2, 3; b) 720° or 12.57 rad, 1080° or 18.85 rad).

3. In question 2, what was the angular displacement for each jump? (Answer: a) 0°; b) 0°).

4. During a biceps curl, the relative angle at the elbow changes from 0 to 160° for each flexion curl. If four flexion curls are accomplished, what is a) the total angular distance and b) the angular displacement at the elbow? (Answer: a) 640°; b) 0°).

5. During an elbow flexion exercise, the relative angle at the elbow was 10° at 0.5 s and 120° at 0.71 s. What was the angular velocity of the elbow? (Answer: 523.81 °/s).

6. In the following diagram, use a protractor to measure:
   a) the relative angle of the knee
   b) the relative angle of the ankle
   c) the relative angle of the elbow
   d) the absolute angle of the trunk
   e) the absolute angle of the thigh
   f) the absolute angle of the upper arm

7. If the angular velocity at 0.47 s was 1.5 rad/s and 2.1 rad/s at 0.51 s, what was the average angular acceleration over this time interval? (Answer: 15.0 rad/s^2).

8. If a skater is rotating during a spin at a constant angular velocity, what is his angular acceleration? (Answer: 0 rad/s^2).

9. There are two styles of place-kicking in football: the traditional "head-on" style and the soccer style. In which style would the kicker achieve a greater linear velocity of the foot? Why?

10. Why would a batter in baseball wish to "choke up" on a bat when facing a pitcher with an outstanding fastball?

Additional Questions

1. Calculate a) the relative angle at the knee and the absolute angles of b) the thigh and c) the leg given the following positions in degree. (Hint : plot these points on a graph before calculating.)
   hip - (1.228, 0.931), knee - (1.122, 0.542), ankle - (0.897, 0.160)
   (Answer: a) 164.75°; b) 74.75°; c) 59.50°).

2. What are the angles in question #1 in radians?
   (Answer: a) 2.86 rad; b) 1.31 rad; c) 1.04 rad).

3. During the support phase of walking, the absolute angle of the thigh has the following angular velocities:

frame	time (s)	angular velocity (rad/s)
38	0.6167	1.033
39	0.6333	1.511
40	0.6500	1.882
41	0.6667	2.190

   Calculate the angular acceleration at frames 39 and 40.
   (Answer: frame 39: 25.42 rad/s^2; frame 40: 20.33 rad/s^2)

4. Sketch the angular velocity and angular acceleration curves using the concepts of slope and local extrema of the following angular position curve.

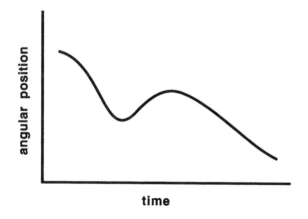

5. A cyclist completes 2.1 cycle revolutions in 1 s. What is a) the angular distance, b) the angular displacement, and c) the angular velocity? (Answer: a) 756°; b) 0°; c) 756 °/s).

6. An individual's arm segment is 0.15 m long and has an angular velocity of 123 °/s. What is the tangential velocity of the wrist? (Answer: 0.32 m/s).

7. An individual is running around a turn with a 12 m radius at a running speed of 4.83 m/s. What is the runner's centripetal acceleration? (Answer: 1.94 m/s^2).

8. Sketch a bowler's arm at an instant on the downswing with the following vectors illustrated on the arm:
   a) angular velocity of the arm
   b) the length of the arm
   c) linear velocity of the wrist
   d) centripetal acceleration

9. A hammer thrower rotates at 14.7 rad/s with an angular acceleration of 6.28 rad/s^2 prior to releasing the hammer. Given a radius (i.e. length of the arm of the athlete plus the length of the chain of the hammer) of 1.5 m, what are the magnitudes of a) the tangential, b) the centripetal accelerations, and c) the resultant acceleration? (Answer: a) 9.42 m/s^2; b) 324.14 m/s^2; c) 324.27 m/s^2).

10. Why must a runner decrease their velocity on an indoor track versus an outdoor track if both tracks have the same turn radius?

References

1. Nordin, M. and Frankel, V.H. (eds.). Biomechanics of the Musculoskeletal System (2nd ed.). Philadelphia: Lea & Febiger, 1979.
2. Winter, D.A. The Biomechanics and Motor Control of Gait. Waterloo, Ont., Canada: University of Waterloo Press, 1987.
3. Plagenhoef, S. Patterns of Human Motion. Inglewood Cliffs, NJ: Prentice-Hall, Inc, 1971.
4. Williams, K.R. Biomechanics of Running. In Exercise and Sports Sciences Review. 13:389–441, 1985.
5. Bates, B.T., James, S.L., Osternig, L.R. Foot function during the support phase of running. Running 24:29, Fall/1978.
6. van Woensel, W. and Cavanagh, P.R. A perturbation study of lower extremity motion during running. International Journal of Sports Biomechanics 8:30–47, 1992.
7. Elliott, B.R. and Blanksby, B.A. A biomechanical analysis of the male jogging action. Journal of Human Movement Studies 5:42–51, 1979.
8. Clarke, T.E., Frederick, E.C., Hamill, C.L. The effects of shoe design parameters of rearfoot control in running. Medicine and Science in Sports and Exercise 15(5):376–381, 1983.
9. Hamill, J., Bates, B.T., Holt, K.G. Timing of lower extremity joint actions during treadmill running. Medicine and Science in Sports and Exercise 24:807–813, 1992.
10. Bates, B.T., Osternig, L.R., Mason, B.R., James, S.L. Functional variability of the lower extremity during the support phase of running. Medicine and Science in Sports and Exercise 11(4):328–331, 1979.
11. Cavanagh, P.R., Pollock, M.L., Landa, J. A biomechanical comparison of good and elite distance runners. In P. Milvy (ed.). The Marathon: Physiological, Medical, Epidemiological, and Psychological Studies. pp. 328–345, New York: New York Acad. Sci., 1977.
12. Hamill, J., Clarkson, P.M., Freedson, P.S., Braun, B. Muscle soreness during running: Biomechanical and physiological considerations. International Journal of Sports Biomechanics 7:125–137, 1990.
13. Nigg, B.M., Luethi, S., Denoth, J., Stacoff, A. Methodological aspects of sport shoe and sport surface analysis. In H. Matsui and K. Kobayashi (eds.). Biomechanics VIII-B. pp. 1041–1052. Champaign, IL: Human Kinetics Publishers, 1983.
14. Hamill, J., Freedson, P.S., Boda, W., Reichsman, F. Effects of shoe type on cardiorespiratory responses and rearfoot control during treadmill running. Medicine and Science in Sports and Exercise 20:515–521, 1987.

Additional Reading

1. Inman, V.T., Ralston, H.J., Tood, F. Human Walking. Baltimore: Williams and Wilkins, 1981.
2. Milliron, M.J., Cavanagh, P.R. Sagittal plane kinematics of the lower extremity during distance running. In P.R. Cavanagh (ed.). Biomechanics of Distance Running. pp. 65-105. Champaign, IL: Human Kinetics Publishing, 1990.
3. Stacoff, A., Kaelin, X., Stuessi, E., Segesser, B. The torsion of the foot in running. International Journal of Sports Biomechanics 5:375-389, 1989.
4. Winter, D.A. Biomechanics and Motor Control of Human Movement (2nd edition). New York: John Wiley and Sons, Inc., 1990.

Glossary

absolute angle:	angle of a segment, measured from the right horizontal, that describes the orientation of the segment in space.
angle:	a figure formed by two lines meeting at a common point called a vertex.
angle-angle diagram:	a graph in which the angle of one segment is plotted as a function of the angle of another segment.
angular acceleration:	the change in angular velocity per unit time.
angular displacement:	the difference between the final angular position and the initial angular position of a rotating body.
angular distance:	the total of all angular changes of a rotating body.
angular kinematics:	the description of angular motion including angular positions, angular velocities, and angular accelerations, without regards to the causes of the motion.
angular motion:	motion about an axis of rotation where different regions of the same object do not move through the same distance in the same time.
angular speed:	the angular distance traveled divided by the time period over which the angular motion occurred.
angular velocity:	the time rate of change of angular displacement.
axis of rotation:	the point about which a body rotates.
centripetal acceleration:	the component of the linear acceleration directed towards the axis of rotation.
degree:	a unit of angular measurement $\frac{1}{360}$ of a revolution.
general motion:	motion that involves both translation and rotation.
instantaneous joint center:	the center of rotation of a joint at any instant in time.
Law of Cosines:	the general case of the Pythagorean Theorem that states: $$a^2 = b^2 + c^2 - 2ab \cos A$$ where a is the length of the side opposite angle A and b and c are the lengths of the other 2 sides in a triangle.
overuse injury:	an injury caused by continual low level stress on a body.
polarity:	the direction of rotation designated as positive or negative.
radial acceleration:	see centripetal acceleration.
radian:	the measure of an angle at the center of a circle described by an arc equal to the length of the radius of the circle (1 rad = 57.3°).
radius of rotation:	the linear distance from the axis of rotation to a point on the rotating body.
relative angle:	the angle formed by the longitudinal axes of two adjacent segments whose vertex is at the joint.

revolution:	a unit of measurement that describes one complete cycle of a rotating body.
Right Hand Rule:	the convention that designates the direction of an angular motion vector; the fingers of the right hand are curled in the direction of the rotation and the right thumb points in the direction of the vector.
rotation:	a motion that occurs when all parts of an object do not undergo the same displacement.
tangent:	the ratio of the side opposite an angle to the adjacent angle in a right angle triangle.
tangential acceleration:	the change in linear velocity per unit time of a body moving along a curved path.
tangential velocity:	the change in linear position per unit time of a body moving along a curved path.
vertex:	the intersection of the two lines that form an angle.

Linear Kinetics

After reading this chapter, the student should be able to:

1. Define force and discuss the characteristics of a force.
2. Compose and resolve forces according to vector operations.
3. State Newton's three laws of motion and their impact on human movement.
4. Differentiate between a contact and a non-contact force.
5. Discuss Newton's law of gravitation and how it affects human movement.
6. Discuss the 6 types of contact forces and how each affects human movement.
7. Represent the external forces acting on the human body on a free-body diagram.
8. Discuss the forces acting on an object as it moves along a curved path.
9. Discuss the relationships between force, pressure, work, energy, and power.

In the previous chapters, linear and angular kinematics or the description of motion were discussed. The motion described was either translatory (linear), rotational (angular), or both (general). In this chapter the concern will be with the causes of motion. For example, why does a runner lean on the curve of a track? What keeps an airplane in the air? Why do golf balls slice or hook? How does a pitcher curve a baseball? The search for the causes of motion date back to antiquity, but answers to some of these questions were suggested by such notables as Aristotle and Galileo. The culmination of these explanations was provided by the great scientist Sir Isaac Newton. In fact, the Laws of Motion described by Newton in his famous book *Principia Mathematica* (1687) form the cornerstone of the mechanics of human movement. The branch of mechanics that deals with the causes of motion is called *kinetics*. Kinetics is concerned with the forces that act on a system. If the motion is translatory, then *linear kinetics* is of concern. The basis for the understanding of the kinetics of linear motion is the concept of force.

Force

According to Newton's principles, objects move when acted upon by a force greater than the resistance to movement provided by the object. A force involves the interaction of two objects and produces a change in the state of motion of an object by pushing or pulling it. The force may produce motion, stop motion, positively or negatively accelerate, or change the direction of the object. In each of the previous cases it should be noted that the acceleration of the object changes or is prevented from changing. A *force*, therefore, may be defined as any interaction, a push or pull, between two objects that can cause an object to accelerate either positively or negatively.

Characteristics of a Force

Forces are vectors and as such have the characteristics of a vector—magnitude and direction. Magnitude represents the amount of force being applied. It is also necessary to state the direction of a force because the force could have a different effect, depending, for example, if the force is pushing in one direction instead of pulling in the other. Vectors, as described in Chapter 8, are usually represented by arrows, with the length of the arrow indicating the magnitude of the force and the arrow head pointing in the direction in which the force is being applied. In the International System (SI) of measurement the unit for force is the newton (N).

Forces have two other equally important characteristics, however; the point of application and the line of action. The *point of application* of a force is that specific point at which the force is applied to an object. In many instances, a force is represented by a point of application at a specific point, although there may be many points of application. For example, the point of application of a muscular force is the center of the muscle's attachment to the bone or the insertion of the muscle. In many cases, the muscle is not attached to a single point on the bone but to many points, such as in the case of the fan-shaped deltoid muscle. In solving mechanical problems, however, it is considered to be attached to a single point. The *line of application* of a force represents a straight line of infinite length in the direction in which the force is acting. A force can be assumed to produce the same acceleration of the object if it acts anywhere along this line of application. The orientation of the line of application is usually given with respect to an X-Y coordinate system. The orientation of the line of application to this system is given as an angular position and is referred to as the *angle of application*. This angle is designated by the Greek letter theta (θ). The four characteristics of a force are illustrated in Figure 10-1.

Composition and Resolution of Forces

Since forces are vectors and have an angle of application, they may be composed into the resultant of several forces or a force may be resolved into its components. That is, a single force vector can be calculated or composed that represents the net effect of all of the forces in the system. Similarly, given the resultant force, the resul-

tant force can be composed or resolved into its horizontal and vertical components. In order to do either, the trigonometric principles presented in Appendix D will be used.

There are several types of force systems, however, that must be defined in order to compose or resolve force systems in which there are multiple forces. Any system of forces acting in a single plane are referred to as *coplanar*, and if they act at a single point they are called *concurrent*. Any set of concurrent coplanar forces may be substituted by a single force, or the resultant, producing the same effect as the multiple forces. The process of finding this single force is called *composition* of force vectors.

When force vectors act along a single line, the system is said to be *collinear*. In this case, vector addition will be used to compose the forces. Consider the force system in Figure 10-2a. The force vectors a, b, and c all act in the same direction and can be replaced by a single force, d, which is the sum of a, b, and c. Thus,

$$d = a + b + c$$
$$= 5\,N + 7\,N + 10\,N$$
$$= 22\,N$$

The force vector d would have the same effect as the other three force vectors. In Figure 10-2b, however, two of the force vectors, a and b, are acting in one direction while the vector, c, is acting in the opposite direction. Thus, the force vector, d, is the algebraic sum of these three force vectors:

$$d = a + b - c$$
$$= 5\,N + 7\,N - 4\,N$$
$$= 8\,N$$

The force vector d still represents the net effect of these force vectors. In both of these examples, sets of collinear force vectors are present.

When the force vectors are not collinear but are coplanar, they may still be composed to determine the resultant force. Graphically, this can be done in exactly the same manner as described in Chapter 8 in the section on adding vectors. Consider Figure 10-2c. The force vectors a and b are not collinear, but may be composed or added to determine their net effect. By placing the arrow of vector a to the tail of vector b, the resultant composed vector c is the distance between the tail of a and the arrow of b. This procedure is illustrated in Figure 10-2d with multiple vectors.

The *resolution* of a force means to break the force into its components, generally with respect to a coordinate system. The coordinate system may be oriented in the vertical and horizontal directions, or it may be oriented

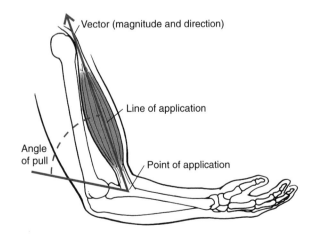

FIGURE 10-1. Characteristics of a force.

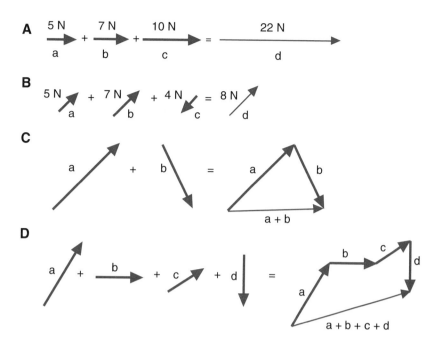

FIGURE 10-2. Force vector addition.

with respect to an anatomical reference such as relative to the long axis of a bone. Generally, however, the single force vector is broken into its vertical and horizontal components. To do this, the angle of application with respect to the coordinate system must be known. In Figure 10-3a, a force vector F acts at an angle of 30° to the horizontal. Using the trigonometric sine and cosine functions (Appendix D), the components of this vector can be determined. Considering the vector and its components to be a right angle triangle as in Figure 10-3b, the horizontal component would be:

$$\cos 30° = \frac{F_x}{F}$$
$$F_x = F * \cos 30°$$
$$F_x = 20 \text{ N} * 0.866$$
$$F_x = 17.32 \text{ N}$$

and the vertical component would be:

$$\sin 30° = \frac{F_y}{F}$$
$$F_y = F * \sin 30°$$
$$F_y = 20 \text{ N} * 0.866$$
$$F_y = 17.32 \text{ N}$$

Note that the components have a magnitude less than the magnitude of the vector F. This is a simple check to determine if the procedure was carried out correctly.

Laws of Motion

The publication of the Principia in 1687 by Sir Isaac Newton (1642–1727) astounded the scientific community of the day. In this book, he introduced his theories of motion and gravity and used these theories to explain a number of phenomena. For this achievement alone, Newton certainly ranks among the greatest thinkers in human history. The three laws of motion that define how many problems in biomechanics are addressed were introduced in this monumental achievement. The statements of these laws are taken from a translation of Newton's Principia (1).

Law I - The Law of Inertia

Every body continues in its state of rest, or of uniform motion in a straight line, unless it is compelled to change that state by forces impressed on it. (1)

The *inertia* of an object is used to describe the resistance of an object to motion. It is directly related to the amount of matter contained in the object. *Mass* is the measure of the amount of matter that constitutes an object and is expressed in kilograms (kg). An object's mass is constant, regardless of where it is measured, so that the mass would be the same whether it was calculated on earth or on the moon. The greater the mass of an object, the greater its inertia, and thus the greater the difficulty in moving it.

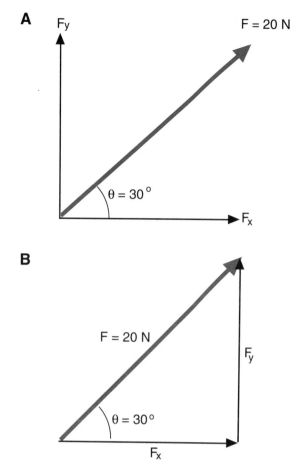

A

B

FIGURE 10-3. Resolution of a force vector into its horizontal and vertical components.

For example, in order to get an object moving, the inertia of the object would have to be overcome. Newton suggested that an object at rest—an object with zero velocity—would remain at rest. This point seems rather obvious. A chair sitting in a room will have a zero velocity because it is not moving. Additionally, an object moving at a constant velocity would continue to do so in a straight line. This concept is not as obvious, since the practical instances of individuals on the earth's surface rarely experience constant velocity motion. If it is noted that a constant velocity results in zero acceleration just the same as zero velocity would, then it can be understood how this law holds for both cases. Therefore, the inertia of these objects would compel them to maintain their status at a constant velocity.

To overcome the inertia of such objects would require a net external force greater than the force of the inertia of the object. For example, if a barbell had a mass of 70 kg,

a force greater than 686.7 N or the acceleration due to gravity (9.81 m/s^2) times 70 kg, must be exerted to lift the bar. If an object experiences an external force, then the object will be either positively or negatively accelerated. To get an object moving, the external force must positively accelerate the object. On the other hand, to stop the object from moving, the external force must negatively accelerate the object.

Law II - Law of Acceleration

The change of motion is proportional to the force impressed and is made in the direction of the straight line in which that force is impressed. (1)

Newton's second law generates an equation that relates force, mass, and acceleration. This relationship is:

$$\text{Force} = \text{mass} * \text{acceleration}$$

or

$$F = ma$$

This equation can also be used to define the unit of force, the newton. By substituting the units for mass and acceleration into the right hand side it can be seen that:

$$\text{newton} = \frac{\text{kg.m}}{\text{s}^2}$$

In this equation, the force is the net force acting on the object in question, that is, the sum of all of the forces involved. In adding up all forces acting on an object, it is necessary to take the direction of the forces into account. If the forces exactly counteract each other, then the net force is zero. If the sum of the forces is zero, the acceleration will also be zero. This case is also described by Newton's first law. If the net force produces an acceleration, the accelerated object will travel in a straight line comparable to the line of action of the net force.

By re-arranging the equation described by Newton's second law, another very important concept in biomechanics can be defined. Acceleration was previously defined as dv/dt. Substituting this expression into the equation of the second law:

$$F = m\frac{dv}{dt}$$
or
$$F = \frac{mdv}{dt}$$

The product of mass and velocity in the numerator of the right hand side of the above equation is known as the *momentum* of an object. Momentum is the quantity of motion of an object. It is generally represented by the letter "p" and has units of kg-m/s. For example, if a football player has a mass of 83 kg and is running at a speed of 4.5 m/s, his momentum would be:

$$p = mass * velocity$$
$$= 83 \text{ kg} * 4.5 \text{ m/s}$$
$$= 373.5 \text{ kg-m/s}$$

Newton's second law can thus be re-stated:

$$F = \frac{dp}{dt}$$

That is, force is equal to the time rate of change of momentum. To change the momentum of an object, an external force must be applied to the object. The momentum may increase or decrease, but in either case, an external force is required.

Law III - Law of Action-Reaction

To every action there is always opposed an equal reaction; or, the mutual action of two bodies upon each other are always equal and directed to contrary parts.

This law illustrates that forces never act in isolation and, in fact, always act in pairs. When two objects interact, the force exerted by object A on object B is counteracted by a force equal and opposite exerted by object B on object A. These forces are equal in magnitude but opposite in direction. In addition, the force—the action—and the counter-force—the reaction—act on different objects. The result is that these two forces cannot cancel each other out since they act on and may have a different effect on the objects. For example, a person landing from a jump exerts a force on the earth, and the earth exerts an equal and opposite force on the person. Because the earth is more massive than the individual, the effect on the individual is greater than their effect on the earth. This example illustrates that while the force and the counterforce are equal, they may not necessarily have comparable results.

Types of Forces

The forces that exist in nature and affect the way humans move may be classified in a number of ways. The most common classification scheme is to describe forces as *contact* or *non-contact* forces (2). A *contact* force involves the actions, pushes or pulls, exerted by one object in direct contact with another object. These are the forces involved, for example, between a bat as it hits the ball, the foot as it impacts the floor, or the punch as it contacts the face. In contrast to contact forces are those that act at a distance. These are called *non-contact* forces. As implied by the name, these are forces that are exerted by objects that are not in direct contact with one another, and in fact may be separated by a considerable distance.

Non-contact Forces

When investigating human movement, the most familiar and important non-contact force is gravity. Any object released from a height will fall freely to the earth's surface pulled by gravity. Another very important contribution of Newton to the understanding of the motion of objects was the Law of Gravitation also presented in Newton's *Principia*. Newton identified gravity as the force that causes objects to fall back to the earth, the moon to orbit the earth, and the planets to revolve about the sun. This law states:

The force of gravity is inversely proportional to the square of the distance between attracting objects and proportional to the product of their masses.

In algebraic terms, the law is described by the following equation:

$$F = \frac{Gm_1m_2}{r^2}$$

where G = universal gravitational constant
m_1 = mass of one object
m_2 = mass of the other object
r = distance between the mass centers of the objects

The constant value G was estimated by Newton and determined accurately by Cavendish in 1798. The value of G is $6.67 * 10^{-11}$ N-m^2/kg^2.

The gravitational attraction of one object of a relatively small size on another object of similar size is extremely small, and therefore can be neglected. In biomechanics, the objects of most concern are the earth, the human body, and projectiles. In these cases the earth's mass is considerable and gravity is a very important force. The attractive force of the earth on an object is called the *weight* of the object. This is stated as:

$$W = F_g = \frac{G\,m_{object}\,M_{earth}}{r^2}$$

If Newton's second law is considered, then:

$$W = m\,a$$

where m is the mass of the individual and a is the acceleration due to gravity. Thus:

$$W = m\,g$$

where g is the acceleration due to gravity. Body weight is thus the product of the individual's mass and the acceleration due to gravity. It is apparent, therefore, that an individual's mass and their body weight are not the same. Body weight is a force and the appropriate units for body weight are newtons. To determine a person's body weight, simply multiply their mass times the acceleration due to gravity (9.81 m/s^2).

Since weight is a force, it also has the attributes of a force. As a vector, it has a line of action and a point of application. An individual's total body weight is considered to have a point of application at the center of mass of the individual, and a line of action from the center of mass of the individual to the center of the earth. Because the earth is so large, this line of action is downward vertically.

Newton's law also suggests that the value for g, the acceleration due to gravity, is dependent on the square of the distance to the center of the earth. Because of the spin on its axis, the earth is not perfectly spherical. The earth is slightly flattened at the poles, resulting in shorter distances to the earth's center at the poles than at the equator. Thus, all points on the earth are not equidistant from its center, and acceleration due to gravity, g, does not have the same value everywhere. The latitude—the position on the earth with respect to the equator—on which a long jump, for example, is performed, can have a significant effect on the distance jumped. Another factor influencing the value of g is altitude. The higher the altitude the lower the value for g. In weighing oneself where a minimal weight was essential, an optimum place for the weigh-in would be on the highest mountain at the earth's equator.

Contact Forces

Because contact forces are those resulting from an interaction of two objects, the number of such forces is considerably greater than the single non-contact force discussed. The following contact forces considered paramount in human movement will be discussed: 1) ground reaction force; 2) joint reaction force; 3) friction; 4) fluid resistance; 5) inertial force; 6) muscle force; and 7) elastic force.

Ground Reaction Force In almost all terrestrial human movement, the individual is acted upon by the ground reaction force (GRF) at some time. This force is the reaction force provided by the surface upon which one is moving. The surface may be a sandy beach, a gymnasium floor, a concrete sidewalk, or a grass lawn. If the individual is swinging from a high bar, the surface of the bar

provides a reaction force. All surfaces provide a reaction force. The ground reaction force is a direct application of Newton's third law of motion concerning action-reaction. The individual pushes against the ground with a force and the ground pushes back against the individual with an equal force in the opposite direction. Note that these forces affect both parties—the ground and the individual—and do not cancel out even though they are equal in magnitude but opposite in direction. It should be noted also that the ground reaction force changes in magnitude, direction, and point of application during the period that the individual is in contact with the surface.

As with all forces, the ground reaction force is a vector and can be resolved into its components. For the purpose of analysis, it is commonly broken down into its components. These components are orthogonal to each other along a three-dimensional coordinate system (Figure 10-4). The components are usually labeled: F_z, vertical (up-down) component; F_y, antero-posterior (forward-backward) component; and F_x, medio-lateral (side-to-side) component. According to the International Society of Biomechanics (ISB), however, standardization in the reporting of three-dimensional data, the components should be labeled as F_y, (vertical); F_x (antero-posterior); and F_z, (medio-lateral). The ISB convention conforms to the reference system for kinematics but is not the most commonly used system. Therefore, the former convention will be used in this book as it is the most often used system. In either case, the antero-posterior and

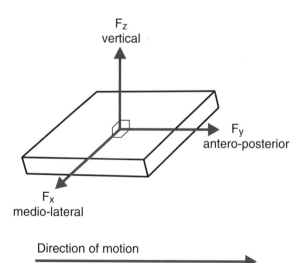

FIGURE 10-4. Ground reaction force components. The origin of the force platform coordinate system is located at the center of the platform.

medio-lateral components are referred to as shear components because they act parallel to the surface of the ground.

Biomechanists measure the ground reaction force components using a *force platform*. A force platform is a very sophisticated measuring scale usually imbedded in the ground, with its surface flush with the surface on which the individual is performing. A typical force platform experimental setup is shown in Figure 10-5. This device can measure the force of the collision of the sole of the individual's foot on the performing surface, or the force of an individual just standing on the platform. Force platforms have been used since the 1930's (3) but became more prominent in biomechanics research in the 1980's. While forces are measured in newtons, ground reaction force data are generally scaled by dividing the force component by the individual's body weight, resulting in units of "times body weight" (BW). In other instances, ground reaction forces may be scaled by dividing the force by body mass, resulting in a unit of newtons per kilogram of body mass (N/kg of body mass).

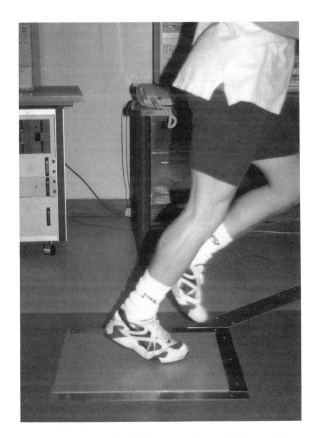

FIGURE 10-5. A typical laboratory force platform set-up.

Ground reaction force data have been used in many studies to investigate a variety of activities. The majority of studies, however, have dealt with the load or impact on the body during landings, either from jumps or during support phase of gait. For example, ground reaction forces have been studied during the support phase of running (4, 5), walking (6), and landings from jumps (7, 8). GRF profiles continually change with time and are generally presented as a function of time. It should be noted that the magnitude of the GRF components for running are much greater than for walking. The magnitude of the ground reaction forces will also vary as a function of locomotor speed (9, 10), increasing with running speed. In addition, the magnitude also increases with increased distance from which the individual lands. That is, the vertical GRF component increases when the individual lands from a height of 1.0 m versus a height of 0.3 m.

The vertical GRF component is much greater in magnitude than the other components, and has received the most attention from biomechanists (Figure 10-6). In walking, the vertical component has a maximum value of 1 to 1.2 BW, and in running, the maximum value can be 3 to 5 BW. The vertical force component in walking has a characteristic bimodal shape. A bimodal shape is one that has two maximum values. The first modal peak occurs during the first half of support and characterizes the portion of support when the total body is lowered after foot contact. The second peak represents the active push against the ground to move into the next step. Figure 10-7 presents a comparison of walking and running vertical ground reaction force component profiles.

In running, the shape of the vertical GRF component is dependent on the footfall pattern of the runner (Figure 10-8). The heel-strike runner's curve has two discernible peaks. The first peak occurs very rapidly after the initial contact and is often referred to as the *passive peak*. The term *passive peak* refers to the fact that this phase is not under muscular control (11). This peak is also referred to as the *impact peak*. The second peak occurs during mid-support and is generally of greater magnitude than the impact peak. Nigg (11) refers to the second peak as the active peak indicating the role the muscles play in the force development to accelerate the body off the ground. The midfoot strike runner has little or no impact peak.

The antero-posterior ground reaction force component also exhibits a characteristic shape similar in both walking and running, but of different magnitude (Figure 10-9). The F_y component reaches magnitudes of 0.15 BW in walking, and up to 0.5 BW in running. During locomotion, this component shows a negative phase during the

A

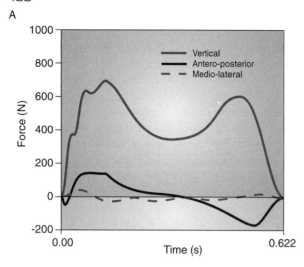

B

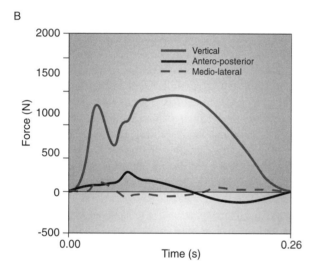

FIGURE 10-6. Ground reaction force components for A) walking and B) running. Note the difference in magnitude between the vertical component and the shear components.

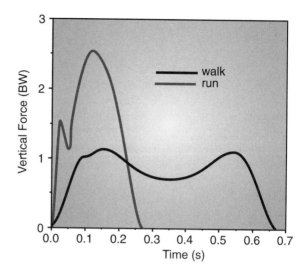

FIGURE 10-7. The vertical ground reaction force component for both walking and running.

A vertical component curve of a single foot contact of an individual landing from a jump is presented in Figure 10-11. In these studies, only the vertical component is presented since it is of much greater magnitude than the other components, and since the major interest in landings has been the effect of vertical loads or impacts on the human body. In the landing curve, the first peak represents the initial ground contact with the forefoot. The second peak is the contact of the heel on the surface. Generally, the second peak is greater than the first peak. Some individuals, however, land in a flat-footed position

first half of support and a positive phase during the second portion of support.

The medio-lateral ground reaction force component is extremely variable and has no consistent pattern from individual to individual. It is very difficult to interpret this force component without a video or film record of the foot contact. Figure 10-10 illustrates both walking and running profiles for the same individual. The great variety in foot placement regarding toeing-in (forefoot adduction) and toeing-out (forefoot abduction) may be a reason for this lack of consistency in the medio-lateral component. The magnitude of the medio-lateral component ranges from 0.01 BW in walking to 0.1 BW in running.

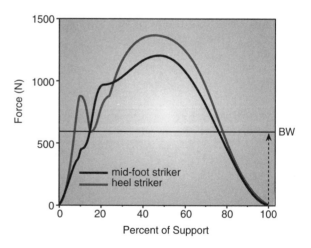

FIGURE 10-8. Vertical ground reaction force component profiles of a runner using a heel-toe footfall pattern and a runner who initially strikes the ground with the mid-foot.

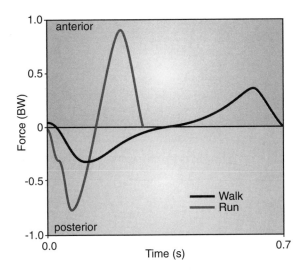

FIGURE 10-9. Antero-posterior (front to back) ground reaction force component for both walking and running.

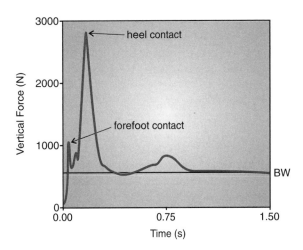

FIGURE 10-11. Vertical ground reaction force component during a landing from a jump.

and have only one impact peak. When the individual comes to rest on the surface, the vertical force curve settles at the individual's body weight. While the magnitude of the vertical component at impact in running is 3 to 5 times body weight, the vertical component in landing can be as much as 11 times body weight, depending on the height from which the person dropped to make contact with the floor (7, 8).

Biomechanists have investigated ground reaction forces to attempt to relate these forces to the kinematics of the lower extremity, particularly to foot function.

Efforts have been made to relate these forces to the rearfoot supination/pronation profiles of runners to identify possible injuries or aid in the design of athletic footwear (12, 13). The use of GRF data for these purposes is probably extrapolating beyond the information provided by the ground reaction forces.

To illustrate this point, a method of calculating the vertical GRF component proposed by Bobbert et al. (14) will be presented. In this method, Bobbert and associates used the kinematically derived values of the accelerations of the centers of mass of each of the body's segments. It should be noted that the vertical GRF component reflects the accelerations of the individual body segments resulting from the motion of the segments. If the vertical forces of all the body segments, including the effect of gravity are summed, the vertical ground reaction force component would be generated. That is:

$$F_z = \sum_{i=1}^{n} m_i \left(a_{zi} - g \right)$$

where F_z is the vertical force component (forces directed upwards are defined as positive), m_i is the mass of the i-th segment, n is the number of segments, a_{zi} is the vertical acceleration of the i-th segment (upward accelerations are defined as positive), and g is the acceleration due to gravity. The antero-posterior GRF component reflects the horizontal (i.e. in the direction of motion) accelerations of the individual body segments. Using similar methods, this force component can be computed as:

$$F_y = \sum_{i=1}^{n} \left(m_i \, a_{yi} \right)$$

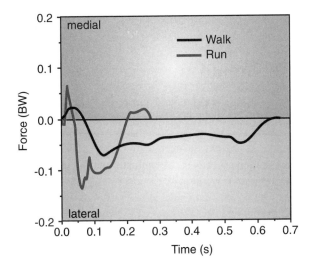

FIGURE 10-10. Medio-lateral (side-to-side) ground reaction force component for both walking and running.

where a_{yi} is the horizontal acceleration of the i-th segment. Similarly, the medio-lateral ground reaction force component reflects the side-to-side accelerations of the individual body segments or:

$$F_x = \sum_{i=1}^{n} \left(m_i\ a_{xi} \right)$$

where a_{xi} is the side-to-side acceleration of the i-th segment. If the center of mass is a single point that represents the mass center of all the body's segments, the vertical component is:

$$F_z = m\ (a_z - g)$$

where m is the total body mass, a_z is the vertical acceleration of the center of mass, and g is the acceleration due to gravity. Similarly, the other components may be represented as the total body mass times the acceleration of the center of mass. That is:

$$F_y = m\ a_y$$
$$F_x = m\ a_x$$

It should be noted, therefore, that the ground reaction force represents the force necessary to accelerate the total body center of gravity (15) and cannot be directly associated with the lower extremity. Caution, then, should be exercised in describing lower extremity function using ground reaction force data.

Since the ground reaction force relates to the motion of the total body center of mass, the antero-posterior force profile can be related to the acceleration profile of the center of mass during support. In chapter 8, a study by Bates et al. (16), illustrating the horizontal velocity pattern of the center of mass during the support phase of the running stride was discussed (Figure 10-12a). When this curve was differentiated, an acceleration curve was generated (Figure 10-12b). This curve has the characteristic shape of the antero-posterior force component in that it has an initial negative acceleration followed by a subsequent positive acceleration. By applying Newton's second law of motion, if each point along this curve was multiplied by the runner's body mass (m), the antero-posterior ground reaction force component would be generated as follows:

$$F_y = m * a_y$$

Conversely, the acceleration curve could be calculated by dividing the F_y force component by the runner's body mass.

$$a_y = \frac{F_y}{m}$$

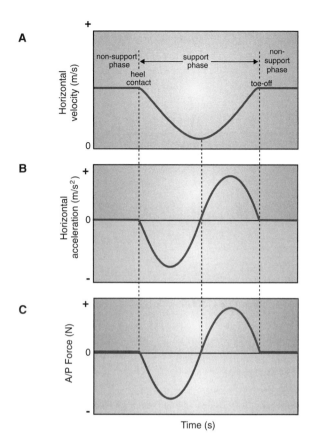

FIGURE 10-12. A) Horizontal velocity of the center of mass of a runner; B) if the velocity curve in A is differentiated, a horizontal acceleration curve of the center of mass is generated; C) multiplying each point along this curve by the runner's body mass, the antero-posterior ground reaction force component is generated.

By either generating the curves using the kinematic procedure or collecting the antero-posterior GRF component, the same conclusion can be reached. The negative portion of the force component indicates a force against the runner serving to brake the runner, that is, to decrease their velocity. The positive portion of the component indicates a force in the direction of motion serving to accelerate the runner, so to increase their velocity. If the running speed is constant, the negative and positive phases will be symmetrical, indicating no loss in velocity. If the negative portion of the curve is greater than the positive portion, then the runner will slow down more than they speed up. Conversely, if the positive portion is greater than the negative, the runner is speeding up.

Joint Reaction Force   In many instances in biomechanical analyses, individual segments are examined either singly or one at a time in a logical order. When this analysis is conducted, the segment is separated at the joints and the forces acting across the joints must be considered. By Newton's third law there must be an equal and opposite force acting at each joint. For example, if one is standing still, the thigh exerts a downward force on the leg across the knee joint. Similarly, the leg exerts an upward force on the thigh of equal magnitude (Figure 10-13). This force is the net force acting across the joint and is referred to as the *joint reaction force*. In most analyses, the magnitude of this force is unknown, but it can be calculated given the appropriate kinematic and kinetic data, in addition to anthropometric data describing the body dimensions.

There has been some confusion as to whether the joint reaction force is the force of the distal bony surface of one segment acting on the proximal bony surface of the contiguous segment. The joint reaction force does not, however, reflect this bone-on-bone force across a joint. The actual *bone-on-bone force* is the sum of the actively contracting muscle forces pulling the joint together and the joint reaction force. Since the force generated by the actively contracting muscles is not known, the bone-on-bone force is difficult to calculate. Very sophisticated calculations, however, have been done to estimate these bone-on-bone forces (17).

Friction   *Friction* is a force acting parallel to the interface of two surfaces that are in contact during the motion or impending motion of one surface as it moves over the other. For example, the weight of a block resting on a horizontal table pulls the block downward pressing it against the table. The table exerts an upward vertical force on the block that is perpendicular or normal to the surface. To move the block horizontally, a horizontal force on the block of sufficient magnitude must be exerted. If this force is too small, the block will not move. In this case the table evidently exerts a horizontal force equal and opposite to the force on the block. This interaction is the frictional force and is due to the bonding of the molecules of the block and the table at the places where the surfaces are in very close contact. Figure 10-14 illustrates this example.

It would appear that the area of contact would influence the force of friction. However, this is not the case. The force of friction is proportional to the normal force between the surfaces, that is:

$$F_f = \mu\,N$$

where μ is the *coefficient of friction* and N is the normal force or the force perpendicular to the surface. The coefficient of friction is calculated by:

$$\mu = \frac{F_f}{N}$$

The coefficient of friction depends on the nature of the interfacing surfaces and is a dimensionless number. The greater the magnitude of the coefficient of friction, the greater the interaction between the molecules of the interfacing surfaces.

Continuing the block and table example, if a steadily increasing force is applied to the block, the table also applies an increasing opposite force resisting the move-

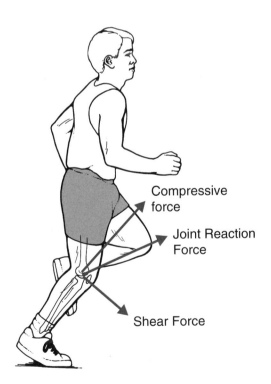

FIGURE 10-13. An illustration of the joint reaction force of the knee with its shear and compressive components.

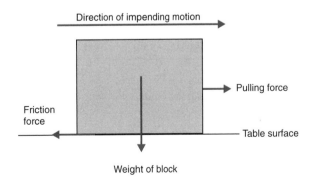

FIGURE 10-14. Illustration of the forces acting on a block that is being pulled across the surface of a table.

ment. At the point where the pulling force is at a maximum and no movement results, the resisting force is called the *maximum static friction force* (F_{sMAX}). Prior to movement, it can be stated that:

$$F_{sMAX} \leq \mu_s N$$

where μ_s is the static coefficient of friction. At some point, however, the force will be sufficiently large and the static friction force cannot prevent the movement of the block. This relationship simply means that, if a block weighing 750 N is standing on a surface with μ_s of 0.5, it will take 50% of the 750 N normal force, or 375 N, of a horizontal force to cause motion between the block and the table. A μ_s of 0.1 would require a horizontal force of 75 N to cause motion and a μ_s of 0.8 would require a 600 N horizontal force. As can be seen, the smaller the coefficient of friction, the less horizontal force required to cause movement.

As the block slides along the surface of the table, molecular bonds are continually made and broken. Thus, once the two surfaces start moving relative to each other, it becomes slightly easier to maintain the motion. The result is a force of sliding friction that opposes the motion. Sliding friction along with rolling friction are types of kinetic friction. *Kinetic friction* is defined as:

$$F_k = \mu_k N$$

where μ_k is the dynamic coefficient of friction or the coefficient of friction during movement. It has been found experimentally that μ_k is less than μ_s and that μ_k depends on the relative speed of the object. At speeds from one cm/s to several m/s, however, μ_k is approximately constant. Figure 10-15 illustrates the friction/external force relationship.

While translational friction is very important in human movement, the evaluation of *rotational friction* must also be considered. Rotational friction is the resistance to rotational or twisting movements. For example, the soles of the shoes of a basketball player accomplishing a pivot interact with the playing surface to resist the turning of the foot. Obviously, the player must be able to accomplish this movement during a game, and thus the rotational friction must allow this motion without influencing the other frictional characteristics of the shoe. A basketball player executing a 180° pivot in a conventional basketball shoe on a wooden floor would have a rotational friction value 4.3 times greater than if this movement was attempted in gym socks (18). The measurement of rotational friction does not yield a coefficient of friction value. The values used to compare rotational friction are based on the measured value of the resistance

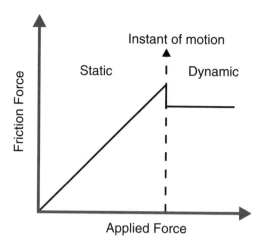

FIGURE 10-15. A theoretical representation of friction force as a function of the applied force. The applied force increases with the friction force until motion occurs.

to rotation and is usually done on a force platform. It should be noted that translational and rotational friction are not independent of each other.

Friction is a complicated but very important influence on human movement. Just to walk across a room requires an appropriate coefficient of friction between the shoe outsoles and the surface of the floor. In everyday activities, one may either try to increase or decrease the coefficient of friction depending on the activity. For example, skaters prefer a fresh ice surface on which to skate, because it has a low coefficient of friction. On the other hand, a golfer wears a glove to increase the coefficient of friction and get a better grip on the club.

Many types of athletes wear cleated shoes to increase the coefficient of friction and get better traction on the playing surface. Valiant (19) suggested that ms equaling 0.8 provides sufficient traction for athletic movements and any coefficient of friction greater than this value may be unsafe. In certain situations cleated shoes may, in fact, result in too much translational and/or rotational friction. This appears to be the case with artificial turf for example. Many injuries such as "turf toe" and anterior cruciate ligament tears have been related to too great a friction force on artificial turf.

When the coefficient of friction is too small, a slip hazard results, but when the coefficient of friction is too great, a trip hazard occurs. In the work place, injuries resulting from slips and falls are numerous and often cause serious injury. Cohen and Compton (20) reported that 50% of 120,682 workers compensation cases in New York state from 1966 to 1970 were caused by slips. In England, Buck (21) cited 1982 statistics in which 14% of all accidents resulted from slips and trips in the manufacturing industry. Thus, the consideration of the coefficient of friction is a very important criteria in the design of any surface on which people perform, whether in the work place or in athletics.

The static coefficient of friction of materials on different surfaces has been successfully measured using a towed sled device (22). This device, illustrated in Figure 10-16, involves an object of a known mass and a force measuring gauge. The moving surface, usually some type of footwear, is placed beneath the mass. The gauge pulls the known mass until it moves and the force is measured by the gauge at the instant of movement. The coefficient of friction can then be calculated using the known mass and the force measured from the gauge. This type of measurement is known as a materials test measure; it does not involve human subjects.

It is difficult to measure the static or kinetic coefficient of friction accurately without sophisticated equipment. Both, however, may be measured with a force platform. The shear components, F_y and F_x, are, in fact, the frictional forces in the antero-posterior and medio-lateral directions, respectively. If the normal force is known, then the dynamic coefficient of friction may be estimated. Generally, this is done using the vertical (F_z) component as the normal force. Thus, the coefficient of friction can be determined by:

$$\mu = \frac{F_y}{F_z}$$

Several researchers have devised instruments to measure both translational and rotational friction. A device developed in the Nike Sports Science Laboratory is one

such instrument (Figure 10-17). It has been used to measure the friction characteristics of many types of athletic shoes.

In general, the magnitude of the coefficient of friction depends on the types of materials constituting the surfaces in contact and the nature of those surfaces. For example, a rubber-soled shoe would have a higher coefficient of friction on a wood gymnasium floor than a leather-soled shoe. The relative roughness or smoothness of the contacting surfaces also affects the coefficient of friction. Intuitively, a rough surface would have a higher coefficient of friction than a smooth surface. The addition of lubricants, moisture, or dust to a surface will also greatly affect frictional characteristics. To determine the coefficient of friction, all of these factors must be considered.

Fluid Resistance   In many activities, human motion is affected by the fluid in which the activities are performed. Both air—a gas, and water—a liquid, are considered fluids. Thus, the motion of a runner is affected by the movement of air and that of a swimmer by the water or the air/water interface. Projectiles, whether human or objects, are also affected by air. For example, anyone who has ever driven a golf ball into the wind will understand the effects of air on the golf ball.

The two properties of a fluid that most affect objects as they pass through it are the fluid's *density* and *viscosity*. Density is defined as the mass per unit volume. Generally, the more dense the fluid, the more resistance it presents to the object. The density of air is particularly affected by humidity, temperature, and pressure. Viscosity is a measure of the fluid's resistance to flow.

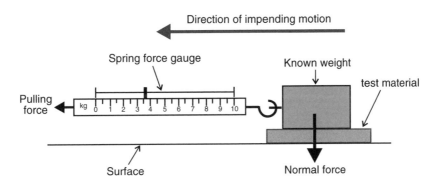

FIGURE 10-16. An illustration of a towed sled device that measures the applied force. The known weight constitutes the normal force and the spring gauge measures the resistant horizontal force to the known weight. The coefficient of friction is computed by the ratio of the spring gauge force/known weight.

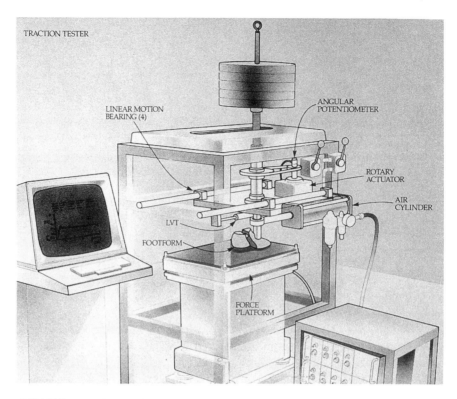

FIGURE 10-17. Device for mechanically evaluating translational and rotational friction characteristics of athletic footwear outsoles. This device was developed at the Nike Sports Research Laboratory.

For example, water is more viscous than air, with the result that water resistance is greater than air resistance. Gases such as air become more viscous as the air temperature rises.

As an object passes through a fluid, it disturbs the fluid. This is true for both air and water. The degree that the fluid is disturbed is dependent on the density and viscosity of the fluid. The greater the disturbance of the fluid, the greater the energy that is transmitted from the object to the fluid. This transfer of energy from the object to the fluid is called *fluid resistance*. The resultant fluid resistance force can be resolved into two components *lift* and *drag* (Figure 10-18).

Drag Force Component - drag is a component of the fluid resistance force that always acts to oppose the motion. The direction of the drag component is always directly opposite to the direction of the velocity vector and acts to retard the motion of the object through the fluid. In most instances drag is synonymous with "air resistance". The magnitude of the drag component may be determined by:

$$F_{drag} = \frac{1}{2} C_d \, A \, \rho \, v^2$$

where C_d is a constant known as the coefficient of drag, A is the projected frontal area of the object, the Greek letter rho (ρ) is the fluid viscosity, and v is the relative velocity of the object, that is, the velocity of the object relative to the fluid. The magnitude of the drag component is a function of the nature of the fluid, the nature and shape of the object, and the velocity of the object through the fluid.

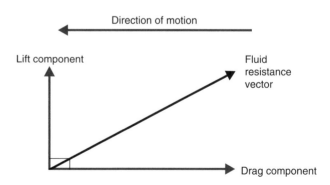

FIGURE 10-18. Fluid resistance vector with its lift and drag components.

There are two types of drag that must be considered. Drag as a result of friction between the object's surface and the fluid is referred to as *surface drag* or *viscous drag*. When an object moves through a fluid, the fluid interacts with the surface of the object, literally sticking to its surface. The resulting fluid layer is called the *boundary layer*. The fluid in the boundary layer is slowed down relative to the object as it passes it by. It results in the object pushing on the fluid and the fluid pushing on the object in the opposite direction. This interaction causes friction between the fluid in the boundary layer and the object's surface. This "fluid friction" opposes the motion of the object through the fluid. A fluid with a high viscosity will generate a high drag component. In addition, the size of the object becomes more important if more surface is exposed to the fluid.

From the formula for drag force, it can be seen that the drag force increases as a function of the velocity squared. The relative velocity of the fluid as it passes by the object actually determines how the object will interact with the fluid. At lower movement velocities of the object, the fluid passes the object in uniform layers of differing speed, with the slowest moving layers closest to the surface of the object. This is called *laminar flow* (Figure 10-19a). Laminar flow occurs when the object is small and smooth and the velocity is small. The drag force consists almost entirely of surface or friction drag. On every object, however, there are points, called the *points of separation*, at which the fluid separates from the object. That is, the fluid does not completely follow the contours of the shape of the object. As the object moves faster through the fluid, the fluid also moves faster by the object. When this occurs, the points of separation move forward on the object and the fluid separates from the contours of the object closer to the front of the object. Thus, at relatively high velocities, the fluid does not maintain a laminar flow and *separated flow* occurs (Figure 10-19b). Separated flow is also referred to as partially turbulent flow. In this instance, the fluid is unable to contour to the shape of the object and the boundary layer separates from the surface. This produces turbulence behind the object. In power-boating, for example, the turbulence behind the boat is called the "wake". The wake of an object moving through a fluid is a low pressure region. Separated flow occurs at a small relative velocity if the object is large and has a rough surface, or under any conditions in which the fluid does not stick to the surface of the object.

When the fluid contacts the front of the object, an area of relatively high pressure is formed. The turbulent area behind the object is an area of pressure lower than the

A. Laminar Flow

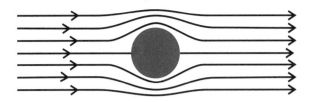

B. Separated Flow

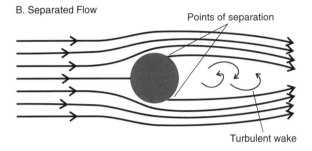

Points of separation

Turbulent wake

C. Turbulent Flow

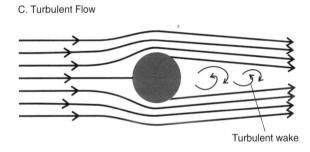

Turbulent wake

FIGURE 10-19. An illustration of: A) laminar flow; B) separated flow; and C) turbulent flow about a sphere.

pressure at the front of the object. The greater the turbulence, the lower the pressure behind the object. The net pressure differential between the front and back of the object retards the object's movement through the fluid. With increasing flow velocity, the point at which the boundary layer separates from the surface of the object moves further to the front of the object, resulting in an even greater pressure differential and greater resistance. Drag resulting from this pressure differential is called *form drag*. In partially turbulent flow, both form and friction drag occur. As the wake increases, however, form drag dominates.

As the relative velocity of the object and the fluid increases, the whole boundary layer becomes turbulent. This type of fluid flow is called *turbulent flow* (Figure 10-19c). Interestingly, the turbulence in the boundary layer actually moves the point of separation towards the back of the object, reducing the ability of the boundary layer to separate from the object. The net result is a reduction in the drag force.

Moving from partially turbulent flow to fully turbulent flow, thus decreasing the drag force, can be accomplished by streamlining and smoothing the surface of the object moving through the fluid. Contrary to what one might expect, putting dimples on a golf ball or seams on a baseball, thereby roughening the surface, actually helps in this transition.

Moving from laminar to partially turbulent flow can be forestalled, and thus drag minimized, by assuming a streamlined shape or making the surface smoother, or both. Athletes such as sprinters, cyclists, swimmers, or skiers generally wear very smooth suits during their events. A significant 10% reduction of drag occurs when a speed skater wears a smooth body suit (23). Wearing smooth clothing also prevents such things as long hair, laces, or loose fitting clothing from increasing drag (24). Kyle (25) reported that loose clothing or thick long hair could raise the total drag from 2 to 8%. He calculated that a 6% drop in air resistance can result in a long jump distance increase from 3 to 5 cm.

Streamlining the shape of the object involves decreasing the projected frontal area. The projected frontal area of the object is the area of the surface that might come in contact with the fluid flow. Athletes in sports in which air resistance must be minimized manipulate this frontal area constantly. For example, an ice speed skater has a number of body positions that can be assumed during a race. When a skater has their arms hanging down in front, this presents a greater frontal area than an "arms back" racing position. The frontal areas in these speed skating positions are 42.21 m^2 and 38.71 m^2, respectively (26). Similarly, a ski racer will assume a tuck position to minimize the frontal area rather than the posture of a recreational skier. To decrease the drag component, a more streamlined position must be assumed. Streamlining helps minimize the pressure differential and thus minimize the form drag on the object. For example, the drag coefficient for a standing human figure is 0.92 while it is 0.80 for a runner and 0.70 for a skier in a low crouch (25).

In many instances, equipment has been designed to minimize fluid resistance. New bicycle designs, solid rear wheels on racing bicycles, clothing for skiers, swimmers, runners and cyclists, bent poles for downhill skiers, new helmet designs, etc. have all contributed to help these athletes in their events. Research on streamlining body positions has also greatly aided athletes in many sports such as cycling, speed skating, and sprint running (26).

Lift Force Component - lift is the component of the fluid resistance force vector that acts perpendicular to the drag component. Thus, it also acts at right angles to the direction of motion. While there is always a drag force component, the lift component only occurs under special circumstances. That is, lift occurs only if the object is spinning or is not perfectly symmetrical. The lift force component is one of the most significant forces in aerodynamics. This is the force that helps airplanes fly. Contrary to what the name suggests, this force component is not always gravity-opposing.

The lift force component is produced by any break in the symmetry of the air flow about an object. This can be shown in an object having an asymmetrical shape, a flat object being tilted thus inclined to the air flow, or by a spinning object. The effect makes the air flowing over one side of the object follow a different path than the air flowing over the other side. The result of this differential air flow is a lower air pressure on one side of the object and a higher air pressure on the other side. This pressure differential causes the object to move towards the side that has the lower pressure. This is illustrated in Figure 10-20 illustrating an airfoil (the cross section of an airplane wing). The air flowing over the top of the airfoil moves at a higher speed than the air flowing beneath it. A principle first stated by Daniel Bernoulli in 1738 relates the speed of air flow to the pressure exerted by the fluid. *Bernoulli's Principle* states that pressure is inversely proportional to the velocity of the fluid and is expressed mathematically as :

$$P \propto \frac{1}{v}$$

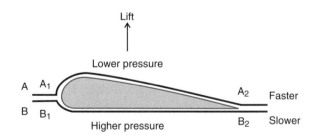

FIGURE 10-20. An illustration of Bernoulli's Principle in creating lift on an airfoil. Molecule of air, A and B, move from A1 to A2 and B1 to B2 respectively in the same amount of time, but the distance from A1 to A2 is greater. Thus the velocity of A is greater than the velocity of B, creating a lower pressure on the top of the airfoil than underneath.

where P is the pressure and v is the velocity of the fluid. Thus, when the velocity of a moving fluid increases, the pressure exerted by the fluid decreases, and vice versa. The result is that the airfoil develops a lift force in the direction of the lower pressure.

The lift force component is used in many instances in human movement. It is used on the wings of airplanes and as "spoilers" on cars, but it is also used in swimming. Lift can be generated by the swimmer's hands as they move through the water. The swimmer's hands resemble airfoils and the swimmer pitches or orients the hands as they are pulled through the water. Pitching the hand puts the lift component in the desired direction of the swim (27, 28). The lift component thus may also contribute to the forward motion of the swimmer.

The lift force also contributes to the curved flight of a spinning ball that is critical in baseball and, most maddeningly, in golf. The spin on the ball results in the air flowing faster on one side of the ball and slower on the other side, thus creating a pressure differential on either side of the ball. Consider the spinning ball in Figure 10-21. Side A of the ball is spinning against the air flow, causing the boundary layer to slow down on that side. On side B, however, because it is moving in the same direction as the air flow, the boundary layer speeds up. By Bernoulli's principle, this results in a pressure differential on either side of the ball. This is comparable to the pressure differential about the airfoil. The ball, therefore, is deflected laterally towards the direction of the spin or the side on which there is a lower pressure area. This effect was first described by Gustav Magnus in 1852, and is known as the *magnus effect*. Baseball pitchers have mastered the art of putting just enough spin on the ball to curve its path successfully. Many golfers try not to put a

sideways spin on the ball, thus avoiding slicing or hooking the ball. They do, however, try to put backspin on the golf ball. The backspin, because of the Magnus Effect, creates a pressure differential between the top and the bottom of the golf ball, with the lower pressure on the top. The golf ball gains lift, and thus distance.

Inertial Force   According to Newton's first law, an object's resistance to motion is due to its inertia. In many instances in human movement, one segment can exert a force on another segment, causing a movement in that segment that is not due to muscle action. When this occurs, an *inertial force* has been generated. Generally a more proximal segment exerts an inertial force on a more distal segment. For example, during the swing phase of running, the ankle is plantar flexed at take-off and slightly dorsiflexed at touchdown. The ankle is relaxed during the swing phase and, in fact, the muscle movement about this joint is very nearly zero, indicating little muscle activity. The leg also swings through, however, and exerts an inertial force on the foot segment, causing the foot to move to the dorsiflexed position. Similarly, the thigh segment exerts an inertial force on the leg.

Muscle Force   When a force was defined, it was noted that a force constituted a push or pull that results in a change in velocity. A muscle, however, can only generate a pulling or tensile force and therefore has only unidirectional capability. The biceps brachii, for example, pulls on its insertion on forearm to flex the elbow. To extend the elbow, the triceps brachii must also exert a pull on its insertion on the forearm. Thus, the movements at any joint must be accomplished by opposing pairs of muscles. It should be noted that gravity also assists in the motion of segments.

In most instances in biomechanical analyses, it is assumed that a muscle force acting across a joint is a net force. That is, the force of individual muscles acting across a joint cannot be taken into consideration. Generally there are a number of muscles acting across any single joint. Each of these muscles would constitute an unknown value. Mathematically, the number of unknown values must have a comparable number of equations. Because there is not a comparable number of equations, there cannot be a solution for each individual muscle force. If a solution was attempted, it would result mathematically in an indeterminate solution; that is, no solution.

It is also assumed that the muscle force acts at a single point. This assumption is again not completely correct since the insertions of muscles are rarely if ever single

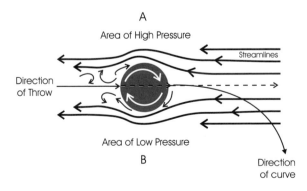

A

Area of High Pressure

Streamlines

Direction of Throw

Area of Low Pressure

B

Direction of curve

FIGURE 10-21. An illustration of the Magnus Effect on a spinning ball. Because of this effect, the ball will curve in the direction of B.

points. Each muscle can be represented as a single force vector that is the resultant of all forces generated by the individual muscle fibers (Figure 10-22). The force vector can be resolved into its components, one component (Fy) acting to cause a rotation at the joint and the other (Fx) acting towards the joint center. If the angle θ (theta) is considered, it can be seen that as θ gets large, such as when the joint is flexing, the rotational component increases while the component acting towards the joint center decreases. This can be assumed because:

$$F_y = F \sin \theta$$

and

$$F_x = F \cos \theta$$

At $\theta = 90°$, the rotational component is a maximum since the sin 90° = 1 while the component acting towards the joint center is zero since cos θ = 0.

The actual muscle force in vivo is very difficult to measure. To do so requires either a mathematical model or the placement of a force measuring device called a force transducer on the tendon of a muscle. Many researchers have developed mathematical models to approximate individual muscle forces (29, 30). In order to do this, however, a number of assumptions must be made, including the direction of the muscle force, point of application, and whether co-contraction occurs or not.

For example, a simple model to determine the peak Achilles tendon force during the support phase of a running stride would use the peak vertical ground reaction force and the point of application of that force. It must be assumed that the line of this force is located a specific distance anterior to the ankle joint and the Achilles tendon is located a specific distance posterior to the ankle joint. In addition, the center of mass of the foot and its location relative to the ankle joint center must be determined. In this rather simple example, the number of anatomical assumptions that must be made is evident.

The second technique, placing a force measuring device on the tendon of a muscle, requires a surgical procedure. Komi et al. (31) and Komi (32) have placed force transducers on the Achilles tendon of individuals. Komi (32) reported peak Achilles tendon forces corresponding to 12.5 BW during running at a speed of 6.0 m/s. While the technique to measure in vivo muscle forces is not well developed for long term studies, it can be used to analyze several very important parameters in muscle mechanics.

Elastic Force When a force is applied to materials, the material undergoes a change in its length. The algebraic statement that reflects this relationship is:

$$F = k \, \Delta s$$

where k is a constant of proportionality and Δs is the change in length. The constant k represents the stiffness or the ability of the material to be compressed or stretched. A stiffer material requires a greater force to compress or stretch it. This relationship is often applied to biological materials and represented in "stress-strain" relationships. This relationship was presented in Chapter 2.

The effect of elastic force can be visualized in an example of a diver on a springboard. The diver uses their body weight as the force to deflect the springboard. The deflected springboard stores an elastic force that is returned as the springboard rebounds to its original state. The result is that the diver is accelerated upwards. A considerable amount of work has been conducted to determine the elasticity of diving springboards used in competition (33, 34).

In most situations, the biological tissue—muscles, tendons, and ligaments—do not exceed their elastic limit. Within this limit, these tissues can store force when they are stretched much like a rubber band does. When the loading force is removed, the elastic force may be returned and, with the muscle force, contribute to the total force of the action. For example, using a pre-stretch prior to a movement will increase the force output by inducing the elastic force potential of the surrounding tissues. There is, however, a time constraint on how long this elastic force can be stored. Attempts to measure the effect of stored elastic force have illustrated that utilizing

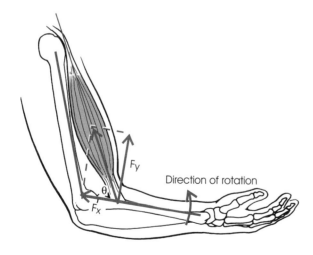

FIGURE 10-22. Representation of a muscle force vector, the angle of pull (θ), and its vertical and horizontal components.

this force can affect oxygen consumption (35). Further estimates of stored elastic force in vertical jumping have been investigated by Komi and Bosco (36) who reported higher jump heights using stored elastic force. It has been suggested by Alexander (37) that elastic force storage is important in the locomotion of both humans and many animals such as kangaroos and ostriches.

Representation of Forces Acting on a System

When one undertakes an analysis of any human movement, it should now be readily apparent that there are a number of forces acting on the system. In order to simplify the problem for a better understanding, a free-body diagram is often used. A *free-body diagram* is a stick figure drawing of the system being analyzed on which the vector representations of the external forces acting on the system are drawn. External forces are those exerted outside the system rather than from inside the system. Thus, internal forces are not represented on a free-body diagram. Before proceeding, however, the term *system* must be defined. In biomechanics, the *system* refers to the total human body or parts of the human body and any other objects that may be important in the analysis. It is critically important to define the system correctly; otherwise, extraneous variables may confound the analysis.

Once the system has been defined, the external forces acting upon the system must be identified and drawn. Figure 10-23 illustrates a free-body diagram of a total body sagittal view of a runner. The external forces acting on the runner are: 1) the ground reaction force; 2) friction; 3) fluid or air resistance; and 4) gravity as reflected in the runner's body weight. Vector representations of the external forces are then drawn on the stick figure at the approximate point of application. If the runner is carrying an implement such as a wrist weight, another force vector representing the weight of this implement must be added to the free-body diagram (Figure 10-24). For the most part, however, the four external forces noted above are the only ones identified on a total body diagram.

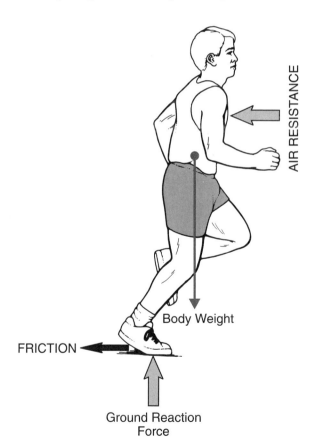

FIGURE 10-23. A free body diagram of a runner with the whole body defined as the system.

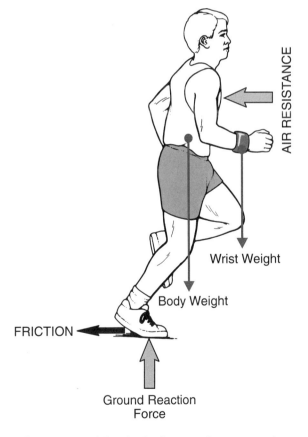

FIGURE 10-24. A free body diagram of a runner-wrist weight system.

When a specific segment and not the total body is defined as the system, the interpretation of what constitutes an external force must be clarified. In drawing a free-body diagram of a particular segment, the segment must be isolated from the rest of the total body. The segment is drawn disconnected from the rest of the body, and all external forces acting on that segment are drawn. The muscle forces that cross the proximal or distal joints of that segment are external to the system and must be classified as external forces. As noted earlier, however, all of the muscles and their forces acting across a joint cannot be identified. An idealized net muscle force, that is, a single force vector, is used to represent the sum total of all muscle forces.

In Figure 10-25, a free-body diagram of the forearm of an individual doing a biceps curl is drawn. There are four external forces acting on this system that must be identified. They are: 1) the net biceps muscle force; 2) the joint reaction force; 3) the force of gravity on the arm represented by the weight of the forearm; and 4) the force of gravity acting on the barbell or the weight of the barbell. These are then drawn as they would act during this movement. In many instances the joint reaction force and the net muscle force are not known but must be calculated. All other forces such as the friction forces in the joints and the forces of the ligaments and tendons are assumed to be negligible. It should be apparent that air resistance is also neglected.

Free-body diagrams are extremely useful tools in biomechanics. Drawing the diagram of the system and identifying the forces acting upon the system define the problem and determine how to undertake the analysis.

Forces Occurring Along a Curved Path

In Chapter 9, linear and angular kinematics were related, using the situations in which an object moved along a curved path. It was determined that the centripetal acceleration acts towards the center of rotation when an object moves along a curved path. Using Newton's second law of motion, F = ma, a formula for the *centripetal force* can be generated. The magnitude of the centripetal or center seeking force is calculated by:

$$F_C = m\, \omega^2\, r$$

where F_C is the centripetal force, m is the mass of the object, ω is the angular velocity, and r is the radius of rotation. Centripetal force may also be defined as:

$$F_C = m\, \frac{v^2}{r}$$

where v is the tangential velocity of the segment.

Newton's third law states that for every action there is an equal and opposite reaction. Thus, for every "center seeking" force there is a "center fleeing" force. The centripetal force acts towards the center of rotation and the *centrifugal force*, the reaction force, acts away from the center of rotation. If an athlete is running around the curve on a track, the centripetal force pulls them towards the center of the curve, while the centrifugal force pulls them away from the center of the curve. It should be noted, however, that the net effect of these forces is zero and thus would not change the direction in which the body is moving. If the centripetal force did not exist, the centrifugal force also would not exist, and so the runner would continue to run in a straight line.

As the runner moves along the curve of the running track, they apply a shear force to the ground, resulting in a shear ground reaction force equal and opposite to the applied force. The shear reaction force constitutes the centripetal force. Figure 10-26 is a free body diagram of the runner as they move along the curved path showing the centripetal force, the vertical reaction force, and the resultant of these two force components. This centripetal force at the runner's foot tends to rotate the runner outwards. To counteract this outward rotation, the runner leans toward the center of the curve. Hamill et al. (38) reported that this shear ground reaction force increased as the radius of rotation decreased.

The resultant of the vertical reaction force and the centripetal force must pass through the center of mass of the runner. If the centripetal force increases, the runner

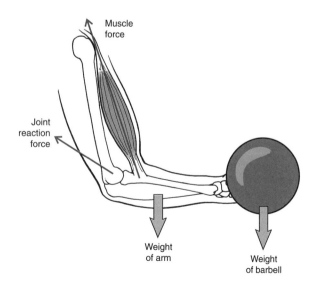

FIGURE 10-25. A free body diagram of the forearm during a biceps curl.

leans more towards the center of rotation, and the resultant vector becomes less vertical. As mentioned in the previous chapter, banked curves on tracks reduce the shear force applied by the runner, and thus reduce the centripetal force. As the centripetal force is reduced, the runner reduces their lean. The resultant force thus acts more vertically as is the case when the runner moves along a straight path.

Special Force Applications

Pressure

Up to this point in the discussion of force, the way a force causes an object to accelerate to achieve a state of motion, has been considered. It should also be discussed, however, how forces, particularly impact forces, are distributed. The concept of pressure is used to describe force distribution. *Pressure* is defined as the force per unit area. That is:

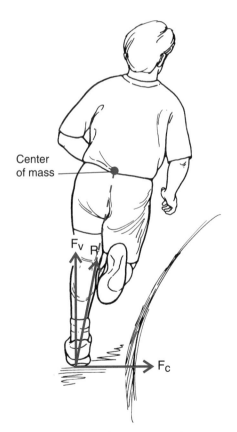

Center of mass —

F_V R

F_C

FIGURE 10-26. A free body diagram of a runner on the curve of a running track. F_C is the centripetal force, F_V is the vertical reaction force and R is the resultant of F_C and F_V.

$$P = \frac{F}{A}$$

where F is a force and A is the area over which the force is applied. Pressure has units of N/m^2. Another unit of pressure often used is the pascal (Pa) or the kilopascal (kPa). One pascal is equal to 1 N/m^2. If an individual with a body weight of 650 N is supported on the soles of their feet with an approximate area of 0.018 m^2, then the pressure on the soles of the feet would be:

$$P = \frac{650 \text{ N}}{0.018 \text{ m}^2}$$
$$= 36.11 \text{ kPa}$$

Another individual with a smaller body weight of 500 N and soles of the feet with an identical area would have a pressure of:

$$P = \frac{500 \text{ N}}{0.018 \text{ m}^2}$$
$$= 27.78 \text{ kPa}$$

If the heavier individual's foot soles had a larger area, 0.020 m^2 for example, the pressure would be 32.50 kPa. The pressure on the soles of the foot of the heavier individual is less than the pressure on the soles of the lighter individual, even though the body weight is different. These pressures appear quite large but imagine if these individuals were females wearing spike heel dress shoes that have much less surface area than ordinary shoes. A more dramatic example would be if these individuals were wearing ice skates that have a distinctly smaller area of contact with the surface than the sole of a normal shoe or a spike heeled ladies' shoe. On the other hand, if these individuals were using skis or snow shoes to walk in deep snow, the pressure would be quite small because of the large area of the skis or snow shoes in contact with the snow. In this way, individuals could walk on the snow without sinking into it.

The concept of pressure is especially important in activities in which a collision or impact results. Generally, when a force of impact is to be minimized, it should be received over as large an area as possible. For example, when landing from a fall, most athletes attempt to initiate a rolling action to spread the impact force over as large an area as possible. In the martial arts, considerable time is spent in learning how to fall correctly, specifically applying the pressure as force per unit area.

A number of sporting activities in which collisions abound have special protective equipment designed to reduce pressure. A few examples are shoulder pads in football and ice hockey; shin pads in ice hockey, field hockey, soccer, and baseball (for the catcher); boxing gloves; and batting helmets in baseball. In all of these examples, the attempt in the design of the protective padding is to spread the impact force over as large an area as possible to reduce the pressure.

With use of a force platform, it is possible to obtain a measure of the *center of pressure*. The center of pressure (COP) is a displacement measure indicating the path of the resultant ground reaction force vector on the force platform. It is equal to the weighted average of the points of application of all the downward acting forces on the force platform. Since the COP is a general measurement, it may be nowhere near the maximal areas of pressure. It does provide, however, a general pattern and has been extensively used in gait analysis. Cavanagh and Lafortune (4) showed different COP patterns for rearfoot and midfoot strikers. Representative COP patterns are illustrated in Figure 10-27. Cavanagh and Lafortune suggested that COP information may be useful in shoe design but these patterns have not been related successfully to foot function during locomotion. Miller (15) noted that COP data provides only very restricted information on the overall pressure distribution on the sole of the foot.

Methods of measuring the localized pressure patterns under the foot or shoe have been developed in recent years. An example of the type of data available on these systems is presented in Figure 10-28. Cavanagh et al. (39) developed such a measuring system and reported distinct local areas of high pressure on the foot throughout the ground contact phase. The greatest pressures were

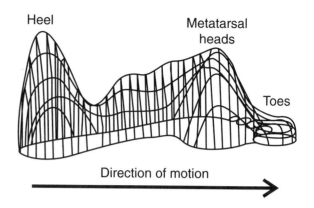

FIGURE 10-28. A three-dimensional representation of the pressure distribution pattern of a normal foot during walking (after Cavanagh, P.R. The biomechanics of running and running shoe problems. In B. Segesser and W. Pforringer (eds.). The Shoe in Sport. pp. 3-15. London: Wolfe Publishing, Ltd., 1989.).

measured at the heel, on the metatarsal heads, and on the hallux. Cavanagh et al. (39) also compared the pressure patterns during running barefoot and with various foam materials attached to the foot. They reported that peak pressures were reduced when wearing the foam materials but the changes in the pressure pattern over the support period were similar. Foti et al. (40), using an in-shoe pressure measuring device, reported that softer midsole shoes distributed the foot-shoe pressure at heel contact during walking better than a hard midsole shoe. The implication is that softer midsole shoes provide a more cushioned "feel" to the wearer.

Work

The term *work* is often used to mean a variety of things. This term is generally used to define anything that causes some mental or physical effort throughout daily activities. In mechanics, however, the term work has a more specific and narrow meaning. Mechanical work is equal to the product of the magnitude of a force applied against an object, and the distance that the object moves in the direction of the force while the force is being applied to the object. For example, in moving an object along the ground, an individual pushes the object with a force parallel to the ground. If the force necessary to move the object was 100 N and the object was moved 1.0 m, the work done would be 100 N-m. The case cited, however, is a very specific one. More generally, work is:

$$W = F * \cos \theta * s$$

where F is the force applied, s is the displacement, and θ is the angle between the force vector and the line of dis-

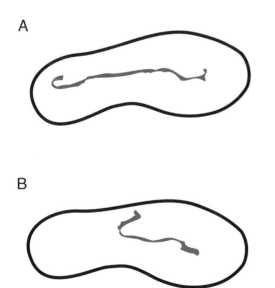

FIGURE 10-27. Center of pressure patterns for the left foot of: a) a heel-toe footfall pattern runner; and b) a midfoot foot-strike pattern runner.

placement. The unit of mechanical work is derived from the product of a force in newtons and a displacement in meters. The most commonly used units are the newton-meter (N-m) and the joule (J). These are equivalent units such that:

$$1 \text{ N-m} = 1 \text{ J}$$

In Figure 10-29a, the force is applied to a block parallel to the line of displacement or at an angle of 0° to the displacement. Since cos 0° =1, then the work done is simply the product of the force and the distance the block is displaced. Thus if the force applied was 50 N and the block was displaced 0.1 m, the mechanical work done would be:

$$W = 50 \text{ N} * \cos 0° * 0.1 \text{m}$$
$$= 50 \text{ N} * 1 * 0.1 \text{ m}$$
$$= 5 \text{ N-m}$$

If the same force were applied at an angle of 30° over the same distance, d, (Figure 10-29b), the work done would be:

$$W = 50 \text{ N} * \cos 30° * 0.1 \text{ m}$$
$$W = 50 \text{ N} * 0.866 * 0.1 \text{ m}$$
$$W = 4.33 \text{ N-m}$$

Therefore, more work is done if the force is applied parallel to the direction of motion than if the force is applied at an angle.

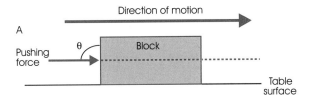

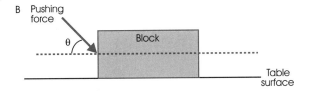

FIGURE 10-29. Illustration of the mechanical work done on a block when: a) a force is applied parallel to the surface (θ = 0° thus cos θ = 1); and b) a force is applied at an angle to the direction of motion (θ = 30° thus cos θ = 0.866).

As the above discussion of work implies, work is done only when the object is moving and its motion is influenced by the applied force. If a force acts on an object and does not cause the object to move, then no mechanical work is done because the distance moved is zero. During an isometric contraction, for example, no work is done since there is no movement. If a weight-lifter holds a 892 N (200 lb) barbell overhead, they are doing no mechanical work. In lifting the barbell overhead, however, mechanical work was done. If they lifted the barbell 1.85 m, the work done would be:

$$W = 892 \text{ N} * 1.85 \text{ m}$$
$$= 1650.2 \text{ J}$$

assuming that they lifted the bar straight up.

Energy

As with the term, work, the definition of the mechanical term, energy, is often misused. Simply stated, *energy* is the capacity to do work. There are many types of energy, some of which are light, heat, nuclear, electrical and mechanical. In biomechanics, however, the main concern is with mechanical energy. The unit of mechanical energy in the metric system is the joule (J). Mechanical energy has two forms—kinetic energy and potential energy.

Kinetic energy (KE) refers to the energy resulting from motion. An object possesses kinetic energy when it is in motion, that is, when it has some velocity. Linear kinetic energy is expressed algebraically as :

$$KE = \frac{1}{2}mv^2$$

where m is the mass of the object and v is the velocity. Since this expression includes the square of the velocity, any change in velocity greatly increases the amount of energy in the object. If the velocity is zero, then the object has no kinetic energy. An approximate value for the kinetic energy of a 625 N runner would be 3600 J while a swimmer of comparable body weight would have a value of 125 J. A moving body must have some energy because a force must be exerted to stop it. To start an object moving, a force must be applied over a distance. Kinetic energy, therefore, is the ability of a moving object to do work resulting from its motion.

Potential energy (PE) is the capacity to do work because of position or form. An object may contain stored energy, for example, simply because of its height or its deformation. In the first case, if a 30 kg barbell is lifted overhead to a height of 2.2 m, 647.5 J of work is done to lift the barbell. That is:

$$W = F * s$$
$$= (30 \text{ kg} * 9.81 \text{ m/s}^2) * 2.2 \text{ m}$$
$$= 647.5 \text{ J}$$

As the barbell is held overhead, it has the potential energy of 647.5 J. The work done to lift it overhead is also the potential energy. It should be noted, however, that the potential energy gradually increases as the bar is lifted. If the bar is lowered, the potential energy decreases. Potential energy is defined algebraically as:

$$PE = m\,g\,h$$

where m is the mass of the object, g is the acceleration due to gravity, and h is the height. The more work done to overcome gravity, therefore, the greater the potential energy.

An object that is deformed may also store potential energy. This type of potential energy has to do with the energy associated with elastic forces. When an object is deformed, the resistance to the deformation increases as the object is stretched. Thus, the force that deforms the object is stored, and may be released as elastic energy. This type of energy is referred to as *strain energy* (SE) and is defined as:

$$SE = \frac{1}{2} k * \Delta x$$

where k is a proportionality constant and Δx is the distance over which the object was deformed. The proportionality constant is dependent on the material deformed and is often called the stiffness constant because it represents the object's ability to store energy. It has already been discussed how certain tissues such as muscles and tendons, and how certain devices such as springboards for diving may store this strain energy and release to aid in human movement. In athletics there are numerous pieces of equipment that achieve such an end. Examples of such are trampolines, bows in archery, and poles in pole vaulting. Perhaps the most sophisticated use of elastic energy storage has been the design of the "tuned" running track at Harvard University. McMahon and Greene (41) analyzed the mechanics of running and the energy interactions between the runner and the track to develop an optimum design for the track surface. In the first season on this new track, an average speed advantage of nearly 3% was observed. Further, it was determined that there was a 93% probability that any given individual will run faster on this new track (42).

It should be intuitive that many instances in human motion can be understood in terms of the interchanges between kinetic and potential energy. The mathematical relationship between the different forms of energy was formulated by the German scientist von Helmholtz (1821–1894). In 1847 he defined what has come to be known as the *law of conservation of energy*. The main point of this law is that energy cannot be created nor destroyed. No machine, including the human machine, can generate more energy than it takes in. It follows, therefore, that the total energy of a closed system is constant since energy does not enter or leave a closed system. A closed system is one that is physically isolated from its surroundings. This point can be phrased mathematically by stating:

$$TE = KE + PE$$

where TE is a constant representing the total energy of the system. In human movement, this only occurs when the object is a projectile, whereby the only external force acting upon it is gravity since fluid resistance is neglected.

Consider the example of a projectile that is projected vertically upwards. At the point of release, the projectile has zero potential energy and a large kinetic energy. As the projectile proceeds vertically upward, the potential energy increases while the kinetic energy decreases, because gravity is slowing down the upward flight. At the peak of the trajectory, the velocity of the projectile is zero and the kinetic energy is zero but the potential energy is a maximum. The total energy of the system, however, did

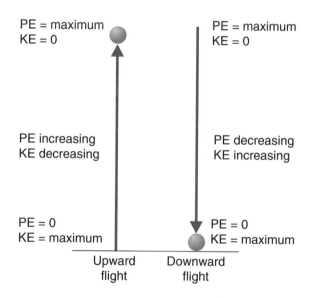

FIGURE 10-30. Changes in potential energy (PE) and kinetic energy (KE) as a ball is projected straight upward vertically and as it falls back to earth.

not change since increases in potential energy resulted in equal decreases in kinetic energy. On the downward flight, the reverse change in the forms of energy occurs. These changes in energy are illustrated in Figure 10-30.

Power

In evaluating the amount of work done by a force, the time over which the force is applied is not taken into account. In the previous example in which the work done (by the weight-lifter to raise the barbell overhead), was calculated, the time it took to raise the barbell was not taken into consideration. Regardless of how long it took to raise the barbell, the amount of work done was 647.5 J. The concept of *power* takes into consideration the work done per unit of time. Power is defined as the rate at which a force does work or:

$$P = \frac{dW}{dt}$$

where W is the work done and dt is the time period in which the work was done. Power has units of watts (W). The change in work is expressed in units of Joules and the change in time in units of seconds. Thus:

$$1\ W = 1\ J/s$$

If power were plotted on a graph as a function of time, the area under the curve would equal the work done.

If the weight-lifter in the previous example raised the bar in 0.5 s, the power developed would be:

$$P = \frac{647.5\ J}{0.5\ s}$$
$$= 1295\ J/s$$
$$= 1295\ W$$

Decreasing the time over which the bar is lifted to 0.35 s would increase the power developed by the weight-lifter to 1850 W. While the work done remains constant, greater power must be developed to do the mechanical work more quickly.

Another definition for power can be developed by re-arranging the formula for power. If the product of the force (F) and the distance over which it was applied (s) is substituted for the mechanical work done, the equation becomes:

$$P = \frac{d(F * s)}{dt}$$

and re-arranging this equation:

$$P = F * \frac{ds}{dt}$$

Since $\frac{ds}{dt}$ was defined in a previous chapter as the velocity in the s-direction, it can be readily seen that:

$$P = F * v$$

where F is the applied force and v is the velocity of the force application.

Power is often confused with force, work, energy, or strength. Power, however, is a combination of force and velocity. In many athletic endeavors, power, or the ability to use the combination of force and velocity, is paramount. One such activity, weight-lifting, has already been discussed, but there are many others, such as shot putting, batting in baseball, or boxing.

Summary

Linear kinetics is the branch of mechanics that deals with the causes of linear motion, or forces. All forces have the characteristics of magnitude, direction, point of application, and line of action. The mechanical laws governing the motion of objects were presented by Sir Isaac Newton and form the basis for the mechanical analysis of human motion. These laws are:

1) Law of inertia - Every body continues in its state of rest or uniform motion in a straight line unless acted upon by an external force.

2) Law of acceleration - the rate change of change of momentum of a body is proportional to the force causing it, and the change takes place in the direction of the force.

3) Law of action-reaction - For every action there is an equal and opposite reaction.

Forces may be categorized as non-contact or contact forces. The most important non-contact force acting during human movement is gravity. The contact forces include the ground reaction force, joint reaction force, friction, fluid resistance, inertial force, muscle force, and elastic force. The ground reaction force is a direct application of Newton's third law and has three components: a vertical component and two shear components acting parallel to the surface of the ground. The joint reaction force is the net force acting across a joint and has components referred to as the compressive and shear components. Friction results from the interaction between two surfaces and is a force acting parallel to the interface of the two surfaces and in a direction opposite to the motion. The coefficient of friction is the quantification of the interaction of the two surfaces. Fluid resistance refers to the transfer of energy from an object to the fluid through which the object is moving. The fluid resistance vector has two components, lift and drag. Drag acts in a direction opposite to the direction of motion and lift is perpendicular to the drag component. An inertial force results from the force applied by one segment on another that is not due to muscle actions. A muscle force is the pull of the muscle on its insertion, resulting in motion at a joint. Muscle forces are generally calculated as net forces, not individual muscle forces, although there are intricate mathematical procedures that can evaluate individual muscle forces. An elastic force results from the rebound of a material to its original length once it has been deformed.

A free-body diagram is a schematic illustration of a system with all external forces represented by vector arrows at their points of application. Internal forces are not presented on free body diagrams. Muscle forces are generally not represented on these diagrams unless the system involves a single segment.

Special force applications include definitions for pressure, mechanical work, energy, and power. Pressure is the force per unit area. Work is the product of the force applied and the distance over which the force is applied. Energy, the capacity to do work, has two forms—kinetic and potential. The relationship between work and energy is defined in the Work-Energy Theorem which states that the amount of work done is equal to the change in energy. When the amount of work done is related to the time over which the work was done, the power developed is being evaluated.

Review Questions

1. What are the a) horizontal and b) vertical components of a force with a magnitude of 50 N acting at 34° to the horizontal? (Answer: a) 41.45 N; b) 27.96 N).
2. The horizontal and vertical components of a force are 47.23 N and 28.56 N, respectively. What is the magnitude of the resultant vector? (Answer: 55.19 N).
3. a) An individual has a mass of 72 kg. What is their body weight?
   b) If an individual has a body weight of 555 N, what is their body mass? (Answer: a) 706.32 N; b) 56.57 kg).
4. How much force must be exerted to accelerate a 500 N weight to 6.5 m/s^2? (Answer: 331.29 N).
5. What are the coefficients of friction if the friction forces are a) 80.9 N, b) 25.7 N and c) 100 N and the normal force is 110 N? (Answer: a) 0.74; b) 0.23; c) 0.91).

6. Sketch the vertical ground reaction force component of an individual who jumps off a force platform starting from a deep crouch, simply extending at the knee and ankle joints.
7. Explain how a golfer puts backspin on the golf ball when it is driven off the tee.
8. a) Sketch a free-body diagram of a person who has jumped into the air (i.e. they are a projectile).
   b) Sketch a free-body diagram of the leg-foot system if ankle weights are placed on the ankle.
9. What is the pressure if a 100 N force is applied over an area of 11 cm^2? (Answer: 9.1 N/cm^2).
10. If an individual's thigh exerts a force of 15.1 N at a velocity of 2.5 m/s^2, what is the power generated by the thigh? (Answer: 37.75 W).

Additional Questions

1. Sketch multi-force systems that are: a) co-planar; b) concurrent; and c) collinear.
2. Differentiate between an astronaut's body mass and their body weight when they are: a) standing on the moon; b) traveling through space in the space vehicle.
3. Sketch a series of antero-posterior ground reaction force component curves in which an individual: a) gradually slows down from a running speed of 5 m/s to 2 m/s; b) gradually speeds up from a running speed of 2 m/s to 5 m/s.
4. Select one activity in which the coefficient of friction should be maximized and one in which it should be minimized. Illustrate in each case how the coefficient of friction can affect performance.
5. If the static coefficient of friction of a basketball shoe on a particular playing surface is 0.45 and the normal force is 750 N, what is the horizontal force necessary to cause the shoe to slide? Is this an acceptable situation? (Answer: 337.50 N).

6. A batter hits a pitched ball to the opposite field. Why does the ball curve?
7. An individual lifts a 100 kg weight to a height of 1.87 m. a) How much work was done? b) When the weight is held overhead, what is the potential energy? c) What is the kinetic energy? (Answer: a) 1834.5 N-m; b) 1834.5 N-m; c) 0 N-m).
8. Sketch a free body diagram of a sprinter in the starting blocks just prior to the start of the race.
9. A runner's leg has a total energy of 20 J at one instant in time and a total energy of 47 J later in the stride. How much work was done on the leg? (Answer: 27 N-m).
10. The work calculated at time 1 and time 2 was 155 N-m and 213 N-m, respectively. Calculate the power if the time interval was .023 s. (Answer: 2521.74 W).]

References

1. Cajori, F. Sir Isaac Newton's Mathematical Principles (translated by Andrew Motte in 1729). Berkeley, CA: University of California Press, 1934.

2. Brancazio, P.J. Sports Science. New York: Simon and Schuster, 1984.

3. Elftman, H. Forces and energy changes in the leg during walking. American Journal of Physiology 125:339-356, 1939.

4. Cavanagh, P.R. and Lafortune, M.A. Ground reaction forces in distance running. Journal of Biomechanics 15:397-406, 1980.

5. Clarke, T.E., Frederick, E.C., Cooper, L.B. The effects of shoe cushioning upon ground reaction forces in running. International Journal of Sports Medicine 4:376-381, 1983.

6. Hamill, J., Bates, B.T., Knutzen, K.M. Ground reaction force symmetry during walking and running. Research Quarterly for Exercise and Sport 55:289-293, 1984.

7. Dufek, J.S. and Bates, B.T. Dynamic performance assessment of selected sport shoes on impact forces. Medicine and Science in Sports and Exercise 23:1062-1067, 1991.

8. McNitt-Gray, J.L. Kinematics and impulse characteristics of drop landings from three heights. International Journal of Sports Biomechanics 7:201-224, 1991.

9. Hamill, J., Bates, B.T., Knutzen, K.M., Sawhill, J.A. Variations in ground reaction force parameters at different running speeds. Human Movement Science 2:47-56, 1983.

10. Munro, C.F., Miller, D.I., Fuglevand, A.J. Ground reaction forces in running - a re-examination. Journal of Biomechanics 20:147-155, 1987.

11. Nigg, B.M. External force measurements with sports shoes and playing surfaces. In B.M. Nigg and B. Kerr (eds.). Biomechanical Aspects of Sports Shoes and Playing Surfaces. pp. 11-23. Calgary: University of Calgary, 1983.

12. Hamill, J. and Bates, B.T. A kinetic evaluation of the effects of in vivo loading on running shoes. Journal of Orthopaedic and Sports Physical Therapy 10:47-53, 1988.

13. Hamill, J., Bates, B.T., Knutzen, K.M., Kirkpatrick, G.M. Relationship between selected static and dynamic lower extremity measures. Clinical Biomechanics 4:217-225, 1989.

14. Bobbert, M.F., Schamhardt, H.C., Nigg, B.M. Calculation of vertical ground reaction force estimates during running from positional data. Journal of Biomechanics 24:1095-1105, 1991.

15. Miller, D.I. Ground reaction forces in distance running. In P.R. Cavanagh (ed.). Biomechanics of Distance Running. pp. 203-224. Champaign, IL: Human Kinetics Publishers, 1990.

16. Bates, B.T., Osternig, L.R., Mason, B.R. Variations of velocity within the support phase of running. In J. Terauds and G. Dales (eds.). Science in Athletics. pp. 51-59. Del Mar: Academic Publishers, 1979.

17. Winter, D.A. Biomechanics and Motor Control of Human Movement (2nd Ed.). New York: John Wiley and Sons, Inc., 1990.

18. Valiant, G.A., Cooper, L.B., McGuirk, T. Measurements of the rotational friction of court shoes on an oak hardwood playing surface. Proceedings of the North American Congress on Biomechanics. pp. 295-296, 1986.

19. Valiant, G.A. Ground reaction forces developed on artificial turf. Proceedings of the First World Congress of Science and Medicine in Football. In H. Reilly and A. Lees. (eds.). pp. 143-158. E&FN: London, 1987.

20. Cohen, H.H. and Compton, D.M.J. Fall accident patterns. Professional Safety. June:16-22, 1982.

21. Buck, P.C. Slipping, tripping and falling accidents at work - a national picture. Ergonomics 28:949-958, 1985.

22. Andres, R.O. and Chaffin, D.B. Ergonomic analysis of slip-resistance measurement devices. Ergonomics 28:1065-1079, 1985.

23. van Ingen Schenau, G.J. The influence of air friction in speed skating. Journal of Biomechanics 16:449-453, 1982.

24. Kyle, C.R. and Caizzo, V.L. The effect of athletic clothing aerodynamics upon running speed. Medicine and Science in Sports and Exercise 18:509-513, 1986.

25. Kyle, C.R. The wind resistance of the human figure in sports. Proceedings of the First IOC World Congress on Sports Sciences. pp. 287-288, Colorado Springs, CO, 1989.

26. Walpert, R.A. and Kyle, C.J. Aerodynamics of the human body in sports. XII International Congress of Biomechanics Proceedings. pp. 346-347, 1989.

27. Barthels, K.M. and Adrian, M.J. Three-dimensional spatial hand patterns of skilled butterfly swimmers. In L. Lewillie and J .P. Clarys (eds.). Swimming II. pp. 154-160, Baltimore: University Park Press, 1975.

28. Schleihauf, R.E. A hydrodynamic analysis of swimming propulsion. In J. Terauds and E.W. Bedington (eds.). Swimming III. pp. 70-109. Baltimore: University Park Press, 1979.

29. Scott, S.H. and Winter, D.A. Talocrural and talocalcaneal joint kinematics and kinetics during the stance phase of walking. Journal of Biomechanics 24:743-752, 1991.

30. Hawkins, D. and Hull, M.L. Muscle force as affected by fatigue: mathematical model and experimental verification. Journal of Biomechanics 26:1117-1128, 1993.

31. Komi, P.V., Salonen, M., Jarvinen, M., Kokko, O. In vivo registration of achilles tendon forces in man. I. Methodological development. International Journal of Sports Medicine 8:3-8, Supplement, 1987.

32. Komi, P.V. Relevance of In Vivo force measurements to human biomechanics. Journal of Biomechanics 23:23-34, 1990.

33. Sprigings, E., Stillings, D., Watson, G. Development of a model to represent an aluminum springboard in diving. International Journal of Sport Biomechanics 5:297-307, 1989.

34. Boda, W.L. and Hamill, J. A mechanical model of the Maxiflex "B" springboard. Proceedings of the VIth bian-

nual meeting of the Canadian Society of Biomechanics. pp. 109-110. Quebec City, Canada, 1990.

35. Asmussen, E. and Bonde-Peterson, F. Apparent efficiency and storage of elastic energy in human muscles during exercise. Acta Physiologica Scandanavia 92:537-545, 1974.

36. Komi, P.V. and Bosco, C. Utilization of stored elastic energy in leg extensor muscles by men and women. Medicine and Science in Sports 10:261-265, 1978.

37. Alexander, R. McN. Elastic energy stores in running vertebrates. American Zoologist 24:85-94, 1984.

38. Hamill, J., Murphy, M., Sussman, D.H. The effect of track turns on lower extremity function. International Journal of Sports Biomechanics 3:276-286, 1987.

39. Cavanagh, P.R. The biomechanics of running and running shoe problems. In B. Segesser and W. Pforringer (eds.). The Shoe in Sport. pp. 3-15. London: Wolfe Publishing, Ltd., 1989.

40. Foti, T., Derrick, T., Hamill, J. Influence of footwear on weight-acceptance plantar pressure distribution during walking. In R. Rodano (ed.). Biomechanics in Sports X. pp. 243-246. Milan, Italy: Edi-Ermes, 1992.

41. McMahon, T.A. and Greene, P.R. The influence of track compliance on running. Journal of Biomechanics 12:893-904, 1979.

42. McMahon, T.A. and Greene, P.R. Fast running tracks. Scientific American 239(6):148-163, 1978.

Additional Reading

1. Alexander, R.M. Elastic Mechanisms in Animal Movement. Cambridge: Cambridge University Press, 1988.
2. Frohlich, C. Aerodynamic effects on discus flight. American Journal of Physics 49:1125-1132, 1981.
3. Greene, P.R. Sprinting with banked turns. Journal of Biomechanics 20:667-680, 1980.
4. Pugh, L. The influence of wind resistance in running and walking and the mechanical efficiency of work against horizontal and vertical forces. Journal of Physiology 213:255-276, 1971.
5. Schirer, E.W., Allman, W.F. (eds.). Newton at the Bat - The Science in Sports. New York: Charles Scribner's Sons, 1984.

Glossary

active peak:	the second of the two peaks in a vertical ground reaction force curve during running.
angle of application:	the angle at which a force vector acts.
Bernoulli's Principle:	the relationship between pressure and velocity that states that pressure is inversely proportional to velocity.
bone-on-bone force:	the force at a joint that includes the joint reaction force and the forces due to muscles and ligaments.
boundary layer:	the thin layer of a fluid adjacent to the surface of an object moving through the fluid.
center of pressure:	the point of application of a force.
centrifugal force:	the reaction force to the centripetal force acting away from the center of rotation when an object moves along a curved path.
centripetal force:	the force acting towards the center of rotation resulting when an object moves along a curved path.
coefficient of friction:	the ratio of the friction force to the normal force pressing two surfaces together.
collinear forces:	a force system in which the lines of action of the forces act along the same line.
composition:	the process by which the resultant of two vectors is determined.
concurrent forces:	a force system in which the lines of action of the forces act at the same point.
contact force:	the pushes or pulls, exerted by one object in direct contact with another object.
coplanar forces:	a system of forces in which the lines of action of the forces all act in the same plane.
density:	the mass per unit volume.
energy:	the capacity to do work.
fluid resistance:	the transfer of energy from an object to the fluid through which the object is moving.
force:	an interaction between two objects in the form of a push or pull that may or may not cause motion.
force platform:	a device that measures ground reaction force.
form drag:	a net resistance force caused by a pressure differential between the leading and trailing edges of an object moving through a fluid.
free body diagram:	a diagram in which all force vectors are drawn.

friction:	the force that resists the motion of one surface upon another.
impact peak:	the initial peak on a vertical ground reaction force.
inertia:	the resistance of a body to a change in its state of motion.
joint reaction force:	the force acting across a joint.
kinetic energy:	the ability of a body to do work by virtue of its motion.
kinetics:	the branch of mechanics that deals with forces that act on a system.
kinetic friction:	the friction that exists between two surfaces as they slide on each other.
laminar flow:	at slower flow velocities, the flow of a fluid smoothly over the surface of an object.
Law of Conservation of Energy:	the law that states that energy can neither be created nor destroyed but can only change form.
linear kinetics:	the branch of kinetics that deals with translational motion.
line of action:	the line along which a force acts.
lift force:	a component of the fluid resistance force that acts perpendicular to the direction of motion.
loading peak:	see active peak.
magnus effect:	the curve in the path of a spinning ball caused by a pressure differential on either side of the ball.
mass:	the measure of a body's inertia.
maximum static friction force:	the maximum friction force measured just prior to the impending motion of an object.
non-contact-force:	a force that acts at a distance from an object.
passive peak:	see impact peak.
point of application:	the point at which a force acts.
point of separation:	the point at which the boundary layer separates from an object.
potential energy:	the ability of a body to do work by virtue of its position.
power:	the quantity of work done per unit time.
pressure:	force per unit area.
resolution:	the breakdown of a vector into its horizontal and vertical components.
rotational friction:	the resistance in rotation of one surface upon another.
separated flow:	the type of fluid flow in which the boundary layer separates from the object, creating turbulence and thus resisting the motion of the object.
shear force:	a force that acts parallel to the surface.
strain energy:	the capacity to do work by virtue of the deformation of an object.

surface drag:	the fluid drag force acting on a body resulting from the friction between the surface of the object and the fluid.
system:	a defined set of forces.
turbulent flow:	the type of fluid flow in which the boundary layer becomes so turbulent that the point of separation moves further back on the object, thus reducing drag.
viscous drag:	see surface drag.
viscosity:	the measure of a fluid's resistance to flow.
weight:	the force of the earth's gravitational attraction to a body's mass.
work:	the product of the force applied to a body and the distance through which the force is applied.

Angular Kinetics

After reading this chapter, the student should be able to:
1. Define torque and discuss the characteristics of a torque.
2. Define the concept of center of mass.
3. Discuss the different methods of determining body segment parameters for the calculation of center of mass.
4. Calculate the segment center of mass and the total body center of mass.
5. Discuss the concept of moment of inertia.
6. Differentiate between the three classes of levers.
7. State the angular analogs of Newton's three laws of motion and their impact on human movement.
8. Understand the impact of angular momentum on human motion.
9. Discuss the relationships between torque, angular work, rotational kinetic energy, and angular power.

In the previous chapter, the notion that movement does not occur unless an external force is applied was discussed. Also discussed were the characteristics of a force; two of which were the line of action and the point of application. If the line of action and the point of application of a force are critical, then it would appear that different types of motion might result dependent on these characteristics. For example, a nurse pushing a wheelchair exerts two equal forces, one on each handle. The result is that these particular lines of action and points of application of the two forces cause the wheelchair to move in a straight line. What happens though when the nurse pushes the chair with only one arm, applying a force to only one of the handles? A force is still applied, but the motion is totally different. In fact, the wheelchair will translate and rotate (Figure 11-1). The situation that has just been described actually represents the majority of the types of motion that occur when humans move. It is rare that a force or a system of forces cause pure translation. In fact, the majority of force applications in human movement cause simultaneous translation and rotation.

The branch of mechanics that deals with the causes of motion is called kinetics. The branch of mechanics that deals with the causes of angular motion is called *angular kinetics*.

Torque or Moment of Force

When a force causes a rotation, the rotation occurs about a pivot point, and the line of action of the force must act at a distance from the pivot point. When a force is applied such that it causes a rotation, the product of that force and the perpendicular distance to its line of action is referred to as a *torque* or a *moment of force*. These terms are synonymous and are used interchangeably in the literature. A torque is not a force but merely the effectiveness of a force in causing a rotation. A torque is defined, therefore, as the tendency of a force to cause a rotation about a specific axis. Any discussion of a torque must be with reference to a specific axis. Mathematically, torque is:

$$T = F * r$$

where T is the torque, F is the applied force in newtons, and r is the perpendicular distance in meters of the line of action of the force to the pivot point. Since torque is the product of a force, with units of newtons, and a distance, with units of meters, torque has units of newton-meters

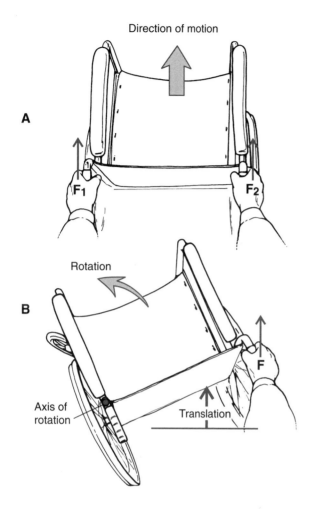

Figure 11-1. In an overhead view of a wheelchair; a) the wheelchair is translated forward by forces F1 and F2 and b) the wheelchair will rotate and translate if only one force, F1, is applied.

Figure 11-2. Representation of the moment arm, line of application, and pivot point constituting a torque.

ever. As noted with angular measurement, a counter-clockwise direction is considered positive (+) while a clockwise direction is negative. Thus, a counter-clockwise torque is positive, while a clockwise torque is negative. Torques in a system may be evaluated through vector operations as has been described for forces. That is, torques may be composed into a resultant torque.

The concept of torque is prevalent in our everyday life. For example, many have probably used a wrench to loosen the nut on a bolt. By applying a force to the wrench, a rotation is produced causing the nut to loosen. Intuitively, the wrench is grasped at the end as in Figure 11-3a. By grasping it in this way, the moment arm is maximized and, for a given force application, the torque is maximized. If the wrench were grasped at its midpoint, the torque would be halved, even though the same force is applied (Figure 11-3b). To achieve the same torque as in the first case, the force must be doubled. Thus, a torque can be increased by increasing the force, increasing the moment arm, or both.

The concept of torque is often used in rehabilitation evaluation. For example, if an individual has injured their elbow, the therapist may use a manual resistance technique to evaluate the joint. The therapist would resist the individual's elbow flexion by exerting a force at the mid-forearm position. This creates a torque which the patient must overcome. As the individual progresses, the therapist may exert approximately the same force level at the wrist instead of the mid-forearm position. By increasing the moment arm while keeping the force constant, the therapist has increased the torque that the patient must overcome. This rather simple technique can be helpful to the therapist in developing a program for the individual's rehabilitation.

Force Couple When a gymnast wishes to execute a twist about a longitudinal axis, they apply not one, but two equal and opposite torques. By applying a backward force with one foot and a forward force with the other, the gymnast creates two torques that cause them to rotate about the longitudinal axis (Figure 11-4). Such a pair of

(N-m). The distance, r, is referred to as the *moment of arm* of the force. This concept is illustrated in Figure 11-2. If the force acts directly through the pivot point or the axis of rotation, then the torque is zero, because the moment arm would be zero regardless of how great the force. In this instance, a pure translational motion would occur. Because the force is not applied through the pivot point, a torque is said to result from an *eccentric force* or literally an off-center force. While an eccentric force primarily causes rotation, it also causes translation.

Torque is a vector quantity and thus has the properties of magnitude and direction. The magnitude is represented by the quantity of the magnitude of the force times the magnitude of the moment arm. The direction is determined by the convention of the Right Hand Rule, how-

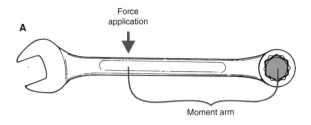

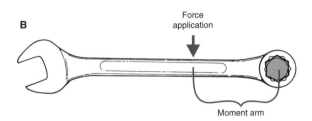

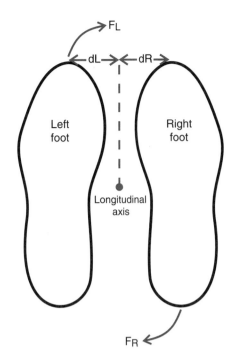

Figure 11-3. An illustration of a wrench with two points of force application. By grasping the wrench at the end (a) a greater torque is generated because the moment arm is greater than in example (b).

Figure 11-4. A torque $F_R * d_R$ is created by the right foot while another torque $F_L * d_L$ is created by the left foot. Since these two torques are equal and in the same angular direction, the force couple will result in a rotation about the longitudinal axis through the center of mass.

forces is called a *force couple*. A force couple is two forces that are equal in magnitude and act in opposite directions. These two forces act at a distance from an axis and produce rotation about the axis. A force couple can be thought of as two moments of force, each creating a rotation about the longitudinal axis of the gymnast. Torques, however, also cause a translation, but because the translation caused by each torque is in the opposite direction, the translation of the body is canceled out. Thus, a force couple causes a pure rotation about an axis with no translation. By placing their feet slightly further apart, the gymnast in Figure 11-4 can increase the moment arm and thus cause a great deal more rotation.

A force couple is calculated by:

$$Force\ Couple = F\ d$$

where F is one of the equal and opposite forces and d is the distance between the forces. While no true force couples exist in the human anatomy, the human body often uses force couples. For example, a force couple would be created when one uses the thumb and forefinger to screw open the top on a jar.

Center of Mass

An individual's body weight is a product of their mass and the acceleration due to gravity. The body weight vec-

tor originates at a point referred to as the *center of gravity*, or the point about which all particles of the body are evenly distributed. The point about which the body's mass is evenly distributed is referred to as the *center of mass*. The terms center of mass and center of gravity are often used synonymously. The center of gravity refers only to the vertical direction because that is the direction in which gravity acts. The more general term is the center of mass.

If the center of mass is the point about which the mass is evenly distributed, then this must also be the balancing point of the body. Thus, the center of mass can be further defined as the point about which the sum of the torques equal zero. That is:

$$\Sigma T_{cm} = 0$$

In Figure 11-5, an object consisting of two point masses is illustrated. Point mass A results in a counterclockwise torque about point C while point B results in a clockwise torque also about point C. If these two torques are equal, then the object is balanced and point C may be considered the center of mass. This does not imply that the mass of these two point masses is the same, but that the torques created by the masses are equal (Figure 11-5).

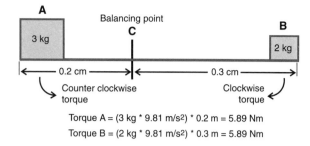

Torque A = (3 kg * 9.81 m/s2) * 0.2 m = 5.89 Nm
Torque B = (2 kg * 9.81 m/s2) * 0.3 m = 5.89 Nm

Figure 11-5. A two-point force system balanced at the center of mass.

The center of mass is a theoretical point whose location may change from instant to instant during a movement. The change in position of the center of mass results from the rapidly changing positions of the body segments during movement. In fact, the center of mass does not necessarily have to be inside the limits of the object. For example, the center of mass of a doughnut is located within the inner hole but outside the physical mass of the doughnut. In the case of a human performer, the positions of the segments can also result in a case in which the center of mass is outside the body. In activities such as high-jumping and pole-vaulting where the body must curl around the bar, the center of mass is certainly outside the limits of the body (1).

Reaction Board Method

A balancing approach may be used to determine the location of the center of mass of an individual in a particular static posture. This method is known as the *reaction board method*. For this technique, a weigh scale and a rigid board measuring approximately 2 m by 0.6 m with angle iron fastened on either end is used. The computation of the location of the center of mass involves summing the torques about the platform support.

A free-body diagram of the set-up is presented in Figure 11-6. One end of the board is placed on the weigh scale with a reaction force of S_y on that end, and R_y on the end in contact with the floor. The center of mass of the board acts through point B a distance b from the non-scale end of the board. The magnitude of S_y for the board alone can be read directly from the scale and acts at a distance s from the non-scale end. The torques about the non-scale end of the board can be calculated. If the scale for the board alone reads 90 N and the board is 2 m long, the torque created by the weight of the board is:

$$\Sigma T = 0$$
$$-S_y * s + B * b = 0$$
$$B * b = S_y * s$$
$$B * b = 90 \text{ N} * 2m$$
$$B * b = 180 \text{ N-m}$$

The subject then may lie supine on the board with their heels directly over the non-scale end and S_{y1}, the weight of the subject and the board, is recorded (Figure

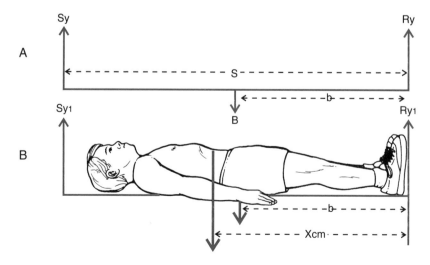

Figure 11-6. Free body diagrams of: a) the reaction board system; and b) the reaction board/person system.

11-6b). Their body weight, W (670 N), acts vertically downward from their center of mass, located a distance x_{cm} from the non-scale end. Once again, the torques acting about the non-scale end can be summed:

$$\Sigma T = 0$$
$$-S_{y1} * s + W * x_{cm} + B * b = 0$$

If the weight of the subject-board system was 424 N, all known quantities can be substituted into this equation to solve for x_{cm}:

$$-424 \text{ N} * 2 \text{ m} + 670 \text{ N} * x_{cm} + 180 \text{ N - m} = 0$$
$$x_{cm} = \frac{848 \text{ N - m} - 180 \text{ N - m}}{670 \text{ N}}$$
$$x_{cm} = 0.99 \text{ m}$$

This indicates that the center of mass of the individual is located 0.99 m from the non-scale end, with respect to the subject's feet. This value is often expressed as a percentage of the individual's height. By orienting the body in different positions, it is evident how the center of mass may change. For example, by simply moving the arms to an overhead position, the center of mass would be expected to be farther from the feet.

Segmental Method

Segment Center of Mass Calculation The reaction board method is an interesting exercise, but not of much use in a biomechanics study. A much more useful approach involves knowledge of the masses and the location of the centers of mass of each of the body's segments. This approach, called the *segmental method* utilizes the x-y coordinates from digitized data, and the above mentioned properties of the segments, to analyze one segment at a time and then calculate the total body center of mass. Before presenting the computations for this method, the source of the information concerning the body segments must be approached. At least three methods for deriving this information have been utilized.

They are: 1) measures based on cadaver studies; 2) mathematical modeling; and 3) radioisotope scanning.

Several researchers have presented formulae that estimate the mass and the location of the center of mass of the various segments based on cadaver studies (2, 3, 4, 5). These researchers have generated regression or prediction equations that make it possible to estimate the mass and the location of the center of mass. Table 11-1 presents the prediction equations from Chandler et al. (4). These predicted parameters are based on known parameters such as the total body weight or the length or circumference of the segment. An example of a regression equation based on Clauser et al. (3) for estimating the segment mass of the leg is as follows:

Leg mass = 0.111 calf circumference + 0.047 tibial height + 0.074 ankle circumference − 4.208

where all dimensions of length are measured in centimeters. The location of the center of mass of the segment is usually presented as a percentage of the segment length from either the proximal or distal end of the segment.

Other researchers have used mathematical models to predict the segment masses and locations of the center of mass. This has been done by representing the individual body segments as regular geometric solids (6, 7, 8, 9). Using this method, the body segments are represented as truncated cones (e.g. upper arm, forearm, thigh, leg, and foot), cylinders (e.g. trunk), or elliptical spheres (e.g. head and hand). One such model, the Hanavan model, is presented in Figure 11-7. Regression equations based on the geometry of these solids requires the input of several measurements for each segment. For example, the thigh segment requires the measurement of the circumference of the upper and lower thigh, the length of the thigh, and the total body mass to estimate the desired parameters for the segment.

Table 11-1. Segment weight prediction equations and location of center of mass.

Segment	Weight (N)	CM Location (%)
head	0.032 BW + 18.70	66.3% from proximal end
trunk	0.532 BW − 6.93	52.2% from proximal end
upper arm	0.022 BW + 4.76	50.7% from proximal end
arm	0.013 BW + 2.41	41.7% from proximal end
hand	0.005 BW + 0.75	51.5% from proximal end
thigh	0.127 BW − 14.82	39.8% from proximal end
leg	0.044 BW − 1.75	41.3% from proximal end
foot	0.009 BW + 2.48	40.0% from proximal end

(From Chandler, R.F., Clauser, C.E., McConville, J.T., Reynolds, H.M., Young, J.W. Investigation of inertial properties of the human body. AMRL Technical Report. pp. 74–137. Wright-Patterson Air Force Base, 1975).

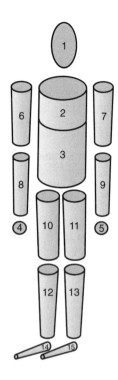

ters. The thigh segment will be used as an example to calculate these parameters. Using Dempster's (2) data, the mass of the thigh of an individual is 0.1 times their total body mass. Thus, if an individual had a total body mass of 75 kg, then the mass of the thigh would be:

$$m_{thigh} = 0.1 * 75 \text{ kg}$$
$$m_{thigh} = 7.5 \text{ kg}$$

Using the same data, the center of mass is located at 43.3% of the length of the thigh measured from the proximal end along the long axis of the segment. Consider the segment endpoint coordinates of the illustration of the thigh in Figure 11-8. The location of the center of mass would be:

$$
\begin{aligned}
x_{cm} &= x_p - (\text{length of the segment in x-direction} * 0.433) \\
&= x_p - ((x_p - x_d) * 0.433) \\
&= 0.45 + ((0.45 - 0.41) * 0.433) \\
&= 0.4327
\end{aligned}
$$

where x_{cm} is the location of the center of mass, x_p is the location of the hip joint, and x_d is the location of the knee joint all in the horizontal direction. Similarly:

$$
\begin{aligned}
y_{cm} &= y_p + (\text{length of the segment in y-direction} * 0.433) \\
&= y_p + ((y_p - y_d) * 0.433) \\
&= 0.78 - ((0.78 - 0.47) * 0.433) \\
&= 0.645
\end{aligned}
$$

Figure 11-7. A representation of the human body using geometric solids (after Miller, D.I. and Morrison, W.E. Prediction of segmental parameters using the Hanavan human body model. Medicine and Science in Sports 7(3):207–212, 1975).

The third method of determining the necessary segment characteristics is the "gamma-scanning" technique suggested by Zatsiorsky and Seluyanov (10). It involves the use of a radioisotope method. Using this method, a measure of a gamma-radiation beam is made before and after it has passed through a segment. In this way one can calculate the mass per unit surface area and thus generate prediction equations to determine the segment characteristics. The following example equation is based on these data and predicts the mass of the leg segment:

$$y = -1.592 + 0.0362 * \text{body mass} + 0.0121 \\ * \text{body height}$$

where y is the leg mass in kilograms. The following equation will predict the location of the center of mass along the longitudinal axis for the same segment:

$$y = -6.05 - 0.039 * \text{body mass} + 0.142 * \text{body height}$$

where y is the location of the center of mass as a percentage of the segment length.

All of these methods have been used in the literature to determine the segment characteristics of the center of mass and all give reasonable estimates of these parame-

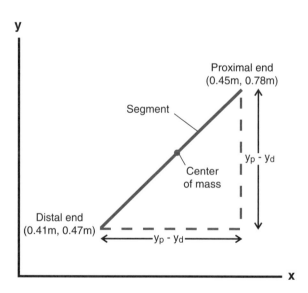

Figure 11-8. An illustration of the thigh segment and the segment endpoint coordinates at an instant in time during a walking stride.

where y_{cm} is the location of the center of mass, y_p is the location of the hip joint, and y_d is the location of the knee joint, all in the vertical direction. The center of mass of the segment is (0.4327, 0.645) in the reference frame. Note that the location of the center of mass must be between the values for the proximal and distal ends of the segments. This procedure must be carried out for each segment, using the coordinates of the joint centers to define the segments.

Total Body Center of Mass Calculation Once the segment center of mass locations have been determined, the total body center of mass can be calculated. Consider the illustration of a hypothetical three segment model in Figure 11-9. The mass and the location of the center of mass of each segment has been previously determined. To determine the horizontal location of the center of mass, the torques about the y-axis are calculated, using the concept that the sum of the torques about the total system center of mass is zero. There are four torques to consider; three created by the segment centers of mass and one by the total system center of mass. Thus:

$$m_1 g x_1 + m_2 g x_2 + m_3 g x_3 = M g x_{cm}$$

where m is the mass of the respective segments, M is the total system mass, g is the acceleration due to gravity, x is the location of the segment centers of mass, and x_{cm} is the location of the system center of mass. Since the term, g, appears in every term in the equation, it can be removed from the equation resulting in:

$$m_1 x_1 + m_2 x_2 + m_3 x_3 = M x_{cm}$$

Substituting the values from Figure 11-8 into this equation, there is only one unknown quantity, x_{cm}. Thus:

$$1(1) + 2(1) + 3(3) = 6 x_{cm}$$
$$x_{cm} = \frac{12}{6}$$
$$x_{cm} = 2$$

Therefore, the center of mass is located 2 units from the y-axis.

Using the same procedure, the vertical location of the system center of mass can be located by determining the torques about the x-axis. Thus, the vertical location of the system center of mass can be calculated by:

$$m_1 y_1 + m_2 y_2 + m_3 y_3 = M y_{cm}$$
$$1(1) + 2(1) + 3(3) = 6 y_{cm}$$
$$y_{cm} = \frac{14}{6}$$
$$y_{cm} = 2.3$$

The center of mass location is 2.3 units from the x-axis. Thus, the center of mass of this total system is (2, 2.3). In the x-direction, the mass is evenly proportioned between the left and right sides and thus the location of the center of mass is basically in the center. This is not the case in the y-direction. The total mass is not proportioned evenly between the top and bottom because the majority of the mass is located towards the top of the system. As such, the y_{cm} location is closer to the majority of the mass of the system. The location of the system center of mass is represented on Figure 11-9 by the intersection of the dashed lines.

The previous example can be used to generalize the procedure for calculating the total body center of mass location. In the previous example, for either the horizontal or vertical position, the products of the coordinate of the segment center of mass and the segment mass for each segment were added and then divided by the total body mass. In algebraic terms:

$$x_{cm} = \frac{\sum_{i=1}^{n} m_i x_i}{M}$$

where x_{cm} is the horizontal location of the total body center of mass, n is the total number of segments, m_i is the mass of the i-th segment, M is the total body mass, and x_i is the horizontal location of the i-th segment center of mass.

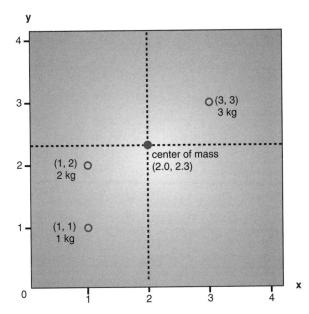

Figure 11.9. Location of the center of mass in a three-point mass system.

Similarly, for the vertical direction:

$$y_{cm} = \frac{\sum\limits_{i=1}^{n} m_i y_i}{M}$$

where y_{cm} is the vertical location of the total body center of mass, n is the total number of segments, m_i is the mass of the i-th segment, M is the total body mass, and y_i is the vertical location of the i-th segment center of mass.

This technique for calculating the total body center of mass is used in many studies in biomechanics. The computation is based on the segment characteristics and the coordinates determined from a kinematic analysis. For most situations, the body is thought of as a 14 segment model (head, trunk, 2 upper arms, 2 lower arms, 2 hands, 2 thighs, 2 legs, and 2 feet). In certain situations in which the actions of the limbs are symmetric, an 8 segment model (head, trunk, upper arm, lower arm, hand, thigh, leg, and foot) will suffice, although the mass of the segments not digitized must be included. As an example of this technique, consider the illustration of the runner in Figure 11-10.

The coordinates of the locations of the segment centers of mass are presented on the illustration of the runner. If it is assumed that the runner has a body mass of 68.2 kg and the segment mass proportions from Dempster (2) are used, the total body center of mass can be calculated. First, the sum for all segments of the product of the segment mass and the x-value for the center of mass of the segment must be calculated. That is:

head = 5.39 * 3.6 = 19.40
trunk = 35.05 * 8.46 = 296.52
left arm = 1.81 * 4.44 = 8.04
right arm = 1.81 * 6.18 = 11.19
left forearm = 1.06 * 3.77 = 3.996
right forearm = 1.06 * 8.49 = 8.999
left hand = 0.41 * 2.50 = 1.03
right hand = 0.41 * 10.91 = 4.47
left thigh = 6.58 * 9.07 = 59.68
right thigh = 6.58 * 12.73 = 83.76
left leg = 3.07 * 8.76 = 26.89
right leg = 3.07 * 16.99 = 52.16
left foot = 0.95 * 10.11 = 9.60
right foot = 0.95 * 19.60 = 18.62

$$\sum\limits_{i=1}^{n} m_i x_i = 604.36$$

The x_{cm} can then be calculated by:

$$x_{cm} = \frac{\sum\limits_{i=1}^{n} m_i x_i}{M} = \frac{604.36}{68.2} = 8.86$$

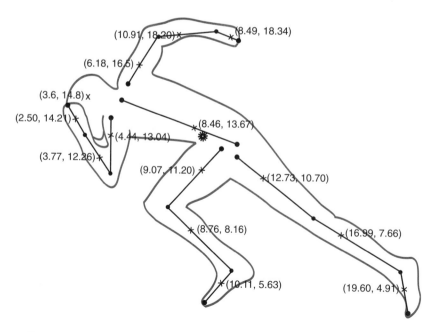

(10.91, 18.20)×   ×(8.49, 18.34)

(6.18, 16.5) ×

(3.6, 14.8) ×

(2.50, 14.2) ×   ×(8.46, 13.67)

×(4.44, 13.04)

(3.77, 12.26) ×

(9.07, 11.20) ×   ×(12.73, 10.70)

(8.76, 8.16) ×   ×(16.99, 7.66)

×(10.11, 5.63)   (19.60, 4.91) ×

Figure 11-10. Segmental centers of mass of a runner at an instant in time. The total body center of mass is designated by an asterisk.

Similarly, calculations can be made for the y_{cm} coordinate. That is:

$$head = 5.39 * 14.8 = 79.77$$
$$trunk = 35.05 * 13.67 = 479.13$$
$$left\ arm = 1.81 * 13.04 = 23.60$$
$$right\ arm = 1.81 * 16.50 = 29.87$$
$$left\ forearm = 1.06 * 12.26 = 12.996$$
$$right\ forearm = 1.06 * 18.34 = 19.44$$
$$left\ hand = 0.41 * 14.21 = 5.83$$
$$right\ hand = 0.41 * 18.20 = 7.46$$
$$left\ thigh = 6.58 * 11.20 = 73.30$$
$$right\ thigh = 6.58 * 10.70 = 70.41$$
$$left\ leg = 3.07 * 8.16 = 25.05$$
$$right\ leg = 3.07 * 7.66 = 23.52$$
$$left\ foot = 0.95 * 5.63 = 5.35$$
$$right\ foot = 0.95 * 4.91 = 4.66$$

$$\sum_{i=1}^{n} m_i x_i = 860.78$$

The y_{cm} can then be calculated by:

$$y_{cm} = \frac{\sum_{i=1}^{n} m_i y_i}{M}$$
$$= \frac{860.78}{68.2}$$
$$= 12.62$$

The total body center of mass coordinates, therefore, are 8.86, 12.62 and are indicated on Figure 11-10 by the asterisk.

Rotation and Leverage

The outcome of a torque is to produce a rotation or pivoting about an axis. If rotations about a fixed point are considered, the effects of the *lever* can be discussed. A lever is a rigid rod that is rotated about a fixed point or axis called the *fulcrum*. A lever consists of a resistance force, an effort force, a bar-like structure, and a fulcrum. In addition, there are two moment or lever arms designated as the effort arm and the resistance arm. The *effort arm* is the perpendicular distance from the line of action of the effort force to the fulcrum. The *resistance arm* is the perpendicular distance from the line of action of the resistance force to the fulcrum. Since both the effort and resistance forces act at a distance from the fulcrum, they create torques about the fulcrum. An anatomical example, such as the forearm segment, can be used to illustrate a lever (Figure 11-11). The long bone of the forearm seg-

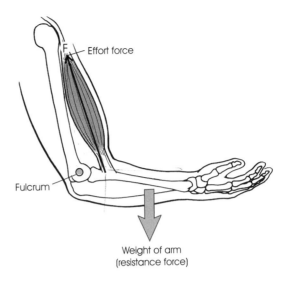

Figure 11.11. An illustration of an anatomical lever.

ment is the rigid bar-like structure and the elbow joint is the fulcrum. The resistance force may be the weight of the segment and possibly an added load carried in the hand or at the wrist. The effort force is produced by the tension developed in the muscles to flex the elbow. Figure 11-12 illustrates several examples of simple machines that are, in effect, different types of levers.

A lever may be evaluated for its mechanical effectiveness by computing its mechanical advantage. *Mechanical advantage* is defined as the ratio of the effort arm to the resistance arm. That is:

$$MA = \frac{effort\ arm}{resistance\ arm}$$

In the construction of a lever, there are three situations that may arise that define the function of the lever. The simplest case is when MA = 1 or when the effort arm equals the resistance arm. In this case, the function of the lever is to alter the direction of motion or balance the lever, but not to magnify either the effort or resistance force. The second case is when the MA > 1, or when the effort arm is greater than the resistance arm. In this case, the torque created by the effort force is magnified by the greater effort arm. Thus, when MA > 1, the lever is said to magnify the effort force. In the third situation, MA < 1, the effort arm is less than the resistance arm. In this case, a much greater effort force is required to overcome the resistance force. The effort force acts over a small distance, however, with the result that the resistance force is moved over a much

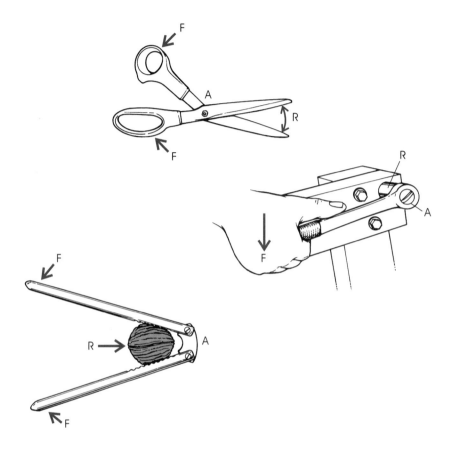

Figure 11-12. Simple machines that are levers.

greater distance in the same amount of time (Figure 11-13). When MA < 1, therefore, velocity or speed of movement is said to be magnified.

There are three classes of levers. In a *first class lever*, the effort force and the resistance force are on opposite sides of the fulcrum. Everyday examples of this lever configuration are the see-saw, the balance scale, and the crow-bar. A first class lever may be configured many different ways and may have a mechanical advantage of 1, greater than 1, or less than 1. First class levers exist in the musculo-skeletal system of the human body. The simultaneous action of the agonist and antagonist muscles acting on opposite sides of a joint create a first class lever. In most instances, however, the first class lever in the human body acts with a mechanical advantage of 1. That is, they act to balance or change the direction of the effort force. An example of the former situation is the action of the splenius muscles acting to balance the head on the

Figure 11-12.(continued) Simple machines that are levers.

atlanto-occipital joint (Figure 11-14). The latter situation, in which the lever acts to simply change the direction of the effort force, is seen in the action of many bony prominences called processes. This type of first class lever is called a *pulley*. One such example is action of the patella in knee joint extension. Here the angle of pull of the quadriceps muscles is altered by the riding action of the patella on the condylar groove of the femur.

In a *second class lever*, the effort force and the resistance force act on the same side of the fulcrum. In this class of lever, the resistance force acts between the fulcrum and the effort force. That is, the resistance force arm is less than the effort arm and thus the mechanical advantage is greater than 1. One example of a second class lever in everyday situations is the wheelbarrow (Figure 11-15). Using the wheelbarrow, effort forces can be applied to act against very significant resistance forces provided by the load carried in the wheelbarrow. There are very few examples of second class levers in the human body, although the act of rising onto the toes is often proclaimed and also disputed as one. This action is used in weight-training and is known as a "calf raise". Since there are so few examples of second class levers in the human body, it is safe to say that humans are not designed to apply great forces via lever systems.

The effort force and the resistance force are also on the same side of the fulcrum in a *third class lever*. In this arrangement, however, the effort force acts between the fulcrum and the line of action of the resistance force. As a result, the effort force arm is less than the resistance force arm and thus the mechanical advantage is less than 1. An example of this type of lever is the shovel when the hand nearest the spade end applies the effort force (Figure 11-16). Therefore, it would appear that a large effort force must be applied to overcome a moderate

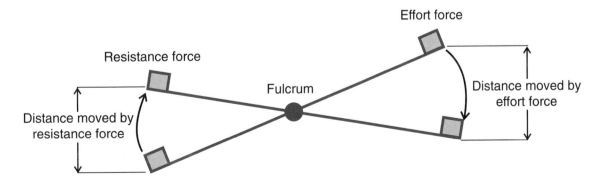

Figure 11-13. A first class lever in which MA < 1, that is, the effort arm is less than the resistance arm. The linear distance moved by the effort force, however, is less than that moved by the resistance force in the same period of time.

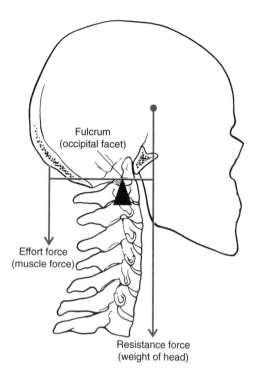

Figure 11-14. An anatomical first class lever in which the weight of the head is the resistance force, the splenius muscles provide the effort force, and the fulcrum is the atlanto-occipital joint.

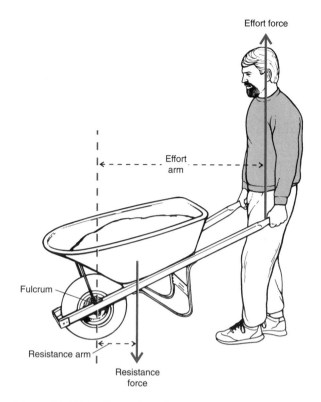

Figure 11-15. An illustration of a wheel-barrow as a second class lever. Note that the resistance force is located between the fulcrum and the effort force. Since the effort arm > resistance arm, MA > 1 and the effort force is magnified.

resistance force. In a third class lever, a large effort force is applied to gain the advantage of increased speed of motion. This is the most prominent type of lever arrangement in the human body, with nearly all joints of the extremities acting as third class levers. It is probably safe to conclude that, from a design standpoint, greater speed of movement exemplified by third class levers appears to be emphasized in the musculo-skeletal system to the exclusion of greater effort force application ability of the second class lever. Figure 11-17 illustrates a third class lever arrangement in the human body.

Moment of Inertia

According to Newton's first law of motion, inertia is an object's tendency to resist a change in velocity. The measure of an object's inertia is its mass. The angular counterpart to mass is the *moment of inertia*. It is a quantity that indicates the resistance of an object to a change in angular motion. Unlike its linear counterpart, mass, the moment of inertia of a body is dependent not only on the mass of the object, but also on the distribution of mass with respect to the axis of rotation. The moment of inertia will also have different values because there are many axes about which

an object may rotate. That is, the moment of inertia is not fixed, but is a changeable quantity. If a gymnast rotating in the air in a lay-out body position is used as an example, the way in which the moment of inertia changes can be illustrated. Suppose the gymnast twists about a longitudinal axis passing through their center of mass. The mass of the gymnast is distributed along and relatively close to this axis (Figure 11-18a). If the gymnast now rotates about a transverse axis through the center of mass, the same mass is distributed much further from the axis of rotation (Figure 11-18b). Since there is a greater mass distribution rotating about the transverse axis than about the longitudinal axis, the moment of inertia is greater in the former case. That is, there is a greater resistance to rotation about the transverse axis than about the longitudinal axis. The gymnast may also alter their mass distribution about an axis by assuming different body positions. They can assume a tuck position, bringing more body mass closer to the transverse axis, thus decreasing the moment of inertia. In multiple aerial somersaults, gymnasts will assume an extreme tuck position by almost placing their head between their knees in an attempt to reduce their moment of inertia. They do

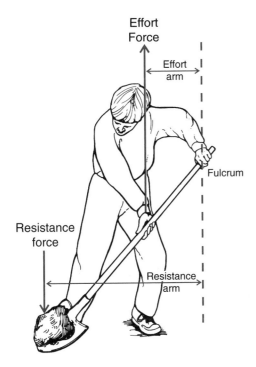

Figure 11-16. An individual using a shovel is a third class lever.

this to provide less resistance to angular acceleration and thus complete the multiple somersaults.

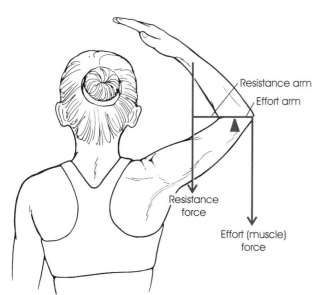

Figure 11-17. The arm held in a flexion position at the elbow is an anatomical third class lever: the resistance force is the weight of the arm, the fulcrum is the elbow joint, and the effort force is provided by the elbow flexor muscles.

The concept of reducing the moment of inertia to enhance angular motion is also seen in running (Figure 11-19). During the swing phase, the foot is brought forward from behind the body to the point on the ground where the next contact will be made. Once the foot leaves the ground, however, the leg flexes considerably at the knee and the foot is raised up close to the buttocks. The effect of this action is to decrease the moment of inertia of the lower extremity relative to a transverse axis through the hip joint. This enables the limb to rotate forward more quickly than would be the case if the lower limb were not flexed. This action is a distinguishing feature of the lower extremity action of sprinters. Figure 11-19 illustrates the change in the moment of inertia of the lower extremity during the recovery action in running.

If all objects are considered to be made up of a number of small particles, each with its own mass and its own distance from the axis of rotation, then the moment of inertia can be represented in mathematical terms by:

$$I = \sum_{i=1}^{n} m_i r_i^2$$

where I is the moment of inertia, n represents the number of particle masses, m_i represents the mass of the i-th particle, and r_i is the distance of the i-th particle from the axis of rotation. That is, the moment of inertia equals the sum of the products of the mass and the distance from the axis of rotation squared of all mass particles comprising the object. A dimensional analysis of the above equation results in units of kilogram-meters squared (kg-m^2) for moment of inertia.

Consider the illustration in Figure 11-20. This hypothetical object is comprised of 5 point masses each with a mass of 0.5 kg. The distances r_1 to r_5 represent the distance from the axis of rotation. The point masses are each 0.1 m apart with the first 0.1 m from the Y-Y axis. Each point mass is 0.1 m from the X-X axis. The moment of inertia about the Y-Y axis is:

$$
\begin{aligned}
I_{Y-Y} &= \sum_{i=1}^{n} m_i r_i^2 \\
&= m_1 r_1^2 + m_2 r_2^2 + m_3 r_3^2 + m_4 r_4^2 + m_5 r_5^2 \\
&= 0.5 \text{ kg} * (0.1 \text{ m})^2 + 0.5 \text{ kg} * (0.2 \text{ m})^2 + 0.5 \text{ kg} * \\
&\qquad (0.3 \text{ m})^2 + 0.5 \text{ kg} * (0.4 \text{ m})^2 + 0.5 \text{ kg} * \\
&\qquad (0.5 \text{ m})^2 \\
&= 0.005 \text{ kg-m}^2 + 0.02 \text{ kg-m}^2 + 0.045 \text{ kg-m}^2 \\
&\qquad + 0.08 \text{ kg-m}^2 + 0.125 \text{ kg-m}^2 \\
&= 0.275 \text{ kg-m}^2
\end{aligned}
$$

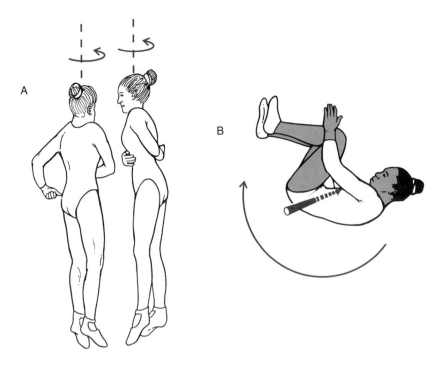

Figure 11-18. The mass distribution of an individual about the longitudinal axis through the total body center of mass (a) and about a transverse axis through the total body center of mass (b).

Fig. 11.19

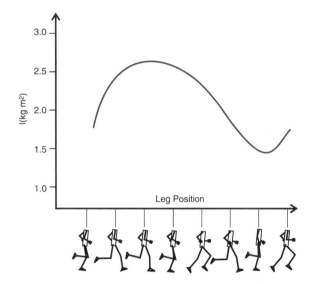

Figure 11-19. A graphical illustration of the changes in the leg's moment of inertia during the stride.

If the axis of rotation is changed to the X-X axis, the moment of inertia of the object about this axis would be:

$$I_{x-x} = \sum_{i=1}^{n} m_i r_i^2$$
$$= m_1 r_1^2 + m_2 r_2^2 + m_3 r_3^2 + m_4 r_4^2 + m_5 r_5^2$$
$$= 0.5 \text{ kg} * (0.1 \text{ m})^2 + 0.5 \text{ kg} * (0.1 \text{ m})^2 + 0.5 \text{ kg} *$$
$$(0.1 \text{ m})^2 + 0.5 \text{ kg} * (0.1 \text{ m})^2$$
$$+ 0.5 \text{ kg} * (0.1 \text{ m})^2$$
$$= 0.005 \text{ kg - m}^2 + 0.005 \text{ kg - m}^2 + 0.005 \text{ kg - m}^2$$
$$+ 0.005 \text{ kg - m}^2 + 0.005 \text{ kg - m}^2$$
$$= 0.025 \text{ kg - m}^2$$

The change in the axis of rotation from the Y-Y axis to the X-X axis thus dramatically reduces the moment of inertia, resulting in less resistance to angular motion about the X-X axis than about the Y-Y axis. If an axis that passes through the geometric center of mass is used, the

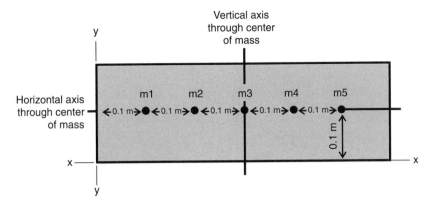

Figure 11-20. A hypothetical five-point mass system.

mass of point 3 would not influence the moment of inertia because the axis passes directly through this point. Thus:

$$I_{cm} = \sum_{i=1}^{n} m_i r_i^2$$
$$= m_1 r_1^2 + m_2 r_2^2 + m_4 r_4^2 + m_5 r_5^2$$
$$= 0.5 \text{ kg} * (0.2 \text{ m})^2 + 0.5 \text{ kg} * (0.1 \text{ m})^2 + 0.5 \text{ kg} *$$
$$(0.1 \text{ m})^2 + 0.5 \text{ kg} * (0.2 \text{ m})^2$$
$$= 0.02 \text{ kg-m}^2 + 0.005 \text{ kg-m}^2 + 0.005 \text{ kg-m}^2$$
$$+ 0.02 \text{ kg-m}^2$$
$$= 0.05 \text{ kg-m}^2$$

From these examples, it should now be clear that the moment of inertia changes according to the axis of rotation.

In the human body, the segments are not as simply constructed as in the above example. Each segment is made up of different tissue types such as bone, muscle, skin, etc., that are not uniformly distributed. The body segments are also irregularly shaped. This means that a segment is not of uniform density and thus it would be impractical to determine the moment of inertia of human body segments using the particle-mass method. Values for the moment of inertia of each body segment have been determined using a number of different methods. Experimentally, moment of inertia values have been obtained in much the same way as the parameters for the center of mass. The values were generated from cadaver studies (2), mathematical modeling (7, 8) and from gamma-scanning techniques (10). Jensen (11) has developed prediction equations specifically for children based on the body mass and height of the child.

It is necessary for a high degree of accuracy to calculate a moment of inertia that is unique for a given individual. Most of the techniques that provide values for segment moments of inertia provide information on the segment *radius of gyration* and from this the moment of inertia may be calculated. The radius of gyration denotes the segment's mass distribution about the axis of rotation and is the distance from the axis of rotation to a point at which the mass can be assumed to be concentrated without changing the inertial characteristics of the segment. Thus, a segment's moment of inertia may be calculated by:

$$I = m(\rho\, l)^2$$

where I is the moment of inertia, m is the mass of the segment, l is the segment length and the Greek letter rho, ρ, is the radius of gyration of the segment as a proportion of the segment length. For example, consider the leg of an individual with a mass and length of 3.6 kg and 0.4 m, respectively. The proportion of the radius of gyration to segment length is 0.302 based on the data of Dempster (2). This information is sufficient to calculate the moment of inertia of the leg about an axis through the center of mass. Thus:

$$I_{cm} = m(\rho_{cm}\, l)^2$$
$$= 3.6 \text{ kg} * (0.302 * 0.4 \text{ m})^2$$
$$= 0.0525 \text{ kg-m}^2$$

The moment of inertia about an axis through the center of mass is 0.0525 kg-m^2. Table 11-2 illustrates the radius of gyration as a proportion of the segment length values from Dempster (2).

Using the radius of gyration technique, the moment of inertia about a transverse axis through the proximal and distal ends of the segment may also be calculated. The

Table 11.2. Radii of gyration as a proportion of segment length about a transverse axis

Segment	Center of Mass	Proximal	Distal
head, neck, trunk	0.503	0.830	0.607
upper arm	0.322	0.542	0.645
arm	0.303	0.526	0.647
hand	0.297	0.587	0.577
thigh	0.323	0.540	0.653
leg	0.302	0.528	0.643
foot	0.475	0.690	0.690

(From Dempster, W.T. Space requirements of the seated operator. WADC Technical Report. pp. 55–159. Wright-Patterson Air Force Base, 1955).

radius of gyration as a proportion of segment length about the proximal end of the leg in the previous example is 0.528. Therefore, the moment of inertia about the proximal end of the segment is calculated as:

$$I_{cm} = m(\rho_{prox}\,l)^2$$
$$= 3.6 \text{ kg} * (0.528 * 0.4 \text{ m})^2$$
$$= 0.161 \text{ kg-m}^2$$

At the distal endpoint of the leg segment, the moment of inertia is:

$$I_{cm} = m(\rho_{dist}\,l)^2$$
$$= 3.6 \text{ kg} * (0.643 * 0.4 \text{ m})^2$$
$$= 0.238 \text{ kg-m}^2$$

The moment of inertia value for any segment is usually given for an axis through the center of mass of the segment. This moment of inertia value is the smallest possible value when compared to any other parallel axis through the segment. For example, in Figure 11-21, three parallel transverse axes are drawn through the leg segment. These axes are through the proximal endpoint, through the center of mass, and through the distal endpoint. Since the mass of the segment is distributed evenly about the center of mass, intuitively then, the moment of inertia about the center of mass axis should be small, and the moments of inertia about the other axes are greater but not equal. This was illustrated in the previous moment of inertia calculations. Since the mass of most segments is distributed closer to the proximal end of the segment, the moment of inertia about the proximal axis is less than about a parallel axis through the distal endpoint.

The moment of inertia, however, can be calculated about any parallel axis, given the moment of inertia about one axis, the mass of the segment, and the perpendicular distance between the parallel axes. This calculation is known as the *Parallel Axis theorem*. Assume the moment of inertia about a transverse axis through the center of mass of a segment is known, and it is necessary to calculate the moment of inertia about a parallel transverse axis through the proximal endpoint. This theorem would state that:

$$I_{prox} = I_{cm} + mr^2$$

where I_{prox} is the moment of inertia about the proximal axis, I_{cm} is the moment of inertia about the center of mass axis, m is the mass of the segment, and r is the perpendicular distance between the two parallel axes. From the calculation in the previous example of the leg, it was

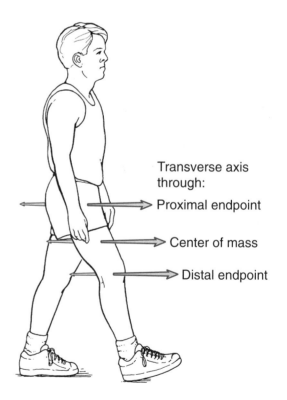

Transverse axis through:

Proximal endpoint

Center of mass

Distal endpoint

Figure 11-21. Parallel transverse axes through the proximal, distal, and center of mass point of the thigh.

determined that the moment of inertia about an axis through the center of mass was 0.0525 kg-m^2. If the center of mass is located 43.3% of the length of the segment from the proximal end, the moment of inertia about a parallel axis through the proximal endpoint of the segment can be calculated. If the length of the segment is 0.4 m, the distance between the proximal end of the segment and the center of mass is:

$$d = 0.433 * 0.4 \text{ m}$$
$$= 0.173 \text{ m}$$

The moment of inertia about the proximal endpoint, then, is:

$$I_{prox} = I_{cm} + mr^2$$
$$= 0.0525 \text{ kg-m}^2 + 3.6\text{kg} * (0.173 \text{ m})^2$$
$$= 0.161 \text{ kg-m}^2$$

This value is the same as was calculated using the radius of gyration/segment length proportion from Dempster's data (2). It can be seen that the moment of inertia about an axis through the center of mass is less than the moment of inertia about any other parallel axis through any other point on the segment.

Angular Momentum

In the previous chapter, linear momentum was defined as the quantity of linear motion. The angular analog of linear momentum is *angular momentum* and is defined as the quantity of angular motion. Since these concepts are analogs, the algebraic formula of angular momentum can be derived from the linear case. If linear momentum is the product of mass and velocity, then angular momentum of any rigid body must be the product of moment of inertia and angular velocity. Thus:

$$H = I\omega$$

where H is the angular velocity, I is the moment of inertia, and ω is the angular velocity. The units for angular momentum can be determined by a unit analysis. That is:

$$H = \text{kg - m}^2 * \text{rad/s}$$
$$H = \frac{\text{kg - m}^2}{\text{s}}$$

Note that radians are dimensionless and disappear from the final unit of angular momentum. Angular momentum is a vector and the direction of the vector is determined by the Right Hand Rule. Once again, counter-clockwise rotations are positive (+) while clockwise rotations are negative (−).

When gravity is the only external force acting on an object, as in projectile motion, the angular momentum generated at take-off remains constant for the duration of the flight. This principle is known as the *conservation of angular momentum*. Angular momentum is conserved during flight because the body weight vector, acting through the total body center of gravity, creates no torque because the moment arm is zero.

Consider the angular momentum of a gymnast performing an aerial somersault about a transverse axis through their total body center of mass (Figure 11-22). The quantity of angular momentum is determined by the torque applied over time at the point of take-off. During the flight phase, angular momentum does not change. The gymnast may manipulate their moment of inertia, however, to spin faster or slower about the transverse axis. At take-off, the gymnast is in a "lay-out" position with a relatively large moment of inertia and a relatively small angular velocity or rate of spin. As the gymnast assumes a tuck position, the moment of inertia decreases and the angular velocity increases accordingly because the quantity of angular momentum is constant. When the gymnast has completed the necessary rotation and is preparing to land, they "open up" or assume a "lay-out" position, increasing their moment of inertia and slowing down their rate of spin. If these actions are done successfully, the gymnast will land on their feet.

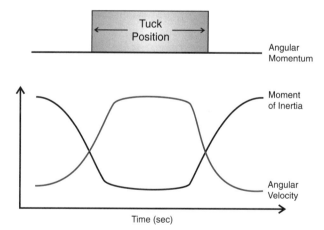

Figure 11-22. The angular momentum, moment of inertia, and angular velocity of a diver completing an aerial somersault. Throughout the aerial portion of the dive, the total body angular momentum of the diver is constant. When the diver is in a lay-out position, the moment of inertia decreases and the angular velocity increases proportionately. During the tuck portion of the dive, the angular velocity increases and the moment of inertia decreases proportionately.

To this point only angular momentum about a single axis has been discussed. Angular momentum about one axis may be transferred to another axis. This occurs in many activities in which the body is a projectile. While the total angular momentum is constant, it may be transferred, for example, from a transverse axis through the center of mass to a longitudinal axis through the center of mass. For example, a diver may twist about the longitudinal and initiate actions to then somersault about the transverse axis. This dive is known as a "full twisting 1 and ½ somersault" dive. Researchers have investigated arm and hip movements to accomplish this change in angular momentum (12, 13, 14, 15). Other activities that use the principles of transferring angular momentum are freestyle skiing and gymnastics.

Rotations may be initiated in midair even when the total body angular momentum is zero. These are called "zero momentum rotations". A prime example of this is the action of a cat when dropped from an upside-down position. The cat initiates a zero momentum rotation and always lands on its feet. As the cat begins to fall, it arches its back or "pikes" to create two body sections, a front and a hind section, and two distinct axes of rotation (Figure 11-23a). The cat's front legs are brought close to its head, decreasing the moment of inertia of the front section and the upper trunk is rotated 180° (Figure 11-23b). The cat extends its hind limbs and rotates the hind section in the opposite direction to counteract the rotation of the front segment. Since the moment of inertia of the hind section is greater than that of the front section, the angular distance that the hind section moves is relatively quite small. To complete the rotation, the cat brings the hind legs and tail into line with its trunk and rotates the back section about an axis through the hind section (Figure 11-23c). The reaction of the front portion of the cat to the hind section rotation is small because the cat creates a large moment of inertia by extending its front legs. Finally, the cat has rotated sufficiently to land upright on its four paws (Figure 11-23d). The use of such actions in sports such as diving and track and field has received considerable attention (12, 16).

During human movement, multiple segments rotate. When this happens, each individual segment has an angular momentum about the segment's center of mass, and also about the total body center of mass. The angular momentum of a segment about its own center of mass is referred to as the *local angular momentum* of the segment. The angular momentum of a segment about the total body center of mass is referred to as the *remote angular momentum* of the segment. A segment's total angular momentum is made up of both local and remote aspects (Figure 11-24). Expressed algebraically:

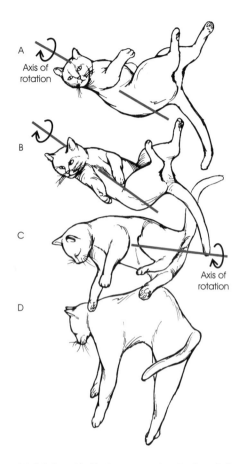

Figure 11-23. A cat initiating a rotation in the air in the absence of an external torque.

$$H_{total} = H_{local} + H_{remote}$$

If the total angular momentum of an individual is calculated, the local aspects of each segment and the remote aspects of each segment must be included. Therefore:

$$H_{total} = \sum_{i=1}^{n} H_{local} + \sum_{i=1}^{n} H_{remote}$$

where i represents each segment and n is the total number of segments.

Local angular momentum is expressed as:

$$H_{local} = I_{cm}\omega$$

where H_{local} is the local angular momentum of the segment, I_{cm} is the moment of inertia about an axis through the segment center of mass, and ω is the angular velocity of the segment about an axis through the segment center of mass. The remote aspect of angular momentum is calculated as:

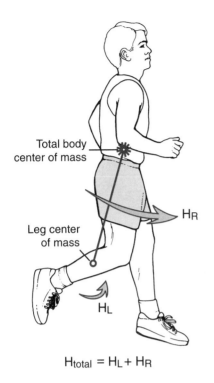

Figure 11-24. An illustration of the local (H_L) and remote (H_R) angular momenta of the leg segment.

$$H_{remote} = md\omega'^2$$

where H_{remote} is the remote angular momentum, m is the mass of the segment, d is the distance from the segment center of mass to the total body center of mass, and ω' is the angular velocity of the segment about an axis through the total body center of mass (17). Figure 11-25 illustrates the proportion of local and remote angular momentum of the total angular momentum in a forward 2 ½ rotation dive. In this instance, the remote angular momentum makes up a greater proportion of the total angular momentum than does the local angular momentum.

This technique for calculating total body angular momentum has been used in a number of biomechanical studies. The sport of diving, for example, has been an area of study concerning the angular momentum requirements of many types of dives. Hamill and associates (18) found that the angular momentum of tower divers increased as the number of rotations required in the dive increased (Figure 11-26). Values as great as 70 kg-m^2/s have been reported for springboard dives (19). The inclusion of a twisting movement with a multiple rotation dive further increases the angular momentum requirements (20).

Hinrichs (21) analyzed the motion of the upper extremities during running by considering angular momentum. He used a three-dimensional analysis to determine angular momentum about the three cardinal axes through the total body center of mass. Hinrichs reported that the arms made a meaningful contribution only about the vertical longitudinal axis. The arms generated alternating positive and negative angular momenta, and tended to cancel out the opposite angular momentum pattern of the legs. These findings are illustrated in Figure 11-27. The upper portion of the trunk was found to rotate in con-

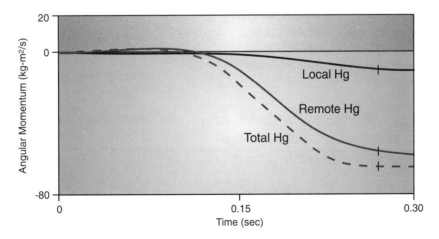

Figure 11-25. The relationship of total body angular momentum, total local angular momentum, and total remote angular momentum of a diver during a forward $2^{1}/_{2}$ rotation dive. The hash-mark denotes the instant of take-off. (after Hamill, J., Ricard, M.D., Golden, D.M. Angular momentum in multiple rotation non-twisting dives. International Journal of Sports Biomechanics 2:78–87, 1986).

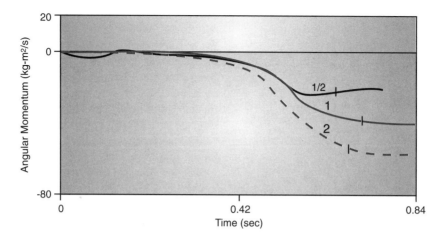

Figure 11-26. Profiles of back dives with $1/2$, 1, and 2 rotations depicting the build-up of angular momentum on the platform. The hash-marks denote the instants of take-off. (after Hamill, J., Ricard, M.D., Golden, D.M. Angular momentum in multiple rotation non-twisting dives. International Journal of Sports Biomechanics 2:78–87, 1986).

junction with the arms while the lower portion of the trunk rotated in conjunction with the legs.

Angular Analogs to Newton's Laws of Motion

The linear case of Newton's laws of motion was presented in the previous chapter. These laws can be restated to represent the angular analogs. It should be noted that each linear quantity has a corresponding angular analog. For example, the angular analog of force is torque, of mass is moment of inertia, of acceleration is angular acceleration. By noting these analogs, they can be directly substituted into the linear laws to create the angular analogs.

First Law

A rotating body will continue in a state of uniform angular motion unless acted on by an external torque.

This law re-states the principle of the conservation of angular momentum. Thus, an object will continue spinning with a constant angular momentum unless acted on by an external torque. As mentioned in the previous section, this principle enables divers and gymnasts, for example, to accomplish aerial maneuvers by manipulating their moments of inertia and angular velocities because they have a constant angular momentum.

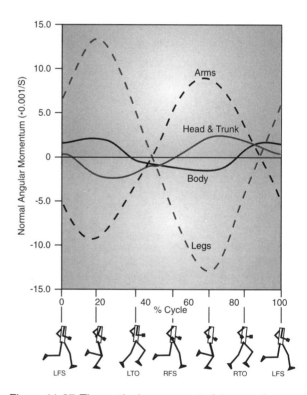

Figure 11-27. The vertical component of the angular momentum of the arms, head and trunk, legs, and total body of a runner at a medium running speed. (after Hinrichs, R.N. Upper extremity function in running. II: Angular momentum considerations. International Journal of Sports Biomechanics 3:242–263, 1987).

Second Law

An external torque will produce an angular acceleration of a body that is proportional to and in the direction of the torque and inversely proportional to the moment of inertia of the body.

This law may be stated algebraically as:

$$T = I\alpha$$

where T is the external torque, I is the moment of inertia of the object, and α is the angular acceleration of the object. For example, if an individual abducts the arm from the body to a horizontal position, the torque at the shoulder results in an angular acceleration of the arm. The greater the moment of inertia of the arm about an axis through the shoulder, the less the angular acceleration of the segment.

The second law may also be expressed to relate torque and angular momentum. The expression of this relationship is once again analogous to the linear case. That is, the sum of the external torques is equal to the time rate of change in angular momentum. That is

$$T = I\alpha$$
$$T = I\frac{\Delta\omega}{\Delta t}$$
$$T = \frac{I\Delta\omega}{\Delta t}$$

where $I\Delta\omega$ is the angular momentum.

Third Law

For every torque exerted by one body on another body, there is an equal and opposite torque exerted by the latter body on the former.

Generally, the torque generated by one body part to rotate that part results in a counter-torque by another body part. In activities such as long jumping in track and field, this concept applies. For example, the long jumper swings their legs forward and upward in preparation for landing. To counteract this lower body torque, the remainder of the body moves forward and downward, producing a torque equal and opposite to the lower body torque. While the torque and counter-torques are equal and opposite, the angular velocity of these two body portions is different because the moments of inertia are different.

When equal torques are applied to different bodies, however, the resulting angular acceleration may not be the same because of the difference in the moments of inertia of the respective bodies. That is, the effect of one body on another may be greater than the latter body on the former. For example, when a second baseman in

baseball executes a pivot during a double play, they jump into the air and throw the ball to first base. In throwing the ball, the muscles in the second baseman's throwing arm create a torque as the arm follows through. The rest of the body must then counter this torque. These torques are equal and opposite but have a much different effect on the respective segments. While the arm undergoes a large angular acceleration, the body angularly accelerates much less because the moment of inertia of the body is greater than that of the arm.

Special Torque Applications

In angular motion, there are special torque applications comparable to the force applications in the linear case. Many of the angular applications are analogs of the linear case and have similar definitions.

Angular Work

Mechanical *angular work* is defined as the product of the magnitude of the torque applied against an object, and the angular distance that the object rotates in the direction of the torque while the torque is being applied. Expressed algebraically:

$$\text{Angular work} = T * \Delta\theta$$

where T is the torque applied and $\Delta\theta$ is the angular distance. Since torque has units of N-m and angular distance has units of radians, the units for angular work are newton-meters (N-m) or joules (J), the same units as in the linear case. For example, if a 40.5 N-m torque is applied over a rotation of 0.79 radians, the work done in rotation is:

$$\begin{aligned} \text{Work} &= T * d\theta \\ &= 40.5 \text{ N-m} * 0.79 \text{ rad} \\ &= 32.0 \text{ N-m} \end{aligned}$$

Thus 32.0 N-m of work is said to be done by the torque, T.

When a muscle contracts and produces tension to move a segment, a torque is produced at the joint and the segment is moved through some angular displacement. Mechanical angular work is done by the muscles that rotated the segment. To differentiate between the kinds of muscle actions, however, angular work done by muscles is characterized as either positive or negative work. *Positive work* is associated with concentric muscle actions or actions in which the muscle is shortening as it creates tension. For example, if a weight-lifter performs a biceps curl on a barbell, the phase in which they flex at the elbows to bring the barbell up is the concentric phase (Figure 11-28a). During this motion, the flexor muscles of the weight-lifter do work on the barbell. *Negative work*, on the other hand, is associated with eccentric muscle actions or actions in which the muscle is lengthening as it creates tension. In a biceps curl, when the weight-lifter is lowering the barbell, resisting the pull of gravity, the flexor muscles are doing negative work (Figure 11-28b). In this instance, the barbell is doing work on the muscles. While it has been found that positive work requires a greater metabolic expenditure than negative work, no direct relationship has been reported

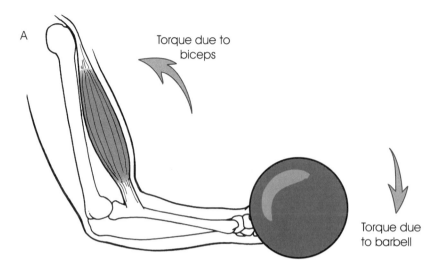

Torque due to biceps > torque due to barbell

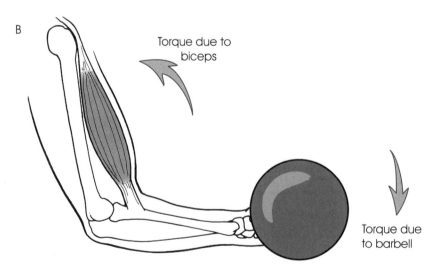

Torque due to biceps < torque to barbell

Figure 11-28. An illustration of positive muscular work (a) and negative muscular work (b) during a biceps curl of a barbell.

between mechanical work of muscles and physiological work.

Energy

In the previous chapter, translational kinetic energy was defined in terms of mass and velocity. Rotational kinetic energy may also be defined similarly, using the angular analogs of mass and velocity, moment of inertia and angular velocity. Thus, *rotational kinetic energy* is defined algebraically as:

$$RKE = \frac{1}{2} I\omega^2$$

where RKE is the rotational kinetic energy, I the moment of inertia and ω the angular velocity. Thus, when the total energy of a system is defined, the rotational kinetic energy must be added to the translational kinetic energy and the potential energy. Total mechanical energy is therefore defined as:

Total Energy = Translational Kinetic Energy + Potential Kinetic Energy + Rotational Kinetic Energy

or

TE = TKE + PE + RKE

In the discussion on angular kinematics, it was noted that single segments undergo large angular velocities during running. Thus, if a single segment was considered, it might be intuitively expected that the rotational kinetic energy might influence the total energy of the segment more significantly than the translational kinetic energy. Winter et al. (22), for example, hypothesized that angular contributions of the leg would be important to the changes in total segment energy in running. Williams (23) offered the following example to illustrate that this is not the case.

Consider the illustration of a theoretical model of a lower extremity segment undergoing both translational and rotational movements in Figure 11-29. If the leg segment is considered, the linear velocity of the center of mass of the leg is:

$$v_{LEGcm} = \omega r$$

where ω is the angular velocity of the leg and r is the distance from the knee to the center of mass of the leg. Using the following body segment values of the leg: moment of inertia = 0.0393 kg-m^2, mass = 3.53 kg, and r = 0.146 m, the magnitude of the rotational kinetic energy is:

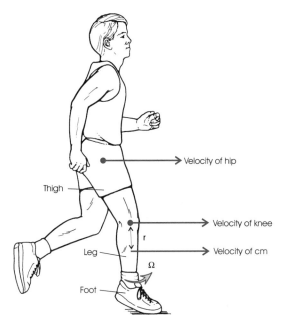

Figure 11-29. Theoretical representation of the translational and rotational velocities of the thigh and leg during running (after Williams, K.R. A Biomechanical and Physiological Evaluation of Running Efficiency. Unpublished doctoral dissertation, The Pennsylvania State University, 1980).

$$RKE = \frac{1}{2} I\omega^2$$
$$= 0.05 * 0.0393 * \omega^2$$
$$= 0.0192 \, \omega^2$$

The translational kinetic energy of the leg under the same circumstances is:

$$TKE = \frac{1}{2} mv^2$$

where m and v are the mass and the linear velocity of the leg. Substituting the expression, ωr, into this equation for the linear velocity, v, the equation becomes:

$$TKE = \frac{1}{2} m(\omega r)^2$$
$$= 0.5 * 3.53 * (0.146 * \omega)^2$$
$$= 0.0376 \, \omega^2$$

Since ω^2 is the same in both answers, the two types of energy can be evaluated, based on the values of 0.0192 times ω^2 for the rotational energy, and 0.0376 times ω^2 for the translational energy. Thus, the translational kinetic energy is almost double that of the rotational kinetic energy. In fact, Williams stated that the rotational kinetic energy of most segments will be much less than the translational kinetic energy. Figure 11-30 illustrates the relationship between the energy components of the foot during a walking stride. In this figure, the magnitude of the total energy is almost completely made up of the translational kinetic energy.

Angular Power   In the previous chapter, power was defined in the linear case as the work done per unit time or the product of force and velocity. *Angular power* may be similarly defined as:

$$\text{Power} = \frac{dW}{dt}$$

where dW is the angular work done and dt is the time in which the work was done. Angular power may also be defined as in the linear case, using the angular analogs of force and velocity, torque and angular velocity. Angular power is the angular work done per unit time and is calculated as the product of torque and angular velocity:

$$\text{Angular Power} = T * \omega$$

where T is the torque applied in N-m and ω is the angular velocity in rad/s. Angular power thus has units of N-m/s or watts. The concept of angular power is often used to describe mechanical muscle power. Muscle power is determined by calculating the net torque of the muscle acting across the joint and the angular velocity of the joint. This computation will be expanded upon in the next chapter.

Summary

Torque or the moment of force results when the line of action does not pass through the center of mass of an object. Torques cause rotational motion of an object about an axis. Torques are vectors and thus must be considered in terms of magnitude and direction. The Right Hand Rule defines whether the torque is positive (counter-clockwise rotation) or negative (clockwise rotation).

The concept of torque can be used to define the center of mass of an object. The sum of the torques about the center of mass of an object equals zero. That is:

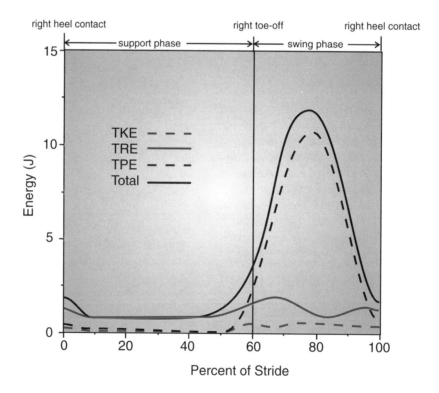

Figure 11-30. The relationship between total energy (TOTAL), rotational kinetic energy (RKE), translational kinetic energy (TKE), and potential energy (PE) of the foot during a walking stride.

$$\Sigma T = 0$$

This relationship defines the center of mass as the balancing point of the object.

The center of mass may be calculated using either the reaction board method or the segmental method. The reaction board method can only be used in circumstances in which the object is static. The use of the segmental method requires the computation of body segment parameters, such as the location of the center of mass and the segment proportion of the total body mass. The location of a segment center of mass requires the coordinates of both the proximal and distal ends of the segment and is defined as:

$$s_{cm} = s_{prox} - (\text{ds} * \% \text{ length from segment's proximal end})$$
$$= s_{prox} - ((s_{prox} - s_{distal}) * \%L)$$

where s_{cm} is the location of the segment center of mass, s_{prox} is the coordinate point of the proximal end of the segment, s_{distal} is the coordinate point of the distal end of the segment, ds is the length of the segment, and %L is the location of the segment center of mass as a proportion of the segment length from the proximal end of the segment. Locating the total body center of mass with the segmental method also uses the concept that the sum of the torques about the center of mass is zero. This computation uses the formula:

$$s_{cm} = \frac{\sum_{i=1}^{n} m_i s_i}{M}$$

where s_{cm} is the location of the total body center of mass, n is the number of segments, m_i is the mass of the i-th segment, s_i is the location of the segment center of mass, and M is the total body center of mass.

A lever is a simple machine that has a balancing point called the fulcrum, and two forces—an effort force and a resistance force. The mechanical advantage (MA) of a lever is defined as the ratio of the effort arm to the resistance arm. Levers can either magnify force (MA > 1), magnify speed of rotation (MA < 1), or change the direction of pull (MA = 1). There are three classes of levers based on the relationship of the effort and resistance forces to the fulcrum. In the human body, however, the third class lever magnifying the speed of movement predominates. Most of the levers in the extremities are third class levers.

Moment of inertia is the angular analog of mass. The moment of inertia of an object is the resistance to angular motion and depends on the axis about which the object is rotating. The calculation of the segment moment of inertia can be calculated using the radius of gyration as:

$$I_{cm} = m(\rho_{cm} L)^2$$

where I_{cm} is the moment of inertia about a transverse axis through the center of mass, m is the mass of the segment, ρ_{cm} is a proportion describing the ratio of the radius of gyration about the center of mass to the segment length, and L is the segment length.

Angular momentum is the angular analog of linear momentum and refers to the quantity of rotation of an object. The angular momentum of a multi-segment body must be understood in terms of local and remote angular momentum. Local angular momentum is the angular momentum of a segment about its own center of mass. The remote angular momentum is the angular momentum of a segment about the total body center of mass. The total segment angular momentum is the sum of the segment's local and remote aspects. The total body angular momentum is the sum of the local and remote aspects of all segments. The angular analog of Newton's first law of motion is a statement of the law of conservation of angular momentum.

Newton's Laws of Motion can be re-stated from the linear case to their angular analogs. The angular analogs of these laws are:

1) A rotating body will continue in a state of uniform angular motion unless acted on by an external torque.

2) An external torque will produce an angular acceleration of a body that is proportional to and in the direction of the torque, and inversely proportional to the moment of inertia of the body.

3) For every torque exerted by one body on another body, there is an equal and opposite torque exerted by the latter body on the former.

There are special torque applications that include angular work, rotational kinetic energy, and angular power. These concepts may be developed by substituting the appropriate angular equivalent in the linear case. Angular work is defined as:

$$\text{Angular work} = T * d\theta$$

where T is the torque applied and $d\theta$ is the angular distance over which the torque was applied. Angular work can be used to define the action of muscles as concentric (positive angular work) or eccentric (negative angular work). Rotational kinetic energy of a segment is defined as:

$$\text{RKE} = \frac{1}{2} I\omega^2$$

where I is the moment of inertia of the segment about its center of mass and ω is the angular velocity of the segment. Angular power is defined as either the rate of doing angular work:

$$\text{Angular Power} = \frac{dW}{dt}$$

where W is the mechanical work done and dt is the time period over which the work is done or the product of torque and angular velocity:

$$\text{Angular Power} = T * \omega$$

where T is the torque and ω is the angular velocity.

Review Questions

1. What type of motion does a force exerted through the center of mass cause? If the force is directed through the axis of rotation, what type of motion does it cause?
2. If a force of 110 N is exerted at a distance of 0.3 m from the axis of rotation, what is the resulting moment of force? (Answer: 33.0 N-m).
3. If a force of 200 N acting 0.34 m from the axis of rotation is balanced by another force of 185 N, what is the moment arm of the second force? (Answer: 0.37 m).
4. There are several techniques for doing sit-ups. Using the concept of torque, explain how one could change the technique to make a sit-up more difficult.
5. Explain a diver's movements in completing a 1½ somersault dive in terms of their moment of inertia.
6. Sketch the relative values of both angular velocity and moment of inertia of a diver accomplishing a 1½ somersault dive.
7. Give examples of household devices which utilize the concepts of first, second, and third class levers.
8. What is the function of a lever with a mechanical advantage = 1, > 1, or < 1? Illustrate each example using a first class lever arrangement.
9. Calculate the rotational energy of a segment given: mass of the segment = 3.72 kg; moment of inertia = 0.065 kg-m^2; ω = 5.26 rad/s. (Answer: 0.90 J).
10. If the net moment at a joint is -33.92 N-m and the angular velocity at the same instant in time is -2.46 rad/s, what is the angular power at that joint? (Answer: 83.44 W).

Additional Questions

1. Explain the force applications necessary to accomplish:
   a) a rotation about the longitudinal axis
   b) a rotation about the transverse axis including a degree of translation
   c) when both a) and b) occur simultaneously

2. An individual is seated while doing a knee flexion-extension exercise on a weight machine. Sketch the thigh and leg at the following instants in time: a) flexed at 90°; b) flexed at 45°; c) completely extended. Draw in the moment arms for all torques in this system at each instant in time.

3. Calculate the center of mass of the following 3-point system with masses of 2 kg, 3.5 kg, and 1.25 kg located respectively at the coordinates (1, 2), (2.2, 3.5), and (4,3). (Answer: 2.18, 2.96).

4. What are the coordinates of the center of mass of a segment with the proximal end (2.3, 4.1) and the distal end (3.5, 5.2) if the center of mass is located 39% from the proximal end of the segment? (Answer: 2.77, 4.53).

5. Calculate the center of mass in each segment given the information in the following table. Assume that the proximal and distal endpoints define each segment and that the top of the spine and crotch define the trunk.

Body Segment	% of total body mass	Segment CM location
head and neck	7.90	0.50 from base of neck
trunk	51.40	0.45 from crotch
arm	2.65	0.436 from proximal end
forearm	1.55	0.43 from proximal end

(Answer: head and neck – 8.0, 15.3; trunk – 11.0, 10.55; left arm – 11.17, 13.78; left forearm – 14.25, 15.45; left hand – 17.30, 16.40; right arm – 10.39, 14.40; right forearm –- 11.24, 17.79; right hand –12.05, 20.96; left thigh – 14.50, 8.35; left leg – 18.53, 9.36; left foot – 20.88, 12.65; right thigh – 10.47, 7.87; right leg – 7.12, 4.76; right foot – 4.49, 1.99).

6. Calculate the individual segment masses for the table in question #5 if the individual has a body mass of 55 kg. (Answer: head and neck – 4.35 kg; trunk– 28.27 kg; arm – 1.46 kg; foreman – 0.85 kg; hand – 0.33 kg; thigh – 5.31 kg; leg – 2.48 kg; foot – 0.77 kg).

7. Calculate the total body center of mass of the individual using the individual segment data from question #5 if they have a body mass of 55 kg. (Answer: 11.37, 10.53).

8. What is the moment of inertia of a segment about a transverse axis through the center of mass with a length of 0.43 m, a mass of 3.72 kg, and a radius of gyration/segment length of 0.302? (Answer: 0.063 kg-m^2).

9. If the center of mass of the segment in the previous question is 43.3% of the length of the segment from the proximal end, what would the relative magnitudes of the moments of inertia be about a transverse axis through: a) the proximal end of the segment; and b) distal end of the segment? (Answer: a) 0.19 kg-m^2; b) 0.28 kg-m^2).

10. A torque of 35 N-m results in a rotation through 0.46 radians in 0.7 s. How much: a) angular work was done; and b) what was the power developed? (Answer: a) 16.1 J; b) 23.0 W).

Additional Reading

1. Frolich, C. The physics of somersaulting and twisting. Scientific American, 242:154-164, 1980.
2. Townend, M.S. Mathematics in Sport. London: Ellis Horwood Limited, 1984.
3. Tricker, R.A.R. and Tricker, B.J.K. The Science of Movement. New York: American Elsevier Publishing Company, Inc., 1967.
4. Yeadon, M.R. : Theoretical models and their application to aerial movement. In B. van Gheluwe and J. Atha (eds.). Current Research In Sports Biomechanics. pp. 86-106. New York: Karger, 1987.

References

1. Hay, J.G. Straddle or Flop. Athletic Journal 55:83-85, 1975.
2. Dempster, W.T. Space requirements of the seated operator. WADC Technical Report. pp. 55-159. Wright-Patterson Air Force Base, 1955.
3. Clauser, C.E., McConville, J.T., Young, J.W. Weight, volume, and center of mass of segments of the human body. AMRL Technical Report. pp. 69-70. Wright-Patterson Air Force Base, 1969.
4. Chandler, R.F., Clauser, C.E., McConville, J.T., Reynolds, H.M., Young, J.W. Investigation of inertial properties of the human body. AMRL Technical Report. pp. 74-137. Wright-Patterson Air Force Base, 1975.
5. Miller, D.I. and Morrison, W.E. Prediction of segmental parameters using the Hanavan human body model. Medicine and Science in Sports 7:207-212, 1975.
6. Amar, J. The Human Motor. London: G. Routledge and Sons, 1920.
7. Hanavan, E.P. A mathematical model of the human body. AMRL Technical Report 64-102, Wright-Patterson Air Force Base, 1964.
8. Hatze, H. A mathematical model for the computational determination of parameter values of anthropomorphic segments. Journal of Biomechanics 13:833-843, 1980.
9. Hatze, H. Estimation of myodynamic parameter values from observations on isometrically contracting muscle groups. European Journal of Applied Physiology 46:325-338, 1981.
10. Zatsiorsky, V. and Seluyanov, V. The mass and inertia characteristics of the main segments of the human body. In H. Matsui and K. Kobayashi (eds.). Biomechanics VIII-B. pp. 1152-1159, Champaign, IL: Human Kinetics Publishers, 1983.
11. Jensen, R.K. The growth of children's moment of inertia. Medicine and Science in Sports and Exercise 18:440-445, 1987.
12. Frolich, C. Do springboard divers violate angular momentum conservation? American Journal of Physics 47:583-592,1979.
13. van Gheluwe, B. A biomechanical simulation model for airborne twists in backward somersaults. Human Movement Studies 7:1-22, 1981.
14. Yeadon, M.R. and Atha, J. The production of a sustained aerial twist during a somersault without the use of asymmetrical arm action. In D.A. Winter et al. (eds.). Biomechanics IX-B. pp. 395-400, Champaign, IL: Human Kinetics Publishers, 1985.
15. Yeadon, M.R. Twisting techniques used in springboard diving. Proceedings of the First IOC World Congress on Sport Sciences. pp. 307-308, 1989.
16. Dyson, G. The Mechanics Of Athletics. London: University of London Press,1973.
17. Hay, J.G., Wilson, B.D., Dapena, J., Woodworth, G.G. A computational technique to determine the angular momentum of a human body. Journal of Biomechanics 10:269-277, 1977.
18. Hamill, J., Ricard, M.D., Golden, D.M. Angular momentum in multiple rotation non-twisting platform dives. International Journal of Sports Biomechanics 2:78-87, 1986.
19. Miller, D.I. and Munro, C.F.: Body segment contributions to height achieved during the flight of a springboard dive. Medicine and Science in Sports and Exercise 16:234-242, 1984.
20. Sanders, R.H. and Wilson, B.D. Angular momentum requirements of the twisting and non-twisting forward 1 1/2 somersault dive. International Journal of Sports Biomechanics 3:47-53, 1990.
21. Hinrichs, R.N.: Upper extremity function in running. II: Angular momentum considerations. International Journal of Sports Biomechanics 3:242-263, 1987.
22. Winter, D.A., Quanbury, A.O., Reimer, G.D. Analysis of instantaneous energy of normal gait. Journal of Biomechanics 9:253-257, 1976.
23. Williams, K.R. A Biomechanical and Physiological Evaluation of Running Efficiency. Unpublished Doctoral dissertation, The Pennsylvania State University, 1980.

Glossary

angular kinetics:	the branch of mechanics that deals with the causes of rotations.
angular momentum:	the quantity of angular motion determined by the product of the object's angular velocity and its moment of inertia.
angular power:	the time rate of change of angular work determined by the product of the torque and the angular velocity.
angular work:	the product of torque applied to an object and the angular distance over which the torque is applied.
center of gravity:	the point at which all of the body's mass seems to be concentrated; the balance point of the body; the point about which the sum of the torques equals zero.
center of mass:	a balance point of a body; the point about which all of the mass particles of the body is evenly distributed.
conservation of angular momentum:	the concept stating that the angular momentum is constant unless the object is acted on by an external force.
eccentric force:	a force that is not applied through the center of mass of an object.
effort arm:	see moment arm.
effort force:	a force applied to a lever causing movement of the lever.
first class lever:	a lever in which the fulcrum is between the effort force and the resistance force.
force couple:	two forces that are equal in magnitude, act in opposite directions at a distance from an axis of rotation, and produce rotation with no translation.
fulcrum:	the axis of rotation of a lever.
lever:	a mechanism for doing work that consists of a fulcrum and two eccentric forces.
local angular momentum:	the angular momentum of a body segment about its own center of mass.
mechanical advantage:	the ratio of the effort arm to the resistance arm of a lever.
moment:	see torque.
moment arm:	the perpendicular distance of the line of action of a force to the axis of rotation.
moment of force:	see torque.
moment of inertia:	the resistance of a body to angular acceleration.
negative work:	the work done on a system when the loading torque is greater than the torque exerted by the muscle.

parallel axis theorem:	a theorem stating the relationship between the moment of inertia about an axis through the body's center of mass (I_{cm}) and any other parallel axis (I_{axis}) such that:

$$I_{axis} = I_{cm} + mr^2$$

where m is the mass of the body and r is the perpendicular distance between the axes.

positive work:	the work done by a system when the torque exerted by a muscle is greater than the torque of the external load.
radius of gyration:	a measure of the distribution of a body's mass about an axis of rotation.
reaction board method:	a static method of calculating the total body center of mass.
remote angular momentum:	the angular momentum of a segment about the total body center of mass.
resistance arm:	see moment arm.
resistance force:	a force that resists the effort force in a lever.
rotational kinetic energy:	the capacity to do angular work; the product of ½ the moment of inertia and the angular velocity squared.
second class lever:	a lever in which the resistance force acts between the fulcrum and the effort force.
segmental method:	a method of calculating the total body center of mass of a multi-segment body by summing the product of the locations of the centers of mass of the segments and the mass of the respective segment and dividing by the total body mass.
third class lever:	a lever in which the effort force acts between the fulcrum and the line of action of the resistance force.
torque:	the product of the magnitude of a force and the perpendicular distance from the line of action of the force to the axis of rotation.

CHAPTER 12

Types of Mechanical Analysis

Student Objectives

After reading this chapter, the student should be able to:

1. Define and conduct a static analysis on a single joint motion.
2. Define stability and discuss its effect on human movement.
3. Define and conduct a dynamic analysis on a single joint motion.
4. Discuss the concept of muscle power.
5. Define the impulse-momentum relationship.
6. Define the work-energy relationship.
7. Discuss the concepts of internal and external work.

In Chapters 8 and 9, it was suggested that human movement may be considered in terms of kinematics. A kinematic approach involved the description of the motion in terms of position, velocity, and acceleration of either individual segments or the total body, without regard to the cause of the movement. In Chapters 10 and 11, the kinetics of motion were discussed. A kinetic approach involves the forces that cause motion. A cause-effect relationship was implied by characterizing kinetics as the cause of motion and kinematics as the product or the effects of the forces. Newton's laws of motion, with particular emphasis on the second law of motion ($F = ma$), provided the link between the cause and effect and essentially form the basis for all further analyses of human movement in biomechanics.

To fully understand the underlying nature of motion, however, the cause of the movement and not merely the description of the outcome must be understood. That is, the forces that cause motion must be fully understood. Although there are multiple conceptual forms that can describe the relationship between the kinematics and the kinetics of a movement, there are three general approaches to this problem that may be used in biomechanical analyses. These approaches can be categorized as: 1) the effect of a force at an instant in time; 2) the effect of a force applied over a period of time; and 3) the effect of a force applied over a distance (1). No one of these methods can be considered better or worse than any other. The choice of which relationship to use simply depends upon which method will best answer the question asked. By using the appropriate analytical technique, however, it should be possible to investigate the forces causing motion more effectively.

The Effects of a Force at an Instant in Time

When considering the effects of a force and the resulting acceleration at an instant in time, Newton's second law of motion must be considered:

$$\Sigma F = ma$$

Two situations, however, based upon the magnitude of the resulting acceleration can be defined. In the first situation, the resulting acceleration will have a zero value. This is the branch of mechanics known as *statics*. In the second case, the resulting acceleration is a non-zero value. This area of study is known as *dynamics*. Using either type of

analysis in the appropriate situation, both the forces acting at a joint and the moments about the joint caused by
the muscles may be determined.

Static Analysis

The static case is devoted to systems at rest or systems
moving at a constant velocity. In both of these situations,
the acceleration is zero. When the acceleration of a system is zero, the system is said to be in *equilibrium*. A system is in equilibrium when, as stated in Newton's first
law, it remains at rest or it is in motion at a constant
velocity.

In translational motion, when a system is in equilibrium, all forces that are acting on the system have their
combined effect cancelled and the resultant is zero. That
is, the sum of all forces acting on the system must total
zero. This is expressed algebraically as:

$$\Sigma F_{system} = 0$$

The forces in this equation can be further expressed in
terms of the 2-dimensional x-y components as:

$$\Sigma F_x = 0$$

and

$$\Sigma F_y = 0$$

Here the sum of the forces in the horizontal (x) direction
must equal zero and the sum of the forces in the vertical
(y) direction must equal zero.

The static case is simply a particular example of
Newton's second law and can be described in terms of a
cause-effect relationship. Note that the left hand side of
these equations describe the cause of the motion and the
right hand side the product or the result of the motion.
Because all forces in the system are in balance, there is
no acceleration. If the forces were not in balance, some
acceleration would occur.

Figure 12-1 presents a free body diagram of a *linear
force system* in which a 100 N box rests on a table.
Gravity acts to pull the box downward on the table with a
force of 100 N. Because the box does not move vertically, an equal and opposite force must act to support the
box. A coordinate system will be defined in which the
upward direction is positive and the downward direction
is negative. The weight of the box, acting downward, will
thus be given a negative sign. There are no horizontal
forces acting in this example. Thus:

$$\Sigma F_y = 0$$
$$-100 \text{ N} + R_y = 0$$
$$R_y = 100 \text{ N}$$

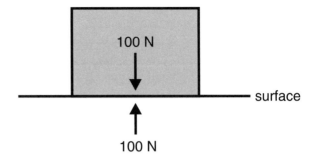

Figure 12-1. A free body diagram of a box on a table. The
box is in equilibrium because there are no horizontal
forces and the sum of the vertical forces is zero.

where R_y is the reaction force to the weight of the box.
Since the weight of the box acts negatively, the reaction
force must act in a positive direction or in the opposite
direction to the weight of the box.

Consider a situation in which there is a system with
multiple forces acting. In Figure 12-2, a tug-of-war is
presented as a linear force system. In this example, the
two contestants on the right balance the three contestants
on the left. The contestants on the left exert forces of 50
N, 150 N, and 300 N, respectively. These forces can be
considered to act in a negative horizontal direction.
Assuming a static situation, the reaction force to produce
equilibrium can be calculated. Thus:

$$\Sigma F_x = 0$$
$$-50 \text{ N} - 150 \text{ N} - 300 \text{ N} + R_x = 0$$
$$R_x = 50 \text{ N} + 150 \text{ N} + 300 \text{ N}$$
$$R_y = 500 \text{ N}$$

The two contestants on the right must exert a reaction
force of 500 N in the positive direction for a state of equilibrium to exist.

The linear force system previously presented is a relatively simple examples of the static case, but in many
instances in human motion, the forces are non-parallel. In
Figure 12-3a, there are two non-parallel forces acting
upon a rigid body, in addition to the weight of the
rigid body. For this system to be in equilibrium, a third
force must act through the intersection of the two non-
parallel forces. The free body diagram in Figure 12-3b
illustrates that the horizontal component of F_3, acting in
a positive direction, must counterbalance the sum of the
horizontal components of the non-parallel forces F_1 and
F_3. Also, the vertical components F_1, F_3, must be coun-
terbalanced by the weight of the rigid body and by the
vertical component of F_3. If F_1 = 100 N, the components of F_1 are:

Figure 12-2. Individuals in a tug-of-war. The system is in equilibrium because the sum of the forces in the horizontal direction is zero. No movement to the left or the right can occur.

$F_{1x} = F_1 \cos 30°$
$\quad = 100\ N * \cos 30°$
$\quad = 86.6\ N$
$F_{1y} = F_1 \sin 30°$
$\quad = 100\ N * \sin 30°$
$\quad = 50.0\ N$

and if $F_2 = 212.13\ N$, the components of F_2 are:

$F_{2x} = F_2 \cos 45°$
$\quad = 212.13\ N * \cos 45°$
$\quad = 150\ N$
$F_{2y} = F_2 \sin 45°$
$\quad = 212.13\ N * \sin 45°$
$\quad = 150\ N$

In the vertical direction, F_{1y} and F_{3y} are considered to act upwards in a positive direction and F_{2y} to act down-

A

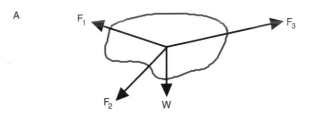

B

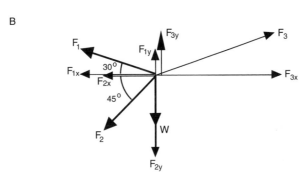

Figure 12-3. A force system in which the sum of the forces is zero (a). A free body diagram of the force system shows the horizontal and vertical components of all forces (b).

wards in a negative direction. The weight of the rigid body, 50 N, also acts in a negative vertical direction. Thus:

$$\Sigma F_y = 0$$
$$F_{3y} + F_{1y} - F_{2y} - W = 0$$
$$F_{3y} = -F_{1y} + F_{2y} + W$$
$$F_{3y} = -50 \text{ N} + 150 \text{ N} + 50 \text{ N}$$
$$F_{3y} = 150 \text{ N}$$

F_{3y} must have a magnitude of 150 N to maintain the system in equilibrium in the vertical direction. In the horizontal direction, F_{3x} is considered to act in a positive direction to the right and F_{1x} and F_{2x} to act in a negative direction to the left. Thus:

$$\Sigma F_x = 0$$
$$F_{3x} - F_{1x} - F_{2x} = 0$$
$$F_{3x} - 86.6 \text{ N} - 150 \text{ N} = 0$$
$$F_{3x} = 86.6 \text{ N} + 150 \text{ N}$$
$$F_{3x} = 236.6 \text{ N}$$

To balance the two non-parallel forces in the horizontal direction, a force of 236.6 N is required. The resultant force F_3 can be determined using the Pythagorean Relationship as:

$$F_3 = \sqrt{F_{3x}^2 + F_{3y}^2}$$
$$F_3 = \sqrt{236.6^2 + 150^2}$$
$$F_3 = 280 \text{ N}$$

The forces F_1, F_2, and the weight of the rigid body, are counteracted by the force F_3, keeping the total system in equilibrium.

A second condition that determines if a system is in equilibrium occurs when the forces in the system are not concurrent. *Concurrent forces* do not coincide at the same point and thus cause rotation about some axis. These rotations, all sum to zero, however, and as this is a static case, no rotation occurs. That is, the sum of the moments of force in the system must sum to equal zero. Stated algebraically, therefore:

$$\Sigma M_{system} = 0$$

Previously, a convention was suggested in which moments causing a counter-clockwise rotation were considered to be positive, while clockwise torques were considered negative. To satisfy this condition of equilibrium, therefore, the sum of the counter-clockwise moments must equal the sum of the clockwise moments and no angular acceleration can occur.

Consider the free body diagram in Figure 12-4 in which a first class lever system is described. On the left side of the fulcrum, the individual A weighs 670 N and is 2.3 m from the axis of rotation. This individual would cause a counter-clockwise or positive moment. Individual B on the right side weighs 541 N and is 2.85 m from the fulcrum, and would cause a clockwise or a negative moment. For this system to be in equilibrium, the clockwise moment must equal the counter-clockwise moment. Thus,

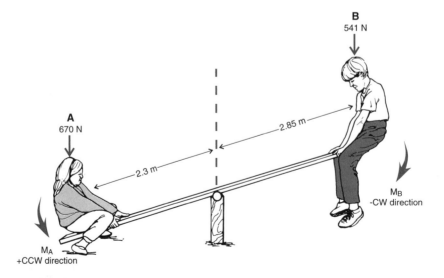

Figure 12-4. A first class lever—a see-saw.

$$M_A = 670 \text{ N} * 2.3 \text{ m}$$
$$M_A = 1541 \text{ N-m}$$

and

$$M_B = 541 \text{ N} * 2.85 \text{ m}$$
$$M_B = 1541 \text{ N-m}$$

For this system to be in equilibrium:

$$\Sigma M = 0$$
$$M_A - M_B = 0$$
$$M_A = M_B$$

Since M_A is positive and M_B is negative and their magnitudes are equal, these moments cancel each other and no rotation occurs. When the individual on the left pushes up and with no further external forces, this system will come to rest in a balanced position.

In many situations in the static analysis of human movement, however, there are multiple torques acting in the system. In Figure 12-5a, the diagram of the forearm of an individual holding a barbell is illustrated. To determine the action at the elbow joint, it would be helpful to know the moment caused by the muscles about the elbow joint. If the elbow joint is considered to be the axis of rotation, there are two negative moments in this system. One negative moment is a result of the weight of the forearm and hand acting through the center of mass of the forearm-hand system, and the other is the result of the weight of the barbell. The counter-clockwise or positive moment is a result of the muscle force acting across the elbow joint. The net moment of the muscles must be equal to the two negative moments for the system to be in equilibrium. Thus,

$$\Sigma M = 0$$
$$M_{Fm} - M_{arm-hand} - M_{barbell} = 0$$

Consider the free body diagram of this system illustrated in Figure 12-5b. The muscle moment that is necessary to keep the system in equilibrium can be calculated given the information in the free body diagram. The forearm-hand complex weighs 45 N and the center of mass is located 0.15 m from the elbow joint. The weight of the barbell is 420 N and the center of mass of the barbell is 0.40 m from the elbow joint. The moment due to the weight of the arm-hand is:

$$M_{arm-hand} = (45 \text{ N} * .15 \text{ m}) = 6.75 \text{ N-m}$$

and the moment due to the barbell is:

$$M_{barbell} = (420 \text{ N} * .40 \text{ m}) = 168.0 \text{ N-m}$$

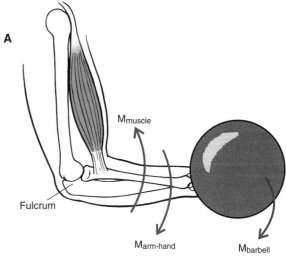

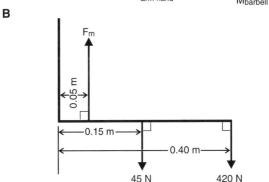

Figure 12-5. An illustration of the forearm during an instant in time of a biceps curl (a) and a free body diagram of the system (b).

The moment due to the muscle force can then be calculated:

$$M_{Fm} - M_{arm-hand} + M_{barbell} = 0$$
$$M_{Fm} - (45 \text{ N} * .15 \text{ m}) - (420 \text{ N} * .40 \text{ m}) = 0$$
$$M_{Fm} = (45 \text{ N} * .15 \text{ m}) + (420 \text{ N} * .40 \text{ m})$$
$$M_{Fm} = 6.75 \text{ N-m} + 168.0 \text{ N-m}$$
$$M_{Fm} = 174.75 \text{ N-m}$$

The muscle must create a torque of 174.75 N-m to counteract the weight of the forearm-hand and the weight of the barbell. Note that this muscle moment cannot be directly attributed to any one muscle that crosses the joint. The muscle moment calculated is the net sum of all muscle actions involved. In this case, since the arm is being held in flexion, it can be stated that the net muscle moment is due to the action of the elbow flexors but it cannot be said exactly which elbow flexors are most involved. In fact, it might be surmised that since this is a static posture, there may be considerable co-contraction occurring.

To determine the muscle action, the net torque about the joint must be considered. The net torque is the sum of all torques acting at the joint, in this case the elbow. Thus:

$$M_{elbow} = M_{Fm} - M_{arm-hand} - M_{barbell}$$
$$M_{elbow} = 174.75 - 6.75 - 168.0$$
$$M_{elbow} = 0$$

The net moment about the elbow joint is zero, indicating that the muscle action must be isometric.

If the moment arm of the elbow flexors was estimated to be a distance of 0.05 m from the elbow joint, the muscle force would be:

$$F_m = \frac{M_{Fm}}{0.05 \text{ m}}$$
$$F_m = \frac{174.75 \text{ N-m}}{0.05 \text{ m}}$$
$$F_m = 3495 \text{ N}$$

It can be seen that the muscle force must be considerably greater than the other two forces because the moment arm for the muscle is very small compared to the moment arms for the forearm-hand or the barbell.

Consider Figure 12-6. In this free body diagram, the arm is placed in a similar posture as in Figure 12-5. The barbell held in the hand, however, now weighs 100 N, and the measured muscle torque at the elbow is 180 N-m. All other measures in this situation are the same as in the previous example. Thus, evaluating the net moment at the elbow:

$$M_{elbow} = M_{Fm} + M_{arm-hand} + M_{barbell}$$
$$M_{elbow} = 180 \text{ N-m} + (45 \text{ N} * .15 \text{ m}) +$$
$$\qquad (100 \text{ N} * .40 \text{ m})$$
$$M_{elbow} = 180 \text{ N-m} - 6.75 \text{ N-m} - 40.0 \text{ N-m}$$
$$M_{elbow} = 133.25 \text{ N-m}$$

Since the net moment at the elbow is positive, the rotation is counter-clockwise, and the muscle action, therefore, is a flexor action.

In the previously described situations, the arm was parallel to the ground, and the moment arms from the axis of rotation were simply the distances measured along the segment to the lines of action of the forces. As the arm flexes or extends at the elbow joint, however, the moment arms change. In Figure 12-7, the moment arms for the muscle force are illustrated with the arm in three positions. With the elbow extended, the moment arm is rather small. As flexion occurs at the elbow, the moment arm increases until the arm is parallel to the ground.

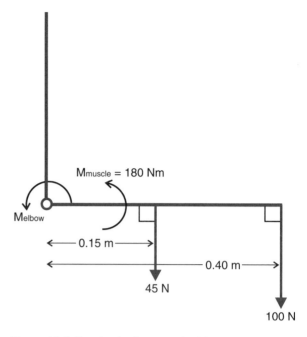

Figure 12-6. Free body diagram of a biceps curl at an instant in time when the forearm is parallel to the horizontal.

When flexion continues past this point, the moment arm becomes smaller once again. The magnitude of the moment arm of the muscle force, therefore, depends on how much flexion or extension occurs at a joint. The change in moment arm during flexion and extension of the limb is also true for the moment arms for both the weight of the forearm-hand and for anything held in the hand.

In Figure 12-8, the arm is held at some angle θ below the horizontal. The value d describes the distance from the axis of rotation to the center of mass of the arm. The moment arm for the arm weight, however, is the distance a. The lines a and d form the sides of a right angle triangle with the line of action of the force. Therefore, with the angle θ, the cosine function can be used to calculate the length a. Thus:

$$\cos \theta = \frac{a}{d}$$
$$a = d * \cos \theta$$

Consider that the arm has a 0° angle when it is parallel to the ground. As the arm extends, the angle increases until it is 90° at full extension. The cosine of 0° is 1 and as the angle becomes greater, the cosine gets smaller until at 90° the cosine is zero. If the angle θ becomes greater as the arm is extended, the moment arm should become correspondingly smaller because the distance d does not

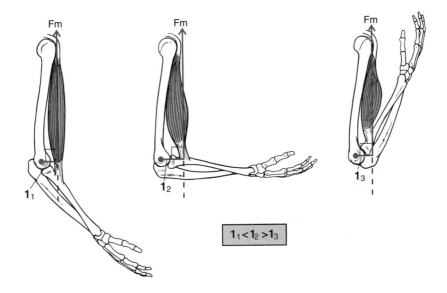

$1_1 < 1_2 > 1_3$

Figure 12-7. The change in moment of the moment arm of the biceps muscle throughout the range of motion. As the elbow flexes from an extended position, the moment arm becomes longer. As the arm continues to flex from the horizontal forearm position, the moment arm becomes shorter.

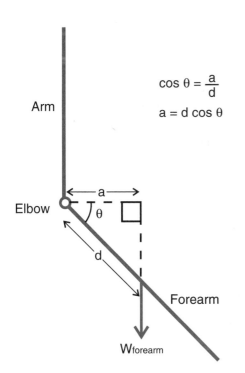

$$\cos \theta = \frac{a}{d}$$

$$a = d \cos \theta$$

Figure 12-8. The cosine of the angle of inclination of the forearm is used to calculate the moment arm when the forearm is not parallel to the horizontal.

change. When the angle θ is 90°, the arm is fully extended and the moment arm is zero because the line of action of the force due to the weight of the arm passes through the axis of rotation.

Consider the example of an individual performing a biceps curl with a weight. The arm is positioned 25° below the horizontal, so that the elbow is slightly extended. This is illustrated in Figure 12-9a. The corresponding free body diagram is presented in Figure 12-9b. Knowing the distance from the axis of rotation to the center of mass and the angle at which the arm is positioned, the moment arm for the weight of the arm can be calculated. Thus:

$$a = 0.15 \text{ m} * \cos 25°$$
$$a = 0.15 \text{ m} * 0.9063$$
$$a = 0.14 \text{ m}$$

Similarly, the moment arm for the weight of the barbell held in the hand can be calculated:

$$b = 0.40 \text{ m} * \cos 25°$$
$$b = 0.40 \text{ m} * 0.9063$$
$$b = 0.36 \text{ m}$$

Notice that the moment arms are smaller than they would be if the arm was held parallel to the ground.

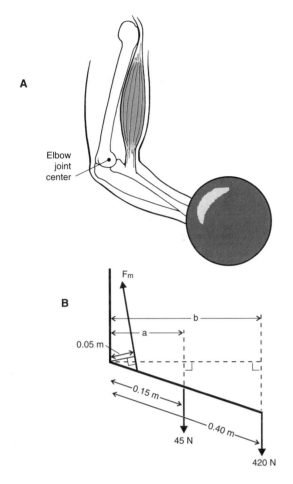

$$F_m = \frac{6.3 \text{ N-m} + 151.2 \text{ N-m}}{0.05 \text{ m}}$$

$$F_m = 3150 \text{ N}$$

Note that the muscle must exert a force much greater than the weight of the arm and the barbell, because the moment arm for the muscle is relatively small. It would appear that our body is at a great disadvantage when it comes to producing large moments of force about a joint. Most of our joints are arranged as third class levers, however, indicating that range of motion is magnified. Muscles can, therefore, exert large forces, but only over very short periods of time.

Static Equilibrium, Stability, and Balance   The concept of stability is closely related to that of equilibrium. *Stability* may be defined in much the same way in which equilibrium is defined; that is, as the resistance to both linear and angular acceleration. The ability of an individual to assume and maintain a stable position is referred to as *balance*. Even in a stable or balanced position, an individual may experience external forces.

If a body is in a state of static equilibrium and is slightly displaced by a force, the object may return to its original position, may continue to move away from its original position, or may stop and assume a new position. If the object is displaced as a result of work done by a force and returns to its original position, it is said to be in a state of *stable equilibrium*. If the object is displaced and tends to increase its displacement, it is in a state of *unstable equilibrium*. A state of *neutral equilibrium* exists if the object is displaced by a force and returns to the position from which it was displaced.

Figure 12-10 illustrates these states of equilibrium. In Figure 12-10a, stable equilibrium is exemplified by a ball on a concave surface. When it is displaced along one side of the surface by some force, it will return to its original position. In Figure 12-10b, an example of unstable equilibrium is presented when a ball on a convex surface is displaced by a force. The ball will come to rest in a new position, not its original position. Figure 12-10c illustrates neutral equilibrium. In this case, when a ball is placed on a flat surface and a force is applied, the ball will move to a new position to which it was displaced.

The human body with its multiple segments is certainly more complex than a ball, but it can assume the different states of equilibrium. An individual doing a head-stand, for example, is in a position of unstable equilibrium. On the other hand, a child sitting on a swing is in a state of stable equilibrium.

There are several factors that will determine the stability of an object. The first factor depends on where the

Figure 12-9. Diagram of the forearm at an instant in time during a biceps curl (a) and the free body diagram of this position (b). Note that the arm is inclined below the horizontal.

This problem may be solved for the muscle force using the same principles of a static analysis that were discussed previously. That is:

$$\Sigma M = 0$$

If, by convention, the moment due to the muscle is considered to be positive and the moments due to the weight of the arm and the weight of the barbell negative, then:

$$M_{Fm} - M_A - M_B = 0$$

Substituting all known values into this equation and rearranging it results in:

$$(F_m * 0.05 \text{ m}) = (45 \text{ N} * 0.14 \text{ m}) + (420 \text{ N} * 0.36 \text{ m})$$

This equation can now be solved to determine the muscle force necessary to maintain this position in a static posture. Thus:

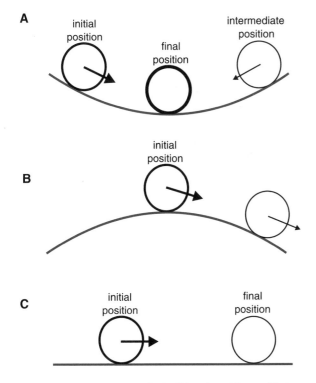

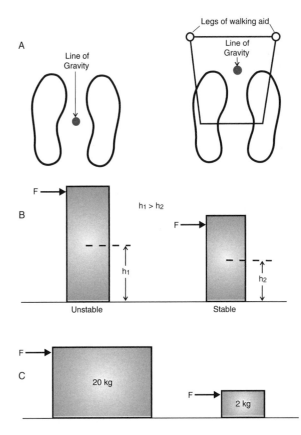

Figure 12-10. Examples of a ball in: a) stable equilibrium; b) unstable equilibrium; and c) neutral equilibrium.

Figure 12-11. Illustrations of the factors that influence the stability of an object: a) increasing the base of support; b) decreasing the height of the center of mass; and c) increasing the mass of the system.

line of gravity falls with respect to the base of support. An object will be most stable if the line of gravity is in the geometric center of the base of support. Increasing the area of the base of support generally increases the stability. It should be noted, however, that a body may be stable in one direction and not in another. For example, spreading one's feet apart increases the area of the base of support and makes the individual stable if pushed in the transverse plane. It does not, however, help the stability in the antero-posterior plane. Increasing the base of support to allow the line of gravity to fall within the base of support may be illustrated by the example of an individual using a "walker" to aid in locomotion. Figure 12-11a illustrates this case. The base of support is produced by the positions of the individual's legs and the legs of the "walker". The "walker" increases the base of support and the individual positions themselves such that the line of action of their center of mass is in the geometric center of this base.

The stability of an object is also inversely proportional to the height of the center of mass. That is, an object with a low center of mass will tend to be very stable when compared to an object with a high center of mass (Figure 12-11b). If the two objects in Figure 12-11b undergo the

same angular displacement as a result of the forces indicated, the line of gravity of the center of mass of the object on the left will move outside the limit of the base of support sooner than that of the object on the right. Therefore, the work done to dislodge the object on the left would be less than for the object on the right. In football, for example, defensive linemen crouch in a three-point stance to keep their center of mass low. This enhances their stability and they are less likely to be moved by the offensive linemen.

The last factor influencing stability is the mass of the object. By referring to the equations of motion, it should be obvious that the greater the mass of an object, the greater the stability of that object (Figure 12-11c). Newton's second law states that the force applied to an object is proportional to the mass of the object and its acceleration. Thus, it takes a proportionately greater force to move an object with a greater mass. Moving a piano, for example, is extremely difficult because of its mass. Many sports such as wrestling and judo, where stability is critical, take body mass into consideration by dividing the contestants into weight divisions because of the disproportionate stability of heavier individuals.

Applications of Statics   Although the techniques of static analysis have been formally presented in this chapter, the principles of static analysis have been used previously in Chapter 11 to calculate the position of the total body center of mass using the reaction board method. It would be appropriate to return to this section in Chapter 11 and review the reaction board technique in terms of a static analysis.

It would appear that static analyses are limited in their usefulness since they describe situations in which no motion or motion at a constant velocity occurs. The static analysis of muscle forces and moments, however, has been used extensively in ergonomics even though the task may involve some movement. The evaluation of work place tasks, such as lifting and manual materials handling, has been examined in considerable detail with static analyses. Using a static analysis to determine an individual's static or isometric strength is widely accepted in determining the ability of the person's lifting ability. Many researchers have suggested that this static evaluation should be used as a pre-employment screening evaluation for applicants for manual materials handling tasks (2, 3, 4). Lind et al. (5) reported that the static endurance in a manual materials handling task is influenced by the posture of the individual performing the task. A model to evaluate static strength evaluations of jobs has been developed by Garg and Chaffin (6). A free body diagram of one such lifting model is presented in Figure 12-12 (7).

Static analysis techniques have also been used in the clinical rehabilitation setting. Quite often no movement of a body or a body segment is desirable and thus a static evaluation may be undertaken. For example, placing a patient in traction demands that a static force system be implemented. Many bracing systems for skeletal problems, such as scoliosis or genu valgum (knock-knee), use a static force system to counteract the forces causing the problem (Figure 12-13). Static analyses have been used in the calculation of muscle forces and have been performed by multiple researchers on many joints (8, 9, 10).

Dynamic Analysis

A static analysis may be used to evaluate the forces on the human body when there are little significant linear or angular accelerations (11). When there are significant linear and/or angular accelerations, however, a dynamic

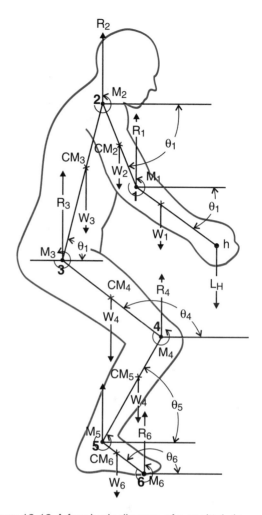

Figure 12-12. A free body diagram of a sagittal view static lifting model showing the reactive forces and moments of forces at the joints (after Chaffin, D.B. and Andersson, G.B.J. Occupational Biomechanics (2nd Edition). New York: John Wiley & Sons, Inc., 1991).

analysis must be undertaken. A dynamic analysis should be used, therefore, when the accelerations are non-zero. The equations of motion for a dynamic analysis were derived from Newton's second law of motion and expanded by the famous Swiss mathematician, Leonhard Euler (1707–1783). The equations of motion for a two-dimensional case are based on:

$$\Sigma F = ma \text{ for the linear case}$$

and

$$\Sigma T = I\alpha \text{ for the rotational case}$$

The linear acceleration may be broken down into its horizontal (x) and vertical (y) components. In a two-

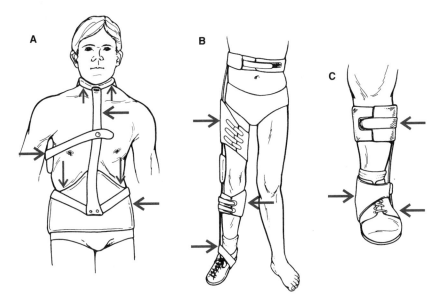

Figure 12-13. Bracing systems that illustrate static force systems: a) a neck brace; b) a three-point pressure brace to correct genu valgum; and c) a brace to correct a foot deformity.

dimensional system, angular acceleration occurs about the z-axis. If $\alpha = 0$, the motion is purely linear. If $a_x = 0$ and $a_y = 0$, the motion is purely rotational. If $\alpha = a_x = a_y = 0$, then a static case exists.

As in the static 2-dimensional analysis, there are three independent equations that are used in a dynamic 2-dimensional analysis. These are:

$$\Sigma F_x = ma_x$$
$$\Sigma F_y = ma_y$$
$$\Sigma T_{cm} = I_{cm}\alpha$$

in which x and y represent the horizontal and vertical coordinate directions, respectively, a is the acceleration of the center of mass, m is the body center of mass, I is the moment of inertia about the center of mass, and α is the angular acceleration about the z-axis through the center of mass. Forces acting on a body may be considered to be muscular, gravitational, contact, or inertial. The gravitational forces are the weights of each of the segments. The contact forces can be reactions, forces with another segment, the ground, or an external object. The inertial forces are ma_x and ma_y while $I\alpha$ is the inertial torque. Using the equations of dynamic motion, the forces and torques acting on a segment can be calculated.

In moving from a static analysis to the dynamic analysis, the problem becomes more difficult. In the static case, no accelerations were present. In the dynamic case, both linear and angular accelerations and the inertial

properties of the body segments resisting these accelerations must be considered. In addition, there is a substantial increase in the work done to collect the data necessary to conduct a dynamic analysis. Since the forces and moments that cause the motion are determined by evaluating the resulting motion itself, a technique called an *inverse dynamics* approach will be used. This approach calculates the forces and moments based on the accelerations of the object instead of measuring the forces directly.

In using the inverse dynamics approach, the system under consideration must be determined. The system is usually defined as a series of segments. The analysis on a series of segments is generally conducted beginning with the most distal segment and proceeds proximally up to the next segment and so on. There are several assumptions that must be made when using this approach. The body is considered to be a rigid-body linked system with frictionless pin joints. Each link, or segment, has a fixed mass and a center of mass located at a fixed point. Lastly, the moment of inertia about any axis of each segment remains constant.

As was stated previously, the dynamic case is certainly more complicated than the static case. As a result only a limited example of a single segment will be presented. Consider Figure 12-14, in which a free body diagram of the foot of an individual during the swing phase of the gait cycle is presented. These data constitute a single frame of

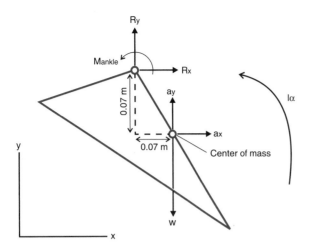

mass of foot = 1.16 kg, I = 0.0096 kg-m^2
α = −14.66 rad/s^2
a_x = −1.35 m/s^2, a_y = 7.65 m/s^2

Figure 12-14. Free body diagram of the foot segment during the swing phase of a walking stride. Convention dictates that moments and forces at the proximal joint are positive as indicated on the free body diagram.

video data and the appropriate inertial properties of the foot segment calculated as described in Chapter 11.

During the swing phase of gait, no external forces other than gravity act on the foot. It can be seen that the only forces acting on the foot are the horizontal and vertical components of the joint reaction force and the weight of the foot acting through the center of mass. The joint reaction force components can be computed by using the first two equations defining the dynamic analysis. First, the horizontal joint reaction force may be defined from:

$$\Sigma F_x = ma_x$$

Since there are no horizontal forces other than the horizontal joint reaction force, this equation becomes:

$$R_x = ma_x$$

If the mass of the foot is 1.16 kg and the horizontal acceleration of the center of mass of the foot is −1.35 m/s^2, the horizontal reaction force is:

$$R_x = 1.16 \text{ kg} * -1.35 \text{ m/s}^2$$
$$R_x = -1.57 \text{ N}$$

Next, the vertical joint reaction force component may be defined from:

$$\Sigma F_y = ma_y$$

There is, however, another vertical force other than the vertical joint reaction force. This force is the weight of the foot itself and so the vertical forces are described as:

$$R_y - W = ma_y$$

If the vertical acceleration of the center of mass of the foot is 7.65 m/s^2, then:

$$R_y = (1.16 \text{ kg} * 7.65 \text{ m/s}^2) + (1.16 \text{ kg} * 9.81 \text{ m/s}^2)$$
$$R_y = 20.3 \text{ N}$$

To determine the net moment acting at the ankle joint, all moments acting on the system must be evaluated. If the center of mass of the foot is considered as the axis of rotation, there are three moments acting on this system, two as a result of the joint reaction forces and the net ankle moment itself. Because the joint reaction forces and their moment arms, the moment of inertia of the foot about an axis through the center of mass, and the angular acceleration of the foot are known, the net ankle moment can be calculated. Thus:

$$\Sigma M_{cm} = I_{cm}\alpha$$
$$M_{ankle} - M_{Rx} - M_{Ry} = I_{cm}\alpha$$

where M_{ankle} is the net muscle moment at the ankle, M_{Rx} is the moment resulting from the horizontal reaction force at the ankle, M_{Ry} is the moment resulting from the vertical reaction force at the ankle, and $I_{cm} \alpha$ is the product of the moment of inertia of the foot and the angular acceleration of the foot. In the general equation of moments acting on the foot, the moments M_{Rx} and M_{Ry} will cause clockwise rotation of the foot about the center of mass of the foot. By convention, clockwise rotations are negative and thus these moments are negative. Substituting in the appropriate values from Figure 12-14 and those calculated in the previous equations and rearranging the equation:

$$M_{ankle} = (0.0096 \text{ kg-m}^2 * -14.66 \text{ rad/s}^2) +$$
$$(.07 \text{ m} * -1.57 \text{ N}) + (.07 \text{ m} * 20.3 \text{ N})$$
$$M_{ankle} = -0.141 \text{ N-m} - 0.110 \text{ N-m} + 1.421 \text{ N-m}$$
$$M_{ankle} = 1.17 \text{ N-m}$$

The net moment at this instant in time is positive, resulting in a counter-clockwise rotation. A counter-clockwise rotation of the foot indicates dorsiflexor activity. As with the static case, the exact muscles acting cannot be determined from this type of analysis. Thus, it cannot be stated whether the muscle activity is dorsiflexor concentric or dorsiflexor eccentric in nature.

It was stated that a dynamic analysis usually proceeds from the more distal joints to the more proximal joints.

The data from this calculation of the foot segment analysis would then be used in the calculations on the next segment—the leg—to calculate the net muscle moment at the knee. The calculation would then continue to the thigh to calculate the hip moment. The analysis is conducted for each joint at each instant in time of the movement under consideration to create a profile of the net muscle moments for the complete movement.

Applications of Dynamics

Dynamic analyses have been used in many biomechanical studies on a number of activities to determine the net moments at various joints. These activities include cycling (12), asymmetric load carrying (13), weight-lifting (14), and throwing (15). The joint moments of force of the lower extremity during locomotion, however, have been most widely researched (16, 17, 18). Winter (17) stated that the resultant moment of force provided powerful diagnostic information when comparing injured gait to non-injured gait.

Figure 12-15 illustrates the net muscle moments of force at the hip, knee, and ankle during a walking stride. In this example, the joint moments are presented in units of N•m but may also be scaled by dividing the moment by the subject's body mass, resulting in units of N-m/kg of body mass. The positive moment at the ankle (Figure 12-15c) just prior to the foot leaving the ground is generally the greatest in magnitude of the lower extremity joint moments of force. Over a number of trials, the moment of force at the ankle is quite consistent during walking, while the knee and hip show greater variability (19). Figure 12-16 illustrates the net joint moments of force for the hip, knee, and ankle during running. The moments of force in running are greater in magnitude than those in walking. The lower extremity moments of force increase in magnitude with increases in locomotor speed. Cavanagh et al. (20), and Mann and Sprague (21) reported considerable variability in the magnitudes of the moments of force between subjects running at the same speed. For slow running this vari-

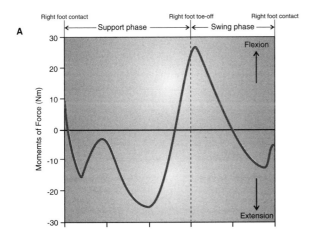

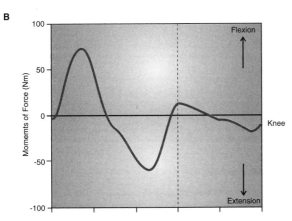

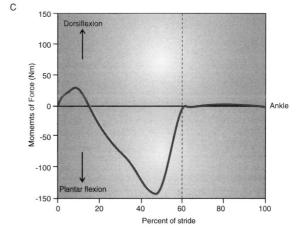

Figure 12-15. Illustrations of the moments of force at the A—hip, B—knee, and C—ankle during a single walking stride. The transition from support to swing is indicated at the dashed line.

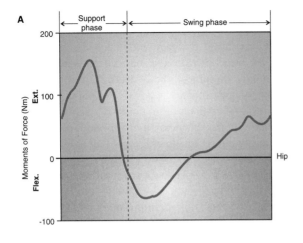

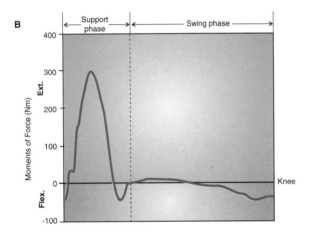

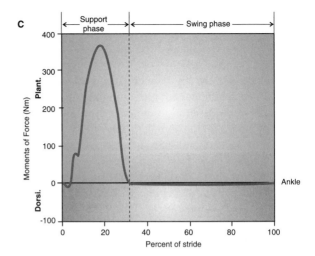

Figure 12-16. Illustration of the moments of force at the A—hip, B—knee, and C—ankle during a single running stride. The transition from support to swing is indicated at the dashed line.

ability is generally smallest at the ankle and greatest at the hip.

Muscle Power The net muscle moment describes the net muscle activity across a joint and does not represent any one particular muscle that crosses the joint. It also does not take into account the situation in which bi-articulate muscles may be acting. This net muscle activity at a joint is simply described as flexor or extensor actions, but it cannot be ascertained if the muscle activity is concentric or eccentric. The net muscle moment can, however, be used in conjunction with the angular velocity of the joint to determine the concentric or eccentric nature of the muscular action. In Chapter 10, concentric actions of muscles were related to the positive work of muscles and eccentric actions as the negative work of muscles. Because the work done by muscles is rarely constant with time, the concept of muscle power should be used. *Muscle power* is the time rate of change of work and is defined as the product of the net muscle moment and the joint angular velocity. It is expressed algebraically as:

$$P_m = M_j * \omega_j$$

where P_m is the muscle power in units of watts (W), M_j is the net muscle moment in N-m, and ω_j is the joint angular velocity in rad/s.

Muscle power may be either positive or negative. Note that since power is the time rate of change of work, the area under the power-time curve is the work done. Thus, a positive muscle power represents positive work or a net concentric action, while negative muscle power represents negative work or a net eccentric action. For example, if M_j and ω_j are either both positive or both negative, muscle power will be positive. If they have opposite polarity, then muscle power will be negative. Consider Figure 12-17, in which the positive and negative work possibilities of the elbow joint are illustrated. In Figure 12-17a, M_j and ω_j are positive, indicating a flexor moment with the arm moving in a flexor direction. The resulting muscle power is positive, indicating a concentric action of the elbow flexors. In Figure 12-17b, both M_j and ω_j are negative, resulting in a positive muscle power or a concentric action. The arm is extending, indicating the muscle action is a concentric action of the elbow extensors. In Figure 12-17c, M_j is positive or a flexor moment, but the arm has a negative ω_j indicating extension. In this case, an external force is causing the arm to extend while the elbow flexors resist. This results in an eccentric action of the elbow flexors and is verified by the negative muscle power. Figure 12-17d illustrates the case in which an eccentric action of the elbow extensors result.

Figure 12-18 illustrates the angular velocity-time, net muscle moment-time, and the muscle power-time pro-files during an elbow flexion followed by elbow extension. In the flexion phase of the movement, the net muscle moment initially is positive, but becomes negative as the arm becomes more flexed. The initial portion, therefore, results in a positive power or a concentric contraction of the flexor muscles. In the latter portion of the flexor phase, the power is negative, indicating an eccentric muscle contraction. The eccentric contraction of the elbow extensor muscles occurs to decelerate the limb. Because the power is negative at this point, however, it does not mean that there is no flexor muscle activity. It simply means that the predominate activity is extensor. In the extension portion of the movement, the situation is reversed. In the initial portion of extension, the muscle activity is concentric extensor activity. In the latter portion, the muscle power profile indicates flexor concentric activity once again to decelerate the limb.

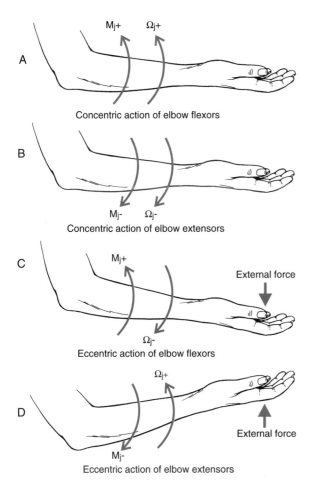

Figure 12-17. The definition of positive and negative power at the elbow joint: a) and b) are situations that result in positive power; c) and d) are situations that result in negative power.

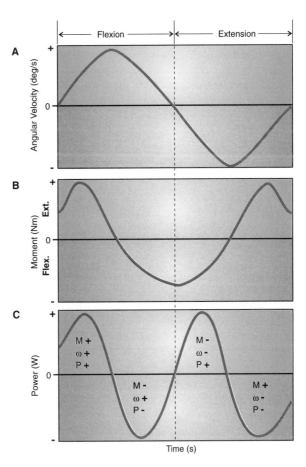

Figure 12-18. Moment of force, angular velocity, and muscle power profiles during an elbow flexion-extension motion.

Figure 12-19 presents the muscle power-time profile at the knee joint during the support phase of walking. The initial portion of the support begins at touchdown and lasts until mid-support. At touchdown, the knee is in a flexed position and continues to flex until mid-support. The knee moment during this period indicates an extensor (negative) moment. Because the knee joint is flexing and the net muscle moment indicates extensor activity, the net muscle moment and the angular velocity must have an opposite polarity. Thus, the resulting muscle power profile during this period is negative, indicating an eccentric action of the knee extensors. The next portion of the support period lasts from mid-support to just before toe-off. In this portion, the knee is extending. The knee moment is still positive, however, indicating an extensor moment. The polarities of the net muscle moment and the angular velocity are the same. The result is a positive power phase until just prior to toe-off. The positive muscle power during this portion of support indicates that the knee extensors are acting concentrically. The positive and negative portions of muscle power profiles can be analyzed in a similar manner to give insight into the actions of muscles during the activity.

The analysis of net muscle moments of force and muscle power have been widely used in research in biomechanics. Winter and Robertson (22), and Robertson and Winter (23) have investigated the power requirements in walking. Robertson (24) described the power functions of the leg muscles during running to identify any common characteristics among a group of runners. Gage (25) reported on the use of the moment of force and the muscle power profiles as a pre- and post-operative comparison. The analysis of muscle power in the lower extremity during locomotion, therefore, appears to be a powerful research and diagnostic tool.

The Effect of a Force Applied Over a Period of Time

For motion to occur, forces must be applied over a period of time. By manipulating the equation describing Newton's second law of motion, a very important physical relationship in human movement can be generated that describes the concept of forces acting over a period of time. This relationship relates the momentum of an object to the force and the time over which the force acts.

This relationship is derived from Newton's second law which states:

$$F = m * a$$

Since $a = \dfrac{dv}{dt}$, this equation can be rewritten as:

$$F = m * \frac{dv}{dt}$$

and further:

$$F = \frac{d(m * v)}{dt}$$

If each side is multiplied by dt to remove the fraction on the right hand side of the equation, the resulting equation is:

$$F * dt = d(m * v)$$

or

$$F * dt = mv_{final} - mv_{initial}$$

The left hand side of this equation, the product of $F * dt$, is a quantity known as an *impulse* and has units of N • s (newton • seconds). Impulse is the measure of what is required to change the motion of an object. Figure 12-20 illustrates the vertical component of the ground reaction force of a single footfall of a runner. This figure represents a force applied over a period of time. Impulse may be expressed graphically as the area under a force-time curve. The right hand side of the equation describes a change in momentum. Thus, when a force is applied to an object over a period of time, a change in the momentum of the object occurs. The derived equation describes what is known as the *impulse-momentum relationship*.

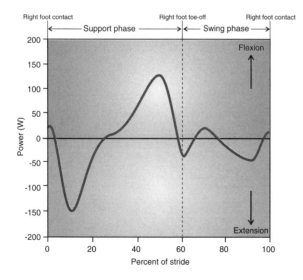

Figure 12-19. Power of the muscles at the knee joint during a single walking stride.

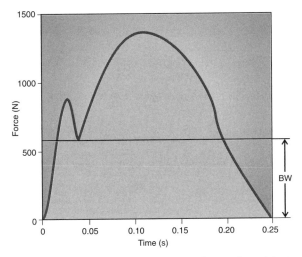

Figure 12-20. Vertical ground reaction force of a subject running at 5 m/s. The vertical impulse is the shaded area under the force-time curve.

Consider an example of an individual with a mass of 65 kg jumping from a squat position into the air. The velocity of the person at the initiation of the jump is zero. By a video analysis, it can be determined that the velocity of the center of mass at take-off was 3.4 m/s. The impulse can then be calculated as:

$$F * dt = mv_{final} - mv_{initial}$$
$$F * dt = 65 \text{ kg} * 3.4 \text{ m/s} - 65 \text{ kg} * 0 \text{ m/s}$$
$$F * dt = 221 \text{ kg-m/s}$$

If it is assumed that the force application took place over 0.2 s. The average force applied, therefore, would be:

$$F * 0.2 \text{ s} = 221 \text{ kg - m/s}$$
$$F = \frac{221 \text{ kg - m/s}}{0.2 \text{ s}}$$
$$F = 1105 \text{ N}$$

The nature of the force applied and the time over which it is applied will determine how the momentum of the object is changed. In order to change the momentum of an object, a greater force can be applied over a short period of time, or a smaller force over a longer period of time. Of course, the tactic used depends on the situation. For example, in landing from a jump, the performer must change their momentum from some initial value to zero. At floor impact, the greater the impulse, the greater the change in momentum that must be accomplished. If an individual lands with little knee flexion (locked knees), then the impact force occurs over a very short period of time. If, however, the individual lands and flexes the joints of the lower extremity, they reduce the impulse by extending the time over which the impact force acts. An

example of "soft" and "rigid" landings is presented in Figure 12-21. This example is taken from a study by DeVita and Skelly (26) and it shows the large peak force in the "rigid" landing as compared to the smaller force in the "soft" landing. Both appear to take place over a relatively comparable period of time, resulting in a greater impulse in the rigid landings.

Previously, a study by Bates et al. (27) was discussed. This study illustrated the horizontal velocity pattern of the center of mass during the support phase of the running stride. It was determined that a runner's horizontal velocity decreased from touchdown until mid-stance and then increased until toe-off. When the velocity curve was differentiated, an acceleration curve was generated and confirmed that the horizontal velocity of the center of mass varied over the support phase. By applying Newton's second law of motion and multiplying each point along this curve by the runner's body mass (m), the antero-posterior ground reaction force component was calculated:

$$F_{a/p} = m * a_{a/p}$$

This force component has a negative portion followed by a positive portion (Figure 12-22). By applying the impulse-momentum relationship, it can again be confirmed that the runner does indeed slow down during the first portion of support and speed up in the latter portion.

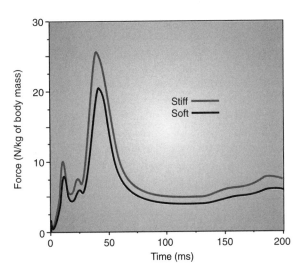

Figure 12-21. Vertical ground reaction force component for soft and stiff landings. Stiff landings had 23% greater linear impulse than the soft landing (after DeVita, P. and Skelly, W.A. Effect of landing stiffness on joint kinetics and energetics in the lower extremity. Medicine and Science in Sports and Exercise, 24:108–115, 1992).

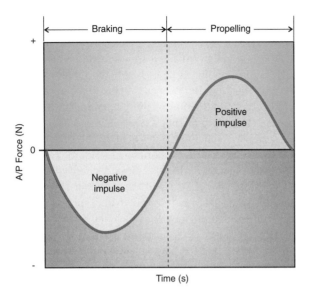

Figure 12-22. Antero-posterior ground reaction force component illustrating the braking and propulsion impulses.

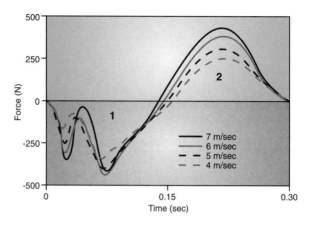

Figure 12-23. Changes in the antero-posterior ground reaction force component as a function of running speed. Area 1 is the braking impulse and area 2 is the propulsion impulse (after Hamill, J., Bates, B.T., Sawhill, J.A., Knutzen, K.M. Variations in ground reaction force parameters at different running speeds. Human Movement Science, 2:47–56, 1983).

The area under the negative portion of the force component or the negative impulse serves to slow the runner down, that is, to change their incoming velocity to some lesser velocity value. The positive area or the positive impulse of the component serves to accelerate the runner, that is, to change their velocity from some lesser value to some outgoing velocity. If the positive change in velocity equals the negative change in velocity, the individual is running at a constant speed. Figure 12-23 illustrates the changes in the braking and propelling impulses across a range of running speeds (28). In many instances in the laboratory when collecting ground reaction force data, the ratio of the negative impulse to the positive impulse is checked to determine if the runner is at a constant velocity, speeding up, or slowing down. It should be understood, however, that, even if the individual is maintaining a constant running speed, the ratio of the positive/negative impulse is rarely 1.0 for any given support period of a stride. The average ratio over a number of footfalls approaches the ratio of 1.0, however.

One interesting research application for the use of the impulse-momentum relationship has been in the investigation of vertical jumping. This area of research has used the force platform to determine the ground reaction forces during the vertical jump. Remember that the ground reaction force reflects the force acting on the center of mass of the individual. Thus, researchers have used the impulse-momentum relationship on data collected from the force platform to determine the parameters nec-

essary to investigate the height of the center of mass above its starting point during vertical jumping (29, 30). Figure 12-24a shows the vertical ground reaction force profile of an individual starting at rest on the platform and jumping into the air. This type of jump is called a counter movement jump because the subject flexes at the knees and then swings the arms up as the knees are extended during the jump. Figure 12-24b illustrates a squat jump in which the subject begins in a squat (knees flexed) position and simply forcefully extends the knees to jump into the air.

In both cases, the impulse-momentum relationship can be used to calculate the peak height of the center of mass above the initial height during the jump. Consider the vertical ground reaction force curve of a counter jump in Figure 12-25. The vertical force in the intitial portion of the curve is the individual's body weight. If the body weight line is extended to the instant of take-off, the area beneath the curve describes the body weight impulse (BW_{imp}). It may be calculated as an integral:

$$BW_{imp} = \int_{t_i}^{t_f} BW \, dt$$

where t_i to t_f represents the time interval when the subject is standing still on the force platform until the instant of take-off. The total area under the force-time curve until the subject leaves the force platform may be designated as $Total_{imp}$ and can be calculated as an integral:

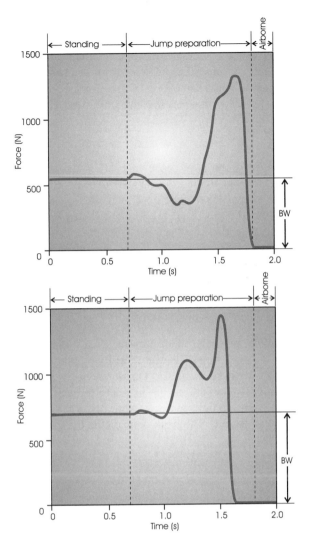

Figure 12-24. The vertical ground reaction force component of two types of vertical jumps: a) the counter-movement jump and b) the squat jump.

$$Total_{imp} = \int_{t_i}^{t_{jump}} F_z \, dt$$

where t_i to t_{jump} represents the time interval when the subject is on the platform until the subject leaves the platform during the jump. The impulse that propelled the subject into the air can then be determined by:

$$Jump_{imp} = Total_{imp} - BW_{imp}$$

The impulse-momentum relationship can then be formulated using the impulse that the subject generated to perform the jump. Thus:

$$Jump_{imp} = m(v_f - v_i)$$

where v_i is the initial velocity of the center of mass and v_f is the take-off velocity of the center of mass. Since $v_i = 0$, then:

$$Jump_{impv} = m(v_f - 0)$$

By substituting the impulse of the jump ($Jump_{imp}$) and the subject's body mass (m) into this equation, the velocity of the center of mass at take-off for the vertical jump can be calculated. The height of the center of mass during the jump is then calculated based on the projectile equations elaborated upon in Chapter 8. Thus:

$$Height_{cm} = \frac{v_m^2}{2g}$$

This calculation has proved to be in good agreement with values calculated from high-speed film data. Komi and Bosco (29) reported an error of 2.0% from the computation on the force platform.

Consider the force-time profile of a counter-movement jump in Figure 12-25. The body weight impulse, the rectangle formed by the body weight and the time of force application, was calculated to be:

$$BW_{imp} = \int_{t_0}^{t_{1.5}} BW \, dt$$

$$BW_{imp} = 620.80 \text{ N.s}$$

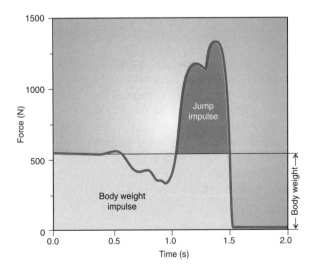

Figure 12-25. The vertical ground reaction force component of a counter-movement jump illustrating the body weight impulse and the jump impulse.

The total impulse, that is the total force application from time = 0.0 s to time = 1.5 s including the body weight, is:

$$Total_{imp} = \int_{t_0}^{t_{1.5}} F_z \, dt$$
$$Total_{imp} = 772.81 \text{ N.s}$$

The jump impulse, therefore, is:

$$Jump_{imp} = Total_{imp} - BW_{imp}$$
$$Jump_{imp} = 772.81 \text{ N.s} - 620.80 \text{ N.s}$$
$$Jump_{imp} = 152.01 \text{ N.s}$$

Substituting this value into the impulse-momentum relationship, it is possible to solve for the velocity of the center of mass at take-off. Thus, with the body mass of the jumper of 56.2 kg, the velocity of take-off is:

$$Jump_{imp} = m(v_f - v_i)$$
$$152.01 \text{ N.s.} = 56.2 \text{ kg} (v_f - 0)$$
$$v_f = \frac{152.01 \text{ N.s}}{56.2 \text{ kg}}$$
$$v_f = 2.70 \text{ m/s}$$

The height of the center of mass during the jump can be calculated by:

$$Height_{cm} = \frac{v_{cm}^2}{2g}$$
$$Height_{cm} = \frac{(2.70 \text{ m/s})^2}{2 * 9.81 \text{ m/s}^2}$$
$$Height_{cm} = 0.373 \text{ m}$$

Therefore, in this particular jump, the center of mass was elevated 37.3 cm above the initial height of the center of mass.

This calculation technique was used by Dowling and Vamos (31) to identify the kinetic and temporal factors related to vertical jump performance. They found a large variability in the patterns of force application between the subjects that made it difficult to identify the characteristics of a good performance. Interestingly, they reported that a high maximum force was necessary but not sufficient for a good performance. They concluded that the pattern of force application was the most important factor in vertical jump performance.

Effect of a Force Applied Over a Distance

Mechanical work was previously defined as the product of a force applied to an object and the distance that the object moved during the force application. When an object is moved, mechanical work is said to have been done on the object. Thus, if no movement occurs, no mechanical work is done. Energy was also previously defined as the capacity to do work. Intuitively, therefore, some relationship exists between work and energy. This very useful relationship is called the *work-energy theorem*. This theorem states that the work done is equal to the change in energy. That is:

$$W = \Delta E$$

where W is the work done and ΔE is the change in energy. That is, for mechanical work to be done, a change in the energy level must occur. The change in energy refers to all types of energy in the system whether it is kinetic, potential, chemical, heat, or light, etc. For example, if a trampolinist weighs 780 N and is at a peak height of 2 m above a trampoline bed, their potential energy based on the height above the trampoline bed is:

$$PE = 780 \text{ N} * 2 \text{ m}$$
$$PE = 1560 \text{ J}$$

Assuming no horizontal movement, at impact their kinetic energy would also be 1560 J while the potential energy would be zero. The kinetic energy would have this initial value and, as the bed deforms, the potential strain energy would increase while the kinetic energy would go to zero. The work done on the trampoline bed would be:

$$W = \Delta KE$$
$$W = 1560 \text{ J} - 0 \text{ J}$$
$$W = 1560 \text{ J}$$

It should be noted that this value of 1560 J is also the value of the potential strain energy of the trampoline bed and, as the bed reformed, would change from this value to zero and constitute the work done by the trampoline bed on the trampolinist.

To evaluate the work done, the energy level of a system must be evaluated at different instants in time. It should be noted that this change in energy represents the work done on the system. For example, a system has an energy level of 26.3 J at position 2 and 13.1 J at position 1. The work done would be:

$$W = \Delta E$$
$$W = 26.3 \text{ J} - 13.1 \text{ J}$$
$$W = 13.2 \text{ J}$$

Because this work has a positive value, it is considered positive work and work is said to be done by the system in the form of concentric muscle actions. On the other hand, if the energy level was 22.4 J at position 1 and 14.5 J at position 2, then:

$$W = \Delta E$$
$$W = 14.5\,J - 22.4\,J$$
$$W = -7.9\,J$$

Now the work done is negative and work is said to have been done on the system. That is, the muscles are undergoing eccentric contractions.

The work-energy relationship has been a very useful relationship in biomechanics to analyze human motion. Researchers have used this analytical method to determine the work done during a number of movements; however, the greatest use of this technique has been in the area of locomotion. The calculation of the total work done resulting from the motion of all of the body's segments is called *internal work*. This calculation has been utilized by many researchers, particularly those studying locomotion (32, 33, 34, 35, 36).

For a single segment, the work done on the segment is:

$$W_s = \Delta KE + \Delta RE + \Delta PE$$
$$W_s = \Delta\left(\frac{1}{2}mv^2\right) + \Delta\left(\frac{1}{2}I\omega^2\right) + \Delta(mgh)$$

where W_s is the work done on the segment, ΔKE is the change in linear kinetic energy of the center of mass of the segment, ΔRE is the change in rotational kinetic energy about the segment center of mass, and ΔPE is the change in potential energy of the segment center of mass. The work done on the total body is the sum of the work done on all segments and is calculated as:

$$W_b = \sum_{i=1}^{n} W_{si}$$

where W_b is the total body work and W_{si} is the work done on the i-th segment. One major limitation of this calculation is that it does not account for all of the energy of a segment, and thus that of the total body. For example, the strain energy due to the deformation of tissue is not considered.

External work, on the other hand, may be defined as the work done by a body on an object. For example, the work done by the body to elevate the total body center of mass while walking up an incline is considered to be external work. *External work* is often calculated during an inclined treadmill walk or run as:

External work = BW * Treadmill Speed * Percent Grade * Duration

where BW is the body weight of the subject walking or running on the treadmill, Percent Grade is the incline of the treadmill, and Duration is the length of time of the walk or run. The product of the treadmill speed, percent grade, and the duration of exercise is the total vertical distance traveled. If the percent grade is zero, that is, the walking surface is level, then the vertical distance traveled is zero. Thus, walking on a level surface results in no external work being done.

Internal work of the total body for a particular movement is derived by summing the changes in the segment energies over a period of time. That is, the change in energy at each instant in time is summed for the length of time that the movement lasts. Usually, this period of time in locomotion studies is the time for one stride. The analysis of the mechanical work done is extremely valuable as a global parameter of the body's behavior without a detailed knowledge of the motion. There are, however, a variety of algorithms that may be used to calculate work (32, 33, 34, 35, 36, 37). These models have incorporated factors to quantify such parameters as positive and negative work, the effect of muscle elastic energy, and the amount of negative work attributable to muscular sources. The major differences among these algorithms is that they differ in terms of how energy is transferred within a segment and between segments. Energy transfer within a segment refers to changes in energy from one form of energy to another, as in the change from potential to kinetic. Energy transfer between segments refers to the exchange of the total energy of a segment from one segment to another. Presently, there is no consensus as to which model is most appropriate. Thus, values of mechanical work for a single running stride may range between 532 W to 1775 W (39).

If it is considered that all energy is transferred between segments, the point in the stride at which the transfer occurs can be illustrated. Figure 12-26 graphically illustrates the magnitude of between segment energy transfer during the running stride. It can be seen that the magnitude of energy transfer decreases during support and increasing from mid-stance to a maximum after toe-off (39).

When work is calculated over a period of time such as a locomotor stride, the result is often presented as power with units of watts. These power values have generally been scaled to body mass resulting in units of watts/kg of body mass. Hintermeister and Hamill (40) investigated the relationship between mechanical power and energy expenditure. They reported that mechanical power significantly influences energy expenditure independent of the running speed. Using several algorithms representing different methods of energy transfer, however, mechanical power explained at best only 56% of the variance in energy expenditure.

By noting the law of conservation of energy, this rather weak relationship between mechanical work and

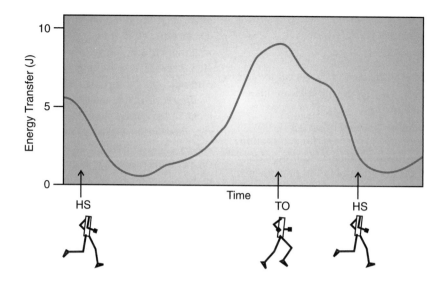

Figure 12-26. The magnitude of between segment transfer during a single running stride (after Williams, K.R. A biomechanical and physiological evaluation of running efficiency. Unpublished Doctoral dissertation, Pennsylvania State University, 1980).

energy expenditure can be explained. Note that this law states that energy is neither created nor destroyed, but may be changed from one form to another. That is, potential energy may be changed to kinetic energy or heat energy. It should be obvious, however, that not all of the energy can or is used to perform mechanical work. In fact, the majority of the available energy of the muscle will go to maintain the metabolism of the muscle and only about 25% of the energy is used for mechanical work. Thus, when using the work-energy theorem to determine the mechanical work of the body, not all of the energy in the system is accounted for.

Summary

The analysis of human motion is usually conducted using one of three techniques: 1) the effect of a force at an instant in time (F = ma); 2) the effect of a force applied over a period of time (impulse-momentum relationship); and 3) the effect of a force applied over a distance (work-energy theorem).

In the first technique, the analysis may be a static case (when a = 0) or a dynamic case (when a ≠ 0). The static 2-dimensional case is determined using the following equations:

$$\Sigma F_x = 0 \text{ for the horizontal component}$$
$$\Sigma F_y = 0 \text{ for the vertical component}$$
$$\Sigma T_z = 0$$

while the 2-dimensional dynamic case uses the following equations:

$$\Sigma F_x = ma_x \text{ for the horizontal component}$$
$$\Sigma F_y = ma_y \text{ for the vertical component}$$
$$\Sigma T_z = I\alpha$$

The purpose in both cases is to determine the net muscle moment about a joint. The dynamic case gives rise to the concept of muscle power which is the product of the muscle moment and the angular velocity:

$$P_m = M_j * \omega_j$$

Since work is the integral of power, positive work implies that the muscle action was concentric, while negative power implies an eccentric muscle action.

The impulse-momentum relationship relates the force applied over a period of time to the change in momentum:

$$F * dt = mv_{final} - mv_{initial}$$

The left hand side of the equation (F * dt) is the impulse, while the right hand side ($mv_{final} - mv_{initial}$) describes the change in momentum. Impulse is defined as the area under the force-time curve and thus is equal to the change in momentum. This type of analysis has been used in research to evaluate the jump height of the center of mass in vertical jumping in association with the equations of constant acceleration.

In the third type of analysis, mechanical work is calculated via the change in mechanical energy. That is:

$$W = \Delta KE + \Delta RE + \Delta PE$$

where KE is the translational kinetic energy, RE is the rotational kinetic energy, and PE is the potential energy.

Work can be calculated for either a single segment or for the total body. When this is done on a segment-by-segment basis, internal work or the work done on the segments by the muscles to move the segments is calculated.

Review Questions

1. Give an example of a system in equilibrium in which there is: a) no motion, and b) motion.
2. Write the equations of motion for a two-dimensional static analysis and for a two-dimensional dynamic analysis.
3. Consider the following free body diagram. Given the mass of the leg and foot is 4.88 kg, the distance from the knee joint to the center of mass of the leg/foot system is 0.146 m, the weight of the barbell is 100 N, and the distance from the knee joint to the center of mass of the barbell is 0.447 m, solve, using a static analysis, for the muscle torque that will place this system in equilibrium. (Answer: 51.69 N-m)

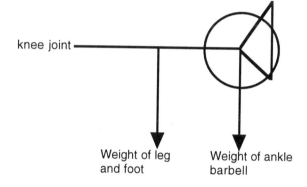

knee joint

Weight of leg and foot

Weight of ankle barbell

4. Give examples of an acrobat in states of stable, unstable, and neutral equilibrium.
5. If the muscle power at a joint is positive, what type of work will be done?

6. If the muscle power is negative, what type of work will be done?
7. Give examples in which the impulse-momentum relationship is used to bring an object to a stop.
8. Calculate the impulse in the following graph. (Answer: 19.25 N.s)

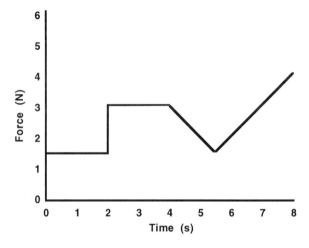

9. If an individual is walking up a hill, what type of work must they be doing? Why?
10. If a person is walking on a level treadmill, what is the external work done? What is the total work done in one walking cycle?

Additional Questions

1. Consider the following free body diagram. Using a static analysis, solve for the horizontal and vertical forces of C that will maintain this system in equilibrium if A = 500 N, B = 350 N, and W = 100 N. (Answer: Cx = 680.5 N, Cy = 97.5 N)

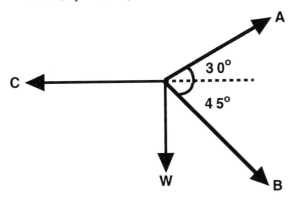

2. If the leg is placed at an angle of 27° below the horizontal, calculate the moment arm of the torque caused by the weight of the leg, given the distance to the center of mass of the leg is 0.15 m. (Answer: 0.13 m).

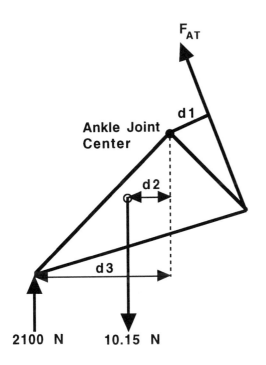

3. Consider the following free body diagram. Using a static analysis, solve for Achilles tendon force which will place this system in equilibrium if d1= 0.05 m, d2= 0.05 m, and d3 = 0.15 m. (Answer: 6289.8 N).

4. Consider the following free body diagram. Using a static analysis, solve for the moment at the elbow if d1 = 0.035 m; d2 = 0.13 m; and d3 = 0.43 m. What is the net muscle force? (Answer: 3048 N).

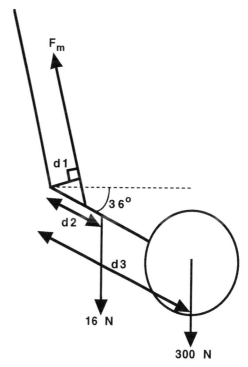

5. Give examples in which a static analysis could be used in place of a dynamic analysis to calculate net muscle moments. Give examples in which a dynamic analysis must be used to calculate net muscle moments.

6. Consider the information listed below. Draw the free diagram body of the foot segment. Using a dynamic analysis, what is the joint moment at the ankle? (Answer: 0.07 N-m)

ankle: 1.25m, 1.013m	ax: 2.40 m/s
fifth metatarsal: 1.26m, 1.00 m	ay: 1.09 m/s
heel: 1.24 m, 1.010 m	α: −21.60 rad/s²
center of mass: 1.255 m, 1.0195 m	

7. Calculate the height of the center of mass above its starting height during a squat jump based on the following information: Body weight = 500 N; Total vertical force = 588 N; Time of force application = 1.2 s (Answer: 0.22 m).

8. Calculate the height of the center of mass above its starting height during a jump based on the following hypothetical graph. (Answer: 0.31 m)

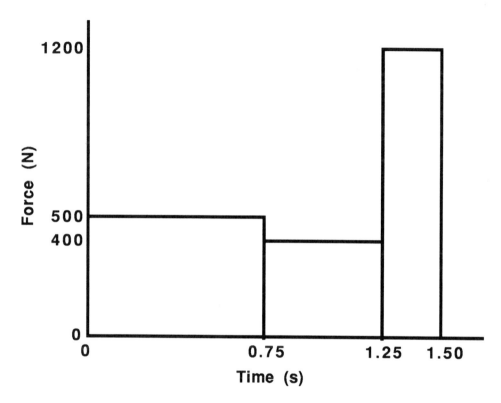

9. What is total energy of the segment given the following information? (Answer: 26.43 J)
   mass : 3.65 kg
   g : 9.81 m/s^2
   vx : – 0.25 m/s
   vy : 2.426 m/s
   moment of inertia : 0.06395 kg-m^2
   angular velocity : 5.2 rad/s^2
   height of center of mass : 0.41 m

10. What is the work done on the segment given the following information? (Answer: 10.2 J)

	TKE	RKE	PE
Frame 2	12.6	1.0	15.0
Frame 100	8.1	0.3	10.0

Additional Reading

1. Bobbert, M.F., van Ingen Schenau, G.J. Coordination in vertical jumping. Journal of Biomechanics 21:249-262, 1988.

2. Putnam, C.A. and Kozey, J.W. Substantive issues in running. In C.L. Vaughn (ed.). Biomechanics of Sport. pp. 1-33. Boca Raton, FL: CRC Press, Inc., 1989.

3. van Ingen Schenau, G.J. and Cavanagh, P.R. Power equations in endurance sports. Journal of Biomechanics 23:865-881, 1990.

4. Winter, D.A. Mechanical power in human movement: Generation, absorption, and transfer. In B. van Gheluwe and J. Atha (eds.). Current Research in Sports Biomechanics. pp. 34-45. New York: Karger, 1987.

References

1. Miller, D.I., Nelson, R.C. Biomechanics of Sport. Philadelphia: Lea & Febiger, 1973.
2. Chaffin, D.B., Herrin, G.D., Keyserling, W.M., Foulke, J.A. Pre-employment strength testing in selecting workers for materials handling jobs. Cincinnati: National Institute for Occupational Safety and Health. Publication, CDC-99-74-62, 1977.
3. Pytel, J.L., Kamon, E. Dynamic strength test as a predictor for maximal and acceptable lifting. Ergonomics 24:663-672, 1981.
4. Kamon, E., Kiser, D., Pytel, J. Dynamic and static lifting capacity and muscular strength of steelmill workers. American Industrial Hygiene Association Journal 43:853-857, 1982.
5. Lind, A.R., Burse, R., Rochelle, R.H., Rinehart, J.S., Petrofsky, J.S. Influence of posture on isometric fatigue. Journal of Applied Physiology 45:270-274, 1978.
6. Garg, A. and Chaffin, D.B. A biomechanical computerized simulation of human strength. Transactions of the American Institute of Industrial Engineers 7:1-15, 1975.
7. Chaffin, D.B. and Andersson, G.B.J. Occupational Biomechanics (2nd Edition). New York: John Wiley & Sons, Inc., 1991.
8. Olsen, V.L., Smidt, G.L., Johnston, R.C. The maximum torque generated by the eccentric, isometric and concentric contractions of the hip abductor muscles. Physical Therapy 52:149-158, 1972.
9. Reilly, D.T., and Martens, M. Experimental analysis of the quadriceps muscle force and the patello-femoral joint reaction force for various activities. Acta Orthopaedia Scandinavia 43:126-137, 1972.
10. Ghista, D.N. and Roaf, R. Orthopaedic Mechanics: Procedures and Devices, Vol. 2. New York: Academic Press, 1981.
11. Ayoub, M.M. and Mital, A. Manual Materials Handling. London: Taylor & Francis Inc., 1989.
12. Redfield, R. and Hull, M.L. On the relation between joint moments and pedalling rates at constant power in cycling. Journal of Biomechanics 19:317-324, 1986.
13. Devita, P., Hong, D.M., Hamill, J. Effects of asymmetric load carrying on the biomechanics of walking. Journal of Biomechanics 24:1119-1129, 1991.
14. Lander, J.E. Effects of center of mass manipulation on the performance of the squat exercise. Unpublished Doctoral dissertation, University of Oregon, 1984.
15. Feltner, M.E. and Dapena, J. Dynamics of the shoulder and elbow joints of the throwing arm during the baseball pitch. International Journal of Sports Biomechanics 2: 235-259, 1986.
16. Cavanagh, P.R. and Gregor, R.J. Knee joint torque during the swing phase of normal treadmill walking. Journal of Biomechanics 8:337-344, 1976.
17. Winter, D.A. Moments of force and mechanical power in jogging. Journal of Biomechanics 16:91-97, 1983.
18. Winter, D.A. Kinematic and kinetic patterns in human gait: variability and compensating effects. Human Movement Science 3:51-76, 1984.
19. Winter, D.A. The Biomechanics and Motor Control of Human Gait. Waterloo, Ontario, Canada: University of Waterloo Press, 1987.
20. Cavanagh, P.R., Pollock, M.L., Landa, J. A biomechanical comparison of elite and good distance runners. Annals of N.Y. Academy of Sciences. pp. 301-328, 1977.
21. Mann, R. and Sprague, P. Kinetics of sprinting. In J. Terauds (ed.) Biomechanics of Sports. Del Mar, CA: Academic Publishers, 1982.
22. Winter, D.A. and Robertson, D.G.E. Joint torque and energy patterns in normal gait. Biological Cybernetics 142:137-142, 1978.
23. Robertson, D.G.E. and Winter, D.A. Mechanical energy generation, absorption, and transfer amongst segments during walking. Journal of Biomechanics 13:845-854, 1980.
24. Robertson, D.G.E. Functions of the leg muscles during the stance phase of running. In B. Jonsson (ed.). Biomechanics X-B. pp. 1021-1027. Champaign, IL: Human Kinetics Publishers, 1987.
25. Gage, J.R. Millions of bits of data: How can we use it to treat cerebral palsy? Proceedings of the Second North American Congress on Biomechanics. pp. 291-294, 1992.
26. DeVita, P. and Skelly, WA. Effect of landing stiffness on joint kinetics and energetics in the lower extremity. Medicine and Science in Sports and Exercise 24:108-115, 1992.
27. Bates, B.T., Osternig, L.R., Mason, B.R. Variations of velocity within the support phase of running. In J. Terauds and G. Dales (eds.). Science in Athletics. Del Mar: Academic Publishers, 1979.
28. Hamill, J., Bates, B.T., Sawhill, J.A., Knutzen, K.M. Variations in ground reaction force parameters at different running speeds. Human Movement Science 2:47-56, 1983.
29. Komi, P.V. and Bosco, C. Utilization of stored elastic energy in leg extensor muscles by men and women. Medicine and Science in Sports 10:261-265, 1978.
30. Bosco, C. and Komi, P.V. Potentiation of the mechanical behavior of the human skeletal muscle through prestretching. Acta Physiologica Scandinavia 106:467-474, 1979.
31. Dowling, J.J. and Vamos, L. Identification of kinetic and temporal factors related to vertical jump performance. Journal of Applied Biomechanics 9:95-110, 1993.
32. Norman, R.W., Sharratt, M.T., Pezzack, J.C., Noble, E.G. Re-examination of the mechanical efficiency of horizontal treadmill running. In P.V. Komi (ed.). Biomechanics V-B. pp. 87-93, Baltimore: University Park Press, 1976.
33. Winter, D.A. A new definition of mechanical work done in human movement. Journal of Applied Physiology 46: 79-83, 1979.
34. Pierrynowski, M.R., Winter, D.A., Norman, R.W. Mechanical energy transfer in treadmill walking. Ergonomics 24: 1-14, 1980.

35. Williams, K.R. and Cavanagh P.R. A model for the calculation of mechanical power during distance running. Journal of Biomechanics 16: 115-128, 1983.

36. White, S.C. and Winter, D.A. : Mechanical power analysis of the lower limb musculature in race walking. International Journal of Sports Biomechanics 1:15-24, 1985.

37. Cavagna, G.A., Thys, H., Zamboni, A. The sources of external work in level walking and running. Journal of Physiology 262:639-657, 1976.

38. Shorten, M.R., Wootton, S.A., Williams, C. Mechanical energy changes and the oxygen cost of running. Engineering in Medicine 10:213-217, 1981.

39. Williams, K.R. A biomechanical and physiological evaluation of running efficiency. Unpublished Doctoral dissertation, The Pennsylvania State University, 1980.

40. Hintermeister, R.A. and Hamill, J. Mechanical power and energy in level treadmill running. Proceedings of the North American Congress Of Biomechanics II. pp. 213-214. Chicago, IL, August, 1992.

Glossary

balance:	the ability of an individual to assume and maintain a stable position.
dynamic analysis:	a calculation of the forces and moments when there are significant linear and/or angular accelerations.
dynamics:	the branch of mechanics in which the system being studied undergoes an acceleration.
equilibrium:	the state of a system whose acceleration is unchanged.
external work:	the work done by a body on another body.
impulse:	the product of the magnitude of a force and its time of application; the area under a force-time curve.
impulse-momentum relationship:	the relationship stating that the impulse is equal to the change in momentum.
impulsive force:	see impulse.
internal work:	the total work done resulting from the motion of all of the body's segments.
inverse dynamics:	an analytical approach calculating the forces and moments based on the accelerations of the object.
muscle power:	the product of the net muscle moment and the angular velocity of the joint.
neutral equilibrium:	the state of a body in which the body will remain in a location if displaced from another location.
stability:	the resistance to a disturbance in the body's equilibrium.
stable equilibrium:	the state of a body in which the body will return to its original location if it is displaced.
statics:	the branch of mechanics in which the system being studied undergoes no acceleration.
unstable equilibrium:	the state of a body in which the body continues to increase its displacement if it is displaced.
work-energy theorem:	the relationship between work and energy stating that the work done is equal to the change in energy.

Ligaments

Ligaments

Ligaments Supporting the Shoulder Complex

Ligament	Insertion	Action
Acromioclavicular	acromion process TO clavicle	prevents separation of clavicle and scapula
Coracoacromial	corocoid process TO acromion process	forms arch over shoulder
Coracoclavicular- Trapezoid; Conoid	coracoid process TO the clavicle	maintains relationship between scapula and clavicle; prevents anterior and posterior scapula movements; prevents upward and downward movements of clavicle on scapula
Coracohumeral	coracoid process TO greater and lesser tuberosity on humerus	checks upward displacement of humeral head; checks external rotation; supports weight of arm
Costoclavicular	clavicle TO 1st rib	checks clavicle elevation, anterior and posterior, and lateral movement; supports weight of upper extremity
Glenohumeral—Inferior; Middle; Superior	upper and anterior edge of glenoid TO over, in front and below humeral head	taut with external rotation and abduction; prevents anterior dislocation of humerus
Interclavicular	clavicle TO clavicle	checks motion of clavicle; supports weight of upper extremity
Sternoclavicular— Anterior, Posterior	clavicle TO sternum	prevents anterior and posterior dislocation; supports weight of upper extremity
Transverse Ligament	across the bicipital groove	keeps biceps tendon in the groove

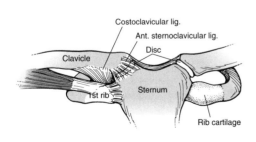

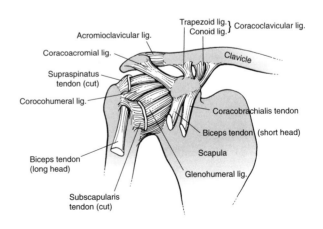

Ligaments Supporting the Elbow and Radioulnar Joints

Ligament	Insertion	Action
Annular	anterior margin of radial notch TO posterior margin of radial notch	surrounds and supports the head of the radius; maintains radius up in the joint
Radial collateral	lateral epicondyle TO annular ligament	supports lateral joint
Ulnar collateral—Posterior; Transverse; Anterior	medial epicondyle; olecranon process TO coronoid process	supports medial joint; resists valgus forces

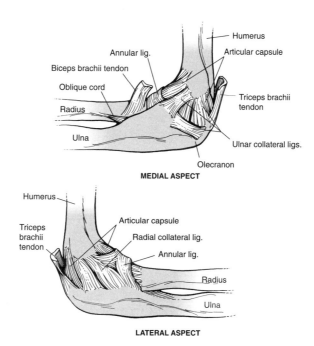

MEDIAL ASPECT

LATERAL ASPECT

Ligaments Supporting the Wrist and Fingers

Ligament	Insertion	Action
Collateral	phalanx TO phalanx; sides of MP, PIP, and DIP joints	supports sides of fingers; prevents varus/valgus forces
Dorsal Intercarpal	first row of carpals TO second row of carpals	keeps carpals together
Deep Transverse	MP of finger TO MP of adjacent finger	taut in finger flexing, disallowing abduction
Dorsal Radiocarpal	lower end of radius TO scaphoid; lunate; triquetrum	connects radius to carpals; supports posterior side of wrist
Palmar Intercarpal	scaphoid TO lunate; lunate TO triquetrum	keeps carpals together
Palmer Plates	across the anterior joint of MP, PIP, DIP	supports anterior MP, PIP, and DIP joint
Palmar Radiocarpal	lower radius TO scaphoid; lunate; triquetrum	connects radius to carpals; supports anterior side of wrist
Radial Collateral	Radius TO scaphoid; trapezium	supports lateral side of wrist; resists valgus forces
Ulnar Collateral	Ulna TO pisiform; triquetrum	supports medial side of wrist; resists varus forces

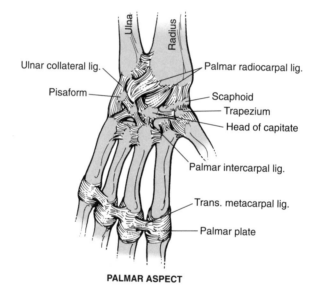

PALMAR ASPECT

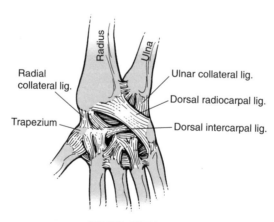

DORSAL ASPECT

Ligaments Stabilizing the Pelvis and Hip

Ligament	Insertion	Action
dorsal sacroiliac	posterior, inferior spine of ilium TO pelvic surface of sacrum	maintain relationship sacrum and ilium
iliofemoral	anterior, inferior iliac spine TO intertrochanteric line of femur	supports anterior hip; resists in movements of extension, internal rotation, external rotation
interosseus (SI)	tuberosity of ilium TO tuberosity of sacrum	prevents downward displacement of sacrum due to body weight
ischiofemoral	posterior acetabulum TO iliofemoral ligament	resists in movements of adduction and internal rotation
ligament of the head	acetabular notch and transverse ligament TO pit on head of femur	transmits vessels to the head of femur; no mechanical function
pubic	transverse fiber from body of pubis TO body of pubis	maintains relationship between right and left pubic bones
pubofemoral	pubic part of acetabulum; superior rami TO intertrochanteric line	resists in movements of abduction and external rotation
ventral sacroiliac	thin; pelvic surface of sacrum TO pelvic surface of ilium	maintains the relationship between the sacrum and the ilium

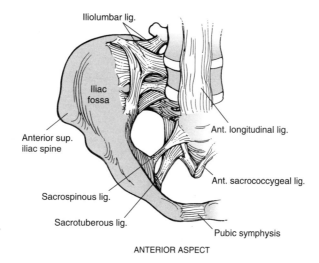

ANTERIOR ASPECT

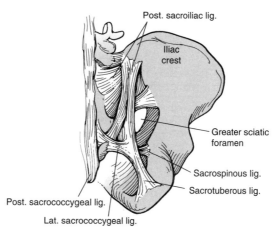

POSTERIOR ASPECT

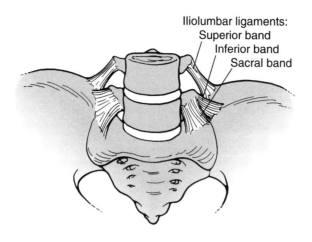

Iliolumbar ligaments:
Superior band
Inferior band
Sacral band

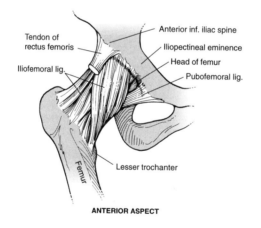

Tendon of
rectus femoris

Iliofemoral lig.

Anterior inf. iliac spine

Iliopectineal eminence

Head of femur

Pubofemoral lig.

Femur

Lesser trochanter

ANTERIOR ASPECT

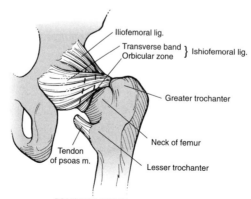

Iliofemoral lig.

Transverse band
Orbicular zone } Ishiofemoral lig.

Greater trochanter

Neck of femur

Tendon
of psoas m.

Lesser trochanter

POSTERIOR ASPECT

Ligaments Stabilizing the Knee Joint

Ligament	Insertion	Action
anterior cruciate	anterior intercondylar area of tibia TO medial surface of lateral condyle	prevents anterior tibial displacement; resists extension, internal rotation, flexion
arcuate	lateral condyle of femur TO head of fibula	reinforces back of capsule
coronary	meniscus TO tibia	holds menisci to tibia
medial collateral	medial epicondyle of femur TO medial condyle of tibia and medial meniscus	resists valgus forces; taut in extension; resists in internal and external rotation
lateral collateral	lateral epicondyle of femur TO head of fibula	resists varus forces; taut in extension
patellar	inferior patella TO tibial tuberosity	transfers force from quadriceps to tibia
posterior cruciate	posterior spine of tibia TO inner condyle of femur	resists posterior tibial movement; resists movements of flexion and rotation
posterior oblique	expansion of semimembranosus muscle	supports posterior, medial capsule
transverse	medial meniscus TO lateral meniscus in front	connects menisci to each other

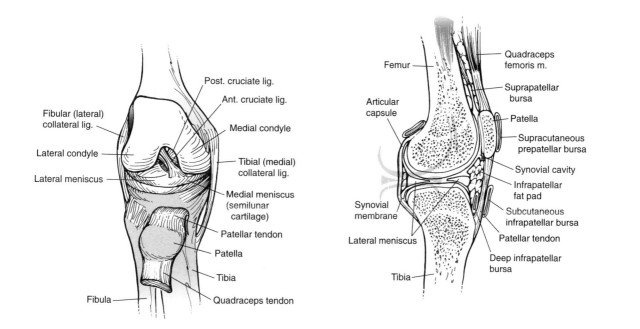

Ligaments Supporting the Foot and Ankle

Ligament	Insertion	Action
anterior talofibular	lateral malleolus TO neck of talus	limits anterior displacement of foot or talar tilt; limits plantar flexion and inversion
anterior talotibial	anterior margin of tibia TO front margin on talus	limits plantar flexion and abduction of foot
calcaneocuboid	calcaneus TO cuboid on dorsal surface	limits inversion of foot
calcaneofibular	lateral malleolus TO tubercle on outer calcaneus	resists backward displacement of foot; resists inversion
deltoid	medial malleolus TO talus, navicular, calcaneus	resists valgus forces to the ankle; limits plantar flexion, dorsiflexion, eversion, and abduction of foot
dorsal (tarsometatarsal)	tarsals TO metatarsals	supports arch; maintains relationship between tarsals and metatarsals
dorsal calcaneocuboid	calcaneus TO cuboid on dorsal side	limits inversion
dorsal talonavicular	neck of talus TO superior surface of navicular	supports talonavicular joint; limits inversion
interosseus (intertarsal)	connects adjacent tarsals	supports arch of foot; supports intertarsal joints
interosseus talocalcaneal	undersurface of talus TO upper surface of calcaneus	limits pronation, supination, abduction, adduction, dorsiflexion, plantar flexion
plantar calcaneocuboid	under surface of calcaneus TO under surface of cuboid	supports the arch
plantar calcaneonavicular	anterior margin of calcaneus TO under surface of navicular	supports the arch; limits abduction
posterior talofibular	inner, back lateral malleolus TO posterior surface of talus	limits plantar flexion, dorsiflexion, inversion; supports lateral ankle
posterior talotibial	tibia TO talus, behind articulating facet	limits plantar flexion; supports medial ankle
talocalcaneal	connecting ant./posterior, medial, lateral talus TO calcaneus	supports subtalar joint

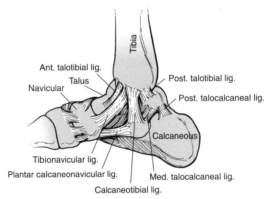

MEDIAL ASPECT

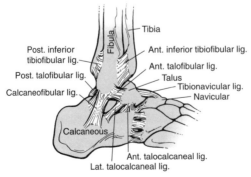

LATERAL ASPECT

Ligaments Supporting the Vertebral Column

Ligament	Insertion	Action
Alar	apex of dens TO medial occipital	limits lateral flexion and rotation of the head; holds dens into atlas
Apical	apex of dens TO front foramen magnum	holds dens into atlas and skull
Anterior longitudinal	sacrum; anterior vertebral body and disc TO above anterior body and disc; atlas	limits hyperextension of the spine; limits forward sliding of vertebrae
Costotransverse	tubercles of ribs TO transverse process of vertebrae	supports rib attachment to the thoracic vertebrae
Cruciform	odontoid bone TO arch of atlas	stabilizes odontoid and atlas; prevents posterior movement of dens in atlas
Iliolumbar	transverse process of L5 TO iliac crest	limits lumbar motion in flexion and rotation
Interspinous	spinous process TO spinous process	limits flexion of trunk; limits shear forces acting on the vertebrae
Intertransverse	transverse process TO transverse process	limits lateral flexion of the trunk
Ligamentum flavum	laminae TO laminae	limits flexion of trunk; assists extension of trunk; creates constant tension on the disc
Ligamentum nuchae	laminae TO laminae in the cervical region; connects with supraspinous ligament	limits cervical flexion; assists in extension; creates constant disc load
Posterior longitudinal	posterior vertebral body and disc TO posterior body and disc of next vertebrae	limits flexion of trunk
Radiate	head of ribs TO body of vertebrae	maintains ribs to thoracic vertebrae
Supraspinous	spinous process TO spinous process of next vertebrae	limits flexion of trunk; resists forward shear force on spine

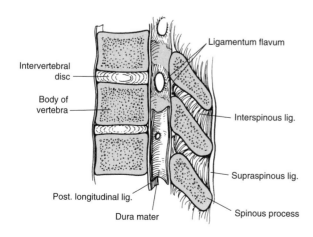

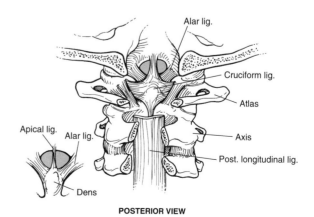

POSTERIOR VIEW

Muscles

Muscles Acting on the Shoulder Complex

Muscle	Insertion	Action	Nerve Supply
Biceps Brachii	supraglenoid tubercle; corocoid process TO radial tuberosity	arm abduction	musculocutaneous nerve; C5 and C6
Coracobrachialis	coracoid process of scapula TO medial surface adjacent to deltoid tuberosity	arm horizontal flexion	musculocutaneous nerve; C6 and C7
Deltoid	lateral third of clavicle; acromion process; spine of scapula TO deltoid tubercle on humerus	arm abduction; arm flexion; arm horizontal flexion; arm horizontal extension; arm internal rotation	axillary nerve; C5 and C6
Infraspinatus	infraspinous fossa TO greater tubercle on humerus	arm external rotation; arm horizontal extension	subscapular nerve; C5 and C6
Latissimus Dorsi	spinous process of thoracic vertebrae 6-12 and lumbar 1-5; lower 3-4 ribs; iliac crest; inferior angle of scapula TO intertubercular groove on humerus	arm internal rotation; arm adduction; arm extension	thoracodorsal nerve; C6, C7, and C8
Levator Scapula	transverse process of cervical vertebrae 1-4 TO superior angle of scapula	shoulder girdle elevation	cervical plexus via C3 and C4; dorsal scapular nerve; C5
Pectoralis Major	clavicle; sternum; ribs 1-6 TO greater tubercle of humerus and intertubercular groove	arm internal rotation; arm horizontal flexion; arm flexion; arm extension	musculocutaneous nerve; C6 and C7
Pectoralis Minor	ribs 3-5 TO corocoid process	shoulder girdle depression; shoulder girdle downward rotation; shoulder girdle protraction	medial anterior thoracic nerve; C8 and T1
Rhomboid	spinous process of C7 and T1-T5 TO medial border of scapula	shoulder girdle retraction; shoulder girdle elevation	dorsal scapular nerve; C5
Serratus Anterior	ribs 1-8 TO underside of scapula along medial border	shoulder girdle protraction; shoulder girdle elevation; shoulder girdle upward rotation	long thoracic nerve; C5, C6, and C7
Subclavius	costal cartilage of rib 1 TO underside of clavicle	shoulder girdle depression	brachial plexus; C5 and C6
Subscapularis	whole underside of scapula TO lesser tubercle on humerus	arm internal rotation	subscapular nerve; C5, C6, and C7
Supraspinatus	supraspinous fossa of scapula TO lesser tubercle of humerus	arm abduction; arm flexion	subscapular nerve; C5
Teres Major	posterior surface of scapula at the inferior angle TO lesser tubercle of humerus	arm internal rotation; arm extension; arm adduction	subscapular nerve; C5 and C6

Muscle	Insertion	Action	Nerve Supply
Teres Minor	lateral border of posterior scapula TO greater tubercle on humerus	arm internal rotation; arm adduction; arm extension	axillary nerve; C5
Trapezius	occipital bone; ligamentum nuchae; spinous process of C7 and T1-T12 TO acromion process; spine of scapula; lateral clavicle	shoulder girdle upward rotation; shoulder girdle elevation; shoulder girdle retraction; arm abduction	accessory nerve-spinal portion of 11th cranial nerve; C3 and C4

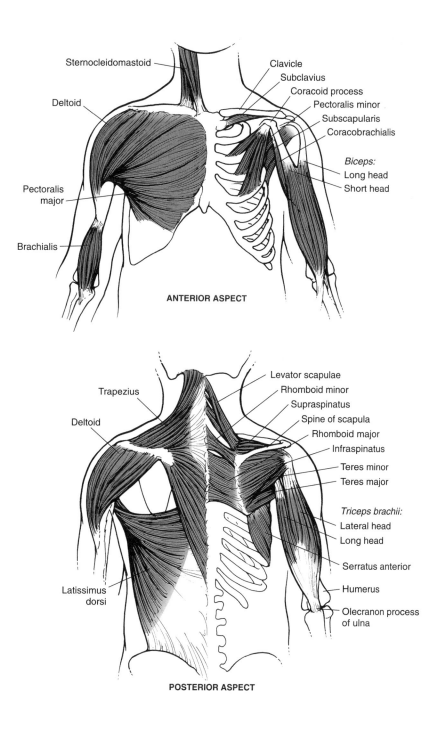

ANTERIOR ASPECT

POSTERIOR ASPECT

Muscles Acting on the Forearm

Muscle	Insertion	Action	Nerve Supply
Anconeus	lateral epicondyle of humerus TO olecarnon process on ulna	forearm extension	radial nerve; C7 and C8
Biceps Brachii	supraglenoid tubercle; corocoid process TO radial tuberosity	forearm flexion; forearm supination	musculocutaneous nerve; C5 and C6
Brachialis	anterior surface of lower humerus TO coronoid process on ulna	forearm flexion	musculocutaneous nerve; C5 and C6
Brachioradialis	lateral supracondylar ridge of humerus TO styloid process of radius	forearm flexion	radial nerve; C5 and C6
Extensor Carpi Radialis Brevis	lateral epicondyle of humerus TO base of 3rd metacarpal	forearm flexion	radial nerve; C6 and C7
Extensor Carpi Radialis Longus	lateral supracondylar ridge of humerus TO base of 2nd metacarpal	forearm flexion	radial nerve; C6 and C7
Extensor Carpi Ulnaris	lateral epicondyle TO base of 5th metacarpal	forearm extension	posterior interosseous branch of radial nerve; C6, C7, and C8
Pronator Quadratus	distal anterior surface of ulna TO distal anterior surface of radius	forearm pronation	anterior interosseous nerve from median nerve; C8 and T1
Pronator Teres	medial epicondyle of humerus and coronoid process on ulna TO midway down on the lateral surface of radius	forearm pronation forearm flexion	median nerve; C5 and C6
Supinator	lateral epicondyle of humerus TO upper, lateral side of radius	forearm supination	posterior interosseus nerve; C6
Triceps Brachii	infraglenoid tubercle on scapula; middle of posterior shaft of humerus; lower shaft of humerus TO olecranon process	forearm extension	radial nerve; C7 and C8

Muscles Acting on the Wrist and Fingers

Muscle	Insertion	Action	Nerve Supply
Abductor Digiti Minimi	pisiform bone TO base of proximal phalanx of the little finger	finger abduction of the little finger	ulnar nerve; C8
Abductor Pollicis Brevis	scaphoid; trapezium TO base of proximal phalanx	thumb abduction	median nerve; C6 and C7
Abductor Pollicis Longus	middle of radius TO radial side of base of 1st metacarpal	thumb abduction	posterior interosseous branch of radial nerve; C6 and C7
Adductor Pollicis	capitate; base of 2nd and 3rd metacarpal TO base of proximal phalanx of thumb	thumb adduction	ulnar nerve; C8 and T1
Dorsal Interossei	between the metacarpals of the four fingers TO the base of the proximal phalanx of 2nd, 3rd, and 4th fingers	finger abduction of index, middle, and ring fingers; finger adduction of middle finger	ulnar nerve; C8 and T1
Extensor Carpi Radialis Brevis	lateral epicondyle of humerus TO base of 3rd metacarpal	hand extension; hand radial flexion	radial nerve; C6 and C7
Extensor Carpi Radialis Longus	lateral supracondylar ridge of humerus TO base of 2nd metacarpal	hand extension; hand radial flexion	radial nerve; C6 and C7
Extensor Carpi Ulnaris	lateral epicondyle of humerus TO base of 5th metacarpal	hand extension; hand ulnar flexion	posterior interosseous branch of radial nerve; C6, C7, and C8
Extensor Digiti Minimi	tendon of extensor digitorum TO proximal phalanx of little finger	finger extension of proximal phalanx of little finger	posterior interosseous branch of radial nerve; C6, C7, and C8
Extensor Digitorum	lateral epicondyle of humerus TO the dorsal hood of the four fingers	finger extension of 1st phalanx; hand extension	posterior interosseous branch of radial nerve; C6, C7, and C8
Extensor Indicis	lower ulna and interosseous membrane TO dorsal hood of index finger	extension of proximal phalanx of index finger	posterior interosseous branch of radial nerve; C6, C7, and C8
Extensor Pollicis Brevis	middle of radius and ulna TO base of proximal phalanx of thumb	thumb extension	posterior interosseous branch of radial nerve; C6 and C7
Extensor Pollicis Longus	middle third of ulna and interosseous membrane TO base of distal phalanx of thumb	thumb extension	posterior interosseous branch of radial nerve; C6, C7, and C8
Flexor Carpi Radialis	medial epicondyle of humerus TO base of 2nd and 3rd metacarpal	hand flexion; hand radial flexion	median nerve; C6 and C7
Flexor Carpi Ulnaris	medial epicondyle TO pisiform; hamate; base of 5th metacarpal	hand flexion; hand ulnar flexion	ulnar nerve; C8 and T1
Flexor Digiti Minimi Brevis	hamate bone TO proximal phalanx of little finger	flexion of the little finger	ulnar nerve; C8
Flexor Digitorum Profundus	anterior, medial ulna TO base of distal phalanx of four fingers	finger flexion-proximal, middle and distal phalanx; hand flexion	anterior interosseous nerve; C8 and T1
Flexor Digitorum Superficialis	medial epicondyle TO base of middle phalanx of four fingers	finger flexion-proximal and middle phalanx; hand flexion	median nerve; C7, C8, and T1
Flexor Pollicis Brevis	trapezium; trapezoid; capitate TO base of proximal phalanx of thumb	thumb flexion	median nerve; ulnar nerve
Flexor Pollicis Longus	middle radius and interosseous membrane TO base of distal phalanx of thumb	thumb flexion	anterior interosseous from median nerve; C8 and T1
Lumbricales	tendon of flexor digitorum profundus TO dorsal hood of the four fingers	finger flexion of proximal phalanx; finger extension of middle and distal phalanx	median nerve; C6 and C7; ulnar nerve; C8
Opponens Digiti Minimi	hamate bone TO 5th metacarpal	finger opposition of little finger	ulnar nerve; C8 and T1

(continued)

Muscles Acting on the Wrist and Fingers (continued)

Muscle	Insertion	Action	Nerve Supply
Opponens Pollicis	trapezium TO 1st metacarpal	thumb opposition	median nerve; C6 and C7
Palmar Interossei	sides of 2nd, 4th, and 5th metacarpal TO base of proximal phalanx of same fingers	finger adduction of index, ring, and little fingers	ulnar nerve; C8 and T1
Palmaris Longus	medial epicondyle TO palmar aponeurosis	hand flexion	median nerve; C6 and C7

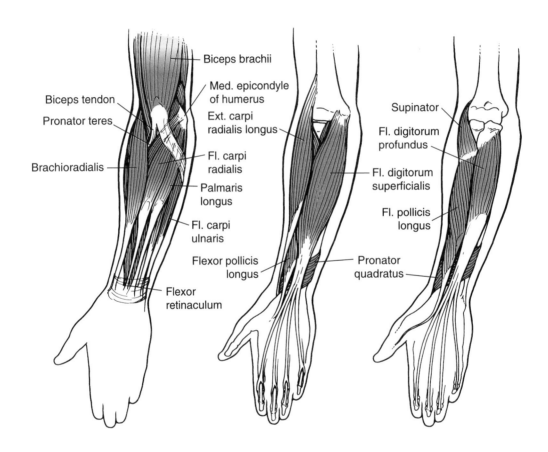

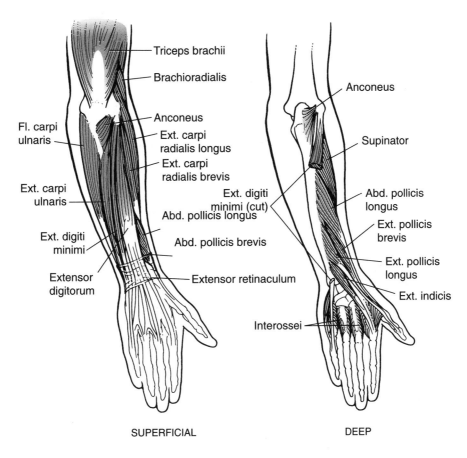

Triceps brachii

Brachioradialis

Fl. carpi
ulnaris

Anconeus

Ext. carpi
radialis longus

Ext. carpi
radialis brevis

Ext. carpi
ulnaris

Ext. digiti
minimi (cut)

Abd. pollicis longus

Ext. digiti
minimi

Abd. pollicis brevis

Extensor
digitorum

Extensor retinaculum

Anconeus

Supinator

Abd. pollicis
longus

Ext. pollicis
brevis

Ext. pollicis
longus

Ext. indicis

Interossei

SUPERFICIAL

DEEP

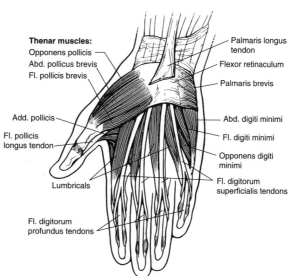

Thenar muscles:
Opponens pollicis
Abd. pollicus brevis
Fl. pollicis brevis

Add. pollicis

Fl. pollicis
longus tendon

Lumbricals

Fl. digitorum
profundus tendons

Palmaris longus
tendon

Flexor retinaculum

Palmaris brevis

Abd. digiti minimi

Fl. digiti minimi

Opponens digiti
minimi

Fl. digitorum
superficialis tendons

Muscles Acting on the Hip Joint

Muscle	Insertion	Action	Nerve Supply
Adductor Brevis	inferior rami of pubis TO upper half of posterior femur	thigh adduction	anterior obturator nerve; L3 and L4
Adductor Longus	inferior rami of pubis TO middle third of posterior femur	thigh adduction; thigh internal rotation	anterior obturator nerve; L3 and L4
Adductor Magnus	anterior pubis and ischial tuberosity TO linea aspera on posterior femur, adductor tubercle	thigh adduction; thigh internal rotation	posterior obturator nerve and sciatic nerve; L3 and L4
Biceps Femoris	ischial tuberosity TO lateral condyle of tibia and head of fubula	thigh extension; shank flexion; shank external rotation	tibial and peroneal portion of sciatic nerve; L5, S1, S2, S3
Gemellus Inferior	ischial tuberosity TO greater trochanter on femur	thigh external rotation	sacral plexus; L4, L5, and S1 sacral nerves
Gemellus Superior	ischial spine TO greater trochanter	thigh external rotation	sacral plexus; L5, S1, and S2 sacral nerves
Gluteus Maximus	posterior ilium, sacrum, and coccyx TO gluteal tuberosity; iliotibial band	thigh extension; thigh external rotation	inferior gluteal nerve; L5, S1, and S2
Gluteus Medius	anterior, lateral ilium TO lateral surface of greater trochanter	thigh abduction; thigh internal rotation	superior gluteal nerve; L4, L5, and S1
Gluteus Minimus	outer, lower ilium TO front of greater trochanter	thigh abduction; thigh internal rotation	superior gluteal nerve; L4, L5, and S1
Gracilis	inferior rami of pubis TO medial tibia (pes anserinus)	thigh adduction; thigh internal rotation; leg flexion; leg internal rotation	anterior obturator nerve; L3 and L4
Iliacus	inner surface of ilium and sacrum TO lesser trochanter	thigh flexion	femoral nerve; L2 and L3
Obturator Internus	sciatic notch and margin of obturator foramen TO greater trochanter	thigh external rotation	sacral plexus; L5, S1 and S2
Obturator Externus	pubis, ischium and margin of obturator foramen TO upper, posterior femur	thigh external rotation	obturator nerve; L3 and L4
Pectineus	pectineal line on pubis TO below lesser trochanter	thigh adduction; thigh flexion	femoral nerve; L2, L3, and L4
Piriformis	anterior, lateral sacrum TO superior greater trochanter	thigh external rotation; thigh abduction	S1, S2, and L5
Psoas	transverse processes, body of L1-L5 and T12 TO lesser trochanter	thigh flexion; trunk flexion	femoral nerve; L1, L2, and L3
Quadratus Femoris	ischial tuberosity TO greater trochanter	thigh external rotation	sacral plexus; L4, L5, and S1
Rectus Femoris	anterior, inferior iliac spine TO patella and tibial tuberosity	thigh flexion; leg extension	femoral nerve; L2, L3, and L4
Sartorius	anterior, superior iliac spine TO medial tibia (pes anserinus)	thigh flexion; thigh external rotation; leg flexion; leg internal rotation	femoral nerve; L2 and L3
Semimembranosus	ischial tuberosity TO medial condyle of tibia	thigh extension; thigh internal rotation; leg flexion; leg internal rotation	tibial portion of sciatic nerve; L5, S1 and S2
Semitendinosus	ischial tuberosity TO medial tibia (pes anserinus)	thigh extension; thigh internal rotation; leg flexion; leg internal rotation	tibial portion of sciatic nerve; L5, S1 and S2
Tensor Fascia Latae	anterior superior iliac spine TO iliotibial tract	thigh flexion; thigh abduction; thigh internal rotation	superior gluteal nerve; L4, L5 and S1

Muscles Acting on the Knee Joint

Muscle	Insertion	Action	Nerve Supply
Biceps Femoris	ischial tuberosity TO lateral condyle of tibia and head of fibula	thigh extension; leg flexion; leg external rotation	tibial and peroneal portion of sciatic nerve; L5, S1, S2, S3
Gastrocnemius	medial and lateral condyles of femur TO calcaneus	leg flexion; foot plantar flexion	tibial nerve; S1 and S2
Gracilis	inferior rami of pubis TO medial tibial (pes anserinus)	thigh adduction; thigh internal rotation; leg flexion; leg internal rotation	anterior obturator nerve; L3 and L4
Popliteus	lateral condyle of femur TO proximal tibia	leg internal rotation; leg flexion	tibial nerve
Rectus Femoris	anterior, inferior iliac spine TO patella and tibial tuberosity	thigh flexion; leg extension	femoral nerve; L2, L3, and L4
Sartorius	anterior, superior iliac spine TO medial tibia (pes anserinus)	thigh flexion; thigh external rotation; leg flexion; leg internal rotation	femoral nerve; L2 and L3
Semimembranosus	ischial tuberosity TO medial condyle of tibia	thigh extension; thigh internal rotation; leg flexion; leg internal rotation	tibial portion of sciatic nerve; L5, S1 and S2
Semitendinosus	ischial tuberosity TO medial tibia (pes anserinus)	thigh extension; thigh internal rotation; leg flexion; leg internal rotation	tibial portion of sciatic nerve; L5, S1 and S2
Vastus Intermedius	anterior, lateral femur TO patella and tibial tuberosity	leg extension	femoral nerve; L2, L3 and L4
Vastus Lateralis	intertrochanteric line; linea aspera TO patella and tibial tuberosity	leg extension	femoral nerve; L2, L3, and L4
Vastus Medialis	linea aspera; trochanteric line TO patella and tibial tuberosity	leg extension	femoral nerve; L2, L3, and L4

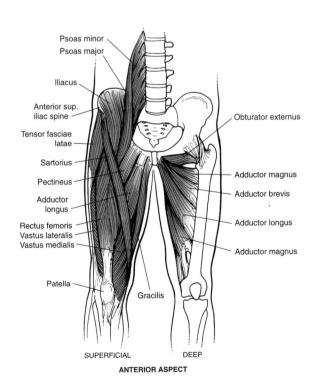

ANTERIOR ASPECT

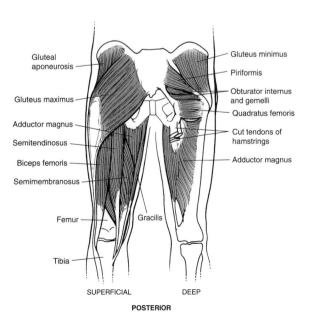

POSTERIOR

Muscles Acting on the Ankle and Foot

Muscle	Insertion	Action	Nerve Supply
Abductor Digiti Minimi	lateral calcaneus TO base of proximal phalanx of 5th toe	little toe abduction	lateral plantar nerve
Abductor Hallucis	medial calcaneus TO medial base of proximal phalanx of 1st toe	big toe abduction	medial plantar nerve
Adductor Hallucis	2nd, 3rd, and 4th metatarsal TO lateral side of proximal phalanx of big toe	big toe adduction	lateral plantar nerve
Dorsal Interossei	sides of metatarsals TO lateral side of proximal phalanx	abduct toes 2-4; adduct 2nd toe; flex proximal phalanx	lateral plantar nerve
Extensor Digitorum Brevis	lateral calcaneus TO proximal phalanx of 1st, 2nd, and 3rd toes	extend toes 1-4	deep peroneal nerve
Extensor Digitorum Longus	lateral condyle of tibia; fibula; interosseus membrane TO dorsal expansion of toes 2-5	foot dorsiflexion; foot eversion; toe 2-5 extension	deep peroneal nerve
Extensor Hallucis Longus	anterior fibula; interosseus membrane TO distal phalanx of big toe	big toe extension; forefoot adduction	deep peroneal nerve
Flexor Digiti Minimi Brevis	5th metatarsal TO proximal phalanx of little toe	little toe flexion	lateral plantar nerve
Flexor Digitorum Brevis	medial calcaneus TO middle phalanx of toes 2-5	toe 2-5 flexion	medial plantar nerve
Flexor Digitorum Longus	posterior tibia TO distal phalanx of toes 2-5	toe 2-5 flexion; foot plantar flexion	tibial nerve
Flexor Hallucis Brevis	cuboid TO medial side of proximal phalanx of big toe	big toe flexion	medial plantar nerve
Flexor Hallucis Longus	lower ⅔rds of posterior fibula and interosseus membrane	big toe flexion; forefoot adduction	tibial nerve
Gastrocnemius	medial and lateral condyles of femur TO calcaneus	leg flexion; foot plantar flexion	tibial nerve; S1 and S2
Lumbricales	tendon of flexor digitorum longus TO base of proximal phalanx of toes 2-5	proximal phalanx flexion, toes 2-5	medial and lateral plantar nerve
Peroneus Brevis	lower, lateral fibula TO 5th metatarsal	foot eversion; foot plantar flexion	superficial peroneal nerve
Peroneus Longus	lateral condyle of tibia and upper, lateral fibula TO 1st cuneiform; lateral 1st metatarsal	foot eversion; foot plantar flexion; forefoot abduction	superficial peroneal nerve
Peroneus Tertius	lower, anterior fibula; interosseous membrane TO base of 5th metatarsal	foot eversion	deep peroneal nerve
Plantar Interossei	medial side of 3-5 metatarsal TO medial side of proximal phalanx of toes 3-5	toe 3-5 abduction	lateral plantar nerve
Plantaris	linea aspera on femur TO calcaneus	foot plantar flexion	tibial nerve
Quadratus Plantae	medial, lateral, inferior calcaneus TO flexor digitorum tendon	toe 2-5 flexion	lateral plantar nerve
Soleus	upper, posterior tibia, fibula, interosseous membrane TO calcaneus	foot plantar flexion	tibial nerve
Tibialis Anterior	upper lateral tibia, interosseous membrane TO medial, plantar surface of 1st cuneiform	foot dorsiflexion; foot inversion	deep peroneal nerve
Tibialis Posterior	upper, posterior tibia, fibula, interosseous membrane TO inferior navicular	foot invertor; foot plantar flexion	tibial nerve

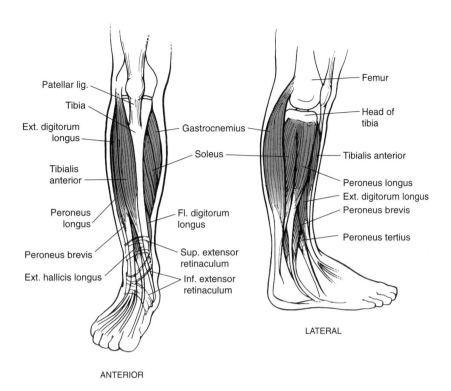

Patellar lig.
Tibia
Ext. digitorum longus
Tibialis anterior
Peroneus longus
Peroneus brevis
Ext. hallicis longus

Gastrocnemius
Soleus
Fl. digitorum longus
Sup. extensor retinaculum
Inf. extensor retinaculum

ANTERIOR

Femur
Head of tibia
Tibialis anterior
Peroneus longus
Ext. digitorum longus
Peroneus brevis
Peroneus tertius

LATERAL

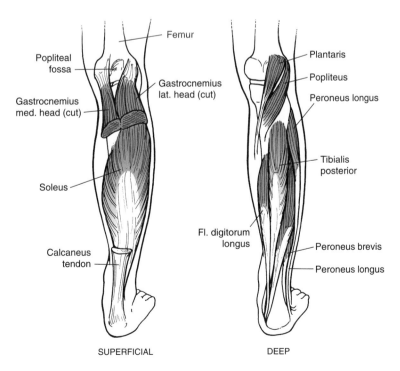

Popliteal fossa
Gastrocnemius med. head (cut)
Soleus
Calcaneus tendon

Femur
Gastrocnemius lat. head (cut)

SUPERFICIAL

Plantaris
Popliteus
Peroneus longus
Tibialis posterior
Fl. digitorum longus
Peroneus brevis
Peroneus longus

DEEP

POSTERIOR ASPECT

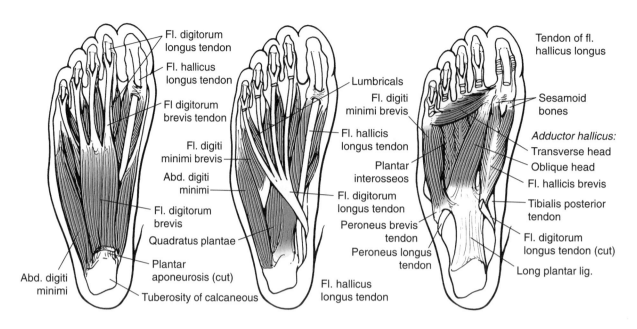

Fl. digitorum longus tendon

Fl. hallicus longus tendon

Fl digitorum brevis tendon

Fl. digiti minimi brevis

Abd. digiti minimi

Fl. digitorum brevis

Quadratus plantae

Plantar aponeurosis (cut)

Abd. digiti minimi

Tuberosity of calcaneous

Lumbricals

Fl. digiti minimi brevis

Fl. hallicis longus tendon

Plantar interosseos

Fl. digitorum longus tendon

Peroneus brevis tendon

Peroneus longus tendon

Fl. hallicus longus tendon

Tendon of fl. hallicus longus

Sesamoid bones

Adductor hallicus:

Transverse head

Oblique head

Fl. hallicis brevis

Tibialis posterior tendon

Fl. digitorum longus tendon (cut)

Long plantar lig.

Muscles Acting on the Vertebral Column

Muscle	Insertion	Action	Nerve Supply
External oblique (ABD)	9-12th ribs alternating with L. Dorsi and S. Ant TO anterior superior spine; pubic tubercle; ant iliac crest	trunk flexion, lateral flexion, rotation to opposite side	intercostal nerve; T7-T12
Iliocostalis (ES) lumborum	sacrum; spinous processes of L1-L5, T11, T12; iliac crest TO lower 6 or 7 ribs	trunk extension, lateral flexion, rotation to same side	spinal nerves; dorsal rami
thoracis	lower 6 ribs TO upper 6 ribs; transverse process of C7	trunk extension, lateral flexion, rotation to same side	spinal nerves; dorsal rami
cervices	3rd-6th ribs TO transverse processes of C4-C6	trunk extension, lateral flexion, rotation to same side	spinal nerves; dorsal rami
Iliopsoas	Bodies of T12, L1-L5; transverse processes of L1-L5; inner surface of ilium and sacrum TO lesser trochanter	thigh flexion; trunk flexion	femoral nerve; ventral rami; L1, L3
Internal oblique (ABD)	iliac crest, lumbar fascia TO ribs 8-10; linea alba	trunk flexion, lateral flexion, rotation to same side	intercostal nerves; T7-T12, L1
Inter-spinales (DP)	spinous process TO spinous process	trunk extension and hyperextension	spinal nerves; dorsal rami
Intertransversarii (DP)	transverse process TO transverse process	trunk extension, lateral flexion	spinal nerves; ventral and dorsal rami
Longissimus (ES) thoracis	posterior and transverse process of L1-L5; thoracolumbar fascia TO transverse process of T1-T12	trunk extension, lateral flexion, rotation to same side	spinal nerves; dorsal rami
cervices	transverse process of T1-T5 TO transverse process of C4-C6	trunk extension, lateral flexion, rotation to same side	spinal nerves; dorsal rami
capitis	transverse process of T1-T5, C4-C7 TO mastoid process	head extension, lateral flexion, rotation	spinal nerves; dorsal rami
Longus capitis	transverse process of C3-C6 TO occipital bone	head and cervical flexion, lateral flexion	cervical nerves; C1-3
Longus cervicis and colli	transverse process of C3-5; bodies of T1-2; bodies of C5-7, T1-3 TO atlas; transverse process of C5-6; bodies of C2-4	cervical flexion and lateral flexion	cervical nerves; C2-7
Multifidus (DP)	sacrum; iliac spine; transverse processes L5-C4 TO spinous process of next vertebrae	trunk extension, lateral flexion, rotation to opposite side	Spinal nerves; dorsal rami
Quadratus lumborum	iliac crest; transverse process of L2-5 TO transverse process of L1-2; last rib	trunk lateral flexion	Thoracic nerves; T12; Lumbar nerves; ventral rami
Rotatores (DP)	Transverse process TO laminae of next vertebrae	trunk extension, rotation to opposite side	spinal nerves; dorsal rami
Scaleni	transverse process of cervical vertebrae TO ribs 1 and 2	cervical lateral flexion, flexion	cervical nerves
Semispinalis capitis	C4-C6 facets; transverse process of C7 TO base of occipital	trunk extension, lateral flexion	cervical nerves; dorsal rami
cervicis	transverse process of T1-T6 TO spinous process of C1-C5	trunk extension, lateral flexion rotation	cervical nerves; dorsal rami
thoracis	transverse processes of T6-T10 TO spinous processes of T1-T4, C6, C7	trunk extension, lateral flexion, rotation	thoracic nerves; dorsal rami
Spinalis (ES) thoracis	spinous processes of L1-L2 and T11-T12 TO spinous process of T1-T8 ligamentum nuchae	trunk extension, lateral flexion	spinal nerves; dorsal rami
cervicis	spinous process of C7 TO spinous process of axis	trunk extension, lateral flexion	spinal nerves; dorsal rami
Splenius capitis	ligamentum nuchae; spinous process of C7, T1-T3 TO mastoid process and occipital bone	cervical extension, lateral flexion, rotation to same side	cervical nerves; dorsal rami

(continued)

Muscles Acting on the Vertebral Column (continued)

Muscle	Insertion	Action	Nerve Supply
Splenius cervicis	spinous process of T3-T6 TO transverse process of C1-C3	cervical extension, lateral flexion, rotation to same side	cervical nerves; dorsal rami
Sternocleido-mastoid	sternum and clavicle TO mastoid process	head and cervical flexion, lateral flexion, rotation to same side	accessory nerve; cranial nerve XI
Transverse abdominus (ABD)	last 6 ribs; iliac crest; inguinal ligament; lumbodorsal fascia TO linea alba; pubic crest	No specific action; increases IAP through compression	intercostal nerves; T7-T12, L1

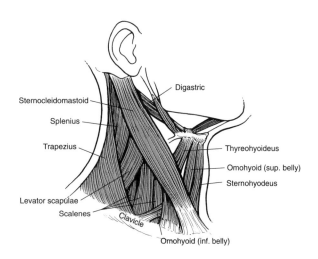

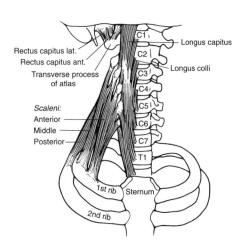

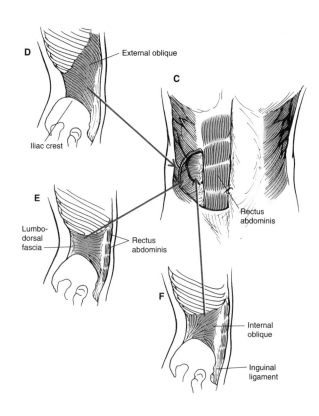

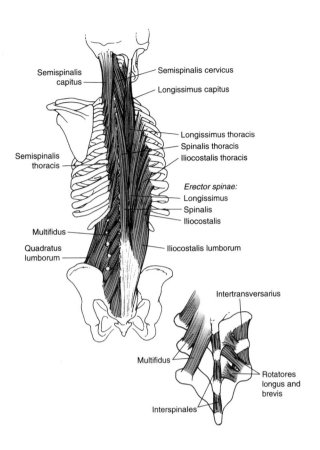

Semispinalis
capitus

Semispinalis cervicus

Longissimus capitus

Longissimus thoracis
Spinalis thoracis
Iliocostalis thoracis

Semispinalis
thoracis

Erector spinae:
Longissimus
Spinalis
Iliocostalis

Multifidus

Iliocostalis lumborum

Quadratus
lumborum

Intertransversarius

Multifidus

Rotatores
longus and
brevis

Interspinales

The Metric System and SI Units

All measurements in the biomechanics literature are expressed in terms of the metric system. This system of measurement uses units that are related to one another by some power of 10. Table C-1 presents the prefixes associated with these powers.

The standard length in the Metric System is the meter. This standard measure was originally indicated by two scratches on a platinum-iridium alloy bar kept at the International Bureau of Weights and Measures in Sevres, France. Table C-2 illustrates the use of these prefixes in common units of length.

Table C.1 Prefixes for the Powers of 10.

Multiple	Prefix	Abbreviation
10^9	giga	G
10^6	mega	M
10^3	kilo	k
10^{-1}	deci	d
10^{-2}	centi	c
10^{-3}	milli	m
10^{-6}	micro	μ
10^{-9}	nano	n

Table C.2 Units of Length.

Metric Unit	Power	Symbol
Kilometer	10^3	km
meter	—	m
decimeter	10^{-1}	dm
centimeter	10^{-2}	cm
millimeter	10^{-3}	mm
micrometer	10^{-6}	μm

The uniform system for the reporting of numerical values is known as the Système International d'Unites or the SI System. This system was developed through international cooperation to standardize the report of scientific information. The base dimensions used in biomechanics are MASS, LENGTH, TIME, TEMPERATURE, ELECTRIC CURRENT, AMOUNT OF SUBSTANCE, and LUMINOUS INTENSITY. The base units in the SI System corresponding to the base dimensions are the kilogram, the meter, the second, and the degree Kelvin. Table C-3 presents the base units of the SI System.

Table C-3 Base Units of Measurement.

Dimension	Unit	Symbol
mass	kilogram	kg
length	meter	m
time	second	s
temperature	degree Kelvin	K
electrical current	ampere	A
amount of substance	mole	mol
luminous intensity	candela	cd

Other units of measurement that are used in biomechanics are derived. These units are presented in Table C-4.

Table C-4 Derived Units of Measurement.

Dimension	Unit	Symbol
acceleration	meters per second squared	m/s^2
angle	radian	rad
area	meter squared	m^2
capacitance	farad	F
concentration	moles per meter cubed	mol/m^3
density	mass per unit volume	kg/m^3
energy	joule	J
impulse	force * time	N.s
lumen	lumen	lm
moment of inertia	kilogram-meters squared	$kg\text{-}m^2$
momentum	kilogram-meters per second	kg-m/s
power	watts	W
pressure	pascal	Pa
resistance	ohm	Ω
speed	meters per second	m/s
torque	newton-meters	N-m
voltage	volt	V
volume	meters cubed	m^3
work	joule	J

Trigonometric Functions

Trigonometry is a branch of mathematics concerned with the measurements of the sides and angles of triangles and their relationships with each other. Many concepts in biomechanics require a knowledge of trigonometry. A triangle is composed of three sides and three angles. The sum of the three angles of a triangle equals 180°. A right angle triangle is a triangle in which one of the angles is a right angle, that is, one of the angles equals 90°. The sum of the remaining angles, therefore, also equals 90°. Consider the triangle with vertices A, B, and C and sides of length AB, AC, and BC in Figure D-1.

The side opposite the right angle, AB, is always the longest side in the right triangle and is referred to as the *hypotenuse*. The other two sides are named according to which of the other angles is under consideration. If angle A is considered , then the side AC is called the *adjacent side* and the side BC is called the *opposite side.* If the angle B is considered (as in the triangle on the right below) then BC is the adjacent side and AC is the opposite side.

Based on the right angle triangle, trigonometric functions may be defined. A mathematical function is a quantity whose value varies and depends on some other quantity or quantities. Trigonometric functions vary with and depend on the values of the two non-right angles in a right angle triangle and the lengths of the sides of the triangle. The trigonometric functions are the ratios of the lengths of the sides of the triangle based on one of the two non-right angles in the triangle. There are six such functions: 1) sine (abbreviation is sin); 2) cosine (abbreviation is cos); 3) tangent (abbreviation is tan); 4) cosecant; 5) secant; and 6) cotangent. In biomechanics, only the first three of these functions are important. The trigonometric functions are thus defined for angle A (Figure D-2a) as:

1) the sin of an angle is defined as the ratio of the side opposite to the angle to the hypotenuse.

$$\sin A = \frac{\text{opposite side}}{\text{hypotenuse}} = \frac{BC}{AB}$$

2) the cos of an angle is defined as the ratio of the side adjacent to the angle to the hypotenuse.

$$\cos A = \frac{\text{adjacent side}}{\text{hypotenuse}} = \frac{AC}{AB}$$

3) the tan of an angle is defined as the ratio of the side opposite to the angle to the side adjacent to the angle.

$$\tan A = \frac{\text{opposite side}}{\text{adjacent side}} = \frac{BC}{AC}$$

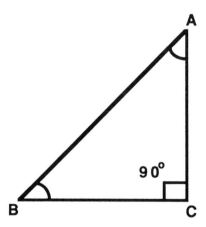

FIGURE D-1. A right angle triangle with the right angle C (90°) and the two non-right angles, A and B. The angles A and B sum to 90°.

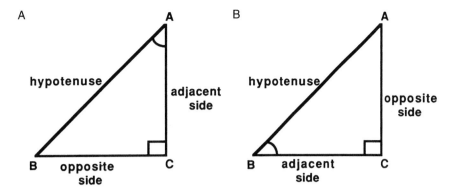

FIGURE D-2. Definitions of the sides of a right angle triangle based on: a) the non-right angle A; and b) the non-right angle B.

Similarly, the ratios for angle B in Figure D-2b may be defined as:

$$\sin B = \frac{\text{opposite side}}{\text{hypotenuse}} = \frac{AC}{AB}$$

$$\cos B = \frac{\text{adjacent side}}{\text{hypotenuse}} = \frac{BC}{AB}$$

$$\tan B = \frac{\text{opposite side}}{\text{adjacent side}} = \frac{AC}{BC}$$

For any angle, the ratios formed by the sides of the right angle triangle will always be the same. For example, the sine of an angle of 32° will always equal 0.5299 regardless of the size of the sides of the triangle. The same is true for the cosine and tangent of the angle. The values for the sine, cosine, and tangent of angles can be presented in tables. Table D-1 presents these values for angles ranging from 0° to 90°.

The values in this table may also be used to determine the angle when the sides of the triangle are known. Consider the triangle in Figure D-3. The length of the hypotenuse, AC is 0.10 m and the length of the side opposite angle C is 0.09 m. The ratio of AB to AC is the sine of angle C. Thus:

$$\sin\ C = \frac{AB}{AC} = \frac{0.03\ m}{0.05\ m} = 0.6$$

If the sine values in Table D-1 are examined, it can be determined that the angle whose sine is 0.6 is approximately 37°. This angle determination is referred to as calculating the arcsin of an angle. Thus:

$$C = \arcsin\frac{AB}{AC}$$

If angle A in Figure D.3 is considered, the ratio of the side AB to the hypotenuse AC is the cosine of the angle. The ratio is still 0.6 and the angle whose cosine is 0.6 is

approximately 53°. This is referred to as the arccosine of an angle.

There are two other very useful trigonometric relationships that are applicable to all triangles—not only right angle triangles. The first of these relationships is the *Law of Sines*. The Law of Sines states that the ratio of the length of any side to the sine of the angle opposite that side is equal to the ratio of any other side to the angle opposite that side. Consider the triangle in Figure E-4.

For this triangle, the Law of Sines can be stated as follows:

$$\frac{A}{\sin\ \alpha} = \frac{B}{\sin\ \beta} = \frac{C}{\sin\ \gamma}$$

The other trigonometric relationship that is applicable to any triangle is the *Law of Cosines*. This relationship states that the square of the length of any side of a triangle is equal to the sum of the squares of the other two sides minus twice the product of the lengths of the other

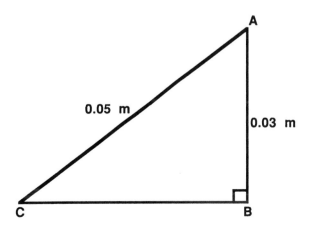

FIGURE D-3. A right triangle with two sides of known lengths but with two unknown angles.

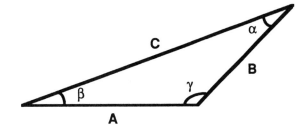

FIGURE D-4. A non-right angle triangle

two sides and the cosine of the angle opposite the original side. Consider the triangle in Figure E-4. For side A of this triangle, the Law of Cosines can be stated as follows:

$$A^2 = B^2 + C^2 - 2BC \cos \alpha$$

Similarly, the Law of Cosines can be stated for sides B and C.

Table D-1 Trigonometric Functions

Angle					Angle				
Degree	Radian	Sine	Cosine	Tangent	Degree	Radian	Sine	Cosine	Tangent
0°	0.000	.0000	.000	.000	46°	0.803	.7193	.6947	1.0355
1°	0.017	.0175	.9998	.0175	47°	0.820	.7314	.6820	1.0724
2°	0.035	.0349	.9994	.0349	48°	0.838	.7431	.6691	1.1106
3°	0.052	.0523	.9986	.0524	49°	0.855	.7547	.6561	1.1504
4°	0.070	.0698	.9976	.0699	50°	0.873	.7660	.6428	1.1918
5°	0.087	.0872	.9962	.0875	51°	0.890	.7771	.6293	1.2349
6°	0.105	.1045	.9945	.1051	52°	0.908	.7880	.6157	1.2799
7°	0.122	.1219	.9925	.1228	53°	0.925	.7986	.6018	1.3270
8°	0.140	.1392	.9903	.1405	54°	0.942	.8090	.5878	1.3764
9°	0.157	.1564	.9877	.1584	55°	0.960	.8192	.5736	1.4281
10°	0.175	.1736	.9848	.1763	56°	0.977	.8290	.5592	1.4826
11°	0.192	.1908	.9816	.1944	57°	0.995	.8387	.5446	1.5399
12°	0.209	.2079	.9781	.2126	58°	1.012	.8480	.5299	1.6003
13°	0.227	.2250	.9744	.2309	59°	1.030	.8572	.5150	1.6643
14°	0.244	.2419	.9703	.2493	60°	1.047	.8660	.5000	1.7321
15°	0.262	.2588	.9659	.2679	61°	1.065	.8746	.4848	1.8040
16°	0.279	.2756	.9613	.2867	62°	1.082	.8829	.4695	1.8807
17°	0.297	.2924	.9563	.3057	63°	1.100	.8910	.4540	1.9626
18°	0.314	.3090	.9511	.3249	64°	1.117	.8988	.4384	2.0503
19°	0.332	.3256	.9455	.3443	65°	1.134	.9063	.4226	2.1445
20°	0.349	.3420	.9397	.3640	66°	1.152	.9135	.4067	2.2460
21°	0.367	.3584	.9336	.3839	67°	1.169	.9205	.3907	2.3559
22°	0.384	.3746	.9272	.4040	68°	1.187	.9272	.3746	2.4751
23°	0.401	.3907	.9205	.4245	69°	1.204	.9336	.3584	2.6051
24°	0.419	.4067	.9135	.4452	70°	1.222	.9397	.3420	2.7475
25°	0.436	.4226	.9063	.4663	71°	1.239	.9455	.3256	2.9042
26°	0.454	.4384	.8988	.4877	72°	1.257	.9511	.3090	3.0777
27°	0.471	.4540	.8910	.5095	73°	1.274	.9563	.2924	3.2709
28°	0.489	.4695	.8829	.5317	74°	1.292	.9613	.2756	3.4874
29°	0.506	.4848	.8746	.5543	75°	1.309	.9659	.2588	3.7321
30°	0.524	.5000	.8660	.5774	76°	1.326	.9703	.2419	4.0108
31°	0.541	.5150	.8572	.6009	77°	1.344	.9744	.2250	4.3315
32°	0.559	.5299	.8480	.6249	78°	1.361	.9781	.2079	4.7046
33°	0.576	.5446	.8387	.6494	79°	1.379	.9816	.1908	5.1446
34°	0.593	.5592	.8290	6745	80°	1.396	.9848	.1736	5.6713
35°	0.611	.5736	.8192	7002	81°	1.414	.9877	.1564	6.3138
36°	0.628	.5878	.8090	7265	82°	1.431	.9903	.1392	7.1154
37°	0.646	.6018	.7986	7536	83°	1.449	.9925	.1219	8.1443
38°	0.663	.6157	.7880	.7813	84°	1.466	.9945	.1045	9.5144
39°	0.681	.6293	.7771	8098	85°	1.484	.9962	.0872	11.430
40°	0.698	.6428	7660	8391	86°	1.501	.9976	.0698	14.301
41°	0.716	.6561	.7547	8693	87°	1.518	.9986	.0523	19.081
42°	0.733	.6691	.7431	9004	88°	1.536	.9994	.0349	28.636
43°	0.750	.6820	.7314	.9325	89°	1.553	.9998	.0175	57.290
44°	0.768	.6947	.7193	.9657	90°	1.571	1.000	.0000	∞
45°	0.785	.7071	.7071	1.0000					

Index

Page numbers in italics refer to figures.